Student Solutions
Manual for Stewart's

SINGLE VARIABLE

SECOND EDITION
CALCULUS

CONCEPTS AND CONTEXTS

Jeffery A. Cole
Anoka-Ramsey Community College

BROOKS/COLE
—————★—————™
THOMSON LEARNING

Australia • Canada • Mexico • Singapore • Spain • United Kingdom • United States

BROOKS/COLE

THOMSON LEARNING

Assistant Editor: *Carol Ann Benedict*
Marketing Team: *Karin Sandberg, Laurie Davidson*
Editorial Assistant: *Molly Nance*
Production Coordinator: *Dorothy Bell*

Cover Design: *Vernon T. Boes*
Cover Photograph: *Erika Ede*
Print Buyer: *Micky Lawler*
Printing and Binding: *Webcom Limited*

For more information about this or any other Brooks/Cole product, contact:
BROOKS/COLE
511 Forest Lodge Road
Pacific Grove, CA 93950 USA
www.brookscole.com
1-800-423-0563 (Thomson Learning Academic Resource Center)

Printed in Canada

10 9 8 7 6 5 4

ISBN 0-534-37923-0

Preface

This *Student Solutions Manual* contains strategies for solving and solutions to selected exercises in the texts *Calculus: Concepts and Contexts, Single Variable, Second Edition,* and Chapters 1–8 of *Calculus: Concepts and Contexts, Second Edition,* by James Stewart. It contains solutions to the odd-numbered exercises in each section, the review sections, the True-False Quizzes, and the Problem Solving sections, as well as solutions to all the exercises in the Concept Checks.

This manual is a text supplement and should be read along *with* the text. You should read all exercise solutions in this manual because many concept explanations are given and then used in subsequent solutions. All concepts necessary to solve a particular problem are not reviewed for every exercise. If you are having difficulty with a previously covered concept, refer back to the section where it was covered for more complete help.

A significant number of today's students are involved in various outside activities, and find it difficult, if not impossible, to attend all class sessions; this manual should help meet the needs of these students. In addition, it is my hope that this manual's solutions will enhance the understanding of all readers of the material and provide insights to solving other exercises.

Some nonstandard use of notation is used in order to save space. If you see a symbol which you don't recognize, refer to the table of Abbreviations and Symbols on page v.

I appreciate feedback concerning errors, solution correctness or style, and manual style. Any comments may be sent directly to me at the address below, at jcole@an.cc.mn.us, or in care of the publisher: Brooks/Cole Thomson Learning, 511 Forest Lodge Road, Pacific Grove, CA 93950.

I would like to thank Jim Stewart, for his guidance; Dan Clegg, of Palomar College, for his careful assistance with most of the new solutions; Brian Betsill, Stephanie Kuhns, and Kathi Townes, of TECH-arts, for their production services; and Gary Ostedt and Carol Ann Benedict, of Brooks/Cole, for entrusting me with this project as well as for their patience and support.

I dedicate this book to my wife, Joan.

Jeffery A. Cole
Anoka-Ramsey Community College
11200 Mississippi Blvd. NW
Coon Rapids, MN 55433

Abbreviations and Symbols

CD	concave downward
CU	concave upward
D	the domain of f
FDT	First Derivative Test
HA	horizontal asymptote(s)
I	interval of convergence
IP	inflection point(s)
R	radius of convergence
VA	vertical asymptote(s)
$\overset{CAS}{=}$	indicates the use of a computer algebra system.
$\overset{H}{=}$	indicates the use of l'Hospital's Rule.
$\overset{j}{=}$	indicates the use of Formula j in the Table of Integrals in the back endpapers.
$\overset{s}{=}$	indicates the use of the substitution $\{u = \sin x, du = \cos x \, dx\}$.
$\overset{c}{=}$	indicates the use of the substitution $\{u = \cos x, du = -\sin x \, dx\}$.

Contents

6   Applications of Integration   223

7   Differential Equations   255

Functions and Models

 Four Ways to Represent a Function • • • • • • • • •

In exercises requiring estimations or approximations, your answers may vary slightly from the answers given here.

1. (a) The point $(-1, -2)$ is on the graph of f, so $f(-1) = -2$.

(b) When $x = 2$, y is about 2.8, so $f(2) \approx 2.8$.

(c) $f(x) = 2$ is equivalent to $y = 2$. When $y = 2$, we have $x = -3$ and $x = 1$.

(d) Reasonable estimates for x when $y = 0$ are $x = -2.5$ and $x = 0.3$.

(e) The domain of f consists of all x-values on the graph of f. For this function, the domain is $-3 \le x \le 3$, or $[-3, 3]$. The range of f consists of all y-values on the graph of f. For this function, the range is $-2 \le y \le 3$, or $[-2, 3]$.

(f) As x increases from -1 to 3, y increases from -2 to 3. Thus, f is increasing on the interval $[-1, 3]$.

3. From Figure 1 in the text, the lowest point occurs at about $(t, a) = (12, -85)$. The highest point occurs at about $(17, 115)$. Thus, the range of the vertical ground acceleration is $-85 \le a \le 115$. In Figure 11, the range of the north-south acceleration is approximately $-325 \le a \le 485$. In Figure 12, the range of the east-west acceleration is approximately $-210 \le a \le 200$.

5. Yes, the curve is the graph of a function because it passes the Vertical Line Test. The domain is $[-3, 2]$ and the range is $[-2, 2]$.

7. No, the curve is not the graph of a function since for $x = -1$ there are infinitely many points on the curve.

9. The person's weight increased to about 160 pounds at age 20 and stayed fairly steady for 10 years. The person's weight dropped to about 120 pounds for the next 5 years, then increased rapidly to about 170 pounds. The next 30 years saw a gradual increase to 190 pounds. Possible reasons for the drop in weight at 30 years of age: diet, exercise, health problems.

11. The water will cool down almost to freezing as the ice melts. Then, when the ice has melted, the water will slowly warm up to room temperature.

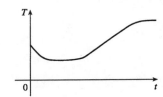

13. Of course, this graph depends strongly on the geographical location!

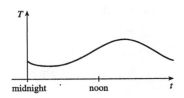

1

15. Height of grass

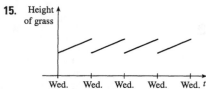

17. (a)

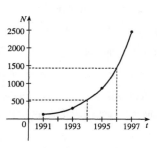

(b) From the graph, we estimate the number of cell-phone subscribers in Malaysia to be about 540 in 1994 and 1450 in 1996.

19. $f(x) = 3x^2 - x + 2$.

$f(2) = 3(2)^2 - 2 + 2 = 12 - 2 + 2 = 12$.

$f(-2) = 3(-2)^2 - (-2) + 2 = 12 + 2 + 2 = 16$.

$f(a) = 3a^2 - a + 2$.

$f(-a) = 3(-a)^2 - (-a) + 2 = 3a^2 + a + 2$.

$f(a+1) = 3(a+1)^2 - (a+1) + 2 = 3(a^2 + 2a + 1) - a - 1 + 2 = 3a^2 + 6a + 3 - a + 1 = 3a^2 + 5a + 4$.

$2f(a) = 2 \cdot f(a) = 2(3a^2 - a + 2) = 6a^2 - 2a + 4$.

$f(2a) = 3(2a)^2 - (2a) + 2 = 3(4a^2) - 2a + 2 = 12a^2 - 2a + 2$.

$f(a^2) = 3(a^2)^2 - (a^2) + 2 = 3(a^4) - a^2 + 2 = 3a^4 - a^2 + 2$.

$[f(a)]^2 = [3a^2 - a + 2]^2 = (3a^2 - a + 2)(3a^2 - a + 2)$

$\quad = 9a^4 - 3a^3 + 6a^2 - 3a^3 + a^2 - 2a + 6a^2 - 2a + 4 = 9a^4 - 6a^3 + 13a^2 - 4a + 4$.

$f(a+h) = 3(a+h)^2 - (a+h) + 2 = 3(a^2 + 2ah + h^2) - a - h + 2 = 3a^2 + 6ah + 3h^2 - a - h + 2$.

21. $f(x) = x - x^2$, so

$f(2+h) = 2 + h - (2+h)^2 = 2 + h - (4 + 4h + h^2) = 2 + h - 4 - 4h - h^2 = -(h^2 + 3h + 2)$,

$f(x+h) = x + h - (x+h)^2 = x + h - x^2 - 2xh - h^2$, and

$\dfrac{f(x+h) - f(x)}{h} = \dfrac{x + h - x^2 - 2xh - h^2 - x + x^2}{h} = \dfrac{h - 2xh - h^2}{h} = \dfrac{h(1 - 2x - h)}{h} = 1 - 2x - h$.

23. $f(x) = x/(3x - 1)$ is defined for all x except when $0 = 3x - 1 \quad \Leftrightarrow \quad x = \frac{1}{3}$, so the domain is $\left\{ x \in \mathbb{R} \mid x \neq \frac{1}{3} \right\} = \left(-\infty, \frac{1}{3} \right) \cup \left(\frac{1}{3}, \infty \right)$.

25. $f(t) = \sqrt{t} + \sqrt[3]{t}$ is defined when $t \geq 0$. These values of t give real number results for $\sqrt{t}$, whereas any value of t gives a real number result for $\sqrt[3]{t}$. The domain is $[0, \infty)$.

27. $h(x) = 1 / \sqrt[4]{x^2 - 5x}$ is defined when $x^2 - 5x > 0 \quad \Leftrightarrow \quad x(x - 5) > 0$. Note that $x^2 - 5x \neq 0$ since that would result in division by zero. The expression $x(x - 5)$ is positive if $x < 0$ or $x > 5$. (See Appendix A for methods for solving inequalities.) Thus, the domain is $(-\infty, 0) \cup (5, \infty)$.

29. $y = \frac{1}{2}t - 1$ is the equation of a line with slope $\frac{1}{2}$ and y-intercept -1. The domain of the function $f(t) = \frac{1}{2}t - 1$ is $\mathbb{R}$, or $(-\infty, \infty)$.

31. $G(x) = \dfrac{3x + |x|}{x}$. Since $|x| = \begin{cases} x & \text{if } x \geq 0 \\ -x & \text{if } x < 0 \end{cases}$, we have

$$G(x) = \begin{cases} \dfrac{3x + x}{x} & \text{if } x > 0 \\[2mm] \dfrac{3x - x}{x} & \text{if } x < 0 \end{cases} = \begin{cases} \dfrac{4x}{x} & \text{if } x > 0 \\[2mm] \dfrac{2x}{x} & \text{if } x < 0 \end{cases} = \begin{cases} 4 & \text{if } x > 0 \\ 2 & \text{if } x < 0 \end{cases}$$

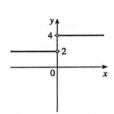

Note that G is not defined for $x = 0$. The domain is $(-\infty, 0) \cup (0, \infty)$.

33. $f(x) = \begin{cases} x & \text{if } x \leq 0 \\ x + 1 & \text{if } x > 0 \end{cases}$

Domain is $\mathbb{R}$, or $(-\infty, \infty)$.

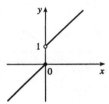

35. $f(x) = \begin{cases} x + 2 & \text{if } x \leq -1 \\ x^2 & \text{if } x > -1 \end{cases}$

Note that for $x = -1$, both $x + 2$ and x^2 are equal to 1. Domain is $\mathbb{R}$.

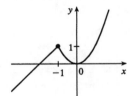

37. Recall that the slope m of a line between the two points (x_1, y_1) and (x_2, y_2) is $m = \dfrac{y_2 - y_1}{x_2 - x_1}$ and an equation of the line connecting those two points is $y - y_1 = m(x - x_1)$. The slope of this line segment is $\dfrac{-6 - 1}{4 - (-2)} = -\dfrac{7}{6}$, so an equation is $y - 1 = -\frac{7}{6}(x + 2)$. The function is $f(x) = -\frac{7}{6}x - \frac{4}{3}$, $-2 \leq x \leq 4$.

39. We need to solve the given equation for y. $x + (y - 1)^2 = 0 \Leftrightarrow (y - 1)^2 = -x \Leftrightarrow y - 1 = \pm\sqrt{-x} \Leftrightarrow y = 1 \pm \sqrt{-x}$. The expression with the positive radical represents the top half of the parabola, and the one with the negative radical represents the bottom half. Hence, we want $f(x) = 1 - \sqrt{-x}$. Note that the domain is $x \leq 0$.

41. For $-1 \leq x \leq 2$, the graph is the line with slope 1 and y-intercept 1, that is, the line $y = x + 1$. For $2 < x \leq 4$, the graph is the line with slope $-\frac{3}{2}$ and x-intercept 4 [which corresponds to the point $(4, 0)$], so $y - 0 = -\frac{3}{2}(x - 4) = -\frac{3}{2}x + 6$. So the function is $f(x) = \begin{cases} x + 1 & \text{if } -1 \leq x \leq 2 \\ -\frac{3}{2}x + 6 & \text{if } 2 < x \leq 4 \end{cases}$

43. Let the length and width of the rectangle be L and W. Then the perimeter is $2L + 2W = 20$ and the area is $A = LW$. Solving the first equation for W in terms of L gives $W = \dfrac{20 - 2L}{2} = 10 - L$. Thus, $A(L) = L(10 - L) = 10L - L^2$. Since lengths are positive, the domain of A is $0 < L < 10$. If we further restrict L to be larger than W, then $5 < L < 10$ would be the domain.

45. Let the length of a side of the equilateral triangle be x. Then by the Pythagorean Theorem, the height y of the triangle satisfies $y^2 + \left(\frac{1}{2}x\right)^2 = x^2$, so that $y^2 = x^2 - \frac{1}{4}x^2 = \frac{3}{4}x^2$ and $y = \frac{\sqrt{3}}{2}x$. Using the formula for the area A of a triangle, $A = \frac{1}{2}(\text{base})(\text{height})$, we obtain $A(x) = \frac{1}{2}(x)\left(\frac{\sqrt{3}}{2}x\right) = \frac{\sqrt{3}}{4}x^2$, with domain $x > 0$.

47. Let each side of the base of the box have length x, and let the height of the box be h. Since the volume is 2, we know that $2 = hx^2$, so that $h = 2/x^2$, and the surface area is $S = x^2 + 4xh$. Thus, $S(x) = x^2 + 4x(2/x^2) = x^2 + (8/x)$, with domain $x > 0$.

49. The height of the box is x and the length and width are $L = 20 - 2x$, $W = 12 - 2x$. Then $V = LWx$ and so $V(x) = (20 - 2x)(12 - 2x)(x) = 4(10 - x)(6 - x)(x) = 4x(60 - 16x + x^2) = 4x^3 - 64x^2 + 240x$. The sides L, W, and x must be positive. Thus, $L > 0 \iff 20 - 2x > 0 \iff x < 10$; $W > 0 \iff 12 - 2x > 0 \iff x < 6$; and $x > 0$. Combining these restrictions gives us the domain $0 < x < 6$.

51. (a)

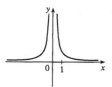

(b) On \$14,000, tax is assessed on \$4000, and 10%(\$4000) = \$400. On \$26,000, tax is assessed on \$16,000, and 10%(\$10,000) + 15%(\$6000) = \$1000 + \$900 = \$1900.

(c) As in part (b), there is \$1000 tax assessed on \$20,000 of income, so the graph of T is a line segment from $(10,000, 0)$ to $(20,000, 1000)$. The tax on \$30,000 is \$2500, so the graph of T for $x > 20,000$ is the ray with initial point $(20,000, 1000)$ that passes through $(30,000, 2500)$.

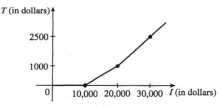

53. (a) Because an even function is symmetric with respect to the y-axis, and the point $(5, 3)$ is on the graph of this even function, the point $(-5, 3)$ must also be on its graph.

(b) Because an odd function is symmetric with respect to the origin, and the point $(5, 3)$ is on the graph of this odd function, the point $(-5, -3)$ must also be on its graph.

55. $f(x) = x^{-2}$

$f(-x) = (-x)^{-2} = \dfrac{1}{(-x)^2} = \dfrac{1}{x^2}$

$= x^{-2} = f(x)$

So f is an even function.

57. $f(x) = x^2 + x$, so $f(-x) = (-x)^2 + (-x) = x^2 - x$. Since this is neither $f(x)$ nor $-f(x)$, the function f is neither even nor odd.

59. $f(x) = x^3 - x$

$f(-x) = (-x)^3 - (-x) = -x^3 + x$

$= -\left(x^3 - x\right) = -f(x)$

So f is odd.

1.2 Mathematical Models · · · · · · · · · · · · · · ·

1. (a) $f(x) = \sqrt[5]{x}$ is a root function with $n = 5$.

(b) $g(x) = \sqrt{1 - x^2}$ is an algebraic function because it is a root of a polynomial.

(c) $h(x) = x^9 + x^4$ is a polynomial of degree 9.

(d) $r(x) = \dfrac{x^2 + 1}{x^3 + x}$ is a rational function because it is a ratio of polynomials.

(e) $s(x) = \tan 2x$ is a trigonometric function.

(f) $t(x) = \log_{10} x$ is a logarithmic function.

3. We notice from the figure that g and h are even functions (symmetric with respect to the y-axis) and that f is an odd function (symmetric with respect to the origin). So (b) $[y = x^5]$ must be f. Since g is flatter than h near the origin, we must have (c) $[y = x^8]$ matched with g and (a) $[y = x^2]$ matched with h.

5. (a) An equation for the family of linear functions with slope 2 is

$y = f(x) = 2x + b$, where b is the y-intercept.

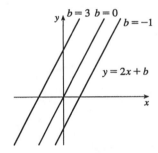

(b) $f(2) = 1$ means that the point $(2, 1)$ is on the graph of f. We can use the point-slope form of a line to obtain an equation for the family of linear functions through the point $(2, 1)$.

$y - 1 = m(x - 2)$, which is equivalent to

$y = mx + (1 - 2m)$ in slope-intercept form.

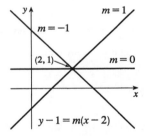

(c) To belong to both families, an equation must have slope $m = 2$, so the equation in part (b), $y = mx + (1 - 2m)$, becomes $y = 2x - 3$. It is the *only* function that belongs to both families.

7. (a)

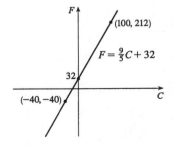

(b) The slope of $\frac{9}{5}$ means that F increases $\frac{9}{5}$ degrees for each increase of $1°$C. (Equivalently, F increases by 9 when C increases by 5 and F decreases by 9 when C decreases by 5.) The F-intercept of 32 is the Fahrenheit temperature corresponding to a Celsius temperature of 0.

9. (a) Using N in place of x and T in place of y, we find the slope to be $\dfrac{T_2 - T_1}{N_2 - N_1} = \dfrac{80 - 70}{173 - 113} = \dfrac{10}{60} = \dfrac{1}{6}$. So a

linear equation is $T - 80 = \frac{1}{6}(N - 173)$ $\Leftrightarrow$ $T - 80 = \frac{1}{6}N - \frac{173}{6}$ $\Leftrightarrow$ $T = \frac{1}{6}N + \frac{307}{6}$ $\left[\frac{307}{6} = 51.1\overline{6}\right]$.

(b) The slope of $\frac{1}{6}$ means that the temperature in Fahrenheit degrees increases one-sixth as rapidly as the number of cricket chirps per minute. Said differently, each increase of 6 cricket chirps per minute corresponds to an increase of 1°F.

(c) When $N = 150$, the temperature is given approximately by $T = \frac{1}{6}(150) + \frac{307}{6} = 76.1\overline{6}°F \approx 76°F$.

11. (a) We are given $\dfrac{\text{change in pressure}}{10 \text{ feet change in depth}} = \dfrac{4.34}{10} = 0.434$. Using P for pressure and d for depth with the point $(d, P) = (0, 15)$, we have the slope-intercept form of the line, $P = 0.434d + 15$.

(b) When $P = 100$, then $100 = 0.434d + 15$ $\Leftrightarrow$ $0.434d = 85$ $\Leftrightarrow$ $d = \frac{85}{0.434} \approx 195.85$ feet. Thus, the pressure is 100 lb/in^2 at a depth of approximately 196 feet.

13. (a) The data appear to be periodic and a sine or cosine function would make the best model. A model of the form $f(x) = a\cos(bx) + c$ seems appropriate.

(b) The data appear to be decreasing in a linear fashion. A model of the form $f(x) = mx + b$ seems appropriate.

Some values are given to many decimal places. These are the results given by several computer algebra systems — rounding is left to the reader.

15. (a)

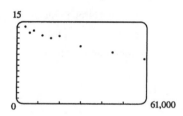

A linear model does seem appropriate.

(b) Using the points $(4000, 14.1)$ and $(60,000, 8.2)$, we obtain

$$y - 14.1 = \frac{8.2 - 14.1}{60,000 - 4000}(x - 4000) \text{ or, equivalently,}$$

$$y \approx -0.000105357x + 14.521429.$$

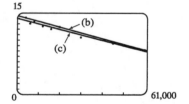

(c) Using a computing device, we obtain the least squares regression line $y = -0.0000997855x + 13.950764$. The following commands and screens illustrate how to find the least squares regression line on a TI-83 Plus. Enter the data into list one (L1) and list two (L2). Press $\boxed{\text{STAT}}$ $\boxed{1}$ to enter the editor.

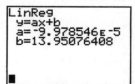

Find the regession line and store it in Y$_1$. Press $\boxed{\text{2nd}}$ $\boxed{\text{QUIT}}$ $\boxed{\text{STAT}}$ $\boxed{\blacktriangleright}$ $\boxed{4}$ $\boxed{\text{VARS}}$ $\boxed{\blacktriangleright}$ $\boxed{1}$ $\boxed{1}$ $\boxed{\text{ENTER}}$.

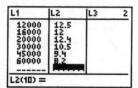
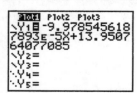

Note from the last figure that the regression line has been stored in Y_1 and that Plot1 has been turned on (Plot1 is highlighted). You can turn on Plot1 from the Y= menu by placing the cursor on Plot1 and pressing ENTER or by pressing 2nd STAT PLOT 1 ENTER.

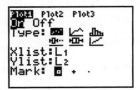

Now press ZOOM 9 to produce a graph of the data and the regression line. Note that choice 9 of the ZOOM menu automatically selects a window that displays all of the data.

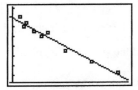

(d) When $x = 25{,}000$, $y \approx 11.456$; or about 11.5 per 100 population.

(e) When $x = 80{,}000$, $y \approx 5.968$; or about a 6% chance.

(f) When $x = 200{,}000$, y is negative, so the model does not apply.

17. (a)

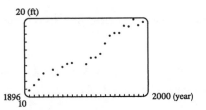

A linear model does seem appropriate.

(b)

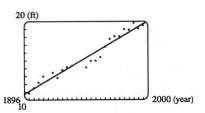

Using a computing device, we obtain the least squares regression line $y = 0.089119747x - 158.2403249$, where x is the year and y is the height in feet.

(c) When $x = 2000$, the model gives $y \approx 20.00$ ft. Note that the actual winning height for the 2000 Olympics is *less than* the winning height for 1996 — so much for that prediction.

(d) When $x = 2100$, $y \approx 28.91$ ft. This would be an increase of 9.49 ft from 1996 to 2100. Even though there was an increase of 8.59 ft from 1900 to 1996, it is unlikely that a similar increase will occur over the next 100 years.

19.

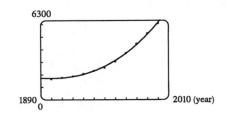

Using a computing device, we obtain the cubic function $y = ax^3 + bx^2 + cx + d$ with $a = 0.0012354312$, $b = -6.722261072$, $c = 12{,}165.08275$, and $d = -7{,}318{,}428.648$. When $x = 1925$, $y \approx 1913$ (million).

1.3 New Functions from Old Functions · · · · · · · · · ·

1. (a) If the graph of f is shifted 3 units upward, its equation becomes $y = f(x) + 3$.

(b) If the graph of f is shifted 3 units downward, its equation becomes $y = f(x) - 3$.

(c) If the graph of f is shifted 3 units to the right, its equation becomes $y = f(x - 3)$.

(d) If the graph of f is shifted 3 units to the left, its equation becomes $y = f(x + 3)$.

(e) If the graph of f is reflected about the x-axis, its equation becomes $y = -f(x)$.

(f) If the graph of f is reflected about the y-axis, its equation becomes $y = f(-x)$.

(g) If the graph of f is stretched vertically by a factor of 3, its equation becomes $y = 3f(x)$.

(h) If the graph of f is shrunk vertically by a factor of 3, its equation becomes $y = \frac{1}{3}f(x)$.

3. (a) (graph 3) The graph of f is shifted 4 units to the right and has equation $y = f(x - 4)$.

(b) (graph 1) The graph of f is shifted 3 units upward and has equation $y = f(x) + 3$.

(c) (graph 4) The graph of f is shrunk vertically by a factor of 3 and has equation $y = \frac{1}{3}f(x)$.

(d) (graph 5) The graph of f is shifted 4 units to the left and reflected about the x-axis. Its equation is
$y = -f(x + 4)$.

(e) (graph 2) The graph of f is shifted 6 units to the left and stretched vertically by a factor of 2. Its equation is
$y = 2f(x + 6)$.

5. (a) To graph $y = f(2x)$ we shrink the graph of f horizontally by a factor of 2.

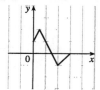

The point $(4, -1)$ on the graph of f corresponds
to the point $\left(\frac{1}{2} \cdot 4, -1\right) = (2, -1)$.

(b) To graph $y = f\left(\frac{1}{2}x\right)$ we stretch the graph of f horizontally by a factor of 2.

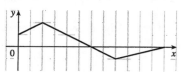

The point $(4, -1)$ on the graph of f corresponds
to the point $(2 \cdot 4, -1) = (8, -1)$.

(c) To graph $y = f(-x)$ we reflect the graph of f about the y-axis.

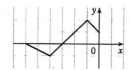

The point $(4, -1)$ on the graph of f corresponds
to the point $(-1 \cdot 4, -1) = (-4, -1)$.

(d) To graph $y = -f(-x)$ we reflect the graph of f about the y-axis, then about the x-axis.

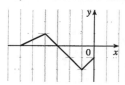

The point $(4, -1)$ on the graph of f corresponds
to the point $(-1 \cdot 4, -1 \cdot -1) = (-4, 1)$.

7. The graph of $y = f(x) = \sqrt{3x - x^2}$ has been shifted 4 units to the left, reflected about the x-axis, and shifted downward 1 unit. Thus, a function describing the graph is

$$y = \underbrace{-1 \cdot}_{\substack{\text{reflect} \\ \text{about} \\ x\text{-axis}}} \underbrace{f(x+4)}_{\substack{\text{shift} \\ 4 \text{ units} \\ \text{left}}} \underbrace{-1}_{\substack{\text{shift} \\ 1 \text{ unit} \\ \text{down}}}$$

This function can be written as

$$y = -f(x+4) - 1 = -\sqrt{3(x+4) - (x+4)^2} - 1 = -\sqrt{3x + 12 - (x^2 + 8x + 16)} - 1$$
$$= -\sqrt{-x^2 - 5x - 4} - 1$$

9. $y = -1/x$: Start with the graph of $y = 1/x$ and reflect about the x-axis.

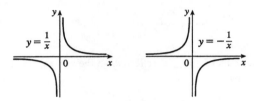

11. $y = \tan 2x$: Start with the graph of $y = \tan x$ and compress horizontally by a factor of 2.

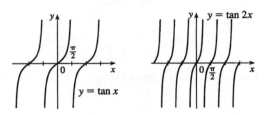

13. $y = \cos(x/2)$: Start with the graph of $y = \cos x$ and stretch horizontally by a factor of 2.

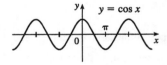
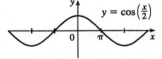

15. $y = \dfrac{1}{x - 3}$: Start with the graph of $y = 1/x$ and shift 3 units to the right.

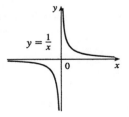
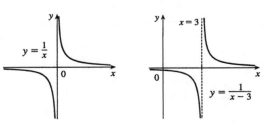

17. $y = \frac{1}{3} \sin\left(x - \frac{\pi}{6}\right)$: Start with the graph of $y = \sin x$, shift $\frac{\pi}{6}$ units to the right, and then compress vertically by a factor of 3.

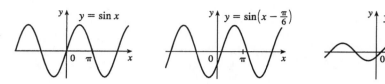

19. $y = 1 + 2x - x^2 = -x^2 + 2x + 1 = -(x^2 - 2x + 1) + 1 + 1 = -(x-1)^2 + 2$: Start with the graph of $y = x^2$, shift 1 unit right, reflect about the x-axis, and then shift 2 units upward.

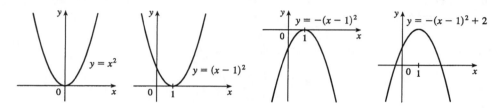

21. $y = 2 - \sqrt{x+1}$: Start with the graph of $y = \sqrt{x}$, reflect about the x-axis, shift 1 unit to the left, and then shift 2 units upward.

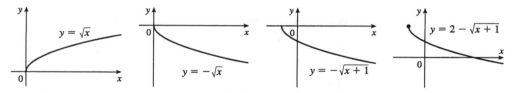

23. $y = |\sin x|$: Start with the graph of $y = \sin x$ and reflect all the parts of the graph below the x-axis about the x-axis.

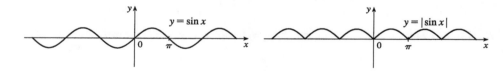

25. This is just like the solution to Example 4 except the amplitude of the curve is $14 - 12 = 2$. So the function is $L(t) = 12 + 2\sin\left[\frac{2\pi}{365}(t - 80)\right]$. March 31 is the 90th day of the year, so the model gives $L(90) \approx 12.34$ h. The daylight time (5:51 A.M. to 6:18 P.M.) is 12 hours and 27 minutes, or 12.45 h. The model value differs from the actual value by $\frac{12.45 - 12.34}{12.45} \approx 0.009$, less than 1%.

27. (a) To obtain $y = f(|x|)$, the portion of the graph of $y = f(x)$ to the right of the y-axis is reflected about the y-axis.

(b) $y = \sin |x|$

(c) $y = \sqrt{|x|}$

$y = \sqrt{x}$

$y = \sqrt{|x|}$

29. Assuming that successive horizontal and vertical gridlines are a unit apart, we can make a table of approximate values as follows.

x	0	1	2	3	4	5	6
$f(x)$	2	1.7	1.3	1.0	0.7	0.3	0
$g(x)$	2	2.7	3	2.8	2.4	1.7	0
$f(x) + g(x)$	4	4.4	4.3	3.8	3.1	2.0	0

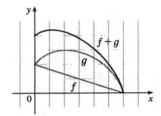

Connecting the points $(x, f(x) + g(x))$ with a smooth curve gives an approximation to the graph of $f + g$. Extra points can be plotted between those listed above if necessary.

31. $f(x) = x^3 + 2x^2$; $g(x) = 3x^2 - 1$. $D = \mathbb{R}$ for both f and g.

$(f + g)(x) = (x^3 + 2x^2) + (3x^2 - 1) = x^3 + 5x^2 - 1$, $D = \mathbb{R}$.

$(f - g)(x) = (x^3 + 2x^2) - (3x^2 - 1) = x^3 - x^2 + 1$, $D = \mathbb{R}$.

$(fg)(x) = (x^3 + 2x^2)(3x^2 - 1) = 3x^5 + 6x^4 - x^3 - 2x^2$, $D = \mathbb{R}$.

$\left(\dfrac{f}{g}\right)(x) = \dfrac{x^3 + 2x^2}{3x^2 - 1}$, $D = \left\{ x \mid x \neq \pm\dfrac{1}{\sqrt{3}} \right\}$ since $3x^2 - 1 \neq 0$.

33. $f(x) = x$, $g(x) = 1/x$

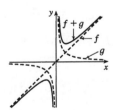

35. $f(x) = \sin x$, $D = \mathbb{R}$; $g(x) = 1 - \sqrt{x}$, $D = [0, \infty)$.

$(f \circ g)(x) = f(g(x)) = f\left(1 - \sqrt{x}\right) = \sin(1 - \sqrt{x})$, $D = [0, \infty)$.

$(g \circ f)(x) = g(f(x)) = g(\sin x) = 1 - \sqrt{\sin x}$. For $\sqrt{\sin x}$ to be defined, we must have

$\sin x \geq 0 \iff x \in [0, \pi], [2\pi, 3\pi], [-2\pi, -\pi], [4\pi, 5\pi], [-4\pi, -3\pi], \ldots$, so

$D = \{x \mid x \in [2n\pi, \pi + 2n\pi]$, where n is an integer$\}$.

$(f \circ f)(x) = f(f(x)) = f(\sin x) = \sin(\sin x)$, $D = \mathbb{R}$.

$(g \circ g)(x) = g(g(x)) = g(1 - \sqrt{x}) = 1 - \sqrt{1 - \sqrt{x}}$, $D = \{x \geq 0 \mid 1 - \sqrt{x} \geq 0\} = [0, 1]$.

37. $f(x) = x + \dfrac{1}{x}$, $D = \{x \mid x \neq 0\}$; $g(x) = \dfrac{x+1}{x+2}$, $D = \{x \mid x \neq -2\}$.

$$(f \circ g)(x) = f(g(x)) = f\left(\frac{x+1}{x+2}\right) = \frac{x+1}{x+2} + \frac{1}{\frac{x+1}{x+2}} = \frac{x+1}{x+2} + \frac{x+2}{x+1}$$

$$= \frac{(x+1)(x+1) + (x+2)(x+2)}{(x+2)(x+1)} = \frac{(x^2 + 2x + 1) + (x^2 + 4x + 4)}{(x+2)(x+1)} = \frac{2x^2 + 6x + 5}{(x+2)(x+1)}$$

Since $g(x)$ is not defined for $x = -2$ and $f(g(x))$ is not defined for $x = -2$ and $x = -1$, the domain of $(f \circ g)(x)$ is $D = \{x \mid x \neq -2, -1\}$.

$$(g \circ f)(x) = g(f(x)) = g\left(x + \frac{1}{x}\right) = \frac{\left(x + \frac{1}{x}\right) + 1}{\left(x + \frac{1}{x}\right) + 2} = \frac{\dfrac{x^2 + 1 + x}{x}}{\dfrac{x^2 + 1 + 2x}{x}} = \frac{x^2 + x + 1}{x^2 + 2x + 1} = \frac{x^2 + x + 1}{(x+1)^2}.$$

Since $f(x)$ is not defined for $x = 0$ and $g(f(x))$ is not defined for $x = -1$, the domain of $(g \circ f)(x)$ is $D = \{x \mid x \neq -1, 0\}$.

$$(f \circ f)(x) = f(f(x)) = f\left(x + \frac{1}{x}\right) = \left(x + \frac{1}{x}\right) + \frac{1}{x + \frac{1}{x}} = x + \frac{1}{x} + \frac{1}{\frac{x^2 + 1}{x}} = x + \frac{1}{x} + \frac{x}{x^2 + 1}$$

$$= \frac{x(x)\left(x^2 + 1\right) + 1\left(x^2 + 1\right) + x(x)}{x\left(x^2 + 1\right)} = \frac{x^4 + x^2 + x^2 + 1 + x^2}{x\left(x^2 + 1\right)}$$

$$= \frac{x^4 + 3x^2 + 1}{x\left(x^2 + 1\right)}, \quad D = \{x \mid x \neq 0\}.$$

$$(g \circ g)(x) = g(g(x)) = g\left(\frac{x+1}{x+2}\right) = \frac{\frac{x+1}{x+2} + 1}{\frac{x+1}{x+2} + 2} = \frac{\frac{x+1+1(x+2)}{x+2}}{\frac{x+1+2(x+2)}{x+2}} = \frac{x+1+x+2}{x+1+2x+4} = \frac{2x+3}{3x+5}.$$

Since $g(x)$ is not defined for $x = -2$ and $g(g(x))$ is not defined for $x = -\frac{5}{3}$, the domain of $(g \circ g)(x)$ is $D = \{x \mid x \neq -2, -\frac{5}{3}\}$.

39. $(f \circ g \circ h)(x) = f(g(h(x))) = f(g(x + 3)) = f((x+3)^2 + 2)$
$$= f(x^2 + 6x + 11) = \sqrt{(x^2 + 6x + 11) - 1} = \sqrt{x^2 + 6x + 10}$$

41. Let $g(x) = x^2 + 1$ and $f(x) = x^{10}$. Then $(f \circ g)(x) = (x^2 + 1)^{10} = F(x)$.

43. Let $g(t) = \cos t$ and $f(t) = \sqrt{t}$. Then $(f \circ g)(t) = \sqrt{\cos t} = u(t)$.

45. Let $h(x) = x^2$, $g(x) = 3^x$, and $f(x) = 1 - x$. Then $(f \circ g \circ h)(x) = 1 - 3^{x^2} = H(x)$.

47. Let $h(x) = \sqrt{x}$, $g(x) = \sec x$, and $f(x) = x^4$. Then $(f \circ g \circ h)(x) = (\sec \sqrt{x})^4 = \sec^4(\sqrt{x}) = H(x)$.

49. (a) $g(2) = 5$, because the point $(2, 5)$ is on the graph of g. Thus, $f(g(2)) = f(5) = 4$, because the point $(5, 4)$ is on the graph of f.

(b) $g(f(0)) = g(0) = 3$

(c) $(f \circ g)(0) = f(g(0)) = f(3) = 0$

(d) $(g \circ f)(6) = g(f(6)) = g(6)$. This value is not defined, because there is no point on the graph of g that has x-coordinate 6.

(e) $(g \circ g)(-2) = g(g(-2)) = g(1) = 4$

(f) $(f \circ f)(4) = f(f(4)) = f(2) = -2$

51. (a) Using the relationship *distance* = *rate* · *time* with the radius r as the distance, we have $r(t) = 60t$.

(b) $A = \pi r^2 \;\;\Rightarrow\;\; (A \circ r)(t) = A(r(t)) = \pi(60t)^2 = 3600\pi t^2$. This formula gives us the extent of the rippled area (in cm^2) at any time t.

53. (a)

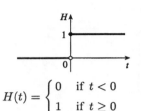

$$H(t) = \begin{cases} 0 & \text{if } t < 0 \\ 1 & \text{if } t \geq 0 \end{cases}$$

(b)

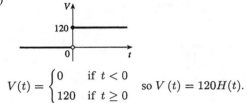

$$V(t) = \begin{cases} 0 & \text{if } t < 0 \\ 120 & \text{if } t \geq 0 \end{cases} \quad \text{so } V(t) = 120H(t).$$

(c)

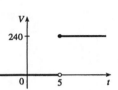

Starting with the formula in part (b), we replace 120 with 240 to reflect the different voltage. Also, because we are starting 5 units to the right of $t = 0$, we replace t with $t - 5$. Thus, the formula is $V(t) = 240H(t - 5)$.

55. (a) By examining the variable terms in g and h, we deduce that we must square g to get the terms $4x^2$ and $4x$ in h. If we let $f(x) = x^2 + c$, then $(f \circ g)(x) = f(g(x)) = f(2x + 1) = (2x + 1)^2 + c = 4x^2 + 4x + (1 + c)$. Since $h(x) = 4x^2 + 4x + 7$, we must have $1 + c = 7$. So $c = 6$ and $f(x) = x^2 + 6$.

(b) We need a function g so that
$f(g(x)) = 3(g(x)) + 5 = h(x) = 3x^2 + 3x + 2 = 3(x^2 + x) + 2 = 3(x^2 + x - 1) + 5$. So we see that
$g(x) = x^2 + x - 1$.

57. We need to examine $h(-x)$.

$$h(-x) = (f \circ g)(-x) = f(g(-x)) = f(g(x)) \quad \text{[because } g \text{ is even]} \quad = h(x)$$

Because $h(-x) = h(x)$, h is an even function.

59. (a) $P = (a, g(a))$ and $Q = (g(a), g(a))$ because Q has the same y-value as P and it is on the line $y = x$.

(b) The x-value of Q is $g(a)$; this is also the x-value of R. The y-value of R is therefore $f(x\text{-value})$, that is, $f(g(a))$. Hence, $R = (g(a), f(g(a)))$.

(c) The coordinates of S are $(a, f(g(a)))$ or, equivalently, $(a, h(a))$.

(d)

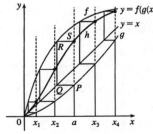

1.4 Graphing Calculators and Computers · · · · · · · ·

1. $f(x) = 10 + 25x - x^3$

(a) $[-4, 4]$ by $[-4, 4]$

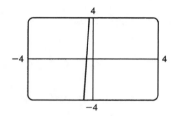

(b) $[-10, 10]$ by $[-10, 10]$

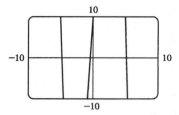

(c) $[-20, 20]$ by $[-100, 100]$

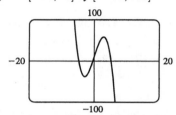

(d) $[-100, 100]$ by $[-200, 200]$

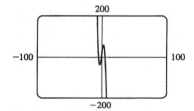

The most appropriate graph is produced in viewing rectangle (c) because the maximum and minimum points are fairly easy to see and estimate.

3. Since the graph of $f(x) = 5 + 20x - x^2$ is a parabola opening downward, an appropriate viewing rectangle should include the maximum point.

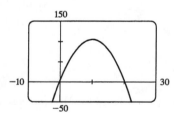

5. $f(x) = \sqrt[4]{81 - x^4}$ is defined when

$81 - x^4 \geq 0 \iff x^4 \leq 81 \iff |x| \leq 3$, so
the domain of f is $[-3, 3]$. Also

$0 \leq \sqrt[4]{81 - x^4} \leq \sqrt[4]{81} = 3$, so the range is $[0, 3]$.

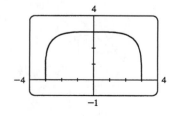

7. The graph of $f(x) = x^2 + (100/x)$ has a vertical asymptote of $x = 0$. As you zoom out, the graph of f looks more and more like that of $y = x^2$.

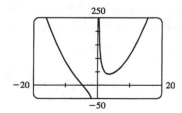

9. $f(x) = \cos(100x)$

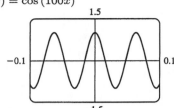

11. $f(x) = \sin(x/40)$

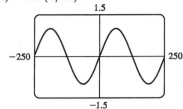

13. $y = 3^{\cos(x^2)}$

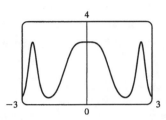

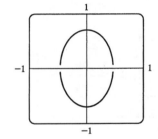

15. We must solve the given equation for y to obtain equations for the upper and lower halves of the ellipse.

$$4x^2 + 2y^2 = 1 \quad \Leftrightarrow \quad 2y^2 = 1 - 4x^2 \quad \Leftrightarrow \quad y^2 = \frac{1 - 4x^2}{2}$$

$$\Leftrightarrow \quad y = \pm\sqrt{\frac{1 - 4x^2}{2}}$$

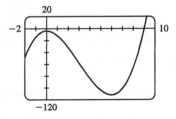

17. From the graph of $f(x) = x^3 - 9x^2 - 4$, we see that there is one solution of the equation $f(x) = 0$ and it is slightly larger than 9. By zooming in or using a root or zero feature, we obtain $x \approx 9.05$.

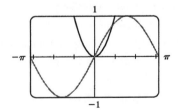

19. We see that the graphs of $f(x) = x^2$ and $g(x) = \sin x$ intersect twice. One solution is $x = 0$. The other solution of $f = g$ is the x-coordinate of the point of intersection in the first quadrant. Using an intersect feature or zooming in, we find this value to be approximately 0.88. Alternatively, we could find that value by finding the positive zero of $h(x) = x^2 - \sin x$.

Note: After producing the graph on a TI-83 Plus, we can find the approximate value 0.88 by using the following keystrokes: 2nd CALC 5 ENTER ENTER 1 ENTER. The "1" is just a guess for 0.88.

21. $g(x) = x^3/10$ is larger than $f(x) = 10x^2$ whenever $x > 100$.

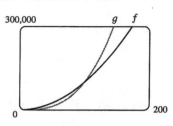

23.

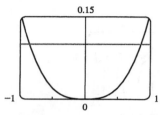

We see from the graphs of $y = |\sin x - x|$ and $y = 0.1$ that there are two solutions to the equation $|\sin x - x| = 0.1$: $x \approx -0.85$ and $x \approx 0.85$. The condition $|\sin x - x| < 0.1$ holds for any x lying between these two values.

25. (a) The root functions $y = \sqrt{x}$, $y = \sqrt[4]{x}$ and $y = \sqrt[6]{x}$

(b) The root functions $y = x$, $y = \sqrt[3]{x}$ and $y = \sqrt[5]{x}$

(c) The root functions $y = \sqrt{x}$, $y = \sqrt[3]{x}$, $y = \sqrt[4]{x}$ and $y = \sqrt[5]{x}$

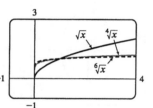

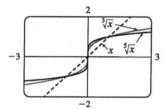

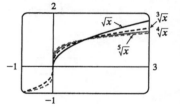

(d) • For any n, the nth root of 0 is 0 and the nth root of 1 is 1; that is, all nth root functions pass through the points $(0,0)$ and $(1,1)$.
  • For odd n, the domain of the nth root function is $\mathbb{R}$, while for even n, it is $\{x \in \mathbb{R} \mid x \geq 0\}$.
  • Graphs of even root functions look similar to that of $\sqrt{x}$, while those of odd root functions resemble that of $\sqrt[3]{x}$.
  • As n increases, the graph of $\sqrt[n]{x}$ becomes steeper near 0 and flatter for $x > 1$.

27. $f(x) = x^4 + cx^2 + x$. If $c < 0$, there are three humps: two minimum points and a maximum point. These humps get flatter as c increases, until at $c = 0$ two of the humps disappear and there is only one minimum point. This single hump then moves to the right and approaches the origin as c increases.

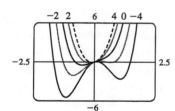

29. $y = x^n 2^{-x}$. As n increases, the maximum of the function moves further from the origin, and gets larger. Note, however, that regardless of n, the function approaches 0 as $x \to \infty$.

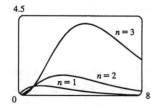

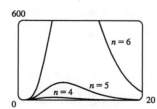

31. $y^2 = cx^3 + x^2$

If $c < 0$, the loop is to the right of the origin, and if c is positive, it is
to the left. In both cases, the closer c is to 0, the larger the loop is.
(In the limiting case, $c = 0$, the loop is "infinite", that is, it doesn't
close.) Also, the larger $|c|$ is, the steeper the slope is on the loopless
side of the origin.

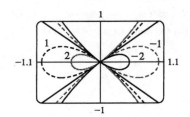

33. The graphing window is 95 pixels wide and we want to start with $x = 0$ and end with $x = 2\pi$. Since there are 94
"gaps" between pixels, the distance between pixels is $\frac{2\pi - 0}{94}$. Thus, the x-values that the calculator actually plots are
$x = 0 + \frac{2\pi}{94} \cdot n$, where $n = 0, 1, 2, \ldots, 93, 94$. For $y = \sin 2x$, the actual points plotted by the calculator are
$\left(\frac{2\pi}{94} \cdot n, \sin\left(2 \cdot \frac{2\pi}{94} \cdot n\right)\right)$ for $n = 0, 1, \ldots, 94$. For $y = \sin 96x$, the points plotted are $\left(\frac{2\pi}{94} \cdot n, \sin\left(96 \cdot \frac{2\pi}{94} \cdot n\right)\right)$
for $n = 0, 1, \ldots, 94$. But

$$\begin{aligned}
\sin\left(96 \cdot \tfrac{2\pi}{94} \cdot n\right) &= \sin\left(94 \cdot \tfrac{2\pi}{94} \cdot n + 2 \cdot \tfrac{2\pi}{94} \cdot n\right) = \sin\left(2\pi n + 2 \cdot \tfrac{2\pi}{94} \cdot n\right) \\
&= \sin(2\pi n)\cos\left(2 \cdot \tfrac{2\pi}{94} \cdot n\right) + \cos(2\pi n)\sin\left(2 \cdot \tfrac{2\pi}{94} \cdot n\right) \quad \text{[Addition formula for the sine]} \\
&= 0 \cdot \cos\left(2 \cdot \tfrac{2\pi}{94} \cdot n\right) + 1 \cdot \sin\left(2 \cdot \tfrac{2\pi}{94} \cdot n\right) = \sin\left(2 \cdot \tfrac{2\pi}{94} \cdot n\right), \quad n = 0, 1, \ldots, 94
\end{aligned}$$

So the y-values, and hence the points, plotted for $y = \sin 96x$ are identical to those plotted for $y = \sin 2x$.
Note: Try graphing $y = \sin 94x$. Can you see why all the y-values are zero?

◆1.5 Exponential Functions • • • • • • • • • • • • •

1. (a) $f(x) = a^x$, $a > 0$

(b) $\mathbb{R}$

(c) $(0, \infty)$

(d) See Figures 4(c), 4(b), and 4(a), respectively.

3. All of these graphs approach 0 as $x \to -\infty$, all of them pass
through the point $(0, 1)$, and all of them are increasing and
approach ∞ as $x \to \infty$. The larger the base, the faster the
function increases for $x > 0$, and the faster it approaches 0 as
$x \to -\infty$.

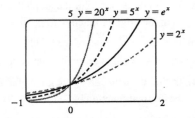

5. The functions with bases greater than 1 (3^x and 10^x) are
increasing, while those with bases less than 1
$\left[\left(\frac{1}{3}\right)^x \text{ and } \left(\frac{1}{10}\right)^x\right]$ are decreasing. The graph of $\left(\frac{1}{3}\right)^x$ is the
reflection of that of 3^x about the y-axis, and the graph of
$\left(\frac{1}{10}\right)^x$ is the reflection of that of 10^x about the y-axis. The
graph of 10^x increases more quickly than that of 3^x for
$x > 0$, and approaches 0 faster as $x \to -\infty$.

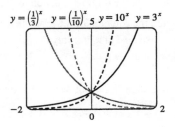

7. We start with the graph of $y = 4^x$ (Figure 3) and then shift 3 units downward. This shift doesn't affect the domain, but the range of $y = 4^x - 3$ is $(-3, \infty)$. There is a horizontal asymptote of $y = -3$.

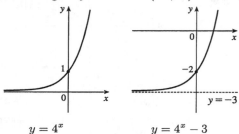

$$y = 4^x \qquad\qquad y = 4^x - 3$$

9. We start with the graph of $y = 2^x$ (Figure 3), reflect it about the y-axis, and then about the x-axis (or just rotate $180°$ to handle both reflections) to obtain the graph of $y = -2^{-x}$. In each graph, $y = 0$ is the horizontal asymptote.

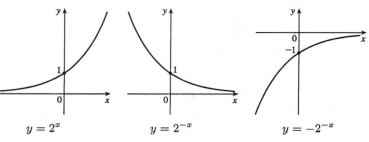

$$y = 2^x \qquad\qquad y = 2^{-x} \qquad\qquad y = -2^{-x}$$

11. We start with the graph of $y = e^x$ (Figure 13), reflect it about the x-axis, and then shift 3 units upward. Note the horizontal asymptote of $y = 3$.

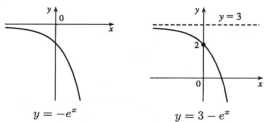

$$y = -e^x \qquad\qquad y = 3 - e^x$$

13. (a) To find the equation of the graph that results from shifting the graph of $y = e^x$ 2 units downward, we subtract 2 from the original function to get $y = e^x - 2$.

(b) To find the equation of the graph that results from shifting the graph of $y = e^x$ 2 units to the right, we replace x with $x - 2$ in the original function to get $y = e^{(x-2)}$.

(c) To find the equation of the graph that results from reflecting the graph of $y = e^x$ about the x-axis, we multiply the original function by -1 to get $y = -e^x$.

(d) To find the equation of the graph that results from reflecting the graph of $y = e^x$ about the y-axis, we replace x with $-x$ in the original function to get $y = e^{-x}$.

(e) To find the equation of the graph that results from reflecting the graph of $y = e^x$ about the x-axis and then about the y-axis, we first multiply the original function by -1 (to get $y = -e^x$) and then replace x with $-x$ in this equation to get $y = -e^{-x}$.

15. Use $y = Ca^x$ with the points $(1, 6)$ and $(3, 24)$. $6 = Ca^1$ $\left(C = \frac{6}{a}\right)$ and $24 = Ca^3$ $\Rightarrow$ $24 = \left(\dfrac{6}{a}\right)a^3$ $\Rightarrow$ $4 = a^2$ $\Rightarrow$ $a = 2$ (since $a > 0$) and $C = 3$. The function is $f(x) = 3 \cdot 2^x$.

17. If $f(x) = 5^x$, then $\dfrac{f(x+h) - f(x)}{h} = \dfrac{5^{x+h} - 5^x}{h} = \dfrac{5^x 5^h - 5^x}{h} = \dfrac{5^x(5^h - 1)}{h} = 5^x\left(\dfrac{5^h - 1}{h}\right)$.

19. 2 ft = 24 in, $f(24) = 24^2$ in = 576 in = 48 ft. $g(24) = 2^{24}$ in = $2^{24}/(12 \cdot 5280)$ mi ≈ 265 mi

21. The graph of g finally surpasses that of f at $x \approx 35.8$.

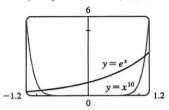

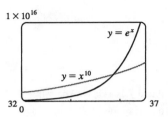

23. (a) Fifteen hours represents 5 doubling periods (one doubling period is three hours).

$100 \cdot 2^5 = 3200$

(b) In t hours, there will be $t/3$ doubling periods. The initial population is 100, so the population y at time t is $y = 100 \cdot 2^{t/3}$.

(c) $t = 20 \implies y = 100 \cdot 2^{20/3} \approx 10{,}159$

(d) We graph $y_1 = 100 \cdot 2^{x/3}$ and $y_2 = 50{,}000$. The two curves intersect at $x \approx 26.9$, so the population reaches 50,000 in about 26.9 hours.

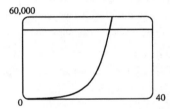

25. An exponential model is $y = ab^t$, where $a = 3.3039025371408 \times 10^{-12}$ and $b = 1.0177407727104$. This model gives $y(1993) \approx 5494$ million and $y(2010) \approx 7409$ million.

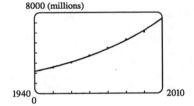

⟨1.6⟩ Inverse Functions and Logarithms • • • • • • • • • • • •

1. (a) See Definition 1.

(b) It must pass the Horizontal Line Test.

3. f is not one-to-one because $2 \neq 6$, but $f(2) = f(6)$.

5. No horizontal line intersects the graph of f more than once. Thus, by the Horizontal Line Test, f is one-to-one.

7. The horizontal line $y = 0$ (the x-axis) intersects the graph of f in more than one point. Thus, by the Horizontal Line Test, f is not one-to-one.

9. The graph of $f(x) = \frac{1}{2}(x + 5)$ is a line with slope $\frac{1}{2}$. It passes the Horizontal Line Test, so f is one-to-one.

Algebraic soution: If $x_1 \neq x_2$, then $x_1 + 5 \neq x_2 + 5 \implies \frac{1}{2}(x_1 + 5) \neq \frac{1}{2}(x_2 + 5) \implies f(x_1) \neq f(x_2)$, so f is one-to-one.

11. $g(x) = |x| \implies g(-1) = 1 = g(1)$, so g is not one-to-one.

13. A football will attain every height h up to its maximum height twice: once on the way up, and again on the way down. Thus, even if t_1 does not equal t_2, $f(t_1)$ may equal $f(t_2)$, so f is not 1-1.

15. f does not pass the Horizontal Line Test,

so f is not 1-1.

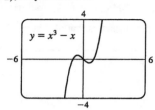

17. Since $f(2) = 9$ and f is 1-1, we know that $f^{-1}(9) = 2$. Remember, if the point $(2, 9)$ is on the graph of f, then the point $(9, 2)$ is on the graph of f^{-1}.

19. First, we must determine x such that $g(x) = 4$. By inspection, we see that if $x = 0$, then $g(x) = 4$. Since g is 1-1 (g is an increasing function), it has an inverse, and $g^{-1}(4) = 0$.

21. We solve $C = \frac{5}{9}(F - 32)$ for F: $\frac{9}{5}C = F - 32$ $\Rightarrow$ $F = \frac{9}{5}C + 32$. This gives us the Fahrenheit temperature F as a function of the Celsius temperature C. $F \geq -459.67$ $\Rightarrow$ $\frac{9}{5}C + 32 \geq -459.67$ $\Rightarrow$ $\frac{9}{5}C \geq -491.67$ $\Rightarrow$ $C \geq -273.15$, the domain of the inverse function.

23. $f(x) = \sqrt{10 - 3x}$ $\Rightarrow$ $y = \sqrt{10 - 3x}$; $y \geq 0$ $\Rightarrow$ $y^2 = 10 - 3x$ $\Rightarrow$ $3x = 10 - y^2$ $\Rightarrow$ $x = -\frac{1}{3}y^2 + \frac{10}{3}$. Interchange x and y: $y = -\frac{1}{3}x^2 + \frac{10}{3}$. So $f^{-1}(x) = -\frac{1}{3}x^2 + \frac{10}{3}$. Note that the domain of f^{-1} is $x \geq 0$.

25. $f(x) = e^{x^3}$ $\Rightarrow$ $y = e^{x^3}$ $\Rightarrow$ $\ln y = x^3$ $\Rightarrow$ $x = \sqrt[3]{\ln y}$. Interchange x and y: $y = \sqrt[3]{\ln x}$. So $f^{-1}(x) = \sqrt[3]{\ln x}$.

27. $y = \ln(x + 3)$ $\Rightarrow$ $x + 3 = e^y$ $\Rightarrow$ $x = e^y - 3$. Interchange x and y: $y = e^x - 3$. So $f^{-1}(x) = e^x - 3$.

29. $y = f(x) = 1 - \dfrac{2}{x^2}$ $\Rightarrow$ $1 - y = \dfrac{2}{x^2}$ $\Rightarrow$ $x^2 = \dfrac{2}{1 - y}$ $\Rightarrow$

$x = \sqrt{\dfrac{2}{1 - y}}$, since $x > 0$. Interchange x and y: $y = \sqrt{\dfrac{2}{1 - x}}$.

So $f^{-1}(x) = \sqrt{\dfrac{2}{1 - x}}$.

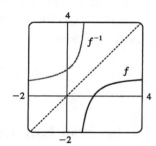

31. The function f is one-to-one, so its inverse exists and the graph of its inverse can be obtained by reflecting the graph of f about the line $y = x$.

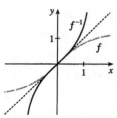

33. (a) It is defined as the inverse of the exponential function with base a, that is, $\log_a x = y$ $\Leftrightarrow$ $a^y = x$.

(b) $(0, \infty)$ (c) $\mathbb{R}$ (d) See Figure 1.

35. (a) $\log_2 64 = 6$ since $2^6 = 64$. (b) $\log_6 \frac{1}{36} = -2$ since $6^{-2} = \frac{1}{36}$.

37. (a) $\log_{10} 1.25 + \log_{10} 80 = \log_{10} (1.25 \cdot 80) = \log_{10} 100 = \log_{10} 10^2 = 2$

(b) $\log_5 10 + \log_5 20 - 3\log_5 2 = \log_5 (10 \cdot 20) - \log_5 2^3 = \log_5 \frac{200}{8} = \log_5 25 = \log_5 5^2 = 2$

39. $2\ln 4 - \ln 2 = \ln 4^2 - \ln 2 = \ln 16 - \ln 2 = \ln \frac{16}{2} = \ln 8$

41. (a) $\log_2 5 = \dfrac{\ln 5}{\ln 2} \approx 2.321928$
(b) $\log_5 26.05 = \dfrac{\ln 26.05}{\ln 5} \approx 2.025563$

43. To graph these functions, we use $\log_{1.5} x = \dfrac{\ln x}{\ln 1.5}$ and

$\log_{50} x = \dfrac{\ln x}{\ln 50}$. These graphs all approach $-\infty$ as $x \to 0^+$, and

they all pass through the point $(1, 0)$. Also, they are all increasing,
and all approach ∞ as $x \to \infty$. The functions with larger bases
increase extremely slowly, and the ones with smaller bases do so
somewhat more quickly. The functions with large bases approach the
y-axis more closely as $x \to 0^+$.

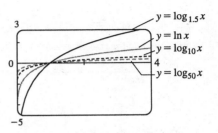

45. 3 ft = 36 in, so we need x such that $\log_2 x = 36$ $\Leftrightarrow$ $x = 2^{36} = 68{,}719{,}476{,}736$. In miles, this is

$68{,}719{,}476{,}736 \text{ in} \cdot \dfrac{1 \text{ ft}}{12 \text{ in}} \cdot \dfrac{1 \text{ mi}}{5280 \text{ ft}} \approx 1{,}084{,}587.7 \text{ mi}$.

47. (a) Shift the graph of $y = \log_{10} x$ five units to the
left to obtain the graph of $y = \log_{10}(x + 5)$.
Note the vertical asymptote of $x = -5$.

$y = \log_{10} x$ $y = \log_{10}(x + 5)$

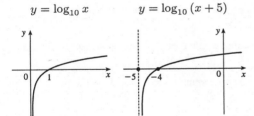

(b) Reflect the graph of $y = \ln x$ about the x-axis
to obtain the graph of $y = -\ln x$.

$y = \ln x$ $y = -\ln x$

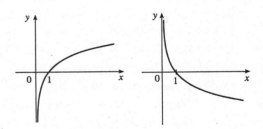

49. (a) $2\ln x = 1$ $\Rightarrow$ $\ln x = \frac{1}{2}$ $\Rightarrow$ $x = e^{1/2} = \sqrt{e}$

(b) $e^{-x} = 5$ $\Rightarrow$ $-x = \ln 5$ $\Rightarrow$ $x = -\ln 5$

51. (a) $2^{x-5} = 3$ $\Leftrightarrow$ $\log_2 3 = x - 5$ $\Leftrightarrow$ $x = 5 + \log_2 3.$ $\Leftrightarrow$

Or: $2^{x-5} = 3$ $\Leftrightarrow$ $\ln(2^{x-5}) = \ln 3$ $\Leftrightarrow$ $(x - 5)\ln 2 = \ln 3$ $\Leftrightarrow$ $x - 5 = \dfrac{\ln 3}{\ln 2}$ $\Leftrightarrow$ $x = 5 + \dfrac{\ln 3}{\ln 2}$

(b) $\ln x + \ln(x - 1) = \ln(x(x - 1)) = 1$ $\Leftrightarrow$ $x(x - 1) = e^1$ $\Leftrightarrow$ $x^2 - x - e = 0$. The quadratic formula
(with $a = 1$, $b = -1$, and $c = -e$) gives $x = \frac{1}{2}(1 \pm \sqrt{1 + 4e})$, but we reject the negative root since the natural
logarithm is not defined for $x < 0$. So $x = \frac{1}{2}(1 + \sqrt{1 + 4e})$.

53. (a) $e^x < 10$ $\Rightarrow$ $\ln e^x < \ln 10$ $\Rightarrow$ $x < \ln 10$ $\Rightarrow$ $x \in (-\infty, \ln 10)$

(b) $\ln x > -1$ $\Rightarrow$ $e^{\ln x} > e^{-1}$ $\Rightarrow$ $x > e^{-1}$ $\Rightarrow$ $x \in (1/e, \infty)$

55. We see that the graph of $y = f(x) = \sqrt{x^3 + x^2 + x + 1}$ is increasing, so f is 1-1. Enter $x = \sqrt{y^3 + y^2 + y + 1}$ and use your CAS to solve the equation for y. Using Derive, we get two (irrelevant) solutions involving imaginary expressions, as well as one which can be simplified to the following:

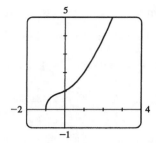

$$y = f^{-1}(x) = -\frac{\sqrt[3]{4}}{6}\left(\sqrt[3]{D - 27x^2 + 20} - \sqrt[3]{D + 27x^2 - 20} + \sqrt[3]{2}\right)$$

where $D = 3\sqrt{3}\sqrt{27x^4 - 40x^2 + 16}$. Maple and Mathematica each give two complex expressions and one real expression, and the real expression is equivalent to that given by Derive. For example, Maple's expression simplifies to $\dfrac{1}{6}\dfrac{M^{2/3} - 8 - 2M^{1/3}}{2M^{1/3}}$, where $M = 108x^2 + 12\sqrt{48 - 120x^2 + 81x^4} - 80$.

57. (a) $n = 100 \cdot 2^{t/3}$ $\Rightarrow$ $\dfrac{n}{100} = 2^{t/3}$ $\Rightarrow$ $\log_2\left(\dfrac{n}{100}\right) = \dfrac{t}{3}$ $\Rightarrow$ $t = 3\log_2\left(\dfrac{n}{100}\right)$. Using formula (10), we can write this as $t = 3 \cdot \dfrac{\ln(n/100)}{\ln 2}$. This function tells us how long it will take to obtain n bacteria (given the number n).

(b) $n = 50{,}000$ $\Rightarrow$ $t = 3\log_2\dfrac{50{,}000}{100} = 3\log_2 500 = 3\left(\dfrac{\ln 500}{\ln 2}\right) \approx 26.9$ hours

59. (a) To find the equation of the graph that results from shifting the graph of $y = \ln x$ 3 units upward, we add 3 to the original function to get $y = \ln x + 3$.

(b) To find the equation of the graph that results from shifting the graph of $y = \ln x$ 3 units to the left, we replace x with $x + 3$ in the original function to get $y = \ln(x + 3)$.

(c) To find the equation of the graph that results from reflecting the graph of $y = \ln x$ about the x-axis, we multiply the original equation by -1 to get $y = -\ln x$.

(d) To find the equation of the graph that results from reflecting the graph of $y = \ln x$ about the y-axis, we replace x with $-x$ in the original equation to get $y = \ln(-x)$.

(e) To find the equation of the graph that results from reflecting the graph of $y = \ln x$ about the line $y = x$, we interchange x and y in the original equation to get $x = \ln y$ $\Leftrightarrow$ $y = e^x$.

(f) To find the equation of the graph that results from reflecting the graph of $y = \ln x$ about the x-axis and then about the line $y = x$, we first multiply the original equation by -1 [to get $y = -\ln x$] and then interchange x and y in this equation to get $x = -\ln y$ $\Leftrightarrow$ $\ln y = -x$ $\Leftrightarrow$ $y = e^{-x}$.

(g) To find the equation of the graph that results from reflecting the graph of $y = \ln x$ about the y-axis and then about the line $y = x$, we first replace x with $-x$ in the original equation [to get $y = \ln(-x)$] and then interchange x and y to get $x = \ln(-y)$ $\Leftrightarrow$ $-y = e^x$ $\Leftrightarrow$ $y = -e^x$.

(h) To find the equation of the graph that results from shifting the graph of $y = \ln x$ 3 units to the left and then reflecting it about the line $y = x$, we first replace x with $x + 3$ in the original equation [to get $y = \ln(x + 3)$] and then interchange x and y in this equation to get $x = \ln(y + 3)$ $\Leftrightarrow$ $y + 3 = e^x$ $\Leftrightarrow$ $y = e^x - 3$.

1.7 Parametric Curves

1.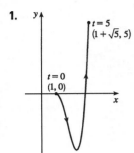

$x = 1 + \sqrt{t}, \quad y = t^2 - 4t, \quad 0 \le t \le 5$

t	0	1	2	3	4	5
x	1	2	$1 + \sqrt{2}$	$1 + \sqrt{3}$	3	$1 + \sqrt{5}$
			2.41	2.73		3.24
y	0	−3	−4	−3	0	5

3.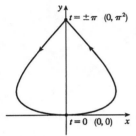

$x = 5 \sin t, \quad y = t^2, \quad -\pi \le t \le \pi$

t	$-\pi$	$-\pi/2$	0	$\pi/2$	π
x	0	−5	0	5	0
y	π^2	$\pi^2/4$	0	$\pi^2/4$	π^2
	9.87	2.47		2.47	9.87

5. (a) $x = 2t + 4, y = t - 1$

t	−3	−2	−1	0	1	2
x	−2	0	2	4	6	8
y	−4	−3	−2	−1	0	1

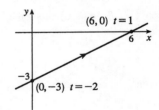

(b) $x = 2t + 4, y = t - 1 \quad \Rightarrow \quad x = 2(y + 1) + 4 = 2y + 6$ or

$y = \frac{1}{2}x - 3$

7. (a) $x = \sqrt{t}, y = 1 - t$

t	0	1	2	3	4
x	0	1	1.414	1.732	2
y	1	0	−1	−2	−3

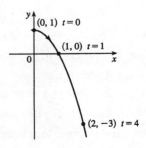

(b) $x = \sqrt{t} \quad \Rightarrow \quad t = x^2. \ y = 1 - t = 1 - x^2.$ Since $t \ge 0, x \ge 0.$

9. (a) $x = \sin \theta$, $y = \cos \theta$, $0 \le \theta \le \pi$.

$x^2 + y^2 = \sin^2 \theta + \cos^2 \theta = 1$.

Since $0 \le \theta \le \pi$, we have $\sin \theta \ge 0$,

so $x \ge 0$.

(b)

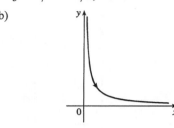

11. (a) $x = e^t$, $y = e^{-t}$.

$y = 1/e^t = 1/x$, $x > 0$

(b)

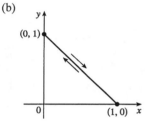

13. (a) $x = \sin^2 \theta$, $y = \cos^2 \theta$. $x + y = \sin^2 \theta + \cos^2 \theta = 1$,

$0 \le x \le 1$. Note that the curve is at $(0, 1)$ whenever $\theta = \pi n$

and is at $(1, 0)$ whenever $\theta = \frac{\pi}{2}n$ for every integer n.

(b)

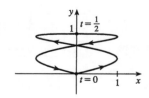

15. $x^2 + y^2 = \cos^2 \pi t + \sin^2 \pi t = 1$, $1 \le t \le 2$, so the particle moves counterclockwise along the circle $x^2 + y^2 = 1$ from $(-1, 0)$ to $(1, 0)$, along the lower half of the circle.

17. $\left(\frac{1}{2}x\right)^2 + \left(\frac{1}{3}y\right)^2 = \sin^2 t + \cos^2 t = 1$, so the particle moves once clockwise along the ellipse $\frac{1}{4}x^2 + \frac{1}{9}y^2 = 1$, starting and ending at $(0, 3)$.

19. We must have $1 \le x \le 4$ and $2 \le y \le 3$. So the graph of the curve must be contained in the rectangle $[1, 4]$ by $[2, 3]$.

21. When $t = 0$ we see that $x = 0$ and $y = 0$, so the curve starts at the origin. As t increases from 0 to $\frac{1}{2}$, the graphs show that y increases from 0 to 1 while x increases from 0 to 1, decreases to 0 and to -1, then increases back to 0, so we arrive at the point $(0, 1)$. Similarly, as t increases from $\frac{1}{2}$ to 1, y decreases from 1 to 0 while x repeats its pattern, and we arrive back at the origin. We could achieve greater accuracy by estimating x- and y-values for selected values of t from the given graphs and plotting the corresponding points.

23. (a) $x = x_1 + (x_2 - x_1)t$, $y = y_1 + (y_2 - y_1)t$, $0 \le t \le 1$. Clearly the curve passes through $P_1(x_1, y_1)$ when $t = 0$ and through $P_2(x_2, y_2)$ when $t = 1$. For $0 < t < 1$, x is strictly between x_1 and x_2 and y is strictly between y_1 and y_2. For every value of t, x and y satisfy the relation $y - y_1 = \dfrac{y_2 - y_1}{x_2 - x_1}(x - x_1)$, which is the equation of the line through $P_1(x_1, y_1)$ and $P_2(x_2, y_2)$.

Finally, any point (x, y) on that line satisfies $\dfrac{y - y_1}{y_2 - y_1} = \dfrac{x - x_1}{x_2 - x_1}$; if we call that common value t, then the given parametric equations yield the point (x, y); and any (x, y) on the line between $P_1(x_1, y_1)$ and $P_2(x_2, y_2)$ yields a value of t in $[0, 1]$. So the given parametric equations exactly specify the line segment from $P_1(x_1, y_1)$ to $P_2(x_2, y_2)$.

(b) $x = -2 + [3 - (-2)]t = -2 + 5t$ and $y = 7 + (-1 - 7)t = 7 - 8t$ for $0 \le t \le 1$.

25. As in Example 5, we let $y = t$ and $x = t - 3t^3 + t^5$ and use a t-interval of $[-2\pi, 2\pi]$.

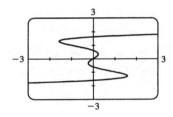

27. The circle $x^2 + y^2 = 4$ can be represented parametrically by $x = 2\cos t$, $y = 2\sin t$; $0 \le t \le 2\pi$. The circle $x^2 + (y - 1)^2 = 4$ can be represented by $x = 2\cos t$, $y = 1 + 2\sin t$; $0 \le t \le 2\pi$. This representation gives us the circle with a counterclockwise orientation starting at $(2, 1)$.

(a) To get a clockwise orientation, we could change the equations to $x = 2\cos t$, $y = 1 - 2\sin t$, $0 \le t \le 2\pi$.

(b) To get three times around in the counterclockwise direction, we use the original equations $x = 2\cos t$, $y = 1 + 2\sin t$ with the domain expanded to $0 \le t \le 6\pi$.

(c) To start at $(0, 3)$ using the original equations, we must have $x_1 = 0$; that is, $2\cos t = 0$. Hence, $t = \frac{\pi}{2}$. So we use $x = 2\cos t$, $y = 1 + 2\sin t$; $\frac{\pi}{2} \le t \le \frac{3\pi}{2}$.

Alternatively, if we want t to start at 0, we could change the equations of the curve. For example, we could use $x = -2\sin t$, $y = 1 + 2\cos t$, $0 \le t \le \pi$.

29. (a) Let $x^2/a^2 = \sin^2 t$ and $y^2/b^2 = \cos^2 t$ to obtain $x = a\sin t$ and $y = b\cos t$ with $0 \le t \le 2\pi$ as possible parametric equations for the ellipse $x^2/a^2 + y^2/b^2 = 1$.

(b) The equations are $x = 3\sin t$ and $y = b\cos t$ for $b \in \{1, 2, 4, 8\}$.

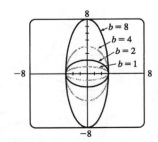

(c) As b increases, the ellipse stretches vertically.

31. The case $\frac{\pi}{2} < \theta < \pi$ is illustrated. C has coordinates $(r\theta, r)$ as in Example 7, and Q has coordinates $(r\theta, r + r\cos(\pi - \theta)) = (r\theta, r(1 - \cos\theta))$ [since $\cos(\pi - \alpha) = \cos\pi\cos\alpha + \sin\pi\sin\alpha = -\cos\alpha$], so P has coordinates $(r\theta - r\sin(\pi - \theta), r(1 - \cos\theta)) = (r(\theta - \sin\theta), r(1 - \cos\theta))$ [since $\sin(\pi - \alpha) = \sin\pi\cos\alpha - \cos\pi\sin\alpha = \sin\alpha$]. Again we have the parametric equations $x = r(\theta - \sin\theta)$, $y = r(1 - \cos\theta)$.

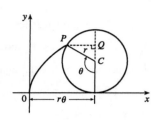

33. It is apparent that $x = |OQ|$ and $y = |QP| = |ST|$. From the diagram, $x = |OQ| = a\cos\theta$ and $y = |ST| = b\sin\theta$. Thus, the parametric equations are $x = a\cos\theta$ and $y = b\sin\theta$. To eliminate θ we rearrange: $\sin\theta = y/b \implies \sin^2\theta = (y/b)^2$ and $\cos\theta = x/a \implies \cos^2\theta = (x/a)^2$. Adding the two equations: $\sin^2\theta + \cos^2\theta = 1 = x^2/a^2 + y^2/b^2$. Thus, we have an ellipse.

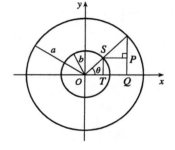

35. $C = (2a\cot\theta, 2a)$, so the x-coordinate of P is $x = 2a\cot\theta$. Let $B = (0, 2a)$. Then $\angle OAB$ is a right angle and $\angle OBA = \theta$, so $|OA| = 2a\sin\theta$ and $A = ((2a\sin\theta)\cos\theta, (2a\sin\theta)\sin\theta)$. Thus, the y-coordinate of P is $y = 2a\sin^2\theta$.

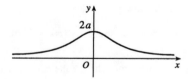

37. $x = t^2$, $y = t^3 - ct$. We use a graphing device to produce the graphs for various values of c with $-\pi \le t \le \pi$. Note that all the members of the family are symmetric about the x-axis. For $c < 0$, the graph does not cross itself, but for $c = 0$ it has a cusp at $(0, 0)$ and for $c > 0$ the graph crosses itself at $x = c$, so the loop grows larger as c increases.

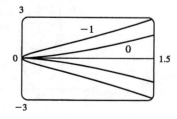

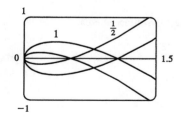

39. Note that all the Lissajous figures are symmetric about the x-axis. The parameters a and b simply stretch the graph in the x- and y-directions respectively. For $a = b = n = 1$ the graph is simply a circle with radius 1. For $n = 2$ the graph crosses itself at the origin and there are loops above and below the x-axis. In general, the figures have $n - 1$ points of intersection, all of which are on the y-axis, and a total of n closed loops.

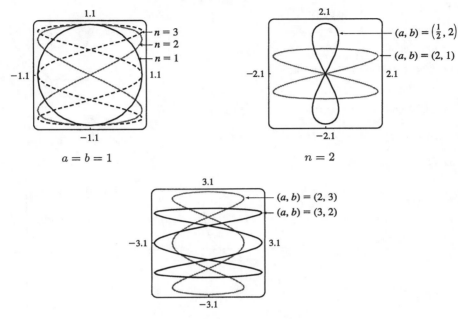

$a = b = 1$

$n = 2$

$n = 3$

 Review

• **CONCEPT CHECK** •

1. (a) A **function** f is a rule that assigns to each element x in a set A exactly one element, called $f(x)$, in a set B. The set A is called the **domain** of the function. The **range** of f is the set of all possible values of $f(x)$ as x varies throughout the domain.

(b) If f is a function with domain A, then its **graph** is the set of ordered pairs $\{(x, f(x)) \mid x \in A\}$.

(c) Use the Vertical Line Test on page 17.

2. The four ways to represent a function are: verbally, numerically, visually, and algebraically. An example of each is given below.

Verbally: An assignment of students to chairs in a classroom (a description in words)

Numerically: A tax table that assigns an amount of tax to an income (a table of values)

Visually: A graphical history of the Dow Jones average (a graph)

Algebraically: A relationship between distance, rate, and time: $d = rt$ (an explicit formula)

3. (a) An **even function** f satisfies $f(-x) = f(x)$ for every number x in its domain. It is symmetric with respect to the y-axis.

(b) An **odd function** g satisfies $g(-x) = -g(x)$ for every number x in its domain. It is symmetric with respect to the origin.

4. A function f is called **increasing** on an interval I if $f(x_1) < f(x_2)$ whenever $x_1 < x_2$ in I.

5. A **mathematical model** is a mathematical description (often by means of a function or an equation) of a real-world phenomenon.

6. (a) Linear function: $f(x) = 2x + 1$, $f(x) = ax + b$

(b) Power function: $f(x) = x^2$, $f(x) = x^a$

(c) Exponential function: $f(x) = 2^x$, $f(x) = a^x$

(d) Quadratic function: $f(x) = x^2 + x + 1$,

$f(x) = ax^2 + bx + c$

(e) Polynomial of degree 5: $f(x) = x^5 + 2$

(f) Rational function: $f(x) = \dfrac{x}{x+2}$, $f(x) = \dfrac{P(x)}{Q(x)}$ where

$P(x)$ and $Q(x)$ are polynomials

7.

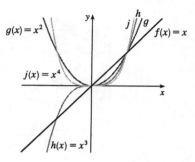

8. (a)

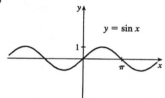

(b)

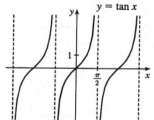

(c)

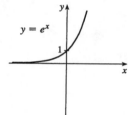

(d)

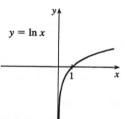

(e)

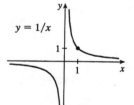

(f)

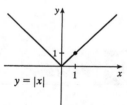

(g)

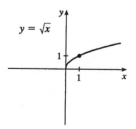

9. (a) The domain of $f + g$ is the intersection of the domain of f and the domain of g; that is, $A \cap B$.

(b) The domain of fg is also $A \cap B$.

(c) The domain of f/g must exclude values of x that make g equal to 0; that is, $\{x \in A \cap B \mid g(x) \neq 0\}$.

10. Given two functions f and g, the **composite** function $f \circ g$ is defined by $(f \circ g)(x) = f(g(x))$. The domain of $f \circ g$ is the set of all x in the domain of g such that $g(x)$ is in the domain of f.

11. (a) If the graph of f is shifted 2 units upward, its equation becomes $y = f(x) + 2$.

(b) If the graph of f is shifted 2 units downward, its equation becomes $y = f(x) - 2$.

(c) If the graph of f is shifted 2 units to the right, its equation becomes $y = f(x - 2)$.

(d) If the graph of f is shifted 2 units to the left, its equation becomes $y = f(x + 2)$.

(e) If the graph of f is reflected about the x-axis, its equation becomes $y = -f(x)$.

(f) If the graph of f is reflected about the y-axis, its equation becomes $y = f(-x)$.

(g) If the graph of f is stretched vertically by a factor of 2, its equation becomes $y = 2f(x)$.

(h) If the graph of f is shrunk vertically by a factor of 2, its equation becomes $y = \frac{1}{2}f(x)$.

(i) If the graph of f is stretched horizontally by a factor of 2, its equation becomes $y = f\left(\frac{1}{2}x\right)$.

(j) If the graph of f is shrunk horizontally by a factor of 2, its equation becomes $y = f(2x)$.

12. (a) A function f is called a *one-to-one function* if it never takes on the same value twice; that is, if $f(x_1) \neq f(x_2)$ whenever $x_1 \neq x_2$. (Or, f is 1-1 if each output corresponds to only one input.) Use the Horizontal Line Test: A function is one-to-one if and only if no horizontal line intersects its graph more than once.

(b) If f is a one-to-one function with domain A and range B, then its *inverse function* f^{-1} has domain B and range A and is defined by

$$f^{-1}(y) = x \quad \Leftrightarrow \quad f(x) = y$$

for any y in B. The graph of f^{-1} is obtained by reflecting the graph of f about the line $y = x$.

13. (a) A **parametric curve** C is the set of points $(x, y) = (f(t), g(t))$ that is traced out as t varies through an interval.

(b) There are several ways to sketch a parametric curve, including creating a table and plotting points, creating an equation in x and y by eliminating the parameter, observing general trends of the equations for x and y, and using a graphing device.

─────────────────── ▲ **TRUE–FALSE QUIZ** ▲ ───────────────────

1. False.   Let $f(x) = x^2$, $s = -1$, and $t = 1$. Then $f(s + t) = (-1 + 1)^2 = 0^2 = 0$, but

$f(s) + f(t) = (-1)^2 + 1^2 = 2 \neq 0 = f(s + t)$.

3. False.   Let $f(x) = x^2$. Then $f(3x) = (3x)^2 = 9x^2$ and $3f(x) = 3x^2$. So $f(3x) \neq 3f(x)$.

5. True.   See the Vertical Line Test.

7. False.   Let $f(x) = x^3$. Then f is one-to-one and $f^{-1}(x) = \sqrt[3]{x}$. But $1/f(x) = 1/x^3$, which is not equal to $f^{-1}(x)$.

9. True.   The function $\ln x$ is an increasing function on $(0, \infty)$.

11. False.   Let $x = e^2$ and $a = e$. Then $\dfrac{\ln x}{\ln a} = \dfrac{\ln e^2}{\ln e} = \dfrac{2\ln e}{\ln e} = 2$ and $\ln \dfrac{x}{a} = \ln \dfrac{e^2}{e} = \ln e = 1$, so in general the statement is false. What *is* true, however, is that $\ln \dfrac{x}{a} = \ln x - \ln a$.

◆ **EXERCISES** ◆

1. (a) When $x = 2$, $y \approx 2.7$. Thus, $f(2) \approx 2.7$.

   (b) $f(x) = 3 \quad \Rightarrow \quad x \approx 2.3, 5.6$

   (c) The domain of f is $-6 \le x \le 6$, or $[-6, 6]$.

   (d) The range of f is $-4 \le y \le 4$, or $[-4, 4]$.

   (e) f is increasing on $[-4, 4]$.

   (f) f is not one-to-one since it fails the Horizontal Line Test.

   (g) f is odd since its graph is symmetric about the origin.

3. (a)

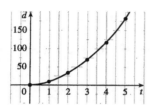

(b) From the graph, we see that the distance is slightly less than 150 feet.

5. $f(x) = \sqrt{4 - 3x^2}$. Domain: $4 - 3x^2 \ge 0 \quad \Rightarrow \quad 3x^2 \le 4 \quad \Rightarrow \quad x^2 \le \frac{4}{3} \quad \Rightarrow \quad |x| \le \frac{2}{\sqrt{3}}$. Range: $y \ge 0$ and $y \le \sqrt{4} \quad \Rightarrow \quad 0 \le y \le 2$.

7. $y = 1 + \sin x$. Domain: $\mathbb{R}$. Range: $-1 \le \sin x \le 1 \quad \Rightarrow \quad 0 \le 1 + \sin x \le 2 \quad \Rightarrow \quad 0 \le y \le 2$.

9. (a) To obtain the graph of $y = f(x) + 8$, we shift the graph of $y = f(x)$ up 8 units.

   (b) To obtain the graph of $y = f(x + 8)$, we shift the graph of $y = f(x)$ left 8 units.

   (c) To obtain the graph of $y = 1 + 2f(x)$, we stretch the graph of $y = f(x)$ vertically by a factor of 2, and then shift the resulting graph 1 unit upward.

   (d) To obtain the graph of $y = f(x - 2) - 2$, we shift the graph of $y = f(x)$ right 2 units (for the "-2" inside the parentheses), and then shift the resulting graph 2 units downward.

   (e) To obtain the graph of $y = -f(x)$, we reflect the graph of $y = f(x)$ about the x-axis.

   (f) To obtain the graph of $y = f^{-1}(x)$, we reflect the graph of $y = f(x)$ about the line $y = x$ (assuming f is one-to-one).

11. $y = -\sin 2x$: Start with the graph of $y = \sin x$, compress horizontally by a factor of 2, and reflect about the x-axis.

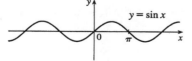
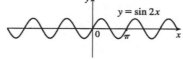
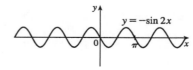

13. $y = (1 + e^x)/2$: Start with the graph of $y = e^x$, shift 1 unit upward, and compress vertically by a factor of 2.

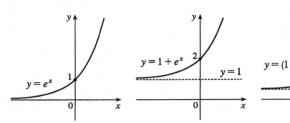

15. $f(x) = \dfrac{1}{x + 2}$: Start with the graph of

$f(x) = 1/x$ and shift 2 units to the left.

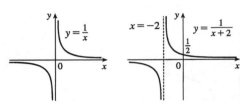

17. (a) The terms of f are a mixture of odd and even powers of x, so f is neither even nor odd.

(b) The terms of f are all odd powers of x, so f is odd.

(c) $f(-x) = e^{-(-x)^2} = e^{-x^2} = f(x)$, so f is even.

(d) $f(-x) = 1 + \sin(-x) = 1 - \sin x$. Now $f(-x) \neq f(x)$ and $f(-x) \neq -f(x)$, so f is neither even nor odd.

19. $f(x) = \ln x, \quad D = (0, \infty); \quad g(x) = x^2 - 9, \quad D = \mathbb{R}$.

$(f \circ g)(x) = f(g(x)) = f(x^2 - 9) = \ln(x^2 - 9)$.

   Domain: $x^2 - 9 > 0 \quad \Rightarrow \quad x^2 > 9 \quad \Rightarrow \quad |x| > 3 \quad \Rightarrow \quad x \in (-\infty, -3) \cup (3, \infty)$

$(g \circ f)(x) = g(f(x)) = g(\ln x) = (\ln x)^2 - 9$.   Domain: $x > 0$, or $(0, \infty)$

$(f \circ f)(x) = f(f(x)) = f(\ln x) = \ln(\ln x)$.   Domain: $\ln x > 0 \quad \Rightarrow \quad x > e^0 = 1$, or $(1, \infty)$

$(g \circ g)(x) = g(g(x)) = g(x^2 - 9) = (x^2 - 9)^2 - 9$.   Domain: $x \in \mathbb{R}$, or $(-\infty, \infty)$

21.

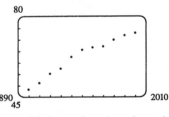

Many models appear to be plausible. Your choice depends on whether you think medical advances will keep increasing life expectancy, or if there is bound to be a natural leveling-off of life expectancy. A linear model, $y = 0.2493x - 423.4818$ gives us an estimate of 77.6 years for the year 2010.

23. We need to know the value of x such that $f(x) = 2x + \ln x = 2$. Since $x = 1$ gives us $y = 2$, $f^{-1}(2) = 1$.

25. (a) $e^{2\ln 3} = (e^{\ln 3})^2 = 3^2 = 9$

(b) $\log_{10} 25 + \log_{10} 4 = \log_{10}(25 \cdot 4) = \log_{10} 100 = \log_{10} 10^2 = 2$

27. (a) After 4 days, $\frac{1}{2}$ gram remains; after 8 days, $\frac{1}{4}$ g; after 12 days, $\frac{1}{8}$ g; after 16 days, $\frac{1}{16}$ g.

(b) $m(4) = \dfrac{1}{2}, m(8) = \dfrac{1}{2^2}, m(12) = \dfrac{1}{2^3}, m(16) = \dfrac{1}{2^4}$. From the pattern, we see that $m(t) = \dfrac{1}{2^{t/4}}$, or $2^{-t/4}$.

(c) $m = 2^{-t/4} \quad \Rightarrow \quad \log_2 m = -t/4 \quad \Rightarrow \quad t = -4\log_2 m$; this is the time elapsed when there are m grams of ^{100}Pd.

(d) $m = 0.01 \quad \Rightarrow \quad t = -4\log_2 0.01 = -4\left(\dfrac{\ln 0.01}{\ln 2}\right) \approx 26.6$ days

29. $f(x) = \ln(x^2 - c)$. If $c < 0$, the domain of f is $\mathbb{R}$. If $c = 0$, the domain of f is $(-\infty, 0) \cup (0, \infty)$. If $c > 0$, the domain of f is $(-\infty, -\sqrt{c}) \cup (\sqrt{c}, \infty)$. As c increases, the dip at $x = 0$ becomes deeper. For $c \geq 0$, the graph has asymptotes at $x = \pm\sqrt{c}$.

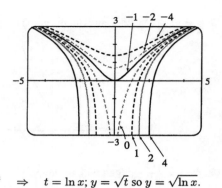

31. (a)

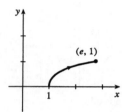

(b) $x = e^t \quad \Rightarrow \quad t = \ln x; y = \sqrt{t}$ so $y = \sqrt{\ln x}$.

$0 \leq t \leq 1 \quad \Rightarrow \quad 0 \leq y \leq 1$ and $1 \leq x \leq e$.

33. We sketch $x = t$, $y = 2t + \ln t$ (the function) and $x = 2t + \ln t$, $y = t$ (its inverse) for $t > 0$.

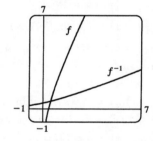

Principles of Problem Solving

1.

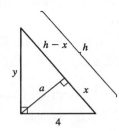

By using the area formula for a triangle, $\frac{1}{2}$ (base) (height), in two ways,

we see that $\frac{1}{2}(4)(y) = \frac{1}{2}(h)(a)$, so $a = \dfrac{4y}{h}$. Since $4^2 + y^2 = h^2$,

$y = \sqrt{h^2 - 16}$, and $a = \dfrac{4\sqrt{h^2 - 16}}{h}$.

3. $|2x - 1| = \begin{cases} 1 - 2x & \text{if } x < \frac{1}{2} \\ 2x - 1 & \text{if } x \geq \frac{1}{2} \end{cases}$ and $|x + 5| = \begin{cases} -x - 5 & \text{if } x < -5 \\ x + 5 & \text{if } x \geq -5 \end{cases}$

Therefore, we consider the three cases $x < -5$, $-5 \leq x < \frac{1}{2}$, and $x \geq \frac{1}{2}$.

If $x < -5$, we must have $1 - 2x - (-x - 5) = 3 \iff x = 3$, which is false, since we are considering $x < -5$.

If $-5 \leq x < \frac{1}{2}$, we must have $1 - 2x - (x + 5) = 3 \iff x = -\frac{7}{3}$.

If $x \geq \frac{1}{2}$, we must have $2x - 1 - (x + 5) = 3 \iff x = 9$.

So the two solutions of the equation are $x = -\frac{7}{3}$ and $x = 9$.

5. $f(x) = |x^2 - 4|x| + 3|$. If $x \geq 0$, then $f(x) = |x^2 - 4x + 3| = |(x - 1)(x - 3)|$.

       Case (i):   If $0 < x \leq 1$, then $f(x) = x^2 - 4x + 3$.

       Case (ii):   If $1 < x \leq 3$, then $f(x) = -(x^2 - 4x + 3) = -x^2 + 4x - 3$.

       Case (iii):   If $x > 3$, then $f(x) = x^2 - 4x + 3$.

This enables us to sketch the graph for $x \geq 0$. Then we use the fact that f is an even function to reflect this part of the graph about the y-axis to obtain the entire graph. Or, we could consider also the cases $x < -3$, $-3 \leq x < -1$, and $-1 \leq x < 0$.

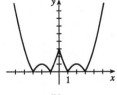

7. $|x| + |y| = 1 + |xy| \iff |xy| - |x| - |y| + 1 = 0 \iff$
$|x||y| - |x| - |y| + 1 = 0 \iff (|x| - 1)(|y| - 1) = 0 \iff x = \pm 1$ or $y = \pm 1$.

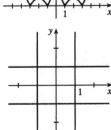

9. $|x| + |y| \leq 1$. The boundary of the region has equation $|x| + |y| = 1$.
In quadrants I, II, III, and IV, this becomes the lines $x + y = 1$,
$-x + y = 1$, $-x - y = 1$, and $x - y = 1$ respectively.

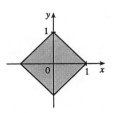

11. $(\log_2 3)(\log_3 4)(\log_4 5)\cdots(\log_{31} 32) = \left(\dfrac{\ln 3}{\ln 2}\right)\left(\dfrac{\ln 4}{\ln 3}\right)\left(\dfrac{\ln 5}{\ln 4}\right)\cdots\left(\dfrac{\ln 32}{\ln 31}\right) = \dfrac{\ln 32}{\ln 2} = \dfrac{\ln 2^5}{\ln 2} = \dfrac{5\ln 2}{\ln 2} = 5$

13. $\ln(x^2 - 2x - 2) \le 0 \;\Rightarrow\; x^2 - 2x - 2 \le e^0 = 1 \;\Rightarrow\; x^2 - 2x - 3 \le 0 \;\Rightarrow\; (x - 3)(x + 1) \le 0 \;\Rightarrow$

$x \in [-1, 3]$. Since the argument must be positive, $x^2 - 2x - 2 > 0 \;\Rightarrow\; \left[x - (1 - \sqrt{3})\right]\left[x - (1 + \sqrt{3})\right] > 0$

$\Rightarrow\; x \in (-\infty, 1 - \sqrt{3}) \cup (1 + \sqrt{3}, \infty)$. The intersection of these intervals is $[-1, 1 - \sqrt{3}) \cup (1 + \sqrt{3}, 3]$.

15. Let d be the distance traveled on each half of the trip. Let t_1 and t_2 be the times taken for the first and second halves of the trip.

For the first half of the trip we have $t_1 = d/30$ and for the second half we have $t_2 = d/60$. Thus, the average speed

for the entire trip is $\dfrac{\text{total distance}}{\text{total time}} = \dfrac{2d}{t_1 + t_2} = \dfrac{2d}{\dfrac{d}{30} + \dfrac{d}{60}} \cdot \dfrac{60}{60} = \dfrac{120d}{2d + d} = \dfrac{120d}{3d} = 40$. The average speed for

the entire trip is 40 mi/h.

17. Let S_n be the statement that $7^n - 1$ is divisible by 6.

- S_1 is true because $7^1 - 1 = 6$ is divisible by 6.

- Assume S_k is true, that is, $7^k - 1$ is divisible by 6. In other words, $7^k - 1 = 6m$ for some positive integer m.
  Then $7^{k+1} - 1 = 7^k \cdot 7 - 1 = (6m + 1) \cdot 7 - 1 = 42m + 6 = 6(7m + 1)$, which is divisible by 6, so S_{k+1} is
  true.

- Therefore, by mathematical induction, $7^n - 1$ is divisible by 6 for every positive integer n.

19. $f_0(x) = x^2$ and $f_{n+1}(x) = f_0(f_n(x))$ for $n = 0, 1, 2, \ldots$.

$f_1(x) = f_0(f_0(x)) = f_0(x^2) = (x^2)^2 = x^4$, $f_2(x) = f_0(f_1(x)) = f_0(x^4) = (x^4)^2 = x^8$,

$f_3(x) = f_0(f_2(x)) = f_0(x^8) = (x^8)^2 = x^{16}, \ldots$. Thus, a general formula is $f_n(x) = x^{2^{n+1}}$.

2 Limits and Derivatives

2.1 The Tangent and Velocity Problems • • • • • • • • •

1. (a) Using $P(15, 250)$, we construct the following table:

t	Q	slope $= m_{PQ}$
5	$(5, 694)$	$\frac{694-250}{5-15} = -\frac{444}{10} = -44.4$
10	$(10, 444)$	$\frac{444-250}{10-15} = -\frac{194}{5} = -38.8$
20	$(20, 111)$	$\frac{111-250}{20-15} = -\frac{139}{5} = -27.8$
25	$(25, 28)$	$\frac{28-250}{25-15} = -\frac{222}{10} = -22.2$
30	$(30, 0)$	$\frac{0-250}{30-15} = -\frac{250}{15} = -16.\overline{6}$

(b) Using the values of t that correspond to the points closest to P ($t = 10$ and $t = 20$), we have

$$\frac{-38.8 + (-27.8)}{2} = -33.3$$

(c) From the graph, we can estimate the slope of the tangent line at P to be

$$\frac{-300}{9} = -33.\overline{3}.$$

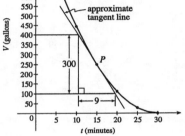

3. For the curve $y = x/(1 + x)$ and the point $P\left(1, \frac{1}{2}\right)$:

(a)

	x	Q	m_{PQ}
(i)	0.5	$(0.5, 0.333333)$	0.333333
(ii)	0.9	$(0.9, 0.473684)$	0.263158
(iii)	0.99	$(0.99, 0.497487)$	0.251256
(iv)	0.999	$(0.999, 0.499750)$	0.250125
(v)	1.5	$(1.5, 0.6)$	0.2
(vi)	1.1	$(1.1, 0.523810)$	0.238095
(vii)	1.01	$(1.01, 0.502488)$	0.248756
(viii)	1.001	$(1.001, 0.500250)$	0.249875

(b) The slope appears to be $\frac{1}{4}$.

(c) $y - \frac{1}{2} = \frac{1}{4}(x - 1)$ or
$y = \frac{1}{4}x + \frac{1}{4}$.

5. (a) At $t = 2$, $y = 40(2) - 16(2)^2 = 16$. The average velocity between times 2 and $2 + h$ is

$$\frac{40(2 + h) - 16(2 + h)^2 - 16}{h} = \frac{-24h - 16h^2}{h} = -24 - 16h, \text{ if } h \neq 0.$$

(i) $h = 0.5$, -32 ft/s   (ii) $h = 0.1$, -25.6 ft/s

(iii) $h = 0.05$, -24.8 ft/s   (iv) $h = 0.01$, -24.16 ft/s

(b) The instantaneous velocity when $t = 2$ (h approaches 0) is -24 ft/s.

7. Average velocity between times 1 and $1 + h$ is

$$\frac{s(1 + h) - s(1)}{h} = \frac{(1 + h)^3/6 - 1/6}{h} = \frac{h^3 + 3h^2 + 3h}{6h} = \frac{h^2 + 3h + 3}{6} \text{ if } h \neq 0.$$

(a) (i) $[1, 3]$: $h = 2$, $\frac{13}{6}$ ft/s   (ii) $[1, 2]$: $h = 1$, $\frac{7}{6}$ ft/s

(iii) $[1, 1.5]$: $h = 0.5$, $\frac{19}{24}$ ft/s   (iv) $[1, 1.1]$: $h = 0.1$, $\frac{331}{600}$ ft/s

(b) As h approaches 0, the velocity approaches $\frac{3}{6} = \frac{1}{2}$ ft/s.

(c) 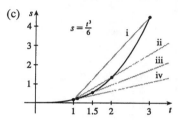   (d)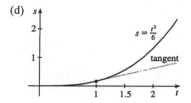

9. For the curve $y = \sin(10\pi/x)$ and the point $P(1, 0)$:

(a)

x	Q	m_{PQ}
2	$(2, 0)$	0
1.5	$(1.5, 0.8660)$	1.7321
1.4	$(1.4, -0.4339)$	-1.0847
1.3	$(1.3, -0.8230)$	-2.7433
1.2	$(1.2, 0.8660)$	4.3301
1.1	$(1.1, -0.2817)$	-2.8173

x	Q	m_{PQ}
0.5	$(0.5, 0)$	0
0.6	$(0.6, 0.8660)$	-2.1651
0.7	$(0.7, 0.7818)$	-2.6061
0.8	$(0.8, 1)$	-5
0.9	$(0.9, -0.3420)$	3.4202

As x approaches 1, the slopes do not appear to be approaching any particular value.

(b)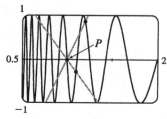

We see that problems with estimation are caused by the frequent oscillations of the graph. The tangent is so steep at P that we need to take x-values much closer to 1 in order to get accurate estimates of its slope.

(c) If we choose $x = 1.001$, then the point Q is $(1.001, -0.0314)$ and $m_{PQ} \approx -31.3794$. If $x = 0.999$, then Q is $(0.999, 0.0314)$ and $m_{PQ} = -31.4422$. The average of these slopes is -31.4108. So we estimate that the slope of the tangent line at P is about -31.4.

2.2 The Limit of a Function • • • • • • • • • • • • • •

1. As x approaches 2, $f(x)$ approaches 5. [Or, the values of $f(x)$ can be made as close to 5 as we like by taking x sufficiently close to 2 (but $x \neq 2$).] Yes, the graph could have a hole at $(2, 5)$ and be defined such that $f(2) = 3$.

3. (a) $f(x)$ approaches 2 as x approaches 1 from the left, so $\lim\limits_{x \to 1^-} f(x) = 2$.

(b) $f(x)$ approaches 3 as x approaches 1 from the right, so $\lim\limits_{x \to 1^+} f(x) = 3$.

(c) $\lim\limits_{x \to 1} f(x)$ does not exist because the limits in part (a) and part (b) are not equal.

(d) $f(x)$ approaches 4 as x approaches 5 from the left and from the right, so $\lim\limits_{x \to 5} f(x) = 4$.

(e) $f(5)$ is not defined, so it doesn't exist.

5. (a) $\lim\limits_{t \to 0^-} g(t) = -1$ (b) $\lim\limits_{t \to 0^+} g(t) = -2$

(c) $\lim\limits_{t \to 0} g(t)$ does not exist because the limits in part (a) and part (b) are not equal.

(d) $\lim\limits_{t \to 2^-} g(t) = 2$ (e) $\lim\limits_{t \to 2^+} g(t) = 0$

(f) $\lim\limits_{t \to 2} g(t)$ does not exist because the limits in part (d) and part (e) are not equal.

(g) $g(2) = 1$ (h) $\lim\limits_{t \to 4} g(t) = 3$

7.

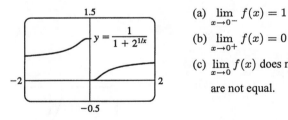

(a) $\lim\limits_{x \to 0^-} f(x) = 1$

(b) $\lim\limits_{x \to 0^+} f(x) = 0$

(c) $\lim\limits_{x \to 0} f(x)$ does not exist because the limits in part (a) and part (b) are not equal.

9. $\lim\limits_{x \to 3^+} f(x) = 4$, $\lim\limits_{x \to 3^-} f(x) = 2$,

     $\lim\limits_{x \to -2} f(x) = 2$, $f(3) = 3$, $f(-2) = 1$

11. For $g(x) = \dfrac{x - 1}{x^3 - 1}$:

x	$g(x)$	x	$g(x)$
0.2	0.806452	1.8	0.165563
0.4	0.641026	1.6	0.193798
0.6	0.510204	1.4	0.229358
0.8	0.409836	1.2	0.274725
0.9	0.369004	1.1	0.302115
0.99	0.336689	1.01	0.330022

It appears that $\lim\limits_{x \to 1} \dfrac{x - 1}{x^3 - 1} = 0.\overline{3} = \frac{1}{3}$.

13. For $f(x) = \dfrac{e^x - 1 - x}{x^2}$:

x	$f(x)$
1	0.718282
0.5	0.594885
0.1	0.517092
0.05	0.508439
0.01	0.501671

x	$f(x)$
-1	0.367879
-0.5	0.426123
-0.1	0.483742
-0.05	0.491770
-0.01	0.498337

It appears that $\displaystyle\lim_{x \to 0} \dfrac{e^x - 1 - x}{x^2} = 0.5 = \tfrac{1}{2}$.

15. (a) From the graphs, it seems that $\displaystyle\lim_{x \to 0} \dfrac{\tan 4x}{x} = 4$.

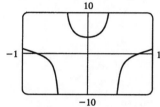
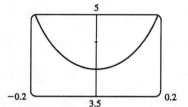

(b)

x	$f(x)$
± 0.1	4.227932
± 0.01	4.002135
± 0.001	4.000021
± 0.0001	4.000000

17. (a) Let $h(x) = (1 + x)^{1/x}$.

x	$h(x)$
-0.001	2.71964
-0.0001	2.71842
-0.00001	2.71830
-0.000001	2.71828
0.000001	2.71828
0.00001	2.71827
0.0001	2.71815
0.001	2.71692

(b)

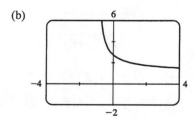

It appears that $\displaystyle\lim_{x \to 0} (1 + x)^{1/x} \approx 2.71828$, which is approximately e.

In Section 3.7 we'll see that the value of the limit is exactly e.

19. For $f(x) = x^2 - (2^x/1000)$:

(a)

x	$f(x)$
1	0.998000
0.8	0.638259
0.6	0.358484
0.4	0.158680
0.2	0.038851
0.1	0.008928
0.05	0.001465

It appears that $\lim\limits_{x \to 0} f(x) = 0$.

(b)

x	$f(x)$
0.04	0.000572
0.02	-0.000614
0.01	-0.000907
0.005	-0.000978
0.003	-0.000993
0.001	-0.001000

It appears that $\lim\limits_{x \to 0} f(x) = -0.001$.

21.

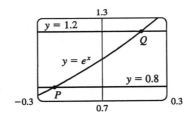

We need to have $0.8 < e^x < 1.2$. From the graph we obtain the approximate points of intersection $P(-0.2231436, 0.8)$ and $Q(0.18232156, 1.2)$. So if x is within 0.182 of 0, then y will be within 0.2 of 1. If we must have e^x within 0.1 of 1, we get $P(-0.1053605, 0.9)$ and $Q(0.09531018, 1.1)$. We would then need x to be within 0.095 of 0.

2.3 Calculating Limits Using the Limit Laws • • • • • • •

1. (a) $\lim\limits_{x \to a} [f(x) + h(x)] = \lim\limits_{x \to a} f(x) + \lim\limits_{x \to a} h(x)$

$$= -3 + 8 = 5$$

(b) $\lim\limits_{x \to a} [f(x)]^2 = \left[\lim\limits_{x \to a} f(x)\right]^2 = (-3)^2 = 9$

(c) $\lim\limits_{x \to a} \sqrt[3]{h(x)} = \sqrt[3]{\lim\limits_{x \to a} h(x)} = \sqrt[3]{8} = 2$

(d) $\lim\limits_{x \to a} \dfrac{1}{f(x)} = \dfrac{1}{\lim\limits_{x \to a} f(x)} = \dfrac{1}{-3} = -\dfrac{1}{3}$

(e) $\lim\limits_{x \to a} \dfrac{f(x)}{h(x)} = \dfrac{\lim\limits_{x \to a} f(x)}{\lim\limits_{x \to a} h(x)} = \dfrac{-3}{8} = -\dfrac{3}{8}$

(f) $\lim\limits_{x \to a} \dfrac{g(x)}{f(x)} = \dfrac{\lim\limits_{x \to a} g(x)}{\lim\limits_{x \to a} f(x)} = \dfrac{0}{-3} = 0$

(g) The limit does not exist, since $\lim\limits_{x \to a} g(x) = 0$ but $\lim\limits_{x \to a} f(x) \neq 0$.

(h) $\lim\limits_{x \to a} \dfrac{2f(x)}{h(x) - f(x)} = \dfrac{2 \lim\limits_{x \to a} f(x)}{\lim\limits_{x \to a} h(x) - \lim\limits_{x \to a} f(x)} = \dfrac{2(-3)}{8 - (-3)} = -\dfrac{6}{11}$

3. $\lim\limits_{x \to 4} \left(5x^2 - 2x + 3\right) = \lim\limits_{x \to 4} 5x^2 - \lim\limits_{x \to 4} 2x + \lim\limits_{x \to 4} 3$ (Limit Laws 2 & 1)

$$= 5 \lim\limits_{x \to 4} x^2 - 2 \lim\limits_{x \to 4} x + 3 \qquad \text{(3 \& 7)}$$

$$= 5 \left(4\right)^2 - 2 \left(4\right) + 3 = 75 \qquad \text{(9 \& 8)}$$

5. $\lim\limits_{t \to -2} (t+1)^9 \left(t^2 - 1\right) = \lim\limits_{t \to -2} (t+1)^9 \lim\limits_{t \to -2} \left(t^2 - 1\right)$ (4)

$$= \left[\lim\limits_{t \to -2} (t+1)\right]^9 \lim\limits_{t \to -2} \left(t^2 - 1\right) \qquad \text{(6)}$$

$$= \left[\lim\limits_{t \to -2} t + \lim\limits_{t \to -2} 1\right]^9 \left[\lim\limits_{t \to -2} t^2 - \lim\limits_{t \to -2} 1\right] \qquad \text{(1 \& 2)}$$

$$= \left[(-2) + 1\right]^9 \left[(-2)^2 - 1\right] = -3 \qquad \text{(8, 7 \& 9)}$$

7. $\lim\limits_{x \to 1} \left(\dfrac{1 + 3x}{1 + 4x^2 + 3x^4}\right)^3 = \left(\lim\limits_{x \to 1} \dfrac{1 + 3x}{1 + 4x^2 + 3x^4}\right)^3$ (6)

$$= \left[\frac{\lim\limits_{x \to 1}(1 + 3x)}{\lim\limits_{x \to 1}(1 + 4x^2 + 3x^4)}\right]^3 \qquad \text{(5)}$$

$$= \left[\frac{\lim\limits_{x \to 1} 1 + 3 \lim\limits_{x \to 1} x}{\lim\limits_{x \to 1} 1 + 4 \lim\limits_{x \to 1} x^2 + 3 \lim\limits_{x \to 1} x^4}\right]^3 \qquad \text{(2, 1, \& 3)}$$

$$= \left[\frac{1 + 3(1)}{1 + 4(1)^2 + 3\,(1)^4}\right]^3 = \left[\frac{4}{8}\right]^3 = \left(\frac{1}{2}\right)^3 = \frac{1}{8} \qquad \text{(7, 8, \& 9)}$$

9. $\lim\limits_{x \to 2} \dfrac{x^2 + x - 6}{x - 2} = \lim\limits_{x \to 2} \dfrac{(x + 3)(x - 2)}{x - 2} = \lim\limits_{x \to 2}(x + 3) = 2 + 3 = 5$

11. $\lim\limits_{x \to 2} \dfrac{x^2 - x + 6}{x - 2}$ does not exist since $x - 2 \to 0$ but $x^2 - x + 6 \to 8$ as $x \to 2$.

13. $\lim\limits_{t \to -3} \dfrac{t^2 - 9}{2t^2 + 7t + 3} = \lim\limits_{t \to -3} \dfrac{(t + 3)(t - 3)}{(2t + 1)(t + 3)} = \lim\limits_{t \to -3} \dfrac{t - 3}{2t + 1} = \dfrac{-3 - 3}{2(-3) + 1} = \dfrac{-6}{-5} = \dfrac{6}{5}$

15. $\lim\limits_{h \to 0} \dfrac{(2 + h)^3 - 8}{h} = \lim\limits_{h \to 0} \dfrac{\left(8 + 12h + 6h^2 + h^3\right) - 8}{h} = \lim\limits_{h \to 0} \dfrac{12h + 6h^2 + h^3}{h}$

$$= \lim\limits_{h \to 0}\left(12 + 6h + h^2\right) = 12 + 0 + 0 = 12$$

17. $\lim\limits_{x \to 7} \dfrac{\sqrt{x + 2} - 3}{x - 7} = \lim\limits_{x \to 7} \dfrac{\sqrt{x + 2} - 3}{x - 7} \cdot \dfrac{\sqrt{x + 2} + 3}{\sqrt{x + 2} + 3} = \lim\limits_{x \to 7} \dfrac{(x + 2) - 9}{(x - 7)\left(\sqrt{x + 2} + 3\right)}$

$$= \lim\limits_{x \to 7} \dfrac{x - 7}{(x - 7)\left(\sqrt{x + 2} + 3\right)} = \lim\limits_{x \to 7} \dfrac{1}{\sqrt{x + 2} + 3} = \dfrac{1}{\sqrt{9} + 3} = \dfrac{1}{6}$$

19. $\lim\limits_{x \to -4} \dfrac{\dfrac{1}{4} + \dfrac{1}{x}}{4 + x} = \lim\limits_{x \to -4} \dfrac{\dfrac{x + 4}{4x}}{4 + x} = \lim\limits_{x \to -4} \dfrac{x + 4}{4x(4 + x)} = \lim\limits_{x \to -4} \dfrac{1}{4x} = \dfrac{1}{4(-4)} = -\dfrac{1}{16}$

21. (a)

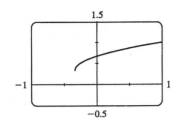

$$\lim_{x \to 0} \frac{x}{\sqrt{1+3x}-1} \approx \frac{2}{3}$$

(b)

x	$f(x)$
−0.001	0.6661663
−0.0001	0.6666167
−0.00001	0.6666617
−0.000001	0.6666662
0.000001	0.6666672
0.00001	0.6666717
0.0001	0.6667167
0.001	0.6671663

The limit appears to be $\frac{2}{3}$.

(c) $\lim_{x \to 0} \left(\dfrac{x}{\sqrt{1+3x}-1} \cdot \dfrac{\sqrt{1+3x}+1}{\sqrt{1+3x}+1} \right) = \lim_{x \to 0} \dfrac{x\left(\sqrt{1+3x}+1\right)}{(1+3x)-1} = \lim_{x \to 0} \dfrac{x\left(\sqrt{1+3x}+1\right)}{3x}$

$\qquad = \frac{1}{3} \lim_{x \to 0} \left(\sqrt{1+3x}+1 \right)$ (Limit Law 3)

$\qquad = \frac{1}{3} \left[\sqrt{\lim_{x \to 0}(1+3x)} + \lim_{x \to 0} 1 \right]$ (1 & 11)

$\qquad = \frac{1}{3} \left(\sqrt{\lim_{x \to 0} 1 + 3 \lim_{x \to 0} x} + 1 \right)$ (1, 3 & 7)

$\qquad = \frac{1}{3} \left(\sqrt{1 + 3 \cdot 0} + 1 \right)$ (7 & 8)

$\qquad = \frac{1}{3}(1+1) = \frac{2}{3}$

23. Let $f(x) = -x^2$, $g(x) = x^2 \cos 20\pi x$ and $h(x) = x^2$. Then $-1 \le \cos 20\pi x \le 1 \Rightarrow$
$-x^2 \le x^2 \cos 20\pi x \le x^2 \Rightarrow f(x) \le g(x) \le h(x)$. So since $\lim_{x \to 0} f(x) = \lim_{x \to 0} h(x) = 0$, by the Squeeze
Theorem we have $\lim_{x \to 0} g(x) = 0$.

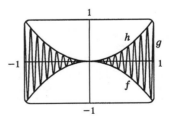

25. $1 \le f(x) \le x^2 + 2x + 2$ for all x. Now $\lim_{x \to -1} 1 = 1$ and

$\lim_{x \to -1} (x^2 + 2x + 2) = \lim_{x \to -1} x^2 + 2 \lim_{x \to -1} x + \lim_{x \to -1} 2 = (-1)^2 + 2(-1) + 2 = 1$. Therefore, by the Squeeze
Theorem, $\lim_{x \to -1} f(x) = 1$.

27. $-1 \le \cos(2/x) \le 1 \Rightarrow -x^4 \le x^4 \cos(2/x) \le x^4$. Since $\lim_{x \to 0} \left(-x^4 \right) = 0$ and $\lim_{x \to 0} x^4 = 0$, we have
$\lim_{x \to 0} \left[x^4 \cos(2/x) \right] = 0$ by the Squeeze Theorem.

29. If $x > -4$, then $|x + 4| = x + 4$, so $\lim\limits_{x \to -4+} |x + 4| = \lim\limits_{x \to -4+} (x + 4) = -4 + 4 = 0$.

If $x < -4$, then $|x + 4| = -(x + 4)$, so $\lim\limits_{x \to -4-} |x + 4| = \lim\limits_{x \to -4-} -(x + 4) = -(-4 + 4) = 0$.

Since the right and left limits are equal, $\lim\limits_{x \to -4} |x + 4| = 0$.

31. Since $|x| = -x$ for $x < 0$, we have $\lim\limits_{x \to 0-} \left(\dfrac{1}{x} - \dfrac{1}{|x|} \right) = \lim\limits_{x \to 0-} \left(\dfrac{1}{x} - \dfrac{1}{-x} \right) = \lim\limits_{x \to 0-} \dfrac{2}{x}$, which does not exist

since the denominator approaches 0 and the numerator does not.

33. (a) (i) If $x \to 1^+$, then $x > 1$ and $g(x) = x - 1$. Thus, $\lim\limits_{x \to 1+} g(x) = \lim\limits_{x \to 1+} (x - 1) = 1 - 1 = 0$.

(ii) If $x \to 1^-$, then $x < 1$ and $g(x) = 1 - x^2$. Thus, $\lim\limits_{x \to 1-} g(x) = \lim\limits_{x \to 1-} (1 - x^2) = 1 - 1^2 = 0$.

Since the left- and right-hand limits of g at 1 are equal, $\lim\limits_{x \to 1} g(x) = 0$.

(iii) If $x \to 0$, then $-1 < x < 1$ and $g(x) = 1 - x^2$. Thus, $\lim\limits_{x \to 0} g(x) = \lim\limits_{x \to 0} (1 - x^2) = 1 - 0^2 = 1$.

(iv) If $x \to -1^-$, then $x < -1$ and $g(x) = -x$. Thus, $\lim\limits_{x \to -1-} g(x) = \lim\limits_{x \to -1-} (-x) = -(-1) = 1$.

(v) If $x \to -1^+$, then $-1 < x < 1$ and $g(x) = 1 - x^2$. Thus,

$$\lim\limits_{x \to -1+} g(x) = \lim\limits_{x \to -1+} (1 - x^2) = 1 - (-1)^2 = 1 - 1 = 0$$

(vi) $\lim\limits_{x \to -1} g(x)$ does not exist because the limits in part (iv) and part (v) are not equal.

(b)

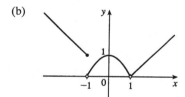

35. (a) (i) $[\![x]\!] = -2$ for $-2 \le x < -1$, so $\lim\limits_{x \to -2+} [\![x]\!] = \lim\limits_{x \to -2+} (-2) = -2$

(ii) $[\![x]\!] = -3$ for $-3 \le x < -2$, so $\lim\limits_{x \to -2-} [\![x]\!] = \lim\limits_{x \to -2-} (-3) = -3$. The right and left limits are different,

so $\lim\limits_{x \to -2} [\![x]\!]$ does not exist.

(iii) $[\![x]\!] = -3$ for $-3 \le x < -2$, so $\lim\limits_{x \to -2.4} [\![x]\!] = \lim\limits_{x \to -2.4} (-3) = -3$.

(b) (i) $[\![x]\!] = n - 1$ for $n - 1 \le x < n$, so $\lim\limits_{x \to n-} [\![x]\!] = \lim\limits_{x \to n-} (n - 1) = n - 1$.

(ii) $[\![x]\!] = n$ for $n \le x < n + 1$, so $\lim\limits_{x \to n+} [\![x]\!] = \lim\limits_{x \to n+} n = n$.

(c) $\lim\limits_{x \to a} [\![x]\!]$ exists $\quad \Leftrightarrow \quad a$ is not an integer.

37. The graph of $f(x) = [\![x]\!] + [\![-x]\!]$ is the same as the graph of $g(x) = -1$ with holes at each integer, since $f(a) = 0$ for any integer a. Thus, $\lim\limits_{x \to 2-} f(x) = -1$ and $\lim\limits_{x \to 2+} f(x) = -1$, so $\lim\limits_{x \to 2} f(x) = -1$.

$f(2) = [\![2]\!] + [\![-2]\!] = 2 + (-2) = 0$.

39. Since $p(x)$ is a polynomial, $p(x) = a_0 + a_1x + a_2x^2 + \cdots + a_nx^n$. Thus, by the Limit Laws,

$$\lim_{x \to a} p(x) = \lim_{x \to a} \left(a_0 + a_1x + a_2x^2 + \cdots + a_nx^n \right)$$

$$= a_0 + a_1 \lim_{x \to a} x + a_2 \lim_{x \to a} x^2 + \cdots + a_n \lim_{x \to a} x^n$$

$$= a_0 + a_1a + a_2a^2 + \cdots + a_na^n = p(a)$$

Thus, for any polynomial p, $\lim_{x \to a} p(x) = p(a)$.

41. Let $f(x) = [\![x]\!]$ and $g(x) = -[\![x]\!]$. Then $\lim_{x \to 3} f(x)$ and $\lim_{x \to 3} g(x)$ do not exist (Example 9) but

$$\lim_{x \to 3} [f(x) + g(x)] = \lim_{x \to 3} ([\![x]\!] - [\![x]\!]) = \lim_{x \to 3} 0 = 0.$$

43. Since the denominator approaches 0 as $x \to -2$, the limit will exist only if the numerator also approaches 0 as
$x \to -2$. In order for this to happen, we need $\lim_{x \to -2} \left(3x^2 + ax + a + 3\right) = 0 \quad \Leftrightarrow$

$3(-2)^2 + a(-2) + a + 3 = 0 \quad \Leftrightarrow \quad 12 - 2a + a + 3 = 0 \quad \Leftrightarrow \quad a = 15$. With $a = 15$, the limit becomes

$$\lim_{x \to -2} \frac{3x^2 + 15x + 18}{x^2 + x - 2} = \lim_{x \to -2} \frac{3(x + 2)(x + 3)}{(x - 1)(x + 2)} = \frac{3(-2 + 3)}{-2 - 1} = -1.$$

2.4 Continuity · · · · · · · · · · · · · · · · ·

1. From Equation 1, $\lim_{x \to 4} f(x) = f(4)$.

3. (a) The following are the numbers at which f is discontinuous and the type of discontinuity at that number:
-4 (removable), -2 (jump), 2 (jump), 4 (infinite).

(b) f is continuous from the left at -2, and continuous from the right at 2 and 4. It is continuous from neither side
at -4.

5.

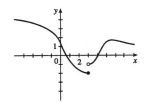

7. (a)

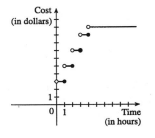

(b) There are discontinuities at $t = 1, 2, 3$, and 4. A person
parking in the lot would want to keep in mind that the
charge will jump at the beginning of each hour.

9. Since f and g are continuous functions,

$$\lim_{x \to 3} [2f(x) - g(x)] = 2 \lim_{x \to 3} f(x) - \lim_{x \to 3} g(x) \quad \text{(by Limit Laws 2 \& 3)}$$
$$= 2f(3) - g(3) \quad \text{(by continuity of } f \text{ and } g \text{ at } x = 3)$$
$$= 2 \cdot 5 - g(3) = 10 - g(3)$$

Since it is given that $\lim_{x \to 3} [2f(x) - g(x)] = 4$, we have $10 - g(3) = 4$, or $g(3) = 6$.

11. $\lim_{x \to -1} f(x) = \lim_{x \to -1} (x + 2x^3)^4 = \left(\lim_{x \to -1} x + 2 \lim_{x \to -1} x^3 \right)^4 = \left[-1 + 2(-1)^3 \right]^4 = (-3)^4 = 81 = f(-1).$

By the definition of continuity, f is continuous at $a = -1$.

13. $f(x) = \ln|x - 2|$ is discontinuous at 2 since $f(2) = \ln 0$ is not defined.

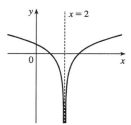

15. $f(x) = \begin{cases} \dfrac{x^2 - x - 12}{x + 3} & \text{if } x \neq -3 \\ -5 & \text{if } x = -3 \end{cases} = \begin{cases} x - 4 & \text{if } x \neq -3 \\ -5 & \text{if } x = -3 \end{cases}$

So $\lim_{x \to -3} f(x) = \lim_{x \to -3} (x - 4) = -7$ and $f(-3) = -5$.

Since $\lim_{x \to -3} f(x) \neq f(-3)$, f is discontinuous at -3.

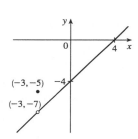

17. $F(x) = \dfrac{x}{x^2 + 5x + 6}$ is a rational function. So by Theorem 5 (or Theorem 7), F is continuous at every number in

its domain, $\{x \mid x^2 + 5x + 6 \neq 0\} = \{x \mid (x + 3)(x + 2) \neq 0\} = \{x \mid x \neq -3, -2\}$ or
$(-\infty, -3) \cup (-3, -2) \cup (-2, \infty)$.

19. By Theorem 5, the polynomial $5x$ is continuous on $(-\infty, \infty)$. By Theorems 9 and 7, $\sin 5x$ is continuous on
$(-\infty, \infty)$. By Theorem 7, e^x is continuous on $(-\infty, \infty)$. By Theorem 4 #4, the product of e^x and $\sin 5x$ is
continuous at all numbers which are in both of their domains, that is, on $(-\infty, \infty)$.

21. By Theorem 5, the polynomial $t^4 - 1$ is continuous on $(-\infty, \infty)$. By Theorem 7, $\ln x$ is continuous on its
domain, $(0, \infty)$. By Theorem 9, $\ln(t^4 - 1)$ is continuous on its domain, which is
$\{t \mid t^4 - 1 > 0\} = \{t \mid t^4 > 1\} = \{t \mid |t| > 1\} = (-\infty, -1) \cup (1, \infty)$.

23. The function $y = 1 \Big/ \left(1 + e^{1/x}\right)$ is discontinuous at $x = 0$ because

the left- and right-hand limits at $x = 0$ are different.

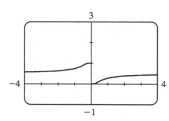

25. Because we are dealing with root functions, $5 + \sqrt{x}$ is continuous on $[0, \infty)$, $\sqrt{x + 5}$ is continuous on $[-5, \infty)$, so

the quotient $f(x) = \dfrac{5 + \sqrt{x}}{\sqrt{5 + x}}$ is continuous on $[0, \infty)$. Since f is continuous at $x = 4$, $\lim\limits_{x \to 4} f(x) = f(4) = \frac{7}{3}$.

27. Because $x^2 - x$ is continuous on $\mathbb{R}$, the composite function $f(x) = e^{x^2 - x}$ is continuous on $\mathbb{R}$, so
$\lim\limits_{x \to 1} f(x) = f(1) = e^{1-1} = e^0 = 1$.

29. $f(x) = \begin{cases} x + 2 & \text{if } x < 0 \\ e^x & \text{if } 0 \le x \le 1 \\ 2 - x & \text{if } x > 1 \end{cases}$

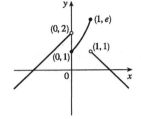

f is continuous on $(-\infty, 0)$ and $(1, \infty)$ since on each of these intervals it is

a polynomial; it is continuous on $(0, 1)$ since it is an exponential. Now

$\lim\limits_{x \to 0^-} f(x) = \lim\limits_{x \to 0^-} (x + 2) = 2$ and $\lim\limits_{x \to 0^+} f(x) = \lim\limits_{x \to 0^+} e^x = 1$, so f is

discontinuous at 0. Since $f(0) = 1$, f is continuous from the right at 0.

Also $\lim\limits_{x \to 1^-} f(x) = \lim\limits_{x \to 1^-} e^x = e$ and $\lim\limits_{x \to 1^+} f(x) = \lim\limits_{x \to 1^+} (2 - x) = 1$, so

f is discontinuous at 1. Since $f(1) = e$, f is continuous from the left at 1.

31. f is continuous on $(-\infty, 3)$ and $(3, \infty)$. Now $\lim\limits_{x \to 3^-} f(x) = \lim\limits_{x \to 3^-} (cx + 1) = 3c + 1$ and

$\lim\limits_{x \to 3^+} f(x) = \lim\limits_{x \to 3^+} (cx^2 - 1) = 9c - 1$. So f is continuous $\Leftrightarrow$ $3c + 1 = 9c - 1$ $\Leftrightarrow$ $6c = 2$ $\Leftrightarrow$ $c = \frac{1}{3}$.

Thus, for f to be continuous on $(-\infty, \infty)$, $c = \frac{1}{3}$.

33. $f(x) = x^3 - x^2 + x$ is continuous on the interval $[2, 3]$, $f(2) = 6$, and $f(3) = 21$. Since $6 < 10 < 21$, there is a

number c in $(2, 3)$ such that $f(c) = 10$ by the Intermediate Value Theorem.

35. $f(x) = x^3 - 3x + 1$ is continuous on the interval $[0, 1]$, $f(0) = 1$, and $f(1) = -1$. Since $-1 < 0 < 1$, there is a

number c in $(0, 1)$ such that $f(c) = 0$ by the Intermediate Value Theorem. Thus, there is a root of the equation

$x^3 - 3x + 1 = 0$ in the interval $(0, 1)$.

37. $f(x) = \cos x - x$ is continuous on the interval $[0, 1]$, $f(0) = 1$, and $f(1) = \cos 1 - 1 \approx -0.46$. Since

$-0.46 < 0 < 1$, there is a number c in $(0, 1)$ such that $f(c) = 0$ by the Intermediate Value Theorem. Thus, there is

a root of the equation $\cos x - x = 0$, or $\cos x = x$, in the interval $(0, 1)$.

39. (a) $f(x) = e^x + x - 2$ is continuous on the interval $[0, 1]$, $f(0) = -1 < 0$, and $f(1) = e - 1 \approx 1.72 > 0$. Since

$-1 < 0 < 1.72$, there is a number c in $(0, 1)$ such that $f(c) = 0$ by the Intermediate Value Theorem. Thus,

there is a root of the equation $e^x + x - 2 = 0$, or $e^x = 2 - x$, in the interval $(0, 1)$.

(b) $f(0.44) \approx -0.007 < 0$ and $f(0.45) \approx 0.018 > 0$, so there is a root between 0.44 and 0.45.

41. (a) Let $f(x) = 100e^{-x/100} - 0.01x^2$.

Then $f(0) = 100 > 0$ and

$f(100) = 100e^{-1} - 100 \approx -63.2 < 0$.

So by the Intermediate Value Theorem, there is a number c in $(0, 100)$ such that $f(c) = 0$. This implies that $100e^{-c/100} = 0.01c^2$.

(b) Using the intersect feature of the graphing device, we find that the root of the equation is $x = 70.347$, correct to three decimal places.

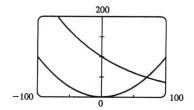

43. $\lim_{h \to 0} \sin(a + h) = \lim_{h \to 0} (\sin a \cos h + \cos a \sin h)$

$\qquad = \lim_{h \to 0} (\sin a \cos h) + \lim_{h \to 0} (\cos a \sin h)$

$\qquad = \left(\lim_{h \to 0} \sin a \right) \left(\lim_{h \to 0} \cos h \right) + \left(\lim_{h \to 0} \cos a \right) \left(\lim_{h \to 0} \sin h \right)$

$\qquad = (\sin a)(1) + (\cos a)(0) = \sin a$

45. If there is such a number, it satisfies the equation $x^3 + 1 = x \iff x^3 - x + 1 = 0$. Let the LHS of this equation be called $f(x)$. Now $f(-2) = -5 < 0$, and $f(-1) = 1 > 0$. Note also that $f(x)$ is a polynomial, and thus continuous. So by the Intermediate Value Theorem, there is a number c between -2 and -1 such that $f(c) = 0$, so that $c = c^3 + 1$.

47. Define $u(t)$ to be the monk's distance from the monastery, as a function of time, on the first day, and define $d(t)$ to be his distance from the monastery, as a function of time, on the second day. Let D be the distance from the monastery to the top of the mountain. From the given information we know that $u(0) = 0$, $u(12) = D$, $d(0) = D$ and $d(12) = 0$. Now consider the function $u - d$, which is clearly continuous. We calculate that $(u - d)(0) = -D$ and $(u - d)(12) = D$. So by the Intermediate Value Theorem, there must be some time t_0 between 0 and 12 such that $(u - d)(t_0) = 0 \iff u(t_0) = d(t_0)$. So at time t_0 after 7:00 A.M., the monk will be at the same place on both days.

2.5 Limits Involving Infinity • • • • • • • • • • • •

1. (a) As x approaches 2 (from the right or the left), the values of $f(x)$ become large.

(b) As x approaches 1 from the right, the values of $f(x)$ become large negative.

(c) As x becomes large, the values of $f(x)$ approach 5.

(d) As x becomes large negative, the values of $f(x)$ approach 3.

3. (a) $\lim_{x \to 2} f(x) = \infty$

(b) $\lim_{x \to -1^-} f(x) = \infty$

(c) $\lim_{x \to -1^+} f(x) = -\infty$

(d) $\lim_{x \to \infty} f(x) = 1$

(e) $\lim_{x \to -\infty} f(x) = 2$

(f) Vertical: $x = -1$, $x = 2$; Horizontal: $y = 1$, $y = 2$

5. $f(0) = 0$, $f(1) = 1$, $\lim\limits_{x \to \infty} f(x) = 0$,

f is odd

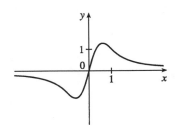

7. $\lim\limits_{x \to 2} f(x) = -\infty$, $\lim\limits_{x \to \infty} f(x) = \infty$,

$\lim\limits_{x \to -\infty} f(x) = 0$, $\lim\limits_{x \to 0^+} f(x) = \infty$,

$\lim\limits_{x \to 0^-} f(x) = -\infty$

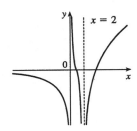

9. If $f(x) = x^2/2^x$, then a calculator gives $f(0) = 0$, $f(1) = 0.5$, $f(2) = 1$, $f(3) = 1.125$, $f(4) = 1$, $f(5) = 0.78125$, $f(6) = 0.5625$, $f(7) = 0.3828125$, $f(8) = 0.25$, $f(9) = 0.158203125$, $f(10) = 0.09765625$, $f(20) \approx 0.00038147$, $f(50) \approx 2.2204 \times 10^{-12}$, $f(100) \approx 7.8886 \times 10^{-27}$.

It appears that $\lim\limits_{x \to \infty} \left(x^2/2^x \right) = 0$.

11. Vertical: $x \approx -1.62$, $x \approx 0.62$, $x = 1$;
Horizontal: $y = 1$

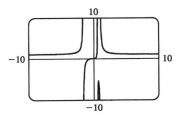

13. $\lim\limits_{x \to -3^+} \dfrac{x + 2}{x + 3} = -\infty$ since the numerator is negative and the denominator approaches 0 from the positive side as $x \to -3^+$.

15. $\lim\limits_{x \to 1} \dfrac{2 - x}{(x - 1)^2} = \infty$ since the numerator is positive and the denominator approaches 0 through positive values as $x \to 1$.

17. $\lim\limits_{x \to (-\pi/2)^-} \sec x = \lim\limits_{x \to (-\pi/2)^-} (1/\cos x) = -\infty$ since $\cos x \to 0^-$ as $x \to (-\pi/2)^-$.

19. Divide both the numerator and denominator by x^3 (the highest power of x that occurs in the denominator).

$$\lim_{x \to \infty} \frac{x^3 + 5x}{2x^3 - x^2 + 4} = \lim_{x \to \infty} \frac{\dfrac{x^3 + 5x}{x^3}}{\dfrac{2x^3 - x^2 + 4}{x^3}} = \lim_{x \to \infty} \frac{1 + \dfrac{5}{x^2}}{2 - \dfrac{1}{x} + \dfrac{4}{x^3}} = \frac{\lim\limits_{x \to \infty} \left(1 + \dfrac{5}{x^2} \right)}{\lim\limits_{x \to \infty} \left(2 - \dfrac{1}{x} + \dfrac{4}{x^3} \right)}$$

$$= \frac{\lim\limits_{x \to \infty} 1 + 5 \lim\limits_{x \to \infty} \dfrac{1}{x^2}}{\lim\limits_{x \to \infty} 2 - \lim\limits_{x \to \infty} \dfrac{1}{x} + 4 \lim\limits_{x \to \infty} \dfrac{1}{x^3}} = \frac{1 + 5(0)}{2 - 0 + 4(0)} = \frac{1}{2}$$

21. First, multiply the factors in the denominator. Then divide both the numerator and denominator by u^4.

$$\lim_{u \to \infty} \frac{4u^4 + 5}{(u^2 - 2)(2u^2 - 1)} = \lim_{u \to \infty} \frac{4u^4 + 5}{2u^4 - 5u^2 + 2} = \lim_{u \to \infty} \frac{\dfrac{4u^4 + 5}{u^4}}{\dfrac{2u^4 - 5u^2 + 2}{u^4}} = \lim_{u \to \infty} \frac{4 + \dfrac{5}{u^4}}{2 - \dfrac{5}{u^2} + \dfrac{2}{u^4}}$$

$$= \frac{\displaystyle\lim_{u \to \infty}\left(4 + \frac{5}{u^4}\right)}{\displaystyle\lim_{u \to \infty}\left(2 - \frac{5}{u^2} + \frac{2}{u^4}\right)} = \frac{\displaystyle\lim_{u \to \infty} 4 + 5 \lim_{u \to \infty}\frac{1}{u^4}}{\displaystyle\lim_{u \to \infty} 2 - 5\lim_{u \to \infty}\frac{1}{u^2} + 2\lim_{u \to \infty}\frac{1}{u^4}} = \frac{4 + 5(0)}{2 - 5(0) + 2(0)}$$

$$= \frac{4}{2} = 2$$

23. $\displaystyle\lim_{x \to \infty}\left(\sqrt{9x^2 + x} - 3x\right) = \lim_{x \to \infty}\frac{\left(\sqrt{9x^2 + x} - 3x\right)\left(\sqrt{9x^2 + x} + 3x\right)}{\sqrt{9x^2 + x} + 3x} = \lim_{x \to \infty}\frac{\left(\sqrt{9x^2 + x}\right)^2 - (3x)^2}{\sqrt{9x^2 + x} + 3x}$

$\qquad = \displaystyle\lim_{x \to \infty}\frac{(9x^2 + x) - 9x^2}{\sqrt{9x^2 + x} + 3x} = \lim_{x \to \infty}\frac{x/x}{\left(\sqrt{9x^2 + x} + 3x\right)/x} = \lim_{x \to \infty}\frac{1}{\sqrt{9 + 1/x} + 3}$

$\qquad = \dfrac{1}{\sqrt{9} + 3} = \dfrac{1}{3 + 3} = \dfrac{1}{6}$

25. $\displaystyle\lim_{x \to \infty}\cos x$ does not exist because, as x increases, $\cos x$ does not approach any one value, but oscillates between 1 and -1.

27. $\displaystyle\lim_{x \to \infty}\frac{x^7 - 1}{x^6 + 1} = \lim_{x \to \infty}\frac{1 - 1/x^7}{(1/x) + (1/x^7)} = \infty$ since $1 - \dfrac{1}{x^7} \to 1$ while $\dfrac{1}{x} + \dfrac{1}{x^7} \to 0^+$ as $x \to \infty$.

Or: Divide numerator and denominator by x^6 instead of x^7.

29. $\displaystyle\lim_{x \to -\infty}\left(x^3 - 5x^2\right) = -\infty$ since $x^3 \to -\infty$ and $-5x^2 \to -\infty$ as $x \to -\infty$.

$\quad$ *Or:* $\displaystyle\lim_{x \to -\infty}\left(x^3 - 5x^2\right) = \lim_{x \to -\infty} x^2(x - 5) = -\infty$ since $x^2 \to \infty$ and $x - 5 \to -\infty$.

31. $\displaystyle\lim_{x \to \infty}\frac{2x^2 + x - 1}{x^2 + x - 2} = \lim_{x \to \infty}\frac{\dfrac{2x^2 + x - 1}{x^2}}{\dfrac{x^2 + x - 2}{x^2}} = \lim_{x \to \infty}\frac{2 + \dfrac{1}{x} - \dfrac{1}{x^2}}{1 + \dfrac{1}{x} - \dfrac{2}{x^2}} = \frac{\displaystyle\lim_{x \to \infty}\left(2 + \frac{1}{x} - \frac{1}{x^2}\right)}{\displaystyle\lim_{x \to \infty}\left(1 + \frac{1}{x} - \frac{2}{x^2}\right)}$

$\qquad = \dfrac{\displaystyle\lim_{x \to \infty} 2 + \lim_{x \to \infty}\frac{1}{x} - \lim_{x \to \infty}\frac{1}{x^2}}{\displaystyle\lim_{x \to \infty} 1 + \lim_{x \to \infty}\frac{1}{x} - 2\lim_{x \to \infty}\frac{1}{x^2}} = \dfrac{2 + 0 - 0}{1 + 0 - 2(0)} = 2$, so $y = 2$ is a horizontal asymptote.

$y = f(x) = \dfrac{2x^2 + x - 1}{x^2 + x - 2} = \dfrac{(2x - 1)(x + 1)}{(x + 2)(x - 1)}$, so

$\displaystyle\lim_{x \to -2^-} f(x) = \infty$, $\displaystyle\lim_{x \to -2^+} f(x) = -\infty$,

$\displaystyle\lim_{x \to 1^-} f(x) = -\infty$, and $\displaystyle\lim_{x \to 1^+} f(x) = \infty$. Thus,

$x = -2$ and $x = 1$ are vertical asymptotes. The

graph confirms our work.

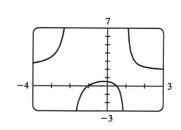

33. (a)

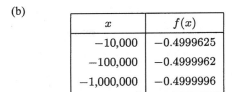

From the graph of $f(x) = \sqrt{x^2 + x + 1} + x$, we estimate the value of $\displaystyle\lim_{x \to -\infty} f(x)$ to be -0.5.

(b)

x	$f(x)$
$-10{,}000$	-0.4999625
$-100{,}000$	-0.4999962
$-1{,}000{,}000$	-0.4999996

From the table, we estimate the limit to be -0.5.

(c) $\displaystyle\lim_{x \to -\infty} \left(\sqrt{x^2 + x + 1} + x\right) = \lim_{x \to -\infty} \left(\sqrt{x^2 + x + 1} + x\right) \left[\frac{\sqrt{x^2 + x + 1} - x}{\sqrt{x^2 + x + 1} - x}\right]$

$\displaystyle = \lim_{x \to -\infty} \frac{(x^2 + x + 1) - x^2}{\sqrt{x^2 + x + 1} - x} = \lim_{x \to -\infty} \frac{(x + 1)\,(1/x)}{\left(\sqrt{x^2 + x + 1} - x\right)(1/x)}$

$\displaystyle = \lim_{x \to -\infty} \frac{1 + (1/x)}{-\sqrt{1 + (1/x) + (1/x^2)} - 1} = \frac{1 + 0}{-\sqrt{1 + 0 + 0} - 1} = -\frac{1}{2}$

Note that for $x < 0$, we have $\sqrt{x^2} = |x| = -x$, so when we divide the radical by x, with $x < 0$, we get

$\dfrac{1}{x}\sqrt{x^2 + x + 1} = -\dfrac{1}{\sqrt{x^2}}\sqrt{x^2 + x + 1} = -\sqrt{1 + (1/x) + (1/x^2)}$.

35. (a) This must be graph IV, the only graph that is always negative to the left of $x = 1$ and positive to the right of $x = 1$.

(b) This is graph III, the only graph with a horizontal asymptote of $y = 1$.

(c) This must be graph II, since it's the only graph that is positive everywhere. [$(x - 1)^2 > 0$ for $x \ne 1$.]

(d) $x^2 - 1 < 0$ if $|x| < 1$. Since the function is the reciprocal of $x^2 - 1$, the graph must be negative for $|x| < 1$ and positive elsewhere. The only graph fitting this description is VI.

(e) $(x - 1)^2 > 0$ if $x \ne 1$, so the sign of y is determined by the sign of the numerator, x. Thus, $y < 0$ if $x < 0$ and $y > 0$ if $x > 0$, as is only the case with graph I.

(f) The graph must have vertical asymptotes at $x = \pm 1$ and an x-intercept at $x = 0$. The only graph fitting this description is V.

37. Let's look for a rational function.

(1) $\displaystyle\lim_{x \to \pm\infty} f(x) = 0 \ \Rightarrow \ $ degree of numerator $<$ degree of denominator

(2) $\displaystyle\lim_{x \to 0} f(x) = -\infty \ \Rightarrow \ $ there is a factor of x^2 in the denominator (not just x, since that would produce a sign change at $x = 0$), and the function is negative near $x = 0$.

(3) $\displaystyle\lim_{x \to 3^-} f(x) = \infty$ and $\displaystyle\lim_{x \to 3^+} f(x) = -\infty \ \Rightarrow \ $ vertical asymptote at $x = 3$; there is a factor of $(x - 3)$ in the denominator.

(4) $f(2) = 0 \ \Rightarrow \ 2$ is an x-intercept; there is at least one factor of $(x - 2)$ in the numerator.

Combining all of this information, and putting in a negative sign to give us the desired left- and right-hand limits, gives us $f(x) = \dfrac{2 - x}{x^2(x - 3)}$ as one possibility.

39. Divide numerator and denominator by the highest power of x in $Q(x)$.

(a) If $\deg P < \deg Q$, then numerator $\to 0$ but denominator doesn't. So $\lim\limits_{x\to\infty} [P(x)/Q(x)] = 0$.

(b) If $\deg P > \deg Q$, then numerator $\to \pm\infty$ but denominator doesn't, so $\lim\limits_{x\to\infty} [P(x)/Q(x)] = \pm\infty$ (depending on the ratio of the leading coefficients of P and Q).

41. $\lim\limits_{x\to\infty} \dfrac{4x-1}{x} = \lim\limits_{x\to\infty} \left(4 - \dfrac{1}{x}\right) = 4$, and $\lim\limits_{x\to\infty} \dfrac{4x^2+3x}{x^2} = \lim\limits_{x\to\infty} \left(4 + \dfrac{3}{x}\right) = 4$. Therefore, by the Squeeze

Theorem, $\lim\limits_{x\to\infty} f(x) = 4$.

43. (a) After t minutes, $25t$ liters of brine with 30 g of salt per liter has been pumped into the tank, so it contains $(5000 + 25t)$ liters of water and $25t \cdot 30 = 750t$ grams of salt. Therefore, the salt concentration at time t will

be $C(t) = \dfrac{750t}{5000 + 25t} = \dfrac{30t}{200 + t} \dfrac{\text{g}}{\text{L}}$.

(b) $\lim\limits_{t\to\infty} C(t) = \lim\limits_{t\to\infty} \dfrac{30t}{200 + t} = \lim\limits_{t\to\infty} \dfrac{30t/t}{200/t + t/t} = \dfrac{30}{0+1} = 30$. So the salt concentration approaches that of the brine being pumped into the tank.

45. (a) If $t = -x/10$, then $x = -10t$ and as $x \to \infty$, $t \to -\infty$. Thus,

$\lim\limits_{x\to\infty} e^{-x/10} = \lim\limits_{t\to-\infty} e^t = 0$ by Equation 8.

(b) $y = e^{-x/10}$ and $y = 0.1$ intersect at $x_1 \approx 23.03$.

If $x > x_1$, then $e^{-x/10} < 0.1$.

(c) $e^{-x/10} < 0.1 \implies -x/10 < \ln 0.1 \implies$

$x > -10\ln \frac{1}{10} = -10\ln 10^{-1} = 10\ln 10$

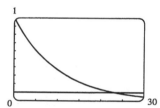

◆ **2.6** **Tangents, Velocities, and Other Rates of Change** • • • • •

1. (a) This is just the slope of the line through two points: $m_{PQ} = \dfrac{\Delta y}{\Delta x} = \dfrac{f(x) - f(3)}{x - 3}$.

(b) This is the limit of the slope of the secant line PQ as Q approaches P: $m = \lim\limits_{x\to3} \dfrac{f(x) - f(3)}{x - 3}$.

3. The slope at D is the largest positive slope, followed by the positive slope at E. The slope at C is zero. The slope at B is steeper than at A (both are negative). In decreasing order, we have the slopes at: D, E, C, A, B.

5. (a) (i) Using Definition 1,

$$m = \lim\limits_{x\to a} \dfrac{f(x) - f(a)}{x - a} \quad \lim\limits_{x\to-3} \dfrac{f(x) - f(-3)}{x - (-3)} = \lim\limits_{x\to-3} \dfrac{(x^2 + 2x) - (3)}{x - (-3)} = \lim\limits_{x\to-3} \dfrac{(x+3)(x-1)}{x+3}$$

$$= \lim\limits_{x\to-3} (x - 1) = -4$$

(ii) Using Equation 2,

$$m = \lim\limits_{h\to0} \dfrac{f(a+h) - f(a)}{h} = \lim\limits_{h\to0} \dfrac{f(-3+h) - f(-3)}{h} = \lim\limits_{h\to0} \dfrac{[(-3+h)^2 + 2(-3+h)] - (3)}{h}$$

$$= \lim\limits_{h\to0} \dfrac{9 - 6h + h^2 - 6 + 2h - 3}{h} = \lim\limits_{h\to0} \dfrac{h(h-4)}{h} = \lim\limits_{h\to0} (h - 4) = -4$$

(b) Using the point-slope form of the equation of a line, an equation of the tangent line is $y - 3 = -4(x + 3)$.
Solving for y gives us $y = -4x - 9$, which is the slope-intercept form of the equation of the tangent line.

(c)

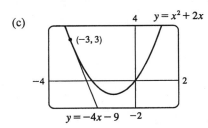

7. Using (1) with $f(x) = \dfrac{x - 1}{x - 2}$ and $P(3, 2)$,

$$m = \lim_{x \to a} \frac{f(x) - f(a)}{x - a} = \lim_{x \to 3} \frac{\frac{x - 1}{x - 2} - 2}{x - 3} = \lim_{x \to 3} \frac{\frac{x - 1 - 2(x - 2)}{x - 2}}{x - 3} = \lim_{x \to 3} \frac{3 - x}{(x - 2)(x - 3)}$$

$$= \lim_{x \to 3} \frac{-1}{x - 2} = \frac{-1}{1} = -1.$$

Tangent line: $y - 2 = -1(x - 3)$ ⇔ $y - 2 = -x + 3$ ⇔ $y = -x + 5$

9. Using (1), $m = \lim_{x \to 1} \dfrac{\sqrt{x} - \sqrt{1}}{x - 1} = \lim_{x \to 1} \dfrac{(\sqrt{x} - 1)(\sqrt{x} + 1)}{(x - 1)(\sqrt{x} + 1)} = \lim_{x \to 1} \dfrac{x - 1}{(x - 1)(\sqrt{x} + 1)} = \lim_{x \to 1} \dfrac{1}{\sqrt{x} + 1} = \dfrac{1}{2}$.

Tangent line: $y - 1 = \frac{1}{2}(x - 1)$ ⇔ $y = \frac{1}{2}x + \frac{1}{2}$.

11. (a) Using (1),

$$m = \lim_{x \to a} \frac{(x^3 - 4x + 1) - (a^3 - 4a + 1)}{x - a} = \lim_{x \to a} \frac{(x^3 - a^3) - 4(x - a)}{x - a}$$

$$= \lim_{x \to a} \frac{(x - a)(x^2 + ax + a^2) - 4(x - a)}{x - a} = \lim_{x \to a} (x^2 + ax + a^2 - 4) = 3a^2 - 4$$

(b) At $(1, -2)$: $m = 3(1)^2 - 4 = -1$, so an equation of the
tangent line is $y - (-2) = -1(x - 1)$ ⇔ $y = -x - 1$.

At $(2, 1)$: $m = 3(2)^2 - 4 = 8$, so an equation of the
tangent line is $y - 1 = 8(x - 2)$ ⇔ $y = 8x - 15$.

(c)

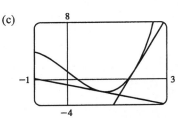

13. (a) Since the slope of the tangent at $t = 0$ is 0, the car's initial velocity was 0.

(b) The slope of the tangent is greater at C than at B, so the car was going faster at C.

(c) Near A, the tangent lines are becoming steeper as x increases, so the velocity was increasing, so the car was
speeding up. Near B, the tangent lines are becoming less steep, so the car was slowing down. The steepest
tangent near C is the one at C, so at C the car had just finished speeding up, and was about to start slowing
down.

(d) Between D and E, the slope of the tangent is 0, so the car did not move during that time.

15. Let $s(t) = 40t - 16t^2$.

$$v(2) = \lim_{t \to 2} \frac{s(t) - s(2)}{t - 2} = \lim_{t \to 2} \frac{(40t - 16t^2) - 16}{t - 2} = \lim_{t \to 2} \frac{-16t^2 + 40t - 16}{t - 2} = \lim_{t \to 2} \frac{-8(2t^2 - 5t + 2)}{t - 2}$$

$$= \lim_{t \to 2} \frac{-8(t - 2)(2t - 1)}{t - 2} = -8 \lim_{t \to 2}(2t - 1) = -8(3) = -24$$

Thus, the instantaneous velocity when $t = 2$ is -24 ft/s.

17. $v(a) = \lim\limits_{h \to 0} \dfrac{s(a + h) - s(a)}{h} = \lim\limits_{h \to 0} \dfrac{4(a + h)^3 + 6(a + h) + 2 - (4a^3 + 6a + 2)}{h}$

$$= \lim_{h \to 0} \frac{4a^3 + 12a^2h + 12ah^2 + 4h^3 + 6a + 6h + 2 - 4a^3 - 6a - 2}{h}$$

$$= \lim_{h \to 0} \frac{12a^2h + 12ah^2 + 4h^3 + 6h}{h} = \lim_{h \to 0} \left(12a^2 + 12ah + 4h^2 + 6\right) = \left(12a^2 + 6\right) \text{ m/s}$$

So $v(1) = 12(1)^2 + 6 = 18$ m/s, $v(2) = 12(2)^2 + 6 = 54$ m/s, and $v(3) = 12(3)^2 + 6 = 114$ m/s.

19. The sketch shows the graph for a room temperature of $72°$ and a refrigerator temperature of $38°$. The initial rate of change is greater in magnitude than the rate of change after an hour.

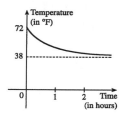

21. (a) (i) $[20, 23]$: $\dfrac{7.9 - 11.5}{3} = -1.2 \,°\text{C/h}$

(ii) $[20, 22]$: $\dfrac{9.0 - 11.5}{2} = -1.25 \,°\text{C/h}$

(iii) $[20, 21]$: $\dfrac{10.2 - 11.5}{1} = -1.3 \,°\text{C/h}$

(b) In the figure, we estimate A to be $(18, 15.5)$ and B as $(23, 6)$. So the slope is

$$\frac{6 - 15.5}{23 - 18} = -1.9 \,°\text{C/h at 8:00 P.M.}$$

23. (a) (i) $[1995, 1997]$: $\dfrac{N(1997) - N(1995)}{1997 - 1995} = \dfrac{2461 - 873}{2} = \dfrac{1588}{2} = 794$ thousand/year

(ii) $[1995, 1996]$: $\dfrac{N(1996) - N(1995)}{1996 - 1995} = \dfrac{1513 - 873}{1} = 640$ thousand/year

(iii) $[1994, 1995]$: $\dfrac{N(1995) - N(1994)}{1995 - 1994} = \dfrac{873 - 572}{1} = 301$ thousand/year

(b) Using the values from (ii) and (iii), we have $\dfrac{640 + 301}{2} = \dfrac{941}{2} = 470.5$ thousand/year.

(c) Estimating A as $(1994, 420)$ and B as $(1996, 1275)$, the slope

at 1995 is $\dfrac{1275 - 420}{1996 - 1994} = \dfrac{855}{2} = 427.5$ thousand/year

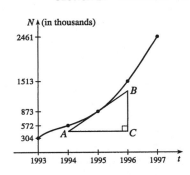

25. (a) (i) $\dfrac{\Delta C}{\Delta x} = \dfrac{C(105) - C(100)}{105 - 100} = \dfrac{6601.25 - 6500}{5} = \$20.25/\text{unit}.$

(ii) $\dfrac{\Delta C}{\Delta x} = \dfrac{C(101) - C(100)}{101 - 100} = \dfrac{6520.05 - 6500}{1} = \$20.05/\text{unit}.$

(b) $\dfrac{C(100 + h) - C(100)}{h} = \dfrac{\left[5000 + 10(100 + h) + 0.05(100 + h)^2\right] - 6500}{h} = \dfrac{20h + 0.05h^2}{h}$

$$= 20 + 0.05h,\ h \neq 0$$

So the instantaneous rate of change is $\displaystyle\lim_{h \to 0} \dfrac{C(100 + h) - C(100)}{h} = \lim_{h \to 0}(20 + 0.05h) = \$20/\text{unit}.$

◆2.7◆ **Derivatives** • • • • • • • • • • • • • • • •

1.

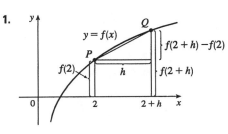

The line from $P(2, f(2))$ to $Q(2 + h, f(2 + h))$

is the line that has slope $\dfrac{f(2 + h) - f(2)}{h}$.

3. $g'(0)$ is the only negative value. The slope at $x = 4$ is smaller than the slope at $x = 2$ and both are smaller than the slope at $x = -2$. Thus, $g'(0) < 0 < g'(4) < g'(2) < g'(-2)$.

5. We begin by drawing a curve through the origin at a slope of 3 to satisfy $f(0) = 0$ and $f'(0) = 3$. Since $f'(1) = 0$, we will round off our figure so that there is a horizontal tangent directly over $x = 1$. Lastly, we make sure that the curve has a slope of -1 as we pass over $x = 2$. Two of the many possibilities are shown.

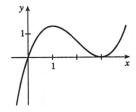

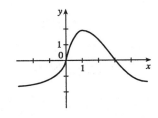

7. Using Definition 2 with $f(x) = 3x^2 - 5x$ and the point $(2, 2)$, we have

$$f'(2) = \lim_{h \to 0} \frac{f(2+h) - f(2)}{h} = \lim_{h \to 0} \frac{[3(2+h)^2 - 5(2+h)] - 2}{h}$$

$$= \lim_{h \to 0} \frac{(12 + 12h + 3h^2 - 10 - 5h) - 2}{h} = \lim_{h \to 0} \frac{3h^2 + 7h}{h} = \lim_{h \to 0} (3h + 7) = 7.$$

So an equation of the tangent line at $(2, 2)$ is $y - 2 = 7(x - 2)$ or $y = 7x - 12$.

9. (a) Using Equation 3 with $F(x) = x^3 - 5x + 1$ and the point $(1, -3)$, we have

$$F'(1) = \lim_{x \to 1} \frac{F(x) - F(1)}{x - 1} = \lim_{x \to 1} \frac{(x^3 - 5x + 1) - (-3)}{x - 1} = \lim_{x \to 1} \frac{x^3 - 5x + 4}{x - 1}$$

$$= \lim_{x \to 1} \frac{(x - 1)(x^2 + x - 4)}{x - 1} = \lim_{x \to 1} (x^2 + x - 4) = -2.$$

So an equation of the tangent line at $(1, -3)$ is $y - (-3) = -2(x - 1)$ $\Leftrightarrow$ $y = -2x - 1$.

Note: Instead of using Equation 3 to compute $F'(1)$, we could have used Definition 2.

(b)

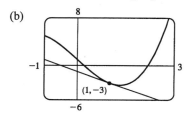

11. (a) $f'(1) = \lim_{h \to 0} \frac{f(1+h) - f(1)}{h} = \lim_{h \to 0} \frac{3^{1+h} - 3^1}{h}$.

So let $F(h) = \dfrac{3^{1+h} - 3}{h}$. We calculate:

(b)

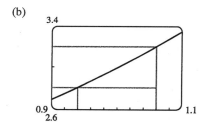

From the graph, we estimate that the slope of the tangent is about

$$\frac{3.2 - 2.8}{1.06 - 0.94} = \frac{0.4}{0.12} \approx 3.3.$$

h	$F(h)$
0.1	3.484
0.01	3.314
0.001	3.298
0.0001	3.296
-0.1	3.121
-0.01	3.278
-0.001	3.294
-0.0001	3.296

We estimate that $f'(1) \approx 3.296$.

13. Use Definition 2 with $f(x) = 3 - 2x + 4x^2$.

$$f'(a) = \lim_{h \to 0} \frac{f(a+h) - f(a)}{h} = \lim_{h \to 0} \frac{[3 - 2(a+h) + 4(a+h)^2] - (3 - 2a + 4a^2)}{h}$$

$$= \lim_{h \to 0} \frac{(3 - 2a - 2h + 4a^2 + 8ah + 4h^2) - (3 - 2a + 4a^2)}{h}$$

$$= \lim_{h \to 0} \frac{-2h + 8ah + 4h^2}{h} = \lim_{h \to 0} \frac{h(-2 + 8a + 4h)}{h} = \lim_{h \to 0} (-2 + 8a + 4h) = -2 + 8a$$

15. $f'(a) = \lim\limits_{h \to 0} \dfrac{f(a+h) - f(a)}{h} = \lim\limits_{h \to 0} \dfrac{\dfrac{2(a+h)+1}{(a+h)+3} - \dfrac{2a+1}{a+3}}{h}$

$= \lim\limits_{h \to 0} \dfrac{(2a+2h+1)(a+3) - (2a+1)(a+h+3)}{h(a+h+3)(a+3)}$

$= \lim\limits_{h \to 0} \dfrac{(2a^2 + 6a + 2ah + 6h + a + 3) - (2a^2 + 2ah + 6a + a + h + 3)}{h(a+h+3)(a+3)}$

$= \lim\limits_{h \to 0} \dfrac{5h}{h(a+h+3)(a+3)} = \lim\limits_{h \to 0} \dfrac{5}{(a+h+3)(a+3)} = \dfrac{5}{(a+3)^2}$

17. $f'(a) = \lim\limits_{h \to 0} \dfrac{f(a+h) - f(a)}{h} = \lim\limits_{h \to 0} \dfrac{\dfrac{1}{\sqrt{(a+h)+2}} - \dfrac{1}{\sqrt{a+2}}}{h}$

$= \lim\limits_{h \to 0} \dfrac{\dfrac{\sqrt{a+2} - \sqrt{a+h+2}}{\sqrt{a+h+2}\sqrt{a+2}}}{h} = \lim\limits_{h \to 0} \left[\dfrac{\sqrt{a+2} - \sqrt{a+h+2}}{h\sqrt{a+h+2}\sqrt{a+2}} \cdot \dfrac{\sqrt{a+2} + \sqrt{a+h+2}}{\sqrt{a+2} + \sqrt{a+h+2}} \right]$

$= \lim\limits_{h \to 0} \dfrac{(a+2) - (a+h+2)}{h\sqrt{a+h+2}\sqrt{a+2}\left(\sqrt{a+2} + \sqrt{a+h+2}\right)}$

$= \lim\limits_{h \to 0} \dfrac{-h}{h\sqrt{a+h+2}\sqrt{a+2}\left(\sqrt{a+2} + \sqrt{a+h+2}\right)}$

$= \lim\limits_{h \to 0} \dfrac{-1}{\sqrt{a+h+2}\sqrt{a+2}\left(\sqrt{a+2} + \sqrt{a+h+2}\right)}$

$= \dfrac{-1}{\left(\sqrt{a+2}\right)^2 \left(2\sqrt{a+2}\right)} = -\dfrac{1}{2(a+2)^{3/2}}$

Note that the answers to Exercises 19–23 are not unique.

19. By Definition 2, $\lim\limits_{h \to 0} \dfrac{(1+h)^{10} - 1}{h} = f'(1)$, where $f(x) = x^{10}$.

21. By Equation 3, $\lim\limits_{x \to 5} \dfrac{2^x - 32}{x - 5} = f'(5)$, where $f(x) = 2^x$.

23. By Definition 2, $\lim\limits_{h \to 0} \dfrac{\cos(\pi + h) + 1}{h} = f'(\pi)$, where $f(x) = \cos x$.

25. $v(2) = f'(2) = \lim\limits_{h \to 0} \dfrac{f(2+h) - f(2)}{h} = \lim\limits_{h \to 0} \dfrac{\left[(2+h)^2 - 6(2+h) - 5\right] - \left[2^2 - 6(2) - 5\right]}{h}$

$= \lim\limits_{h \to 0} \dfrac{(4 + 4h + h^2 - 12 - 6h - 5) - (-13)}{h} = \lim\limits_{h \to 0} \dfrac{h^2 - 2h}{h} = \lim\limits_{h \to 0} (h - 2) = -2 \text{ m/s}$

27. (a) $f'(x)$ is the rate of change of the production cost with respect to the number of ounces of gold produced. Its units are dollars per ounce.

(b) After 800 ounces of gold have been produced, the rate at which the production cost is increasing is $17/ounce. So the cost of producing the 800th (or 801st) ounce is about $17.

(c) In the short term, the values of $f'(x)$ will decrease because more efficient use is made of start-up costs as x increases. But eventually $f'(x)$ might increase due to large-scale operations.

29. (a) $f'(v)$ is the rate at which the fuel consumption is changing with respect to the speed. Its units are (gal/h)/(mi/h).

(b) The fuel consumption is decreasing by 0.05 (gal/h)/(mi/h) as the car's speed reaches 20 mi/h. So if you increase your speed to 21 mi/h, you could expect to decrease your fuel consumption by about 0.05 (gal/h)/(mi/h).

31. $T'(6)$ is the rate of change of the temperature with respect to time when $t = 6$. Its units are $°C/h$. To estimate the value of $T'(6)$, we will average the difference quotients obtained using the times $t = 4$ and $t = 8$.

Let $A = \dfrac{T(4) - T(6)}{4 - 6} = \dfrac{29 - 32}{-2} = 1.5$ and $B = \dfrac{T(8) - T(6)}{8 - 6} = \dfrac{33 - 32}{2} = 0.5$. Then

$T'(6) = \lim\limits_{t \to 6} \dfrac{T(t) - T(6)}{t - 6} \approx \dfrac{A + B}{2} = 1 \, °C/h.$

33. $C'(1980)$ is the rate of change of U.S. cash per capita in circulation with respect to time. To estimate the value of $C'(1980)$, we will average the difference quotients obtained using the years 1970 and 1990.

Let $A = \dfrac{C(1970) - C(1980)}{1970 - 1980} = \dfrac{265 - 571}{-10} = 30.6$ and $B = \dfrac{C(1990) - C(1980)}{1990 - 1980} = \dfrac{1063 - 571}{10} = 49.2$.

Then $C'(1980) = \lim\limits_{t \to 1980} \dfrac{C(t) - C(1980)}{t - 1980} \approx \dfrac{A + B}{2} = 39.9$ dollars per year.

35. Since $f(x) = x \sin(1/x)$ when $x \neq 0$ and $f(0) = 0$, we have

$f'(0) = \lim\limits_{h \to 0} \dfrac{f(0 + h) - f(0)}{h} = \lim\limits_{h \to 0} \dfrac{h \sin(1/h) - 0}{h} = \lim\limits_{h \to 0} \sin(1/h)$. This limit does not exist since $\sin(1/h)$

takes the values -1 and 1 on any interval containing 0. (Compare with Example 4 in Section 2.2.)

2.8 The Derivative as a Function • • • • • • • • • • •

1. It appears that f is an odd function, so f' will be an even function—that is, $f'(-a) = f'(a)$.

(a) $f'(-3) \approx 1.5$ (b) $f'(-2) \approx 1$

(c) $f'(-1) \approx 0$ (d) $f'(0) \approx -4$

(e) $f'(1) \approx 0$ (f) $f'(2) \approx 1$

(g) $f'(3) \approx 1.5$

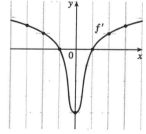

3. (a)$' = $ II, since from left to right, the slopes of the tangents to graph (a) start out negative, become 0, then positive, then 0, then negative again. The actual function values in graph II follow the same pattern.

(b)$' = $ IV, since from left to right, the slopes of the tangents to graph (b) start out at a fixed positive quantity, then suddenly become negative, then positive again. The discontinuities in graph IV indicate sudden changes in the slopes of the tangents.

(c)$' = $ I, since the slopes of the tangents to graph (c) are negative for $x < 0$ and positive for $x > 0$, as are the function values of graph I.

(d)$' = $ III, since from left to right, the slopes of the tangents to graph (d) are positive, then 0, then negative, then 0, then positive, then 0, then negative again, and the function values in graph III follow the same pattern.

Hints for Exercises 5–11: First plot x-intercepts on the graph of f' for any horizontal tangents on the graph of f. Look for any corners on the graph of f—there will be a discontinuity on the graph of f'. On any interval where f has a tangent with positive (or negative) slope, the graph of f' will be positive (or negative). If the graph of the function is linear, the graph of f' will be a horizontal line.

5.

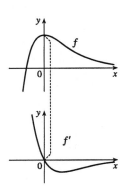

7.

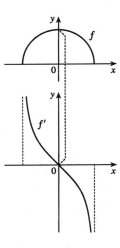

9.

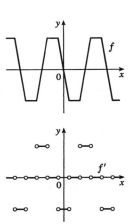

11.

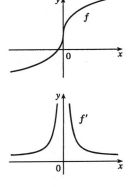

13. It appears that there are horizontal tangents on the graph of M for $t = 1963$ and $t = 1971$. Thus, there are zeros for those values of t on the graph of M'. The derivative is negative for the years 1963 to 1971.

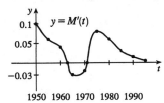

15.

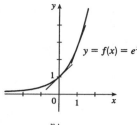

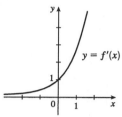

The slope at 0 appears to be 1 and the slope at 1 appears to be 2.7. As x decreases, the slope gets closer to 0. Since the graphs are so similar, we might guess that $f'(x) = e^x$.

17. (a) By zooming in, we estimate that $f'(0) = 0$, $f'\left(\frac{1}{2}\right) = 1$, $f'(1) = 2$, and $f'(2) = 4$.

(b) By symmetry, $f'(-x) = -f'(x)$. So $f'\left(-\frac{1}{2}\right) = -1$, $f'(-1) = -2$, and $f'(-2) = -4$.

(c) It appears that $f'(x)$ is twice the value of x, so we guess that $f'(x) = 2x$.

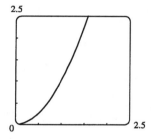

(d) $f'(x) = \lim\limits_{h \to 0} \dfrac{f(x+h) - f(x)}{h} = \lim\limits_{h \to 0} \dfrac{(x+h)^2 - x^2}{h} = \lim\limits_{h \to 0} \dfrac{(x^2 + 2hx + h^2) - x^2}{h}$

$\qquad = \lim\limits_{h \to 0} \dfrac{2hx + h^2}{h} = \lim\limits_{h \to 0} \dfrac{h(2x + h)}{h} = \lim\limits_{h \to 0} (2x + h) = 2x$

19. $f'(x) = \lim\limits_{h \to 0} \dfrac{f(x+h) - f(x)}{h} = \lim\limits_{h \to 0} \dfrac{[4 - 7(x+h)] - (4 - 7x)}{h} = \lim\limits_{h \to 0} \dfrac{(4 - 7x - 7h) - (4 - 7x)}{h}$

$\qquad = \lim\limits_{h \to 0} \dfrac{-7h}{h} = \lim\limits_{h \to 0} (-7) = -7$

Domain of f = domain of f' = $\mathbb{R}$.

21. $f'(x) = \lim\limits_{h \to 0} \dfrac{f(x+h) - f(x)}{h} = \lim\limits_{h \to 0} \dfrac{\left[(x+h)^3 - 3(x+h) + 5\right] - (x^3 - 3x + 5)}{h}$

$\qquad = \lim\limits_{h \to 0} \dfrac{(x^3 + 3x^2h + 3xh^2 + h^3 - 3x - 3h + 5) - (x^3 - 3x + 5)}{h}$

$\qquad = \lim\limits_{h \to 0} \dfrac{3x^2h + 3xh^2 + h^3 - 3h}{h} = \lim\limits_{h \to 0} \dfrac{h(3x^2 + 3xh + h^2 - 3)}{h}$

$\qquad = \lim\limits_{h \to 0} (3x^2 + 3xh + h^2 - 3) = 3x^2 - 3$

Domain of f = domain of f' = $\mathbb{R}$.

23. $g'(x) = \lim\limits_{h \to 0} \dfrac{g(x+h) - g(x)}{h} = \lim\limits_{h \to 0} \dfrac{\sqrt{1 + 2(x+h)} - \sqrt{1+2x}}{h} \left[\dfrac{\sqrt{1+2(x+h)} + \sqrt{1+2x}}{\sqrt{1+2(x+h)} + \sqrt{1+2x}} \right]$

$= \lim\limits_{h \to 0} \dfrac{(1 + 2x + 2h) - (1 + 2x)}{h \left[\sqrt{1+2(x+h)} + \sqrt{1+2x} \right]} = \lim\limits_{h \to 0} \dfrac{2}{\sqrt{1 + 2x + 2h} + \sqrt{1+2x}} = \dfrac{2}{2\sqrt{1+2x}} = \dfrac{1}{\sqrt{1+2x}}$

Domain of $g = \left[-\frac{1}{2}, \infty \right)$, domain of $g' = \left(-\frac{1}{2}, \infty \right)$.

25. $G'(t) = \lim\limits_{h \to 0} \dfrac{G(t+h) - G(t)}{h} = \lim\limits_{h \to 0} \dfrac{\dfrac{4(t+h)}{(t+h)+1} - \dfrac{4t}{t+1}}{h} = \lim\limits_{h \to 0} \dfrac{\dfrac{4(t+h)(t+1) - 4t(t+h+1)}{(t+h+1)(t+1)}}{h}$

$= \lim\limits_{h \to 0} \dfrac{\left(4t^2 + 4ht + 4t + 4h \right) - \left(4t^2 + 4ht + 4t \right)}{h(t+h+1)(t+1)}$

$= \lim\limits_{h \to 0} \dfrac{4h}{h(t+h+1)(t+1)} = \lim\limits_{h \to 0} \dfrac{4}{(t+h+1)(t+1)} = \dfrac{4}{(t+1)^2}$

Domain of G = domain of $G' = (-\infty, -1) \cup (-1, \infty)$.

27. (a) $f'(x) = \lim\limits_{h \to 0} \dfrac{f(x+h) - f(x)}{h} = \lim\limits_{h \to 0} \dfrac{\left[x + h - \left(\dfrac{2}{x+h} \right) \right] - \left[x - \left(\dfrac{2}{x} \right) \right]}{h}$

$= \lim\limits_{h \to 0} \left[\dfrac{h - \dfrac{2}{(x+h)} + \dfrac{2}{x}}{h} \right] = \lim\limits_{h \to 0} \left[1 + \dfrac{-2x + 2(x+h)}{h(x)(x+h)} \right] = \lim\limits_{h \to 0} \left[1 + \dfrac{2h}{h(x)(x+h)} \right]$

$= \lim\limits_{h \to 0} \left[1 + \dfrac{2}{x(x+h)} \right] = 1 + \dfrac{2}{x^2}$

(b) Notice that when f has steep tangent lines, $f'(x)$ is very large. When f is flatter, $f'(x)$ is smaller.

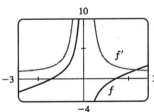

29. (a) $U'(t)$ is the rate at which the unemployment rate is changing with respect to time. Its units are percent per year.

(b) To find $U'(t)$, we use $\lim\limits_{h \to 0} \dfrac{U(t+h) - U(t)}{h} \approx \dfrac{U(t+h) - U(t)}{h}$ for small values of h.

For 1989: $U'(1989) = \dfrac{U(1990) - U(1989)}{1990 - 1989} = \dfrac{5.6 - 5.3}{1} = 0.30$

For 1990: We estimate $U'(1990)$ by using $h = -1$ and $h = 1$, and then averaging the two results to obtain a final estimate.

$h = -1 \quad \Rightarrow \quad U'(1990) \approx \dfrac{U(1989) - U(1990)}{1989 - 1990} = \dfrac{5.3 - 5.6}{-1} = 0.30;$

$h = 1 \quad \Rightarrow \quad U'(1990) \approx \dfrac{U(1991) - U(1990)}{1991 - 1990} = \dfrac{6.8 - 5.6}{1} = 1.20.$

So we estimate that $U'(1990) \approx \frac{1}{2}(0.30 + 1.20) = 0.75$.

t	1989	1990	1991	1992	1993	1994	1995	1996	1997	1998
$U'(t)$	0.30	0.75	0.95	0.05	−0.70	−0.65	−0.35	−0.35	−0.45	−0.40

31. f is not differentiable at $x = -1$ or at $x = 11$ because the graph has vertical tangents at those points; at $x = 4$, because there is a discontinuity there; and at $x = 8$, because the graph has a corner there.

33. As we zoom in toward $(-1, 0)$, the curve appears more and more like a straight line, so f is differentiable at $x = -1$. But no matter how much we zoom in toward the origin, the curve doesn't straighten out—we can't eliminate the sharp point (a cusp). So f is not differentiable at $x = 0$.

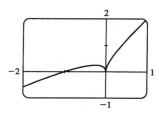

35. $a = f$, $b = f'$, $c = f''$. We can see this because where a has a horizontal tangent, $b = 0$, and where b has a horizontal tangent, $c = 0$. We can immediately see that c can be neither f nor f', since at the points where c has a horizontal tangent, neither a nor b is equal to 0.

37. We can immediately see that a is the graph of the acceleration function, since at the points where a has a horizontal tangent, neither c nor b is equal to 0. Next, we note that $a = 0$ at the point where b has a horizontal tangent, so b must be the graph of the velocity function, and hence, $b' = a$. We conclude that c is the graph of the position function.

39. $f'(x) = \lim\limits_{h \to 0} \dfrac{f(x+h) - f(x)}{h} = \lim\limits_{h \to 0} \dfrac{\left[1 + 4(x+h) - (x+h)^2\right] - \left(1 + 4x - x^2\right)}{h}$

$\qquad = \lim\limits_{h \to 0} \dfrac{\left(1 + 4x + 4h - x^2 - 2xh - h^2\right) - \left(1 + 4x - x^2\right)}{h} = \lim\limits_{h \to 0} \dfrac{4h - 2xh - h^2}{h}$

$\qquad = \lim\limits_{h \to 0} (4 - 2x - h) = 4 - 2x$

$f''(x) = \lim\limits_{h \to 0} \dfrac{f'(x+h) - f'(x)}{h} = \lim\limits_{h \to 0} \dfrac{[4 - 2(x+h)] - (4 - 2x)}{h} = \lim\limits_{h \to 0} \dfrac{-2h}{h} = \lim\limits_{h \to 0} (-2) = -2$

We see from the graph that our answers are reasonable because the graph of f' is that of a linear function and the graph of f'' is that of a constant function.

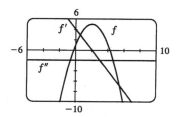

41. $f'(x) = \lim\limits_{h \to 0} \dfrac{f(x+h) - f(x)}{h} = \lim\limits_{h \to 0} \dfrac{\left[2(x+h)^2 - (x+h)^3\right] - \left(2x^2 - x^3\right)}{h}$

$\qquad = \lim\limits_{h \to 0} \dfrac{h\left(4x + 2h - 3x^2 - 3xh - h^2\right)}{h} = \lim\limits_{h \to 0} \left(4x + 2h - 3x^2 - 3xh - h^2\right) = 4x - 3x^2$

$f''(x) = \lim\limits_{h \to 0} \dfrac{f'(x+h) - f'(x)}{h} = \lim\limits_{h \to 0} \dfrac{\left[4(x+h) - 3(x+h)^2\right] - \left(4x - 3x^2\right)}{h}$

$\qquad = \lim\limits_{h \to 0} \dfrac{h(4 - 6x - 3h)}{h} = \lim\limits_{h \to 0} (4 - 6x - 3h) = 4 - 6x$

$$f'''(x) = \lim_{h \to 0} \frac{f''(x+h) - f''(x)}{h} = \lim_{h \to 0} \frac{[4 - 6(x+h)] - (4 - 6x)}{h} = \lim_{h \to 0} \frac{-6h}{h} = \lim_{h \to 0} (-6) = -6$$

$$f^{(4)}(x) = \lim_{h \to 0} \frac{f'''(x+h) - f'''(x)}{h} = \lim_{h \to 0} \frac{-6 - (-6)}{h} = \lim_{h \to 0} \frac{0}{h} = \lim_{h \to 0} (0) = 0$$

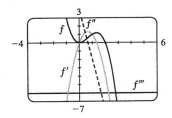

The graphs are consistent with the geometric interpretations of the derivatives because f' has zeros where f has a local minimum and a local maximum, f'' has a zero where f' has a local maximum, and f''' is a constant function equal to the slope of f''.

43. (a) Note that we have factored $x - a$ as the difference of two cubes in the third step.

$$f'(a) = \lim_{x \to a} \frac{f(x) - f(a)}{x - a} = \lim_{x \to a} \frac{x^{1/3} - a^{1/3}}{x - a} = \lim_{x \to a} \frac{x^{1/3} - a^{1/3}}{(x^{1/3} - a^{1/3})(x^{2/3} + x^{1/3}a^{1/3} + a^{2/3})}$$

$$= \lim_{x \to a} \frac{1}{x^{2/3} + x^{1/3}a^{1/3} + a^{2/3}} = \frac{1}{3a^{2/3}} \text{ or } \tfrac{1}{3}a^{-2/3}$$

(b) $f'(0) = \lim_{h \to 0} \frac{f(0+h) - f(0)}{h} = \lim_{h \to 0} \frac{\sqrt[3]{h} - 0}{h} = \lim_{h \to 0} \frac{1}{h^{2/3}}$. This function increases without bound, so the limit does not exist, and therefore $f'(0)$ does not exist.

(c) $\lim_{x \to 0} |f'(x)| = \lim_{x \to 0} \frac{1}{3x^{2/3}} = \infty$ and f is continuous at $x = 0$ (root function), so f has a vertical tangent at $x = 0$.

45. $f(x) = |x - 6| = \begin{cases} -(x - 6) & \text{if } x < 6 \\ x - 6 & \text{if } x \geq 6 \end{cases} = \begin{cases} 6 - x & \text{if } x < 6 \\ x - 6 & \text{if } x \geq 6 \end{cases}$

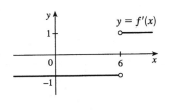

$$\lim_{x \to 6^+} \frac{f(x) - f(6)}{x - 6} = \lim_{x \to 6^+} \frac{|x - 6| - 0}{x - 6} = \lim_{x \to 6^+} \frac{x - 6}{x - 6} = \lim_{x \to 6^+} 1 = 1.$$

But $\lim_{x \to 6^-} \frac{f(x) - f(6)}{x - 6} = \lim_{x \to 6^-} \frac{|x - 6| - 0}{x - 6} = \lim_{x \to 6^-} \frac{6 - x}{x - 6}$

$$= \lim_{x \to 6^-} (-1) = -1$$

So $f'(6) = \lim_{x \to 6} \frac{f(x) - f(6)}{x - 6}$ does not exist. However, $f'(x) = \begin{cases} -1 & \text{if } x < 6 \\ 1 & \text{if } x > 6 \end{cases}$

Another way of writing the answer is $f'(x) = \frac{x - 6}{|x - 6|}$.

47. (a) If f is even, then

$$f'(-x) = \lim_{h \to 0} \frac{f(-x + h) - f(-x)}{h} = \lim_{h \to 0} \frac{f(x - h) - f(x)}{h}$$

$$= -\lim_{h \to 0} \frac{f(x - h) - f(x)}{-h} \quad [\text{let } \Delta x = -h] \quad = -\lim_{\Delta x \to 0} \frac{f(x + \Delta x) - f(x)}{\Delta x} = -f'(x)$$

Therefore, f' is odd.

(b) If f is odd, then

$$f'(-x) = \lim_{h \to 0} \frac{f(-x + h) - f(-x)}{h} = \lim_{h \to 0} \frac{-f(x - h) + f(x)}{h}$$

$$= \lim_{h \to 0} \frac{f(x - h) - f(x)}{-h} \quad [\text{let } \Delta x = -h] \quad = \lim_{\Delta x \to 0} \frac{f(x + \Delta x) - f(x)}{\Delta x} = f'(x)$$

Therefore, f' is even.

49.

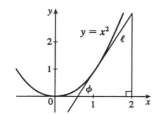

In the right triangle in the diagram, let Δy be the side opposite angle ϕ and Δx the side adjacent angle ϕ. Then the slope of the tangent line ℓ is $m = \Delta y / \Delta x = \tan \phi$. Note that $0 < \phi < \frac{\pi}{2}$. We know (see Exercise 17) that the derivative of $f(x) = x^2$ is $f'(x) = 2x$. So the slope of the tangent to the curve at the point $(1, 1)$ is 2. Thus, ϕ is the angle between 0 and $\frac{\pi}{2}$ whose tangent is 2; that is, $\phi = \tan^{-1} 2 \approx 63°$.

◆2.9◆ Linear Approximations • • • • • • • • • • • • • •

1. (a) If $f(x) = 3^x$, then

$$f'(0) = \lim_{h \to 0} \frac{f(h) - f(0)}{h} = \lim_{h \to 0} \frac{3^h - 1}{h}.$$

Let $y = \dfrac{3^h - 1}{h}$.

From the table, we conclude that $f'(0) \approx 1.0986$.

x	y
0.01	1.1047
0.001	1.0992
0.0001	1.0987

x	y
−0.01	1.0926
−0.001	1.0980
−0.0001	1.0986

(b) An approximate equation of the tangent line is

$y - 1 = 1.0986(x - 0)$, or $y = 1.0986x + 1$.

Thus, the linearization of f at 0 is

$L(x) = 1.0986x + 1$.

$L(0.05) = 1.0986(0.05) + 1 = 1.05493$ and

$L(0.1) = 1.0986(0.1) + 1 = 1.10986$. So we

estimate that $3^{0.05} \approx 1.0549$ and $3^{0.1} \approx 1.1099$.

(c)

The approximations are y-values on the tangent line. Since the tangent line lies *below* the curve, the approximations are *less* than the true values.

3. (a)

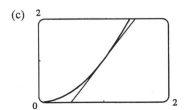

Consider the triangle in the graph. The vertices that lie on the tangent line are $\left(2, \frac{4}{3}\right)$ and $\left(-1, \frac{1}{3}\right)$. Check these points with the linearization in part (b). Thus,

$$f'(1) \approx \frac{\Delta y}{\Delta x} = \frac{\frac{4}{3} - \frac{1}{3}}{2 - (-1)} = \frac{1}{3}$$

(b) Using $f(x) = \sqrt[3]{x}$ and $a = 1$, we have $L(x) = f(1) + f'(1)(x - 1) = 1 + \frac{1}{3}(x - 1) = \frac{1}{3}x + \frac{2}{3}$.

(c)

x	$L(x)$	Calculator
0.5	$0.83\overline{3}$	0.793701
0.9	$0.96\overline{6}$	0.965489
0.99	$0.99\overline{6}$	0.996655
1.01	$1.00\overline{3}$	1.003322
1.1	$1.03\overline{3}$	1.032280
1.5	$1.16\overline{6}$	1.144714
2	$1.33\overline{3}$	1.259921

We see from the table that all of the estimates are overestimates. The most accurate estimates are for $x = 0.99$ and $x = 1.01$, which are the closest values to $x = 1$.

(d) The graph in part (a) shows that the tangent line lies above the curve. That explains why the estimates are overestimates.

5. (a) From Exercise 2.8.17(d) for $f(x) = x^2$, $f'(x) = 2x$, so $f'(1) = 2$.

(b) $L(x) = f(1) + f'(1)(x - 1)$
$L(x) = 1 + 2(x - 1) = 2x - 1$

The estimates using L are all underestimates of the actual function values.

x	$L(x) = 2x - 1$	$f(x) = x^2$
0.9	0.8	0.81
0.95	0.9	0.9025
0.99	0.98	0.9801
1.01	1.02	1.0201
1.05	1.1	1.1025
1.1	1.2	1.21

(c)

Since the tangent line lies under the graph, our underestimate claim in part (b) is supported.

7. As in Example 3, $T(0) = 185$, $T(10) = 172$, $T(20) = 160$, and

$$T'(20) \approx \frac{T(10) - T(20)}{10 - 20} = \frac{172 - 160}{-10} = -1.2 \, °\text{F/min}.$$

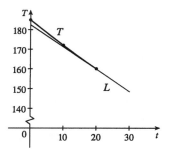

$T(30) \approx T(20) + T'(20)(30 - 20) \approx 160 - 1.2(10) = 148 \, °\text{F}$.

We would expect the temperature of the turkey to get closer to $75 \, °\text{F}$
as time increases. Since the temperature decreased $13 \, °\text{F}$ in the first
10 minutes and $12 \, °\text{F}$ in the second 10 minutes, we can assume that the
slopes of the tangent line are increasing through negative values:

$-1.3, -1.2, \ldots$. Hence, the tangent lines are under the curve and $148 \, °\text{F}$

is an underestimate. From the figure, we estimate the slope of the tangent line at $t = 20$ to be $\frac{184 - 147}{0 - 30} = -\frac{37}{30}$.

Then the linear approximation becomes $T(30) \approx T(20) + T'(20) \cdot 10 \approx 160 - \frac{37}{30}(10) = 147\frac{2}{3} \approx 147.7$.

9. If $C(t)$ represents the cash per capita in circulation in year t, then we estimate $C(2000)$ using a linear

approximation based on the tangent line to the graph of C at $(1990, C(1990)) = (1990, 1063)$.

$$C'(1990) \approx \frac{C(1980) - C(1990)}{1980 - 1990} = \frac{571 - 1063}{-10} = 49.2 \, \frac{\text{dollars}}{\text{year}}.$$

$C(2000) \approx C(1990) + C'(1990)(2000 - 1990) \approx 1063 + 49.2(10) = 1555$.

So our estimate of cash per capita in circulation in the year 2000 is $1555. For the given data, C' is increasing, so
the tangent line approximations are below the curve, indicating that our prediction is an underestimate.

11. Extend the tangent line at the point $(2030, 21)$ to the t-axis.
Answers will vary based on this approximation—we'll use
$t = 1900$ as our t-intercept. The linearization is then

$$P(t) \approx P(2030) + P'(2030)(t - 2030)$$
$$\approx 21 + \frac{21}{130}(t - 2030)$$

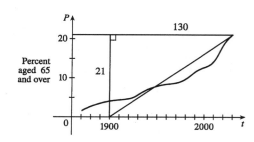

$P(2040) = 21 + \frac{21}{130}(2040 - 2030) \approx 22.6\%$

$P(2050) = 21 + \frac{21}{130}(2050 - 2030) \approx 24.2\%$

These predictions are probably too high since the tangent line lies above the graph at $t = 2030$.

13. (a) The graph shows that $f'(1) = 2$, so $L(x) = f(1) + f'(1)(x - 1) = 5 + 2(x - 1) = 2x + 3$.
$f(0.9) \approx L(0.9) = 4.8$ and $f(1.1) \approx L(1.1) = 5.2$.

(b) From the graph, we see that $f'(x)$ is positive and decreasing. This means that the slopes of the tangent lines are
positive, but the tangents are becoming less steep. So the tangent lines lie *above* the curve. Thus, the estimates in
part (a) are too large.

2.10 What Does f' Say about f? · · · · · · · · · · · · · ·

1. (a) Since $f'(x) > 0$ on $(1, 5)$, f is increasing on this interval. Since $f'(x) < 0$ on $(0, 1)$ and $(5, 6)$, f is decreasing on these intervals.

(b) Since $f'(x) = 0$ at $x = 1$ and f' changes from negative to positive there, f changes from decreasing to increasing and has a local minimum at $x = 1$. Since $f'(x) = 0$ at $x = 5$ and f' changes from positive to negative there, f changes from increasing to decreasing and has a local maximum at $x = 5$.

(c) Since $f(0) = 0$, start at the origin. Draw a decreasing function on $(0, 1)$ with a local minimum at $x = 1$. Now draw an increasing function on $(1, 5)$ and the steepest slope should occur at $x = 3$ since that's where the largest value of f' occurs. Lastly, draw a decreasing function on $(5, 6)$ making sure you have a local maximum at $x = 5$.

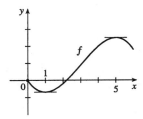

3. The derivative f' is increasing when the slopes of the tangent lines of f are becoming larger as x increases. This seems to be the case on the interval $(2, 5)$. The derivative is decreasing when the slopes of the tangent lines of f are becoming smaller as x increases, and this seems to be the case on $(-\infty, 2)$ and $(5, \infty)$. So f' is increasing on $(2, 5)$ and decreasing on $(-\infty, 2)$ and $(5, \infty)$.

5. If $D(t)$ is the size of the deficit as a function of time, then at the time of the speech $D'(t) > 0$, but $D''(t) < 0$ because $D''(t) = (D')'(t)$ is the rate of change of $D'(t)$.

7. (a) The rate of increase of the population is initially very small, then increases rapidly until about 1932 when it starts decreasing. The rate becomes negative by 1936, peaks in magnitude in 1937, and approaches 0 in 1940.

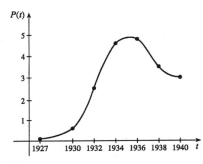

(b) Inflection points (IP) appear to be at $(1932, 2.5)$ and $(1937, 4.3)$. The rate of change of population density starts to decrease in 1932 and starts to increase in 1937. The rates of population increase and decrease have their maximum values at those points.

9. Most students learn more in the third hour of studying than in the eighth hour, so $K(3) - K(2)$ is larger than $K(8) - K(7)$. In other words, as you begin studying for a test, the rate of knowledge gain is large and then starts to taper off, so $K'(t)$ decreases and the graph of K is concave downward.

11. (a) f is increasing where f' is positive, that is, on $(0, 2)$, $(4, 6)$, and $(8, \infty)$; and decreasing where f' is negative, that is, on $(2, 4)$ and $(6, 8)$.

(b) f has local maxima where f' changes from positive to negative, at $x = 2$ and at $x = 6$, and local minima where f' changes from negative to positive, at $x = 4$ and at $x = 8$.

(c) f is concave upward (CU) where f' is increasing, that is, on $(3, 6)$ and $(6, \infty)$, and concave downward (CD) where f' is decreasing, that is, on $(0, 3)$.

(d) There is a point of inflection where f changes from being CD to being CU, that is, at $x = 3$.

(e)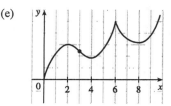

13. The function must be always decreasing and concave downward.

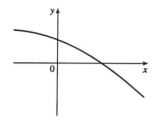

15. Since $f'(x) > 0$ if $x < 2$ and if $x > 2$, f is increasing on $(-\infty, 2)$ and $(2, \infty)$. The graph of f must level off and have a horizontal tangent at $x = 2$ because $f'(2) = 0$.

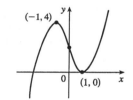

17. $f(-1) = 4$ and $f(1) = 0$ gives us two points to start with. $f'(-1) = f'(1) = 0$ $\Rightarrow$ horizontal tangents at $x = \pm 1$. $f'(x) < 0$ if $|x| < 1$ $\Rightarrow$ f is decreasing on $(-1, 1)$. $f'(x) > 0$ if $|x| > 1$ $\Rightarrow$ f is increasing on $(-\infty, -1)$ and $(1, \infty)$. $f''(x) < 0$ if $x < 0$ $\Rightarrow$ f is concave downward on $(-\infty, 0)$. $f''(x) > 0$ if $x > 0$ $\Rightarrow$ f is concave upward on $(0, \infty)$ and there is an inflection point at $x = 0$.

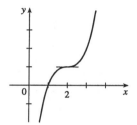

19. First we plot the points which are known to be on the graph: $(2, -1)$ and $(0, 0)$. We can also draw a short line segment of slope 0 at $x = 2$, since we are given that $f'(2) = 0$. Now we know that $f'(x) < 0$ (that is, the function is decreasing) on $(0, 2)$, and that $f''(x) < 0$ on $(0, 1)$ and $f''(x) > 0$ on $(1, 2)$. So we must join the points $(0, 0)$ and $(2, -1)$ in such a way that the curve is concave down on $(0, 1)$ and concave up on $(1, 2)$. The curve must be concave up and increasing on $(2, 4)$ and concave down and increasing toward $y = 1$ on $(4, \infty)$. Now we just need to reflect the curve in the y-axis, since we are given that f is an even function [the condition that $f(-x) = f(x)$ for all x].

21. (a) Since e^{-x^2} is positive for all x, $f'(x) = xe^{-x^2}$ is positive where $x > 0$ and negative where $x < 0$. Thus, f is increasing on $(0, \infty)$ and decreasing on $(-\infty, 0)$.

(b) Since f changes from decreasing to increasing at $x = 0$, f has a minimum value there.

23. (a) To find the intervals on which f is increasing, we need to find the intervals on which $f'(x) = 3x^2 - 1$ is

positive. $3x^2 - 1 > 0 \iff 3x^2 > 1 \iff x^2 > \frac{1}{3} \iff |x| > \sqrt{\frac{1}{3}}$, so

$x \in \left(-\infty, -\sqrt{\frac{1}{3}}\right) \cup \left(\sqrt{\frac{1}{3}}, \infty\right)$. Thus, f is increasing on $\left(-\infty, -\sqrt{\frac{1}{3}}\right)$ and on $\left(\sqrt{\frac{1}{3}}, \infty\right)$. In a similar

fashion, f is decreasing on $\left(-\sqrt{\frac{1}{3}}, \sqrt{\frac{1}{3}}\right)$.

(b) To find the intervals on which f is concave upward, we need to find the intervals on which $f''(x) = 6x$ is

positive. $6x > 0 \iff x > 0$. So f is concave upward on $(0, \infty)$ and f is concave downward on $(-\infty, 0)$.

(c) There is an inflection point at $(0, 0)$ since f changes its direction of concavity at $x = 0$.

25. b is the antiderivative of f. For small x, f is negative, so the graph of its antiderivative must be decreasing. But both a and c are increasing for small x, so only b can be f's antiderivative. Also, f is positive where b is increasing, which supports our conclusion.

27. The graph of F will have a minimum at 0 and a maximum at 2, since $f = F'$ goes from negative to positive at $x = 0$, and from positive to negative at $x = 2$.

29.

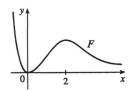

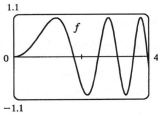

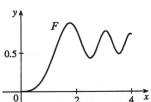

2 Review

————————————— • **CONCEPT CHECK** • —————————————

1. (a) $\lim_{x \to a} f(x) = L$: See Definition 2.2.1 and Figures 1 and 2 in Section 2.2.

(b) $\lim_{x \to a^+} f(x) = L$: See the paragraph after Definition 2.2.2 and Figure 9(b) in Section 2.2.

(c) $\lim_{x \to a^-} f(x) = L$: See Definition 2.2.2 and Figure 9(a) in Section 2.2.

(d) $\lim_{x \to a} f(x) = \infty$: See Definition 2.5.1 and Figure 2 in Section 2.5.

(e) $\lim_{x \to \infty} f(x) = L$: See Definition 2.5.4 and Figure 9 in Section 2.5.

2. In general, the limit of a function fails to exist when the function does not approach a fixed number. For each of the following functions, the limit fails to exist at $x = 2$.

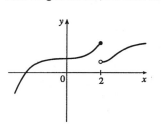

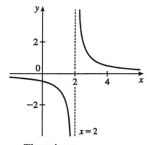

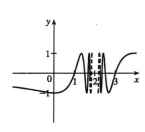

The left- and right-hand limits are not equal.

There is an infinite discontinuity.

There are an infinite number of oscillations.

3. (a)−(g) See the statements of Limit Laws 1−6 and 11 in Section 2.3.

4. See Theorem 3 in Section 2.3.

5. (a) See Definition 2.5.2 and Figures 2−4 in Section 2.5.

(b) See Definition 2.5.5 and Figures 9 and 10 in Section 2.5.

6. (a) $y = x^4$: No asymptote

(b) $y = \sin x$: No asymptote

(c) $y = \tan x$: Vertical asymptotes $x = \frac{\pi}{2} + \pi n$, n an integer

(d) $y = \tan^{-1} x$: Horizontal asymptotes $y = \pm \frac{\pi}{2}$

(e) $y = e^x$: Horizontal asymptote $y = 0$ ($\lim_{x \to -\infty} e^x = 0$)

(f) $y = \ln x$: Vertical asymptote $x = 0$ ($\lim_{x \to 0^+} \ln x = -\infty$)

(g) $y = 1/x$: Vertical asymptote $x = 0$, horizontal asymptote $y = 0$

(h) $y = \sqrt{x}$: No asymptote

7. (a) A function f is continuous at a number a if $f(x)$ gets closer to $f(a)$ as x gets close to a; that is, $\lim_{x \to a} f(x) = f(a)$.

(b) A function f is continuous on the interval $(-\infty, \infty)$ if f is continuous at every real number a. The graph of such a function has no breaks and every vertical line crosses it.

8. See Theorem 2.4.10.

9. See Definition 2.6.1.

10. See the paragraph containing Formula 3 in Section 2.6.

11. (a) The average rate of change of y with respect to x over the interval $[x_1, x_2]$ is $\dfrac{f(x_2) - f(x_1)}{x_2 - x_1}$.

(b) The instantaneous rate of change of y with respect to x at $x = x_1$ is $\lim_{x_2 \to x_1} \dfrac{f(x_2) - f(x_1)}{x_2 - x_1}$.

12. See Definition 2.7.2. The pages following the definition discuss interpretations of $f'(a)$ as the slope of a tangent line to the graph of f at $x = a$ and as an instantaneous rate of change of $f(x)$ with respect to x when $x = a$.

13. See the paragraphs before and after Example 7 in Section 2.8.

14. (a) A function f is differentiable at a number a if its derivative f' exists at $x = a$; that is, if $f'(a)$ exists.

(b) See Theorem 2.8.4. This theorem also tells us that if f is *not* continuous at a, then f is *not* differentiable at a.

15. (a) See the first box in Section 2.10.

(b) See the second box in Section 2.10.

16. (a) When x is near a, the linear approximation to f at a is the approximation $f(x) \approx f(a) + f'(a)(x - a)$.

(b) An antiderivative of a function f is a function F such that $F' = f$.

▲ **TRUE–FALSE QUIZ** ▲

---

1. False.   Limit Law 2 applies only if the individual limits exist (these don't).

3. True.   Limit Law 5 applies.

5. False.   Consider $\lim\limits_{x \to 5} \dfrac{x^2 - 5x}{x - 5}$ or $\lim\limits_{x \to 5} \dfrac{\sin(x - 5)}{x - 5}$. The first limit exists and is equal to 5. By Example 3 in Section 2.2, we know that the latter limit exists (and it is equal to 1).

7. True.   A polynomial is continuous everywhere, so $\lim\limits_{x \to b} p(x)$ exists and is equal to $p(b)$.

9. True.   See Figure 11 in Section 2.5.

11. False.   Consider $f(x) = \begin{cases} 1/(x - 1) & \text{if } x \neq 1 \\ 2 & \text{if } x = 1 \end{cases}$

13. True.   Use Theorem 2.4.8 with $a = 2$, $b = 5$, and $g(x) = 4x^2 - 11$. Note that $f(4) = 3$ is not needed.

15. False.   See the note after Theorem 4 in Section 2.8.

17. False. $\dfrac{d^2y}{dx^2}$ is the second derivative while $\left(\dfrac{dy}{dx}\right)^2$ is the first derivative squared. For example, if $y = x$, then $\dfrac{d^2y}{dx^2} = 0$, but $\left(\dfrac{dy}{dx}\right)^2 = 1$.

◆ **EXERCISES** ◆

---

1. (a) (i) $\lim\limits_{x \to 2+} f(x) = 3$

(ii) $\lim\limits_{x \to -3+} f(x) = 0$

(iii) $\lim\limits_{x \to -3} f(x)$ does not exist since the left and right limits are not equal. (The left limit is -2.)

(iv) $\lim\limits_{x \to 4} f(x) = 2$

(v) $\lim\limits_{x \to 0} f(x) = \infty$

(vi) $\lim\limits_{x \to 2-} f(x) = -\infty$

(vii) $\lim\limits_{x \to \infty} f(x) = 4$

(viii) $\lim\limits_{x \to -\infty} f(x) = -1$

(b) The horizontal asymptotes are $y = 4$ and $y = -1$.

(c) The vertical asymptotes are $x = 0$ and $x = 2$.

(d) f is discontinuous at $x = -3, 0, 2,$ and 4. The discontinuities are jump, infinite, infinite, and removable, respectively.

3. Since the exponential function is continuous, $\lim\limits_{x \to 1} e^{x^3 - x} = e^{1 - 1} = e^0 = 1$.

5. $\displaystyle\lim_{x \to -3} \frac{x^2 - 9}{x^2 + 2x - 3} = \lim_{x \to -3} \frac{(x+3)(x-3)}{(x+3)(x-1)} = \lim_{x \to -3} \frac{x-3}{x-1} = \frac{-3-3}{-3-1} = \frac{-6}{-4} = \frac{3}{2}$

7. $\displaystyle\lim_{h \to 0} \frac{(h-1)^3 + 1}{h} = \lim_{h \to 0} \frac{(h^3 - 3h^2 + 3h - 1) + 1}{h} = \lim_{h \to 0} \frac{h^3 - 3h^2 + 3h}{h} = \lim_{h \to 0} (h^2 - 3h + 3) = 3$

Another solution: Factor the numerator as a sum of two cubes and then simplify.

$$\lim_{h \to 0} \frac{(h-1)^3 + 1}{h} = \lim_{h \to 0} \frac{(h-1)^3 + 1^3}{h} = \lim_{h \to 0} \frac{[(h-1) + 1]\left[(h-1)^2 - 1(h-1) + 1^2\right]}{h}$$

$$= \lim_{h \to 0} \left[(h-1)^2 - h + 2\right] = 1 - 0 + 2 = 3$$

9. $\displaystyle\lim_{r \to 9} \frac{\sqrt{r}}{(r-9)^4} = \infty$ since $(r-9)^4 \to 0$ as $r \to 9$ and $\dfrac{\sqrt{r}}{(r-9)^4} > 0$ for $r \neq 9$.

11. $\displaystyle\lim_{x \to \infty} e^{-3x} = 0$ since $-3x \to -\infty$ as $x \to \infty$ and $\displaystyle\lim_{t \to -\infty} e^t = 0$.

13. 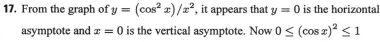 $\displaystyle\lim_{x \to 0} \frac{1 - \sqrt{1 - x^2}}{x} \cdot \frac{1 + \sqrt{1 - x^2}}{1 + \sqrt{1 - x^2}} = \lim_{x \to 0} \frac{1 - (1 - x^2)}{x\left(1 + \sqrt{1 - x^2}\right)} = \lim_{x \to 0} \frac{x^2}{x\left(1 + \sqrt{1 - x^2}\right)} = \lim_{x \to 0} \frac{x}{1 + \sqrt{1 - x^2}} = 0$

15. $\displaystyle\lim_{x \to \infty} \frac{\sqrt{3x^2 - 1}}{x - 1} = \lim_{x \to \infty} \frac{\sqrt{3x^2 - 1}/\sqrt{x^2}}{(x - 1)/x} = \lim_{x \to \infty} \frac{\sqrt{3 - 1/x^2}}{1 - 1/x} = \frac{\sqrt{3}}{1} = \sqrt{3}$

17. From the graph of $y = \left(\cos^2 x\right)/x^2$, it appears that $y = 0$ is the horizontal asymptote and $x = 0$ is the vertical asymptote. Now $0 \leq (\cos x)^2 \leq 1$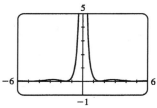

$\Rightarrow \dfrac{0}{x^2} \leq \dfrac{\cos^2 x}{x^2} \leq \dfrac{1}{x^2} \Rightarrow 0 \leq \dfrac{\cos^2 x}{x^2} \leq \dfrac{1}{x^2}$. But $\displaystyle\lim_{x \to \pm\infty} 0 = 0$ and

$\displaystyle\lim_{x \to \pm\infty} \frac{1}{x^2} = 0$, so by the Squeeze Theorem,

$\displaystyle\lim_{x \to \pm\infty} \frac{\cos^2 x}{x^2} = 0$. Thus, $y = 0$ is the horizontal asymptote. $\displaystyle\lim_{x \to 0} \frac{\cos^2 x}{x^2} = \infty$ because $\cos^2 x \to 1$ and $x^2 \to 0$ as

$x \to 0$, so $x = 0$ is the vertical asymptote.

19. Since $2x - 1 \leq f(x) \leq x^2$ for $0 < x < 3$ and $\displaystyle\lim_{x \to 1} (2x - 1) = 1 = \lim_{x \to 1} x^2$, we have $\displaystyle\lim_{x \to 1} f(x) = 1$ by the Squeeze Theorem.

21. (a) $f(x) = \sqrt{-x}$ if $x < 0$, $f(x) = 3 - x$ if $0 \leq x < 3$, $f(x) = (x - 3)^2$ if $x > 3$.

   (i) $\displaystyle\lim_{x \to 0^+} f(x) = \lim_{x \to 0^+} (3 - x) = 3$ (ii) $\displaystyle\lim_{x \to 0^-} f(x) = \lim_{x \to 0^-} \sqrt{-x} = 0$

   (iii) Because of (i) and (ii), $\displaystyle\lim_{x \to 0} f(x)$ does not exist.   (iv) $\displaystyle\lim_{x \to 3^-} f(x) = \lim_{x \to 3^-} (3 - x) = 0$

   (v) $\displaystyle\lim_{x \to 3^+} f(x) = \lim_{x \to 3^+} (x - 3)^2 = 0$ (vi) Because of (iv) and (v), $\displaystyle\lim_{x \to 3} f(x) = 0$.

   (b) f is discontinuous at 0 since $\displaystyle\lim_{x \to 0} f(x)$ does not exist.   (c)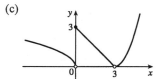

   f is discontinuous at 3 since $f(3)$ does not exist.

23. $f(x) = 2x^3 + x^2 + 2$ is a polynomial, so it is continuous on $[-2, -1]$ and $f(-2) = -10 < 0 < 1 = f(-1)$. So by the Intermediate Value Theorem there is a number c in $(-2, -1)$ such that $f(c) = 0$, that is, the equation $2x^3 + x^2 + 2 = 0$ has a root in $(-2, -1)$.

25. (a) $s = 1 + 2t + t^2/4$. The average velocity over the time interval $[1, 1 + h]$ is

$$\frac{s(1+h) - s(1)}{h} = \frac{1 + 2(1+h) + (1+h)^2/4 - 13/4}{h} = \frac{10h + h^2}{4h} = \frac{10 + h}{4}. \text{ So for the following}$$

intervals the average velocities are:

(i) $[1, 3]$: $(10 + 2)/4 = 3$ m/s

(ii) $[1, 2]$: $(10 + 1)/4 = 2.75$ m/s

(iii) $[1, 1.5]$: $(10 + 0.5)/4 = 2.625$ m/s

(iv) $[1, 1.1]$: $(10 + 0.1)/4 = 2.525$ m/s

(b) When $t = 1$ the velocity is $\lim\limits_{h \to 0} \dfrac{s(1+h) - s(1)}{h} = \lim\limits_{h \to 0} \dfrac{10 + h}{4} = 2.5$ m/s.

27. Estimating the slopes of the tangent lines at $x = 2, 3,$ and 5, we obtain approximate values 0.4, 2, and 0.1. Since the graph is concave downward at $x = 5$, $f''(5)$ is negative. Arranging the numbers in increasing order, we have:
$f''(5) < 0 < f'(5) < f'(2) < 1 < f'(3)$.

29. (a) Estimating $f'(1)$ from the triangle in the graph,

we get $\dfrac{\Delta y}{\Delta x} \approx \dfrac{-0.37}{0.50} = -0.74.$

To estimate $f'(1)$ numerically, we have

$$f'(1) = \lim_{h \to 0} \frac{f(1+h) - f(1)}{h}$$

$$= \lim_{h \to 0} \frac{e^{-(1+h)^2} - e^{-1}}{h} = y$$

From the table, we have $f'(1) \approx -0.736$.

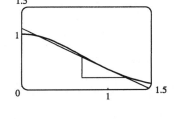

h	y
0.01	-0.732
0.001	-0.735
0.0001	-0.736
-0.01	-0.739
-0.001	-0.736
-0.0001	-0.736

(b) $y - e^{-1} \approx -0.736(x - 1)$ or $y \approx -0.736x + 1.104$

(c) See the graph in part (a).

31. (a) $f'(r)$ is the rate at which the total cost changes with respect to the interest rate. Its units are dollars/(percent per year).

(b) The total cost of paying off the loan is increasing by $1200/(percent per year) as the interest rate reaches 10%. So if the interest rate goes up from 10% to 11%, the cost goes up approximately $1200.

(c) As r increases, C increases. So $f'(r)$ will always be positive.

For Exercise 33, see the hints before Exercise 5 in Section 2.8.

33.

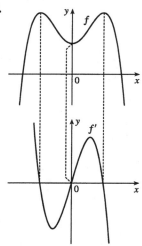

35. (a) $f'(x) = \lim\limits_{h \to 0} \dfrac{f(x+h) - f(x)}{h} = \lim\limits_{h \to 0} \dfrac{\sqrt{3 - 5(x+h)} - \sqrt{3 - 5x}}{h} \dfrac{\sqrt{3 - 5(x+h)} + \sqrt{3 - 5x}}{\sqrt{3 - 5(x+h)} + \sqrt{3 - 5x}}$

$= \lim\limits_{h \to 0} \dfrac{[3 - 5(x+h)] - (3 - 5x)}{h\left(\sqrt{3 - 5(x+h)} + \sqrt{3 - 5x}\right)} = \lim\limits_{h \to 0} \dfrac{-5}{\sqrt{3 - 5(x+h)} + \sqrt{3 - 5x}} = \dfrac{-5}{2\sqrt{3 - 5x}}$

(b) Domain of f: (the radicand must be nonnegative) $3 - 5x \geq 0 \;\Rightarrow\; 5x \leq 3 \;\Rightarrow\; x \in \left(-\infty, \frac{3}{5}\right]$

Domain of f': exclude $\frac{3}{5}$ because it makes the denominator zero; $x \in \left(-\infty, \frac{3}{5}\right)$

(c) Our answer to part (a) is reasonable because $f'(x)$ is always negative and f is always decreasing.

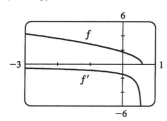

37. f is not differentiable: at $x = -4$ because f is not continuous, at $x = -1$ because f has a corner, at $x = 2$ because f is not continuous, and at $x = 5$ because f has a vertical tangent.

39. (a) Remember that the slope of the tangent for $f(x) = e^x$ at $x = 0$ is 1 (by definition of the number e).

(b) $L(x) = f(0) + f'(0)(x - 0) = 1 + 1(x) = x + 1$

(c)

x	-0.2	-0.1	-0.01	0.01	0.1	0.2
$e^x \approx L(x)$	0.8	0.9	0.99	1.01	1.1	1.2

(d) The graph of $f(x) = e^x$ is concave upward, so its tangent lines are below the graph and thus, the approximations are underestimates. The most accurate estimates are for those closest to $x = 0$, namely, for $e^{-0.01}$ and for $e^{0.01}$.

41. (a) $f'(x) > 0$ on $(-2, 0)$ and $(2, \infty)$ ⟹ f is increasing on those intervals. $f'(x) < 0$ on $(-\infty, -2)$ and $(0, 2)$ ⟹ f is decreasing on those intervals.

(b) $f'(x) = 0$ at $x = -2$, 0, and 2, so these are where local maxima or minima will occur. At $x = \pm 2$, f' changes from negative to positive, so f has local minima at those values. At $x = 0$, f' changes from positive to negative, so f has a local maximum there.

(c) f' is increasing on $(-\infty, -1)$ and $(1, \infty)$ ⟹ (d)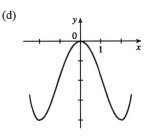

$f'' > 0$ and f is concave upward on those intervals.

f' is decreasing on $(-1, 1)$ ⟹ $f'' < 0$ and f is concave downward on this interval.

43. $f(0) = 0$, $f'(-2) = f'(1) = f'(9) = 0$,

$\lim\limits_{x \to \infty} f(x) = 0$, $\lim\limits_{x \to 6} f(x) = -\infty$,

$f'(x) < 0$ on $(-\infty, -2)$, $(1, 6)$, and $(9, \infty)$,

$f'(x) > 0$ on $(-2, 1)$ and $(6, 9)$,

$f''(x) > 0$ on $(-\infty, 0)$ and $(12, \infty)$,

$f''(x) < 0$ on $(0, 6)$ and $(6, 12)$

45. (a) Using the data closest to $t = 6$, we have $\dfrac{s(8) - s(6)}{8 - 6} = \dfrac{180 - 95}{2} = 42.5$ and

$\dfrac{s(4) - s(6)}{4 - 6} = \dfrac{40 - 95}{-2} = 27.5$. Averaging these two values gives us $\dfrac{42.5 + 27.5}{2} = 35$ ft/s as an estimate for the speed of the car after 6 seconds.

(b)

From the graph, it appears that the inflection point is at $(8, 180)$.

(c) The velocity of the car is at a maximum at the inflection point.

Focus on Problem Solving

1. Let $t = \sqrt[6]{x}$, so $x = t^6$. Then $t \to 1$ as $x \to 1$, so

$$\lim_{x \to 1} \frac{\sqrt[3]{x} - 1}{\sqrt{x} - 1} = \lim_{t \to 1} \frac{t^2 - 1}{t^3 - 1} = \lim_{t \to 1} \frac{(t-1)(t+1)}{(t-1)(t^2 + t + 1)} = \lim_{t \to 1} \frac{t+1}{t^2 + t + 1} = \frac{1+1}{1^2 + 1 + 1} = \frac{2}{3}.$$

Another method: Multiply both numerator and denominator by $(\sqrt{x} + 1)\left(\sqrt[3]{x^2} + \sqrt[3]{x} + 1\right)$.

3. For $-\frac{1}{2} < x < \frac{1}{2}$, we have $2x - 1 < 0$ and $2x + 1 > 0$, so $|2x - 1| = -(2x - 1)$ and $|2x + 1| = 2x + 1$.

Therefore, $\lim_{x \to 0} \dfrac{|2x - 1| - |2x + 1|}{x} = \lim_{x \to 0} \dfrac{-(2x - 1) - (2x + 1)}{x} = \lim_{x \to 0} \dfrac{-4x}{x} = \lim_{x \to 0} (-4) = -4.$

5. Since $[\![x]\!] \le x < [\![x]\!] + 1$, we have $1 \le \dfrac{x}{[\![x]\!]} < 1 + \dfrac{1}{[\![x]\!]}$ for $x \ge 1$. As $x \to \infty$, $[\![x]\!] \to \infty$, so $\dfrac{1}{[\![x]\!]} \to 0$ and

$1 + \dfrac{1}{[\![x]\!]} \to 1$. Thus, $\lim_{x \to \infty} \dfrac{x}{[\![x]\!]} = 1$ by the Squeeze Theorem.

7. f is continuous on $(-\infty, a)$ and (a, ∞). To make f continuous on $\mathbb{R}$, we must have continuity at a. Thus,

$$\lim_{x \to a^+} f(x) = \lim_{x \to a^-} f(x) \quad \Rightarrow \quad \lim_{x \to a^+} x^2 = \lim_{x \to a^-} (x + 1) \quad \Rightarrow \quad a^2 = a + 1 \quad \Rightarrow \quad a = (1 \pm \sqrt{5})/2 \approx 1.618$$

or -0.618.

9. (a) Consider $G(x) = T(x + 180°) - T(x)$. Fix any number a. If $G(a) = 0$, we are done:
Temperature at a = Temperature at $a + 180°$. If $G(a) > 0$, then $G(a + 180°) = T(a + 360°) - T(a + 180°)$
$= T(a) - T(a + 180°) = -G(a) < 0$. Also, G is continuous since temperature varies continuously. So, by the
Intermediate Value Theorem, G has a zero on the interval $[a, a + 180°]$. If $G(a) < 0$, then a similar argument
applies.

(b) Yes. The same argument applies.

(c) The same argument applies for quantities that vary continuously, such as barometric pressure. But one could
argue that altitude above sea level is sometimes discontinuous, so the result might not always hold for that
quantity.

11. Let a be the x-coordinate of Q. Since the derivative of $y = 1 - x^2$ is $y' = -2x$, the slope at Q is $-2a$. But since
the triangle is equilateral, $\overline{AO}/\overline{OC} = \sqrt{3}/1$, so the slope at Q is $-\sqrt{3}$. Therefore, we must have that $-2a = -\sqrt{3}$
$\Rightarrow \quad a = \frac{\sqrt{3}}{2}$. Thus, the point Q has coordinates $\left(\frac{\sqrt{3}}{2}, 1 - \left(\frac{\sqrt{3}}{2}\right)^2\right) = \left(\frac{\sqrt{3}}{2}, \frac{1}{4}\right)$ and by symmetry, P has
coordinates $\left(-\frac{\sqrt{3}}{2}, \frac{1}{4}\right)$.

13. (a) Put $x = 0$ and $y = 0$ in the equation: $f(0) = f(0 + 0) = f(0) + f(0) + 0^2 \cdot 0 + 0 \cdot 0^2 = 2f(0)$. Subtracting $f(0)$ from each side of this equation gives $f(0) = 0$.

(b) $f'(0) = \lim\limits_{h \to 0} \dfrac{f(0 + h) - f(0)}{h} = \lim\limits_{h \to 0} \dfrac{\left[f(0) + f(h) + 0^2 h + 0h^2\right] - f(0)}{h}$

$= \lim\limits_{h \to 0} \dfrac{f(h)}{h} = \lim\limits_{x \to 0} \dfrac{f(x)}{x} = 1$

(c) $f'(x) = \lim\limits_{h \to 0} \dfrac{f(x + h) - f(x)}{h} = \lim\limits_{h \to 0} \dfrac{\left[f(x) + f(h) + x^2 h + xh^2\right] - f(x)}{h}$

$= \lim\limits_{h \to 0} \dfrac{f(h) + x^2 h + xh^2}{h} = \lim\limits_{h \to 0} \left[\dfrac{f(h)}{h} + x^2 + xh\right] = 1 + x^2$

15. $\lim\limits_{x \to a} f(x) = \lim\limits_{x \to a} \left(\frac{1}{2}\left[f(x) + g(x)\right] + \frac{1}{2}\left[f(x) - g(x)\right]\right)$

$= \frac{1}{2} \lim\limits_{x \to a} \left[f(x) + g(x)\right] + \frac{1}{2} \lim\limits_{x \to a} \left[f(x) - g(x)\right]$

$= \frac{1}{2} \cdot 2 + \frac{1}{2} \cdot 1 = \frac{3}{2}$, and

$\lim\limits_{x \to a} g(x) = \lim\limits_{x \to a} \left(\left[f(x) + g(x)\right] - f(x)\right) = \lim\limits_{x \to a}\left[f(x) + g(x)\right] - \lim\limits_{x \to a} f(x) = 2 - \frac{3}{2} = \frac{1}{2}$.

So $\lim\limits_{x \to a}\left[f(x)g(x)\right] = \left[\lim\limits_{x \to a} f(x)\right]\left[\lim\limits_{x \to a} g(x)\right] = \frac{3}{2} \cdot \frac{1}{2} = \frac{3}{4}$.

Another solution: Since $\lim\limits_{x \to a}\left[f(x) + g(x)\right]$ and $\lim\limits_{x \to a}\left[f(x) - g(x)\right]$ exist, we must have

$\lim\limits_{x \to a}\left[f(x) + g(x)\right]^2 = \left(\lim\limits_{x \to a}\left[f(x) + g(x)\right]\right)^2$ and $\lim\limits_{x \to a}\left[f(x) - g(x)\right]^2 = \left(\lim\limits_{x \to a}\left[f(x) - g(x)\right]\right)^2$, so

$\lim\limits_{x \to a}\left[f(x)\,g(x)\right] = \lim\limits_{x \to a} \frac{1}{4}\left(\left[f(x) + g(x)\right]^2 - \left[f(x) - g(x)\right]^2\right)$ (because all of the f^2 and g^2 cancel)

$= \frac{1}{4}\left(\lim\limits_{x \to a}\left[f(x) + g(x)\right]^2 - \lim\limits_{x \to a}\left[f(x) - g(x)\right]^2\right) = \frac{1}{4}\left(2^2 - 1^2\right) = \frac{3}{4}$.

17. We are given that $|f(x)| \leq x^2$ for all x. In particular, $|f(0)| \leq 0$, but $|a| \geq 0$ for all a. The only conclusion is that

$f(0) = 0$. Now $\left|\dfrac{f(x) - f(0)}{x - 0}\right| = \left|\dfrac{f(x)}{x}\right| = \dfrac{|f(x)|}{|x|} \leq \dfrac{x^2}{|x|} = \dfrac{|x^2|}{|x|} = |x| \quad \Rightarrow \quad -|x| \leq \dfrac{f(x) - f(0)}{x - 0} \leq |x|$. But

$\lim\limits_{x \to 0} \left(-|x|\right) = 0 = \lim\limits_{x \to 0} |x|$, so by the Squeeze Theorem, $\lim\limits_{x \to 0} \dfrac{f(x) - f(0)}{x - 0} = 0$. So by the definition of a

derivative, f is differentiable at 0 and, furthermore, $f'(0) = 0$.

Differentiation Rules

 Derivatives of Polynomials and Exponential Functions • • •

1. (a) e is the number such that $\lim\limits_{h \to 0} \dfrac{e^h - 1}{h} = 1$.

(b)

x	$(2.7^x - 1)/x$
-0.001	0.9928
-0.0001	0.9932
0.001	0.9937
0.0001	0.9933

x	$(2.8^x - 1)/x$
-0.001	1.0291
-0.0001	1.0296
0.001	1.0301
0.0001	1.0297

From the tables (to two decimal places), $\lim\limits_{h \to 0} \dfrac{2.7^h - 1}{h} = 0.99$ and $\lim\limits_{h \to 0} \dfrac{2.8^h - 1}{h} = 1.03$. Since $0.99 < 1 < 1.03$, $2.7 < e < 2.8$.

3. $f(x) = 5x - 1 \quad \Rightarrow \quad f'(x) = 5 - 0 = 5$

5. $f(x) = 9x^4 - 3x^2 + 8 \quad \Rightarrow \quad f'(x) = 9(4x^{4-1}) - 3(2x^{2-1}) + 0 = 36x^3 - 6x$

7. $y = x^{-2/5} \quad \Rightarrow \quad y' = -\frac{2}{5}x^{(-2/5)-1} = -\frac{2}{5}x^{-7/5} = -\dfrac{2}{5x^{7/5}}$

9. $G(x) = \sqrt{x} - 2e^x = x^{1/2} - 2e^x \quad \Rightarrow \quad G'(x) = \frac{1}{2}x^{-1/2} - 2e^x = \dfrac{1}{2\sqrt{x}} - 2e^x$

11. $V(r) = \frac{4}{3}\pi r^3 \quad \Rightarrow \quad V'(r) = \frac{4}{3}\pi(3r^2) = 4\pi r^2$

13. $F(x) = (16x)^3 = 4096x^3 \quad \Rightarrow \quad F'(x) = 4096(3x^2) = 12{,}288x^2$

15. $y = 4\pi^2 \quad \Rightarrow \quad y' = 0$ since $4\pi^2$ is a constant.

17. $y = \dfrac{x^2 + 4x + 3}{\sqrt{x}} = x^{3/2} + 4x^{1/2} + 3x^{-1/2} \quad \Rightarrow$

$y' = \frac{3}{2}x^{1/2} + 4(\frac{1}{2})x^{-1/2} + 3(-\frac{1}{2})x^{-3/2} = \frac{3}{2}\sqrt{x} + \dfrac{2}{\sqrt{x}} - \dfrac{3}{2x\sqrt{x}}$

19. $v = t^2 - \dfrac{1}{\sqrt[4]{t^3}} = t^2 - t^{-3/4} \quad \Rightarrow \quad v' = 2t - (-\frac{3}{4})t^{-7/4} = 2t + \dfrac{3}{4t^{7/4}} = 2t + \dfrac{3}{4t\sqrt[4]{t^3}}$

21. $z = \dfrac{A}{y^{10}} + Be^y = Ay^{-10} + Be^y \quad \Rightarrow \quad z' = -10Ay^{-11} + Be^y = -\dfrac{10A}{y^{11}} + Be^y$

23. $f(x) = 2x^2 - x^4 \Rightarrow f'(x) = 4x - 4x^3$.

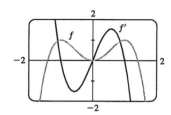

Notice that $f'(x) = 0$ when f has a horizontal tangent and that f' is an odd function while f is an even function.

25. $f(x) = 3x^{15} - 5x^3 + 3 \Rightarrow$
$f'(x) = 45x^{14} - 15x^2$.

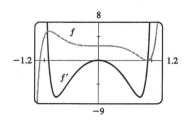

Notice that $f'(x) = 0$ when f has a horizontal tangent, f' is positive when f is increasing, and f' is negative when f is decreasing.

27. $f(x) = x - 3x^{1/3} \Rightarrow$
$f'(x) = 1 - x^{-2/3} = 1 - 1/x^{2/3}$. Note that $f'(x) = 0$ when f has a horizontal tangent, f' is positive when f is increasing, and f' is negative when f is decreasing.

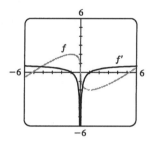

29. (a)

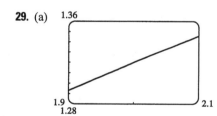

The endpoints of f in this graph are about $(1.9, 1.2927)$ and $(2.1, 1.3455)$. An estimate of $f'(2)$ is
$\frac{1.3455 - 1.2927}{2.1 - 1.9} = \frac{0.0528}{0.2} = 0.264$.

(b) $f(x) = x^{2/5} \Rightarrow f'(x) = \frac{2}{5}x^{-3/5} = 2/\left(5x^{3/5}\right)$.
$f'(2) = 2/\left(5 \cdot 2^{3/5}\right) = 2^{2/5}/5 \approx 0.263902$.

31. $y = f(x) = x + \frac{4}{x} \Rightarrow f'(x) = 1 - \frac{4}{x^2}$. So the slope of the tangent line at $(2, 4)$ is $f'(2) = 0$ and its equation is $y - 4 = 0(x - 2)$ or $y = 4$.

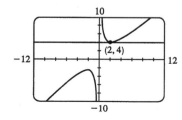

33. $y = f(x) = x + \sqrt{x} \Rightarrow f'(x) = 1 + \frac{1}{2}x^{-1/2}$.
So the slope of the tangent line at $(1, 2)$ is
$f'(1) = 1 + \frac{1}{2}(1) = \frac{3}{2}$ and its equation is
$y - 2 = \frac{3}{2}(x - 1)$ or $y = \frac{3}{2}x + \frac{1}{2}$.

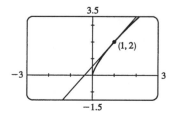

35. (a)

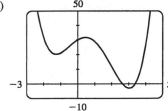

(b)

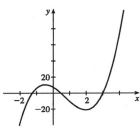

From the graph in part (a), it appears that f' is zero at $x_1 \approx -1.25$, $x_2 \approx 0.5$, and $x_3 \approx 3$. The slopes are negative (so f' is negative) on $(-\infty, x_1)$ and (x_2, x_3). The slopes are positive (so f' is positive) on (x_1, x_2) and (x_3, ∞).

(c) $f(x) = x^4 - 3x^3 - 6x^2 + 7x + 30 \implies$

$f'(x) = 4x^3 - 9x^2 - 12x + 7$

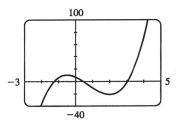

37. $f(x) = x^4 - 3x^3 + 16x \implies f'(x) = 4x^3 - 9x^2 + 16 \implies f''(x) = 12x^2 - 18x$

39. $f(x) = 2x - 5x^{3/4} \implies f'(x) = 2 - \frac{15}{4}x^{-1/4} \implies f''(x) = \frac{15}{16}x^{-5/4}$

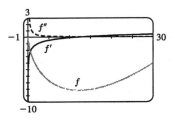

Note that f' is negative when f is decreasing and positive when f is increasing. f'' is always positive since f' is always increasing.

41. (a) $s = t^3 - 3t \implies v(t) = s'(t) = 3t^2 - 3 \implies a(t) = v'(t) = 6t$

(b) $a(2) = 6(2) = 12$ m/s^2

(c) $v(t) = 3t^2 - 3 = 0$ when $t^2 = 1$, that is, $t = 1$ and $a(1) = 6$ m/s^2.

43. $f(x) = 1 + 2e^x - 3x \implies f'(x) = 2e^x - 3$. $f'(x) > 0 \implies 2e^x - 3 > 0 \implies 2e^x > 3 \implies e^x > 1.5 \implies x > \ln 1.5 \approx 0.41$. f is increasing when f' is positive; that is, on $(\ln 1.5, \infty)$.

45. $y = x^3 - x^2 - x + 1$ has a horizontal tangent when $y' = 3x^2 - 2x - 1 = 0$. $(3x + 1)(x - 1) = 0 \iff x = 1$ or $-\frac{1}{3}$. Therefore, the points are $(1, 0)$ and $\left(-\frac{1}{3}, \frac{32}{27}\right)$.

47. $y = 6x^3 + 5x - 3 \implies m = y' = 18x^2 + 5$, but $x^2 \geq 0$ for all x, so $m \geq 5$ for all x.

49.

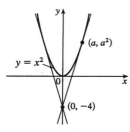

Let (a, a^2) be a point on the parabola at which the tangent line passes through the point $(0, -4)$. The tangent line has slope $2a$ and equation $y - (-4) = 2a(x - 0) \quad \Leftrightarrow \quad y = 2ax - 4$. Since (a, a^2) also lies on the line, $a^2 = 2a(a) - 4$, or $a^2 = 4$. So $a = \pm 2$ and the points are $(2, 4)$ and $(-2, 4)$.

51. $y = f(x) = 1 - x^2 \quad \Rightarrow \quad f'(x) = -2x$, so the tangent line at $(2, -3)$ has slope $f'(2) = -4$. The normal line has slope $-\frac{1}{-4} = \frac{1}{4}$ and equation $y + 3 = \frac{1}{4}(x - 2)$ or $y = \frac{1}{4}x - \frac{7}{2}$.

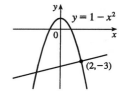

53. $f'(x) = \lim\limits_{h \to 0} \dfrac{f(x + h) - f(x)}{h} = \lim\limits_{h \to 0} \dfrac{\dfrac{1}{x + h} - \dfrac{1}{x}}{h} = \lim\limits_{h \to 0} \dfrac{x - (x + h)}{hx(x + h)}$

$= \lim\limits_{h \to 0} \dfrac{-h}{hx(x + h)} = \lim\limits_{h \to 0} \dfrac{-1}{x(x + h)} = -\dfrac{1}{x^2}$

55. Let $P(x) = ax^2 + bx + c$. Then $P'(x) = 2ax + b$ and $P''(x) = 2a$. $P''(2) = 2 \quad \Rightarrow \quad 2a = 2 \quad \Rightarrow \quad a = 1$. $P'(2) = 3 \quad \Rightarrow \quad 2(1)(2) + b = 3 \quad \Rightarrow \quad 4 + b = 3 \quad \Rightarrow \quad b = -1$.

$P(2) = 5 \quad \Rightarrow \quad 1(2)^2 + (-1)(2) + c = 5 \quad \Rightarrow \quad 2 + c = 5 \quad \Rightarrow \quad c = 3$. So $P(x) = x^2 - x + 3$.

57. (a) At this stage, we would guess that an antiderivative of x^2 must have x^3 in it. Differentiating x^3 gives us $3x^2$, so we know that we must divide x^3 by 3. That gives us $F(x) = \frac{1}{3}x^3$. Checking, we have $F'(x) = \frac{1}{3}(3x^2) = x^2 = f(x)$. Because we can add an arbitrary constant C to F without changing its derivative, we have an infinite number of antiderivatives of the form $F(x) = \frac{1}{3}x^3 + C$.

(b) As in part (a), antiderivatives of $f(x) = x^3$ and $f(x) = x^4$ are $F(x) = \frac{1}{4}x^4 + C$ and $F(x) = \frac{1}{5}x^5 + C$.

(c) Similarly, an antiderivative for $f(x) = x^n$ is $F(x) = \dfrac{1}{n + 1}x^{n+1} + C$, since then

$F'(x) = \dfrac{1}{n + 1}\left[(n + 1)x^n\right] = x^n = f(x)$ for $n \neq -1$.

59. $y = f(x) = ax^2 \quad \Rightarrow \quad f'(x) = 2ax$. So the slope of the tangent to the parabola at $x = 2$ is $m = 2a(2) = 4a$. The slope of the given line, $2x + y = b \quad \Leftrightarrow \quad y = -2x + b$, is seen to be -2, so we must have $4a = -2 \quad \Leftrightarrow \quad a = -\frac{1}{2}$. So when $x = 2$, the point in question has y-coordinate $-\frac{1}{2} \cdot 2^2 = -2$. Now we simply require that the given line, whose equation is $2x + y = b$, pass through the point $(2, -2)$: $2(2) + (-2) = b \quad \Leftrightarrow \quad b = 2$. So we must have $a = -\frac{1}{2}$ and $b = 2$.

61. $y = f(x) = ax^3 + bx^2 + cx + d \quad \Rightarrow \quad f'(x) = 3ax^2 + 2bx + c$. The point $(-2, 6)$ is on f, so $f(-2) = 6 \quad \Rightarrow$ $-8a + 4b - 2c + d = 6$ **(1)**. The point $(2, 0)$ is on f, so $f(2) = 0 \quad \Rightarrow \quad 8a + 4b + 2c + d = 0$ **(2)**. Since there are horizontal tangents at $(-2, 6)$ and $(2, 0)$, $f'(\pm 2) = 0$. $f'(-2) = 0 \quad \Rightarrow \quad 12a - 4b + c = 0$ **(3)** and $f'(2) = 0 \quad \Rightarrow \quad 12a + 4b + c = 0$ **(4)**. Subtracting equation **(3)** from **(4)** gives $8b = 0 \quad \Rightarrow \quad b = 0$. Adding **(1)** and **(2)** gives $8b + 2d = 6$, so $d = 3$ since $b = 0$. From **(3)** we have $c = -12a$, so **(2)** becomes $8a + 4(0) + 2(-12a) + 3 = 0 \quad \Rightarrow \quad 3 = 16a \quad \Rightarrow \quad a = \frac{3}{16}$. Now $c = -12a = -12\left(\frac{3}{16}\right) = -\frac{9}{4}$ and the desired cubic function is $y = \frac{3}{16}x^3 - \frac{9}{4}x + 3$.

63. *Solution 1:* Let $f(x) = x^{1000}$. Then, by the definition of derivative,

$f'(1) = \lim\limits_{x \to 1} \dfrac{f(x) - f(1)}{x - 1} = \lim\limits_{x \to 1} \dfrac{x^{1000} - 1}{x - 1}$. But this is just the limit we want to find, and we know (from the

Power Rule) that $f'(x) = 1000x^{999}$, so $f'(1) = 1000(1)^{999} = 1000$. So $\lim\limits_{x \to 1} \dfrac{x^{1000} - 1}{x - 1} = 1000$.

Solution 2: Note that $\left(x^{1000} - 1\right) = (x - 1)\left(x^{999} + x^{998} + x^{997} + \cdots + x^2 + x + 1\right)$. So

$$\lim_{x \to 1} \frac{x^{1000} - 1}{x - 1} = \lim_{x \to 1} \frac{(x - 1)\left(x^{999} + x^{998} + x^{997} + \cdots + x^2 + x + 1\right)}{x - 1}$$

$$= \lim_{x \to 1} \left(x^{999} + x^{998} + x^{997} + \cdots + x^2 + x + 1\right) = \underbrace{1 + 1 + 1 + \cdots + 1 + 1 + 1}_{1000 \text{ ones}}$$

$$= 1000, \text{ as above.}$$

3.2 The Product and Quotient Rules • • • • • • • • • • •

1. Product Rule: $y = \left(x^2 + 1\right)\left(x^3 + 1\right) \Rightarrow$

$\qquad y' = \left(x^2 + 1\right)\left(3x^2\right) + \left(x^3 + 1\right)(2x) = 3x^4 + 3x^2 + 2x^4 + 2x = 5x^4 + 3x^2 + 2x.$

Multiplying first: $y = \left(x^2 + 1\right)\left(x^3 + 1\right) = x^5 + x^3 + x^2 + 1 \Rightarrow y' = 5x^4 + 3x^2 + 2x$ (equivalent)

3. By the Product Rule, $f(x) = x^2 e^x \Rightarrow f'(x) = x^2 \dfrac{d}{dx}\left(e^x\right) + e^x \dfrac{d}{dx}\left(x^2\right) = x^2 e^x + e^x(2x) = xe^x(x + 2).$

5. By the Quotient Rule, $y = \dfrac{e^x}{x^2} \Rightarrow$

$$y' = \frac{x^2 \dfrac{d}{dx}\left(e^x\right) - e^x \dfrac{d}{dx}\left(x^2\right)}{\left(x^2\right)^2} = \frac{x^2\left(e^x\right) - e^x(2x)}{x^4} = \frac{xe^x(x - 2)}{x^4} = \frac{e^x(x - 2)}{x^3}$$

7. $h(x) = \dfrac{x + 2}{x - 1} \Rightarrow h'(x) = \dfrac{(x - 1)(1) - (x + 2)(1)}{(x - 1)^2} = \dfrac{x - 1 - x - 2}{(x - 1)^2} = -\dfrac{3}{(x - 1)^2}$

9. $H(x) = \left(x^3 - x + 1\right)\left(x^{-2} + 2x^{-3}\right) = \left(x^3 - x + 1\right)\left(x^{-2}\right) + \left(x^3 - x + 1\right)\left(2x^{-3}\right)$

$\qquad = x - x^{-1} + x^{-2} + 2 - 2x^{-2} + 2x^{-3} = 2 + x - x^{-1} - x^{-2} + 2x^{-3} \Rightarrow$

$H'(x) = 0 + 1 - (-1)x^{-2} - (-2)x^{-3} + 2(-3)x^{-4} = 1 + x^{-2} + 2x^{-3} - 6x^{-4}.$

Another method: Use the Product Rule.

11. $y = \dfrac{t^2}{3t^2 - 2t + 1} \Rightarrow$

$$y' = \frac{\left(3t^2 - 2t + 1\right)(2t) - t^2(6t - 2)}{\left(3t^2 - 2t + 1\right)^2} = \frac{2t\left[3t^2 - 2t + 1 - t(3t - 1)\right]}{\left(3t^2 - 2t + 1\right)^2}$$

$$= \frac{2t\left(3t^2 - 2t + 1 - 3t^2 + t\right)}{\left(3t^2 - 2t + 1\right)^2} = \frac{2t(1 - t)}{\left(3t^2 - 2t + 1\right)^2}$$

13. $y = \left(r^2 - 2r\right)e^r \Rightarrow y' = \left(r^2 - 2r\right)\left(e^r\right) + e^r(2r - 2) = e^r\left(r^2 - 2r + 2r - 2\right) = e^r\left(r^2 - 2\right)$

15. $y = \dfrac{v^3 - 2v\sqrt{v}}{v} = v^2 - 2\sqrt{v} = v^2 - 2v^{1/2} \quad \Rightarrow \quad y' = 2v - 2\left(\frac{1}{2}\right)v^{-1/2} = 2v - v^{-1/2}.$

We can change the form of the answer as follows: $2v - v^{-1/2} = 2v - \dfrac{1}{\sqrt{v}} = \dfrac{2v\sqrt{v} - 1}{\sqrt{v}} = \dfrac{2v^{3/2} - 1}{\sqrt{v}}$

17. $f(x) = \dfrac{x}{x + c/x} \quad \Rightarrow$

$f'(x) = \dfrac{(x + c/x)(1) - x(1 - c/x^2)}{\left(x + \frac{c}{x}\right)^2} = \dfrac{x + c/x - x + c/x}{\left(\dfrac{x^2 + c}{x}\right)^2} = \dfrac{2c/x}{\dfrac{(x^2 + c)^2}{x^2}} \cdot \dfrac{x^2}{x^2} = \dfrac{2cx}{(x^2 + c)^2}$

19. $y = 2xe^x \quad \Rightarrow \quad y' = 2(x \cdot e^x + e^x \cdot 1) = 2e^x(x + 1)$. At $(0, 0)$, $y' = 2e^0(0 + 1) = 2 \cdot 1 \cdot 1 = 2$, and an equation of the tangent line is $y - 0 = 2(x - 0)$, or $y = 2x$.

21. (a) $y = f(x) = \dfrac{1}{1 + x^2} \quad \Rightarrow$

$f'(x) = \dfrac{(1 + x^2)(0) - 1(2x)}{(1 + x^2)^2} = \dfrac{-2x}{(1 + x^2)^2}$. So the slope of the

tangent line at the point $\left(-1, \frac{1}{2}\right)$ is $f'(-1) = \dfrac{2}{2^2} = \frac{1}{2}$ and its

equation is $y - \frac{1}{2} = \frac{1}{2}(x + 1)$ or $y = \frac{1}{2}x + 1$.

(b)

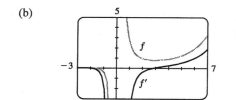

23. (a) $f(x) = \dfrac{e^x}{x^3} \quad \Rightarrow \quad f'(x) = \dfrac{x^3(e^x) - e^x(3x^2)}{(x^3)^2} = \dfrac{x^2 e^x(x - 3)}{x^6} = \dfrac{e^x(x - 3)}{x^4}$

(b)

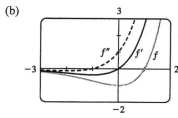

$f' = 0$ when f has a horizontal tangent line, f' is negative when f is decreasing, and f' is positive when f is increasing.

25. (a) $f(x) = (x - 1)e^x \quad \Rightarrow \quad f'(x) = (x - 1)e^x + e^x(1) = e^x(x - 1 + 1) = xe^x$.

$f''(x) = x(e^x) + e^x(1) = e^x(x + 1)$

(b)

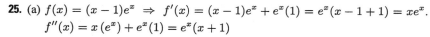

$f' = 0$ when f has a horizontal tangent and $f'' = 0$ when f' has a horizontal tangent. f' is negative when f is decreasing and positive when f is increasing. f'' is negative when f' is decreasing and positive when f' is increasing. f'' is negative when f is concave down and positive when f is concave up.

27. We are given that $f(5) = 1$, $f'(5) = 6$, $g(5) = -3$, and $g'(5) = 2$.

(a) $(fg)'(5) = f(5)g'(5) + g(5)f'(5) = (1)(2) + (-3)(6) = 2 - 18 = -16$

(b) $\left(\dfrac{f}{g}\right)'(5) = \dfrac{g(5)f'(5) - f(5)g'(5)}{[g(5)]^2} = \dfrac{(-3)(6) - (1)(2)}{(-3)^2} = -\dfrac{20}{9}$

(c) $\left(\dfrac{g}{f}\right)'(5) = \dfrac{f(5)g'(5) - g(5)f'(5)}{[f(5)]^2} = \dfrac{(1)(2) - (-3)(6)}{(1)^2} = 20$

29. $f(x) = e^x g(x) \quad \Rightarrow \quad f'(x) = e^x g'(x) + g(x)e^x = e^x [g'(x) + g(x)]$.
$f'(0) = e^0 [g'(0) + g(0)] = 1(5 + 2) = 7$

31. (a) From the graphs of f and g, we obtain the following values: $f(1) = 2$ since the point $(1, 2)$ is on the graph
of f; $g(1) = 1$ since the point $(1, 1)$ is on the graph of g; $f'(1) = 2$ since the slope of the line segment between

$(0, 0)$ and $(2, 4)$ is $\dfrac{4 - 0}{2 - 0} = 2$; $g'(1) = -1$ since the slope of the line segment between $(-2, 4)$ and $(2, 0)$

is $\dfrac{0 - 4}{2 - (-2)} = -1$. Now $u(x) = f(x)g(x)$, so $u'(1) = f(1)g'(1) + g(1)f'(1) = 2 \cdot (-1) + 1 \cdot 2 = 0$.

(b) $v(x) = f(x)/g(x)$, so $v'(5) = \dfrac{g(5)f'(5) - f(5)g'(5)}{[g(5)]^2} = \dfrac{2(-\frac{1}{3}) - 3 \cdot \frac{2}{3}}{2^2} = \dfrac{-\frac{8}{3}}{4} = -\dfrac{2}{3}$

33. Let $P(t)$ be the population and let $A(t)$ be the average annual income at time t, where t is measured in years and
$t = 0$ corresponds to July 1993. Then the total personal income is given by $T(t) = P(t)A(t)$. We wish to find
$T'(0)$. $T'(t) = P(t)A'(t) + A(t)P'(t)$. The term $P(t)A'(t)$ represents the portion of the rate of change of total
income due to the existing population's increasing income. The term $A(t)P'(t)$ represents the
portion of the rate of change of total income due to the increasing population.
$T'(0) = P(0)A'(0) + A(0)P'(0) \approx (3,354,000)(1900) + (21,107)(45,000) = 7,322,415,000$. So the total
personal income was rising at a rate of about \$7.322 billion per year.

35. f is increasing when f' is positive. $f(x) = x^3 e^x \quad \Rightarrow \quad f'(x) = x^3 e^x + e^x (3x^2) = x^2 e^x (x + 3)$. Now $x^2 \geq 0$
and $e^x > 0$ for all x, so $f'(x) > 0$ when $x + 3 > 0$ and $x \neq 0$; that is, when $x \in (-3, 0) \cup (0, \infty)$. So f is
increasing on $(-3, \infty)$.

37. If $y = f(x) = \dfrac{x}{x + 1}$, then $f'(x) = \dfrac{(x + 1)(1) - x(1)}{(x + 1)^2} = \dfrac{1}{(x + 1)^2}$. When $x = a$, the equation of the tangent

line is $y - \dfrac{a}{a + 1} = \dfrac{1}{(a + 1)^2}(x - a)$. This line passes through $(1, 2)$ when $2 - \dfrac{a}{a + 1} = \dfrac{1}{(a + 1)^2}(1 - a) \quad \Leftrightarrow$

$2(a + 1)^2 - a(a + 1) = 1 - a \quad \Leftrightarrow \quad 2a^2 + 4a + 2 - a^2 - a - 1 + a = 0 \quad \Leftrightarrow \quad a^2 + 4a + 1 = 0$.

The quadratic formula gives the roots of this equation as $a = \dfrac{-4 \pm \sqrt{4^2 - 4(1)(1)}}{2(1)} = \dfrac{-4 \pm \sqrt{12}}{2} = -2 \pm \sqrt{3}$,

so there are two such tangent lines. Since

$$f(-2 \pm \sqrt{3}) = \frac{-2 \pm \sqrt{3}}{-2 \pm \sqrt{3} + 1} = \frac{-2 \pm \sqrt{3}}{-1 \pm \sqrt{3}} \cdot \frac{-1 \mp \sqrt{3}}{-1 \mp \sqrt{3}}$$

$$= \frac{2 \pm 2\sqrt{3} \mp \sqrt{3} - 3}{1 - 3} = \frac{-1 \pm \sqrt{3}}{-2} = \frac{1 \mp \sqrt{3}}{2},$$

the lines touch the curve at $A\left(-2 + \sqrt{3}, \frac{1 - \sqrt{3}}{2}\right) \approx (-0.27, -0.37)$ and

$B\left(-2 - \sqrt{3}, \frac{1 + \sqrt{3}}{2}\right) \approx (-3.73, 1.37)$.

We will sometimes use the form $f'g + gf'$ rather than the form $fg' + gf'$ for the Product Rule.

39. (a) $(fgh)' = [(fg)h]' = (fg)'h + (fg)h' = (f'g + fg')h + (fg)h' = f'gh + fg'h + fgh'$

(b) Putting $f = g = h$ in part (a), we have

$$\frac{d}{dx}[f(x)]^3 = (fff)' = f'ff + ff'f + fff' = 3fff' = 3[f(x)]^2 f'(x).$$

(c) $\dfrac{d}{dx}(e^{3x}) = \dfrac{d}{dx}(e^x)^3 = 3(e^x)^2 e^x = 3e^{2x}e^x = 3e^{3x}$

41. For $f(x) = x^2 e^x$, $f'(x) = x^2 e^x + e^x(2x) = e^x(x^2 + 2x)$. Similarly, we have

$$f''(x) = e^x(x^2 + 4x + 2)$$

$$f'''(x) = e^x(x^2 + 6x + 6)$$

$$f^{(4)}(x) = e^x(x^2 + 8x + 12)$$

$$f^{(5)}(x) = e^x(x^2 + 10x + 20)$$

It appears that the coefficient of x in the quadratic term increases by 2 with each differentiation. The pattern for the constant terms seems to be $0 = 1 \cdot 0$, $2 = 2 \cdot 1$, $6 = 3 \cdot 2$, $12 = 4 \cdot 3$, $20 = 5 \cdot 4$. So a reasonable guess is that $f^{(n)}(x) = e^x[x^2 + 2nx + n(n-1)]$.

Proof: Let S_n be the statement that $f^{(n)}(x) = e^x[x^2 + 2nx + n(n-1)]$.

1. S_1 is true because $f'(x) = e^x(x^2 + 2x)$.

2. Assume that S_k is true; that is, $f^{(k)}(x) = e^x[x^2 + 2kx + k(k-1)]$. Then

$$f^{(k+1)}(x) = \frac{d}{dx}\left[f^{(k)}(x)\right] = e^x(2x + 2k) + \left[x^2 + 2kx + k(k-1)\right]e^x$$

$$= e^x[x^2 + (2k+2)x + (k^2 + k)] = e^x[x^2 + 2(k+1)x + (k+1)k]$$

This shows that S_{k+1} is true.

3. Therefore, by mathematical induction, S_n is true for all n; that is, $f^{(n)}(x) = e^x[x^2 + 2nx + n(n-1)]$

for every positive integer n.

43. $\dfrac{d}{dx}(x^{-n}) = \dfrac{d}{dx}\left(\dfrac{1}{x^n}\right) = -\dfrac{nx^{n-1}}{(x^n)^2} = -n \cdot \dfrac{x^{n-1}}{x^{2n}} = -nx^{n-1-2n} = -nx^{-n-1}$

◆3.3 Rates of Change in the Natural and Social Sciences · · ·

1. (a) $s = f(t) = t^3 - 12t^2 + 36t \Rightarrow v(t) = f'(t) = 3t^2 - 24t + 36$

(b) $v(3) = 27 - 72 + 36 = -9$ m/s

(c) The particle is at rest when $v(t) = 0$. $3t^2 - 24t + 36 = 0 \Rightarrow 3(t-2)(t-6) = 0 \Rightarrow t = 2, 6$.

(d) The particle is moving in the positive direction when $v(t) > 0$. $3(t-2)(t-6) > 0 \Leftrightarrow 0 \le t < 2$ or $t > 6$.

(e) Since the particle is moving forward and backward, we need to calculate the distance traveled in the intervals $[0, 2]$, $[2, 6]$, and $[6, 8]$ separately.

(f)

$|f(2) - f(0)| = |32 - 0| = 32$.

$|f(6) - f(2)| = |0 - 32| = 32$.

$|f(8) - f(6)| = |32 - 0| = 32$.

The total distance is $32 + 32 + 32 = 96$ m.

(g) $a(t) = v'(t) = 6t - 24$. $a(3) = 6(3) - 24 = -6$ (m/s)/s or m/s^2.

(h)

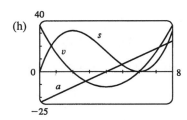

(i) The particle is speeding up when v and a have the same sign. This occurs when $2 < t < 4$ and when $t > 6$. It is slowing down when v and a have opposite signs; that is, when $0 \le t < 2$ and when $4 < t < 6$.

3. (a) $s(t) = t^3 - 4.5t^2 - 7t \quad \Rightarrow \quad v(t) = s'(t) = 3t^2 - 9t - 7 = 5 \quad \Leftrightarrow \quad 3t^2 - 9t - 12 = 0 \quad \Leftrightarrow$
$3(t - 4)(t + 1) = 0 \quad \Leftrightarrow \quad t = 4$ or -1. Since $t \ge 0$, the particle reaches a velocity of 5 m/s at $t = 4$ s.

(b) $a(t) = v'(t) = 6t - 9 = 0 \quad \Leftrightarrow \quad t = 1.5$. The acceleration changes from negative to positive, so the velocity changes from decreasing to increasing. Thus, at $t = 1.5$ s, the velocity has its minimum value.

5. (a) $A(x) = x^2 \quad \Rightarrow \quad A'(x) = 2x$. $A'(15) = 30$ mm^2/mm is the rate at which the area is increasing with respect to the side length as x reaches 15 mm.

(b) The perimeter is $P(x) = 4x$, so
$A'(x) = 2x = \frac{1}{2}(4x) = \frac{1}{2}P(x)$. The figure suggests that if
Δx is small, then the change in the area of the square is
approximately half of its perimeter (2 of the 4 sides) times
Δx. From the figure, $\Delta A = 2x(\Delta x) + (\Delta x)^2$. If Δx is
small, then $\Delta A \approx 2x(\Delta x)$ and so $\Delta A/\Delta x \approx 2x$.

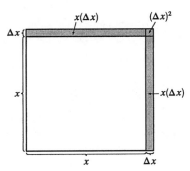

7. (a) Using $A(r) = \pi r^2$, we find that the average rate of change is:

(i) $\dfrac{A(3) - A(2)}{3 - 2} = \dfrac{9\pi - 4\pi}{1} = 5\pi$

(ii) $\dfrac{A(2.5) - A(2)}{2.5 - 2} = \dfrac{6.25\pi - 4\pi}{0.5} = 4.5\pi$

(iii) $\dfrac{A(2.1) - A(2)}{2.1 - 2} = \dfrac{4.41\pi - 4\pi}{0.1} = 4.1\pi$

(b) $A(r) = \pi r^2 \quad \Rightarrow \quad A'(r) = 2\pi r$, so $A'(2) = 4\pi$.

(c) The circumference is $C(r) = 2\pi r = A'(r)$. The figure suggests that if Δr is
small, then the change in the area of the circle (a ring around the outside) is
approximately equal to its circumference times Δr . Straightening out this ring
gives us a shape that is approximately rectangular with length $2\pi r$ and width
Δr, so $\Delta A \approx 2\pi r(\Delta r)$. Algebraically,
$\Delta A = A(r + \Delta r) - A(r) = \pi(r + \Delta r)^2 - \pi r^2 = 2\pi r(\Delta r) + \pi(\Delta r)^2$. So
we see that if Δr is small, then $\Delta A \approx 2\pi r(\Delta r)$ and therefore,
$\Delta A/\Delta r \approx 2\pi r$.

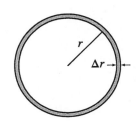

9. $S(r) = 4\pi r^2 \Rightarrow S'(r) = 8\pi r \Rightarrow$

(a) $S'(1) = 8\pi \text{ ft}^2/\text{ft}$ (b) $S'(2) = 16\pi \text{ ft}^2/\text{ft}$ (c) $S'(3) = 24\pi \text{ ft}^2/\text{ft}$

As the radius increases, the surface area grows at an increasing rate. In fact, the rate of change is linear with respect to the radius.

11. The mass is $f(x) = 3x^2$, so the linear density at x is $\rho(x) = f'(x) = 6x$.

(a) $\rho(1) = 6 \text{ kg/m}$ (b) $\rho(2) = 12 \text{ kg/m}$ (c) $\rho(3) = 18 \text{ kg/m}$

Since ρ is an increasing function, the density will be the highest at the right end of the rod and lowest at the left end.

13. The quantity of charge is $Q(t) = t^3 - 2t^2 + 6t + 2$, so the current is $Q'(t) = 3t^2 - 4t + 6$.

(a) $Q'(0.5) = 3(0.5)^2 - 4(0.5) + 6 = 4.75 \text{ A}$

(b) $Q'(1) = 3(1)^2 - 4(1) + 6 = 5 \text{ A}$
The current is lowest when Q' has a minimum. $Q''(t) = 6t - 4 < 0$ when $t < \frac{2}{3}$. So the current decreases when $t < \frac{2}{3}$ and increases when $t > \frac{2}{3}$. Thus, the current is lowest at $t = \frac{2}{3}$ s.

15. (a) To find the rate of change of volume with respect to pressure, we first solve for V in terms of P.

$$PV = C \Rightarrow V = \frac{C}{P} \Rightarrow \frac{dV}{dP} = -\frac{C}{P^2}.$$

(b) From the formula for dV/dP in part (a), we see that as P increases, the absolute value of dV/dP decreases. Thus, the volume is decreasing more rapidly at the beginning.

(c) $\beta = -\dfrac{1}{V}\dfrac{dV}{dP} = -\dfrac{1}{V}\left(-\dfrac{C}{P^2}\right) = \dfrac{C}{(PV)P} = \dfrac{C}{CP} = \dfrac{1}{P}$

17. (a) 1920: $m_1 = \frac{1860 - 1750}{1920 - 1910} = \frac{110}{10} = 11$, $m_2 = \frac{2070 - 1860}{1930 - 1920} = \frac{210}{10} = 21$,

$(m_1 + m_2)/2 = (11 + 21)/2 = 16 \text{ million/year}$

1980: $m_1 = \frac{4450 - 3710}{1980 - 1970} = \frac{740}{10} = 74$, $m_2 = \frac{5280 - 4450}{1990 - 1980} = \frac{830}{10} = 83$,

$(m_1 + m_2)/2 = (74 + 83)/2 = 78.5 \text{ million/year}$

(b) $P(t) = at^3 + bt^2 + ct + d$, where $a = 0.0012354312$, $b = -6.722261072$, $c = 12{,}165.08275$, and $d = -7{,}318{,}428.648$.

(c) $P(t) = at^3 + bt^2 + ct + d \Rightarrow P'(t) = 3at^2 + 2bt + c$

(d) $P'(1920) = 3(0.0012354312)(1920)^2 + 2(-6.722261072)(1920) + 12{,}165.08275$

$\qquad = 14{,}481{,}352/\text{year}$ [smaller than the answer in part (a), but close to it]

$P'(1980) = 75{,}082{,}751/\text{year}$ (smaller, but close)

(e) $P'(1985) = 81{,}337{,}413/\text{year}$, so the rate of growth in 1985 was about 81.3 million/year.

19. (a) $[\text{C}] = \dfrac{a^2kt}{akt + 1} \Rightarrow$

rate of reaction $= \dfrac{d\,[\text{C}]}{dt} = \dfrac{(akt + 1)(a^2k) - (a^2kt)(ak)}{(akt + 1)^2} = \dfrac{a^2k(akt + 1 - akt)}{(akt + 1)^2} = \dfrac{a^2k}{(akt + 1)^2}$

(b) If $x = [\text{C}]$, then $a - x = a - \dfrac{a^2kt}{akt + 1} = \dfrac{a^2kt + a - a^2kt}{akt + 1} = \dfrac{a}{akt + 1}$.

So $k(a - x)^2 = k\left(\dfrac{a}{akt + 1}\right)^2 = \dfrac{a^2k}{(akt + 1)^2} = \dfrac{d\,[\text{C}]}{dt}$ [from part (a)] $= \dfrac{dx}{dt}$.

(c) As $t \to \infty$, $[\text{C}] = \dfrac{a^2kt}{akt + 1} = \dfrac{(a^2kt)/t}{(akt + 1)/t} = \dfrac{a^2k}{ak + (1/t)} \to \dfrac{a^2k}{ak} = a \text{ moles/L}$.

(d) As $t \to \infty$, $\dfrac{d[C]}{dt} = \dfrac{a^2 k}{(akt + 1)^2} \to 0$.

(e) As t increases, nearly all of the reactants A and B are converted into product C. In practical terms, the reaction virtually stops.

21. (a) Using $v = \dfrac{P}{4\eta l}\left(R^2 - r^2\right)$ with $R = 0.01$, $l = 3$, $P = 3000$, and $\eta = 0.027$, we have v as a function of r:

$$v(r) = \frac{3000}{4(0.027)3}\left(0.01^2 - r^2\right). \quad v(0) = 0.\overline{925} \text{ cm/s}, \ v(0.005) = 0.69\overline{4} \text{ cm/s}, \ v(0.01) = 0.$$

(b) $v(r) = \dfrac{P}{4\eta l}\left(R^2 - r^2\right) \ \Rightarrow \ v'(r) = \dfrac{P}{4\eta l}(-2r) = -\dfrac{Pr}{2\eta l}$. When $l = 3$, $P = 3000$, and $\eta = 0.027$, we have

$$v'(r) = -\frac{3000r}{2(0.027)3}. \ v'(0) = 0, \ v'(0.005) = -92.\overline{592} \text{ (cm/s)/cm, and } v'(0.01) = -185.\overline{185} \text{ (cm/s)/cm.}$$

(c) The velocity is greatest where $r = 0$ (at the center) and the velocity is changing most where $r = R = 0.01$ cm (at the edge).

23. (a) $C(x) = 2000 + 3x + 0.01x^2 + 0.0002x^3 \ \Rightarrow \ C'(x) = 3 + 0.02x + 0.0006x^2$

(b) $C'(100) = 3 + 0.02(100) + 0.0006(10,000) = 3 + 2 + 6 = \$11/\text{pair}$. $C'(100)$ is the rate at which the cost is increasing as the 100th pair of jeans is produced. It predicts the cost of the 101st pair.

(c) The cost of manufacturing the 101st pair of jeans is

$$C(101) - C(100) = (2000 + 303 + 102.01 + 206.0602) - (2000 + 300 + 100 + 200)$$
$$= 11.0702 \approx \$11.07$$

25. (a) $A(x) = \dfrac{p(x)}{x} \ \Rightarrow \ A'(x) = \dfrac{xp'(x) - p(x) \cdot 1}{x^2} = \dfrac{xp'(x) - p(x)}{x^2}$. $A'(x) > 0 \ \Rightarrow \ A(x)$ is increasing; that is, the average productivity increases as the size of the work force increases.

(b) $p'(x)$ is greater than the average productivity $\ \Rightarrow \ p'(x) > A(x) \ \Rightarrow \ p'(x) > \dfrac{p(x)}{x} \ \Rightarrow$

$$xp'(x) > p(x) \ \Rightarrow \ xp'(x) - p(x) > 0 \ \Rightarrow \ \frac{xp'(x) - p(x)}{x^2} > 0 \ \Rightarrow \ A'(x) > 0.$$

27. $PV = nRT \ \Rightarrow \ T = \dfrac{PV}{nR} = \dfrac{PV}{(10)(0.0821)} = \dfrac{1}{0.821}(PV)$. Using the Product Rule, we have

$$\frac{dT}{dt} = \frac{1}{0.821}[P(t)V'(t) + V(t)P'(t)] = \frac{1}{0.821}[(8)(-0.15) + (10)(0.10)] \approx -0.2436 \text{ K/min.}$$

29. (a) If the populations are stable, then the growth rates are neither positive nor negative; that is,

$$\frac{dC}{dt} = 0 \text{ and } \frac{dW}{dt} = 0.$$

(b) "The caribou go extinct" means that the population is zero, or mathematically, $C = 0$.

(c) We have the equations $\dfrac{dC}{dt} = aC - bCW$ and $\dfrac{dW}{dt} = -cW + dCW$. Let $dC/dt = dW/dt = 0$, $a = 0.05$, $b = 0.001$, $c = 0.05$, and $d = 0.0001$ to obtain **(1)** $0.05C - 0.001CW = 0$ and **(2)** $-0.05W + 0.0001CW = 0$. Adding 10 times **(2)** to **(1)** eliminates the CW-terms and gives us $0.05C - 0.5W = 0 \ \Rightarrow \ C = 10W$. Substituting $C = 10W$ into **(1)** results in $0.05(10W) - 0.001(10W)W = 0 \ \Leftrightarrow \ 0.5W - 0.01W^2 = 0 \ \Leftrightarrow \ 50W - W^2 = 0 \ \Leftrightarrow \ W(50 - W) = 0 \ \Leftrightarrow \ W = 0$ or 50. Since $C = 10W$, $C = 0$ or 500. Thus, the population pairs (C, W) that lead to stable populations are $(0, 0)$ and $(500, 50)$. So it is possible for the two species to live in harmony.

3.4 Derivatives of Trigonometric Functions • • • • • •

1. $f(x) = x - 3\sin x \Rightarrow f'(x) = 1 - 3\cos x$

3. $g(t) = t^3 \cos t \Rightarrow g'(t) = t^3(-\sin t) + (\cos t) \cdot 3t^2 = 3t^2 \cos t - t^3 \sin t$ or $t^2(3\cos t - t\sin t)$

5. $h(\theta) = \csc\theta + e^\theta \cot\theta \Rightarrow$
$h'(\theta) = -\csc\theta\cot\theta + e^\theta\left(-\csc^2\theta\right) + (\cot\theta)e^\theta = -\csc\theta\cot\theta + e^\theta\left(\cot\theta - \csc^2\theta\right)$

7. $y = \dfrac{\tan x}{x} \Rightarrow \dfrac{dy}{dx} = \dfrac{x\sec^2 x - \tan x}{x^2}$

9. $y = \dfrac{x}{\sin x + \cos x} \Rightarrow$
$\dfrac{dy}{dx} = \dfrac{(\sin x + \cos x) - x(\cos x - \sin x)}{(\sin x + \cos x)^2} = \dfrac{\sin x + \cos x - x\cos x + x\sin x}{\sin^2 x + \cos^2 x + 2\sin x\cos x}$
$= \dfrac{\sin x + \cos x + x\sin x - x\cos x}{1 + \sin 2x}$ or $\dfrac{(1+x)\sin x + (1-x)\cos x}{1 + \sin 2x}$

11. $y = \sec\theta\tan\theta \Rightarrow y' = \sec\theta\left(\sec^2\theta\right) + \tan\theta(\sec\theta\tan\theta) = \sec\theta\left(\sec^2\theta + \tan^2\theta\right)$
Using the identity $1 + \tan^2\theta = \sec^2\theta$, we can write alternative forms of the answer as
$$\sec\theta\left(1 + 2\tan^2\theta\right) \quad \text{or} \quad \sec\theta\left(2\sec^2\theta - 1\right)$$

13. $\dfrac{d}{dx}(\csc x) = \dfrac{d}{dx}\left(\dfrac{1}{\sin x}\right) = \dfrac{(\sin x)(0) - 1(\cos x)}{\sin^2 x} = \dfrac{-\cos x}{\sin^2 x} = -\dfrac{1}{\sin x} \cdot \dfrac{\cos x}{\sin x} = -\csc x\cot x$

15. $\dfrac{d}{dx}(\cot x) = \dfrac{d}{dx}\left(\dfrac{\cos x}{\sin x}\right) = \dfrac{(\sin x)(-\sin x) - (\cos x)(\cos x)}{\sin^2 x} = -\dfrac{\sin^2 x + \cos^2 x}{\sin^2 x} = -\dfrac{1}{\sin^2 x} = -\csc^2 x$

17. $y = \tan x \Rightarrow y' = \sec^2 x \Rightarrow$ the slope of the tangent line at $\left(\frac{\pi}{4}, 1\right)$ is $\sec^2\frac{\pi}{4} = \left(\sqrt{2}\right)^2 = 2$ and an equation is
$y - 1 = 2\left(x - \frac{\pi}{4}\right)$ or $y = 2x + 1 - \frac{\pi}{2}$.

19. (a) $y = x\cos x \Rightarrow y' = x(-\sin x) + \cos x(1) = \cos x - x\sin x$.
So the slope of the tangent at the point $(\pi, -\pi)$ is
$\cos\pi - \pi\sin\pi = -1 - \pi(0) = -1$, and an equation is
$y + \pi = -(x - \pi)$ or $y = -x$.

(b)

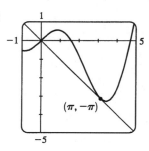

21. (a) $f(x) = 2x + \cot x \Rightarrow f'(x) = 2 - \csc^2 x$

(b)

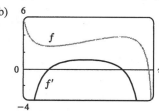

Notice that $f'(x) = 0$ when f has a horizontal tangent.
Also, $f'(x)$ is large negative when the graph of f is
steep.

23. $H(\theta) = \theta\sin\theta \Rightarrow H'(\theta) = \theta(\cos\theta) + (\sin\theta) \cdot 1 = \theta\cos\theta + \sin\theta \Rightarrow$
$H''(\theta) = \theta(-\sin\theta) + (\cos\theta) \cdot 1 + \cos\theta = -\theta\sin\theta + 2\cos\theta$

25. $f(x) = x + 2\sin x$ has a horizontal tangent when $f'(x) = 0 \quad \Leftrightarrow \quad 1 + 2\cos x = 0 \quad \Leftrightarrow \quad \cos x = -\frac{1}{2} \quad \Leftrightarrow$

$x = \frac{2\pi}{3} + 2\pi n$ or $\frac{4\pi}{3} + 2\pi n$, where n is an integer. Note that $\frac{4\pi}{3}$ and $\frac{2\pi}{3}$ are $\pm\frac{\pi}{3}$ units from π. This allows us to

write the solutions in the more compact equivalent form $(2n + 1)\pi \pm \frac{\pi}{3}$, n an integer.

27. $f(x) = x - 2\sin x$, $0 \le x \le 2\pi$. $f'(x) = 1 - 2\cos x$. So $f'(x) > 0 \quad \Leftrightarrow \quad 1 - 2\cos x > 0 \quad \Leftrightarrow$

$-2\cos x > -1 \quad \Leftrightarrow \quad \cos x < \frac{1}{2} \quad \Leftrightarrow \quad \frac{\pi}{3} < x < \frac{5\pi}{3} \quad \Rightarrow \quad f$ is increasing on $\left(\frac{\pi}{3}, \frac{5\pi}{3}\right)$.

29. (a) $x(t) = 8\sin t \quad \Rightarrow \quad v(t) = x'(t) = 8\cos t \quad \Rightarrow \quad a(t) = x''(t) = -8\sin t$

(b) The mass at time $t = \frac{2\pi}{3}$ has position $x\left(\frac{2\pi}{3}\right) = 8\sin\frac{2\pi}{3} = 8\left(\frac{\sqrt{3}}{2}\right) = 4\sqrt{3}$, velocity

$v\left(\frac{2\pi}{3}\right) = 8\cos\frac{2\pi}{3} = 8\left(-\frac{1}{2}\right) = -4$, and acceleration $a\left(\frac{2\pi}{3}\right) = -8\sin\frac{2\pi}{3} = -8\left(\frac{\sqrt{3}}{2}\right) = -4\sqrt{3}$. Since

$v\left(\frac{2\pi}{3}\right) < 0$, the particle is moving to the left. Because v and a have the same sign, the particle is speeding up.

31.

From the diagram we can see that $\sin\theta = x/10 \quad \Leftrightarrow \quad x = 10\sin\theta$. We want to find the

rate of change of x with respect to θ; that is, $dx/d\theta$. Taking the derivative of the above

expression, $dx/d\theta = 10(\cos\theta)$. So when $\theta = \frac{\pi}{3}$,

$$dx/d\theta = 10\cos\frac{\pi}{3} = 10\left(\frac{1}{2}\right) = 5 \text{ ft/rad}$$

33. $\dfrac{d}{dx}(\sin x) = \cos x \quad \Rightarrow \quad \dfrac{d^2}{dx^2}(\sin x) = -\sin x \quad \Rightarrow \quad \dfrac{d^3}{dx^3}(\sin x) = -\cos x \quad \Rightarrow \quad \dfrac{d^4}{dx^4}(\sin x) = \sin x$.

The derivatives of $\sin x$ occur in a cycle of four. Since $99 = 4(24) + 3$, we have

$\dfrac{d^{99}}{dx^{99}}(\sin x) = \dfrac{d^3}{dx^3}(\sin x) = -\cos x$.

35. $y = A\sin x + B\cos x \quad \Rightarrow \quad y' = A\cos x - B\sin x \quad \Rightarrow \quad y'' = -A\sin x - B\cos x$. Substituting these

expressions for y, y', and y'' into the given differential equation $y'' + y' - 2y = \sin x$ gives us

$(-A\sin x - B\cos x) + (A\cos x - B\sin x) - 2(A\sin x + B\cos x) = \sin x \quad \Leftrightarrow$

$-3A\sin x - B\sin x + A\cos x - 3B\cos x = \sin x \quad \Leftrightarrow \quad (-3A - B)\sin x + (A - 3B)\cos x = 1\sin x$, so we

must have $-3A - B = 1$ and $A - 3B = 0$ (since 0 is the coefficient of $\cos x$ on the right side). Solving for A and

B, we add the first equation to three times the second to get $B = -\frac{1}{10}$ and $A = -\frac{3}{10}$.

37. $\displaystyle\lim_{x\to0}\frac{\tan 4x}{x} = \lim_{x\to0}\left(\frac{\sin 4x}{x}\cdot\frac{1}{\cos 4x}\right) = \lim_{x\to0}\left(\frac{4\sin 4x}{4x}\cdot\frac{1}{\cos 4x}\right) = 4\lim_{x\to0}\frac{\sin 4x}{4x}\cdot\lim_{x\to0}\frac{1}{\cos 4x} = 4\cdot1\cdot1 = 4$

39. $\displaystyle\lim_{\theta\to0}\frac{\sin\theta}{\theta + \tan\theta} = \frac{\displaystyle\lim_{\theta\to0}\frac{\sin\theta}{\theta}}{\displaystyle\lim_{\theta\to0}\frac{\theta + \tan\theta}{\theta}} = \frac{1}{\displaystyle\lim_{\theta\to0}\left(1 + \frac{\sin\theta}{\theta}\cdot\frac{1}{\cos\theta}\right)} = \frac{1}{1 + 1\cdot1} = \frac{1}{2}$

41. By the definition of radian measure, $s = r\theta$, where r is the radius of the circle.

By drawing the bisector of the angle θ, we can see that $\sin\dfrac{\theta}{2} = \dfrac{d/2}{r} \quad \Rightarrow \quad d = 2r\sin\dfrac{\theta}{2}$.

So $\displaystyle\lim_{\theta\to0^+}\frac{s}{d} = \lim_{\theta\to0^+}\frac{r\theta}{2r\sin(\theta/2)} = \lim_{\theta\to0^+}\frac{2\cdot(\theta/2)}{2\sin(\theta/2)} = \lim_{\theta\to0}\frac{\theta/2}{\sin(\theta/2)} = 1$. [This is just the reciprocal of the limit

$\displaystyle\lim_{x\to0}\frac{\sin x}{x} = 1$ combined with the fact that as $\theta \to 0$, $\frac{\theta}{2} \to 0$ also.]

The Chain Rule • • • • • • • • • • • • • • • •

1. Let $u = g(x) = 4x$ and $y = f(u) = \sin u$. Then $\dfrac{dy}{dx} = \dfrac{dy}{du}\dfrac{du}{dx} = (\cos u)(4) = 4\cos 4x$.

3. Let $u = g(x) = 1 - x^2$ and $y = f(u) = u^{10}$. Then $\dfrac{dy}{dx} = \dfrac{dy}{du}\dfrac{du}{dx} = (10u^9)(-2x) = -20x(1-x^2)^9$.

5. Let $u = g(x) = \sqrt{x}$ and $y = f(u) = e^u$.

Then $\dfrac{dy}{dx} = \dfrac{dy}{du}\dfrac{du}{dx} = (e^u)\left(\tfrac{1}{2}x^{-1/2}\right) = e^{\sqrt{x}} \cdot \dfrac{1}{2\sqrt{x}} = \dfrac{e^{\sqrt{x}}}{2\sqrt{x}}$.

7. $F(x) = \sqrt[4]{1 + 2x + x^3} = (1 + 2x + x^3)^{1/4} \Rightarrow$

$F'(x) = \tfrac{1}{4}(1 + 2x + x^3)^{-3/4} \cdot \dfrac{d}{dx}(1 + 2x + x^3) = \dfrac{1}{4(1 + 2x + x^3)^{3/4}} \cdot (2 + 3x^2)$

$= \dfrac{2 + 3x^2}{4(1 + 2x + x^3)^{3/4}} = \dfrac{2 + 3x^2}{4\sqrt[4]{(1 + 2x + x^3)^3}}$

9. $g(t) = \dfrac{1}{(t^4 + 1)^3} = (t^4 + 1)^{-3} \Rightarrow g'(t) = -3(t^4 + 1)^{-4}(4t^3) = -12t^3(t^4 + 1)^{-4} = \dfrac{-12t^3}{(t^4 + 1)^4}$

11. $y = \cos(a^3 + x^3) \Rightarrow y' = -\sin(a^3 + x^3) \cdot 3x^2$ [a^3 is just a constant] $= -3x^2 \sin(a^3 + x^3)$

13. $y = e^{-mx} \Rightarrow y' = e^{-mx}\dfrac{d}{dx}(-mx) = e^{-mx}(-m) = -me^{-mx}$

15. $f(x) = xe^{-x^2} \Rightarrow f'(x) = xe^{-x^2}(-2x) + e^{-x^2} \cdot 1 = e^{-x^2}(-2x^2 + 1) = e^{-x^2}(1 - 2x^2)$

17. $G(x) = (3x - 2)^{10}(5x^2 - x + 1)^{12} \Rightarrow$

$G'(x) = (3x - 2)^{10}(12)(5x^2 - x + 1)^{11}(10x - 1) + (5x^2 - x + 1)^{12}(10)(3x - 2)^9(3)$

$= 6(3x - 2)^9(5x^2 - x + 1)^{11}[2(3x - 2)(10x - 1) + 5(5x^2 - x + 1)]$

$= 6(3x - 2)^9(5x^2 - x + 1)^{11}[(60x^2 - 46x + 4) + (25x^2 - 5x + 5)]$

$= 6(3x - 2)^9(5x^2 - x + 1)^{11}(85x^2 - 51x + 9)$

19. $y = e^{x \cos x} \Rightarrow y' = e^{x \cos x} \cdot \dfrac{d}{dx}(x \cos x) = e^{x \cos x}[x(-\sin x) + (\cos x) \cdot 1] = e^{x \cos x}(\cos x - x \sin x)$

21. This function is a quotient raised to a power. To find its derivative, we use the Chain Rule by differentiating the

power first and then multiplying by the derivative of the quotient. $F(y) = \left(\dfrac{y - 6}{y + 7}\right)^3 \Rightarrow$

$F'(y) = 3\left(\dfrac{y - 6}{y + 7}\right)^2 \dfrac{(y + 7)(1) - (y - 6)(1)}{(y + 7)^2} = 3\left(\dfrac{y - 6}{y + 7}\right)^2 \dfrac{13}{(y + 7)^2} = \dfrac{39(y - 6)^2}{(y + 7)^4}$

23. $y = \dfrac{r}{\sqrt{r^2 + 1}} \Rightarrow$

$y' = \dfrac{\sqrt{r^2 + 1}\,(1) - r \cdot \tfrac{1}{2}(r^2 + 1)^{-1/2}(2r)}{\left(\sqrt{r^2 + 1}\right)^2} = \dfrac{\sqrt{r^2 + 1} - \dfrac{r^2}{\sqrt{r^2 + 1}}}{\left(\sqrt{r^2 + 1}\right)^2} = \dfrac{\dfrac{\sqrt{r^2 + 1}\sqrt{r^2 + 1} - r^2}{\sqrt{r^2 + 1}}}{\left(\sqrt{r^2 + 1}\right)^2}$

$= \dfrac{(r^2 + 1) - r^2}{\left(\sqrt{r^2 + 1}\right)^3} = \dfrac{1}{(r^2 + 1)^{3/2}}$ or $(r^2 + 1)^{-3/2}$

(continued)

Another solution: Write y as a product and make use of the Product Rule.

$$y = r(r^2 + 1)^{-1/2} \quad \Rightarrow \quad y' = r \cdot -\tfrac{1}{2}(r^2 + 1)^{-3/2}(2r) + (r^2 + 1)^{-1/2} \cdot 1$$
$$= (r^2 + 1)^{-3/2}\left[-r^2 + (r^2 + 1)^1\right] = (r^2 + 1)^{-3/2}(1) = (r^2 + 1)^{-3/2}$$

The step that students usually have trouble with is factoring out $(r^2 + 1)^{-3/2}$. But this is no different than factoring out x^2 from $x^2 + x^5$; that is, we are just factoring out a factor with the *smallest* exponent that appears on it. In this case, $-\tfrac{3}{2}$ is smaller than $-\tfrac{1}{2}$.

25. Using Formula 5 and the Chain Rule, $y = 2^{\sin \pi x} \quad \Rightarrow$

$$y' = 2^{\sin \pi x}(\ln 2) \cdot \frac{d}{dx}(\sin \pi x) = 2^{\sin \pi x}(\ln 2) \cdot \cos \pi x \cdot \pi = 2^{\sin \pi x}(\pi \ln 2)\cos \pi x$$

27. $y = \cot^2(\sin \theta) = [\cot(\sin \theta)]^2 \quad \Rightarrow$

$$y' = 2\left[\cot(\sin \theta)\right] \cdot \frac{d}{d\theta}\left[\cot(\sin \theta)\right] = 2\cot(\sin \theta) \cdot \left[-\csc^2(\sin \theta) \cdot \cos \theta\right] = -2\cos \theta \cot(\sin \theta) \csc^2(\sin \theta)$$

29. $y = \sin\left(\tan \sqrt{\sin x}\right) \quad \Rightarrow$

$$y' = \cos\left(\tan \sqrt{\sin x}\right) \cdot \frac{d}{dx}\left(\tan \sqrt{\sin x}\right) = \cos\left(\tan \sqrt{\sin x}\right)\sec^2 \sqrt{\sin x} \cdot \frac{d}{dx}(\sin x)^{1/2}$$

$$= \cos\left(\tan \sqrt{\sin x}\right)\sec^2 \sqrt{\sin x} \cdot \tfrac{1}{2}(\sin x)^{-1/2} \cdot \cos x$$

$$= \cos\left(\tan \sqrt{\sin x}\right)\left(\sec^2 \sqrt{\sin x}\right)\left(\frac{1}{2\sqrt{\sin x}}\right)(\cos x)$$

31. $y = \sin(\sin x) \quad \Rightarrow \quad y' = \cos(\sin x) \cdot \cos x$. At $(\pi, 0)$, $y' = \cos(\sin \pi) \cdot \cos \pi = \cos(0) \cdot (-1) = 1(-1) = -1$, and an equation of the tangent line is $y - 0 = -1(x - \pi)$, or $y = -x + \pi$.

33. (a) $y = \dfrac{2}{1 + e^{-x}} \quad \Rightarrow \quad y' = \dfrac{(1 + e^{-x})(0) - 2(-e^{-x})}{(1 + e^{-x})^2} = \dfrac{2e^{-x}}{(1 + e^{-x})^2}$.

(b)
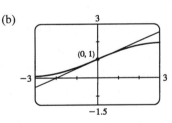

At $(0, 1)$, $y' = \dfrac{2e^0}{(1 + e^0)^2} = \dfrac{2(1)}{(1 + 1)^2} = \dfrac{2}{2^2} = \dfrac{1}{2}$.

So an equation of the tangent line is $y - 1 = \tfrac{1}{2}(x - 0)$ or $y = \tfrac{1}{2}x + 1$.

35. (a) $f(x) = \dfrac{\sqrt{1 - x^2}}{x} \quad \Rightarrow$

$$f'(x) = \frac{x \cdot \tfrac{1}{2}(1 - x^2)^{-1/2}(-2x) - \sqrt{1 - x^2}(1)}{x^2} \cdot \frac{\sqrt{1 - x^2}}{\sqrt{1 - x^2}}$$

$$= \frac{-x^2 - (1 - x^2)}{x^2\sqrt{1 - x^2}} = \frac{-1}{x^2\sqrt{1 - x^2}}$$

(b)
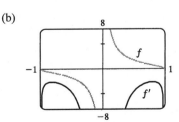

Notice that all tangents to the graph of f have negative slopes and $f'(x) < 0$ always.

37. $F(x) = f(g(x)) \quad \Rightarrow$
$F'(x) = f'(g(x)) \cdot g'(x)$, so $F'(3) = f'(g(3)) \cdot g'(3) = f'(6) \cdot g'(3) = 7 \cdot 4 = 28$. Notice that we did not use $f'(3) = 2$.

39. (a) $h(x) = f(g(x))$ $\Rightarrow$ $h'(x) = f'(g(x)) \cdot g'(x)$, so $h'(1) = f'(g(1)) \cdot g'(1) = f'(2) \cdot 6 = 5 \cdot 6 = 30$.

(b) $H(x) = g(f(x))$ $\Rightarrow$ $H'(x) = g'(f(x)) \cdot f'(x)$, so $H'(1) = g'(f(1)) \cdot f'(1) = g'(3) \cdot 4 = 9 \cdot 4 = 36$.

41. (a) $u(x) = f(g(x))$ $\Rightarrow$ $u'(x) = f'(g(x))g'(x)$. So $u'(1) = f'(g(1))g'(1) = f'(3)g'(1)$. To find $f'(3)$, note that f is linear from $(2, 4)$ to $(6, 3)$, so its slope is $\dfrac{3 - 4}{6 - 2} = -\dfrac{1}{4}$. To find $g'(1)$, note that g is linear from $(0, 6)$ to $(2, 0)$, so its slope is $\dfrac{0 - 6}{2 - 0} = -3$. Thus, $f'(3)g'(1) = \left(-\frac{1}{4}\right)(-3) = \frac{3}{4}$.

(b) $v(x) = g(f(x))$ $\Rightarrow$ $v'(x) = g'(f(x))f'(x)$. So $v'(1) = g'(f(1))f'(1) = g'(2)f'(1)$, which does not exist since $g'(2)$ does not exist.

(c) $w(x) = g(g(x))$ $\Rightarrow$ $w'(x) = g'(g(x))g'(x)$. So $w'(1) = g'(g(1))g'(1) = g'(3)g'(1)$. To find $g'(3)$, note that g is linear from $(2, 0)$ to $(5, 2)$, so its slope is $\dfrac{2 - 0}{5 - 2} = \dfrac{2}{3}$. Thus, $g'(3)g'(1) = \left(\frac{2}{3}\right)(-3) = -2$.

43. $h(x) = f(g(x))$ $\Rightarrow$ $h'(x) = f'(g(x))g'(x)$. So $h'(0.5) = f'(g(0.5))g'(0.5) = f'(0.1)g'(0.5)$. We can estimate the derivatives by taking the average of two secant slopes.

For $f'(0.1)$: $m_1 = \dfrac{14.8 - 12.6}{0.1 - 0} = 22$, $m_2 = \dfrac{18.4 - 14.8}{0.2 - 0.1} = 36$. So $f'(0.1) \approx \dfrac{m_1 + m_2}{2} = \dfrac{22 + 36}{2} = 29$.

For $g'(0.5)$: $m_1 = \dfrac{0.10 - 0.17}{0.5 - 0.4} = -0.7$, $m_2 = \dfrac{0.05 - 0.10}{0.6 - 0.5} = -0.5$. So $g'(0.5) \approx \dfrac{m_1 + m_2}{2} = -0.6$.

Hence, $h'(0.5) = f'(0.1)g'(0.5) \approx (29)(-0.6) = -17.4$.

45. (a) $f(x) = \sqrt{x}$ $\Rightarrow$ $f'(x) = 1/(2\sqrt{x})$, so f is not differentiable at $x = 0$. Since h is differentiable on $[0, \infty)$ and $\sqrt{x}$ is differentiable on $(0, \infty)$, it follows that $G(x) = h(\sqrt{x})$ is differentiable on $(0, \infty)$.

(b) By the Chain Rule, $G'(x) = h'(\sqrt{x})\dfrac{d}{dx}\sqrt{x} = h'(\sqrt{x}) \cdot \dfrac{1}{2\sqrt{x}} = \dfrac{h'(\sqrt{x})}{2\sqrt{x}}$.

47. (a) $F(x) = f(e^x)$ $\Rightarrow$ $F'(x) = f'(e^x)\dfrac{d}{dx}(e^x) = f'(e^x)e^x$

(b) $G(x) = e^{f(x)}$ $\Rightarrow$ $G'(x) = e^{f(x)}\dfrac{d}{dx}f(x) = e^{f(x)}f'(x)$

49. For the tangent line to be horizontal, $f'(x) = 0$. $f(x) = 2\sin x + \sin^2 x$ $\Rightarrow$ $f'(x) = 2\cos x + 2\sin x \cos x = 0$ $\Leftrightarrow$ $2\cos x (1 + \sin x) = 0$ $\Leftrightarrow$ $\cos x = 0$ or $\sin x = -1$, so $x = \frac{\pi}{2} + 2n\pi$ or $\frac{3\pi}{2} + 2n\pi$, where n is any integer. Now $f\left(\frac{\pi}{2}\right) = 3$ and $f\left(\frac{3\pi}{2}\right) = -1$, so the points on the curve with a horizontal tangent are $\left(\frac{\pi}{2} + 2n\pi, 3\right)$ and $\left(\frac{3\pi}{2} + 2n\pi, -1\right)$, where n is any integer.

51. $y = Ae^{-x} + Bxe^{-x}$ $\Rightarrow$
$y' = A(-e^{-x}) + B[x(-e^{-x}) + e^{-x} \cdot 1] = -Ae^{-x} + Be^{-x} - Bxe^{-x} = (B - A)e^{-x} - Bxe^{-x}$ $\Rightarrow$
$y'' = (B - A)(-e^{-x}) - B[x(-e^{-x}) + e^{-x} \cdot 1] = (A - B)e^{-x} - Be^{-x} + Bxe^{-x} = (A - 2B)e^{-x} + Bxe^{-x}$,
so $y'' + 2y' + y = (A - 2B)e^{-x} + Bxe^{-x} + 2[(B - A)e^{-x} - Bxe^{-x}] + Ae^{-x} + Bxe^{-x}$
$= [(A - 2B) + 2(B - A) + A]e^{-x} + [B - 2B + B]xe^{-x} = 0$.

53. The use of D, D^2, ..., D^n is just a derivative notation (see page 161). In general, $Df(2x) = 2f'(2x)$, $D^2f(2x) = 4f''(2x)$, ..., $D^nf(2x) = 2^nf^{(n)}(2x)$. Since $f(x) = \cos x$ and $50 = 4(12) + 2$, we have $f^{(50)}(x) = f^{(2)}(x) = -\cos x$, so $D^{50}\cos 2x = -2^{50}\cos 2x$.

55. $s(t) = 10 + \frac{1}{4}\sin(10\pi t)$ $\Rightarrow$ the velocity after t seconds is
$v(t) = s'(t) = \frac{1}{4}\cos(10\pi t)(10\pi) = \frac{5\pi}{2}\cos(10\pi t)$ cm/s.

57. (a) $B(t) = 4.0 + 0.35 \sin \dfrac{2\pi t}{5.4}$ $\Rightarrow$ $\dfrac{dB}{dt} = \left(0.35 \cos \dfrac{2\pi t}{5.4}\right)\left(\dfrac{2\pi}{5.4}\right) = \dfrac{0.7\pi}{5.4} \cos \dfrac{2\pi t}{5.4} = \dfrac{7\pi}{54} \cos \dfrac{2\pi t}{5.4}$

(b) At $t = 1$, $\dfrac{dB}{dt} = \dfrac{7\pi}{54} \cos \dfrac{2\pi}{5.4} \approx 0.16$.

59. $s(t) = 2e^{-1.5t} \sin 2\pi t$ $\Rightarrow$
$v(t) = s'(t) = 2\left[e^{-1.5t}(\cos 2\pi t)(2\pi) + (\sin 2\pi t)e^{-1.5t}(-1.5)\right] = 2e^{-1.5t}(2\pi \cos 2\pi t - 1.5 \sin 2\pi t)$

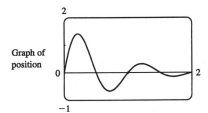

Graph of position

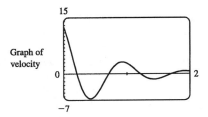

Graph of velocity

61. By the Chain Rule, $a(t) = \dfrac{dv}{dt} = \dfrac{dv}{ds}\dfrac{ds}{dt} = \dfrac{dv}{ds}\,v(t) = v(t)\,\dfrac{dv}{ds}$. The derivative dv/dt is the rate of change of the velocity with respect to time (in other words, the acceleration) whereas the derivative dv/ds is the rate of change of the velocity with respect to the displacement.

63. (a) Using a calculator or CAS, we obtain the model $Q = ab^t$ with $a = 100.0124369$ and $b = 0.000045145933$. We can change this model to one with base e and exponent $\ln b$: $Q = ae^{t \ln b} = 100.0124369 e^{-10.00553063t}$.

(b) $Q'(t) = ab^t \ln b$. $Q'(0.04) \approx -670.63\ \mu$A. The result of Example 2 in Section 2.1 was $-670\ \mu$A.

65. $x = t \sin t$, $y = t \cos t$. To find the value of t that corresponds to the point $(0, -\pi)$, we can solve $0 = t \sin t$, which says that either $t = 0$ or $\sin t = 0$. If $\sin t = 0$, then $t = 0, \pm \pi, \ldots$. Since $y = -\pi = t \cos t$, we see that $t = \pi$ is the desired value. $\dfrac{dy}{dt} = \cos t - t \sin t$, $\dfrac{dx}{dt} = \sin t + t \cos t$, and $\dfrac{dy}{dx} = \dfrac{dy/dt}{dx/dt} = \dfrac{\cos t - t \sin t}{\sin t + t \cos t}$. When $t = \pi$, $(x, y) = (0, -\pi)$ and $\dfrac{dy}{dx} = \dfrac{-1}{-\pi} = \dfrac{1}{\pi}$, so an equation of the tangent is $y + \pi = \dfrac{1}{\pi}(x - 0)$ or $y = \dfrac{1}{\pi}x - \pi$.

67. (a) $x = t^2$, $y = t^3 - 3t$ $\Rightarrow$ $\dfrac{dy}{dx} = \dfrac{dy/dt}{dx/dt} = \dfrac{3t^2 - 3}{2t}$. At the point $(3, 0)$,

$x = 3$ $\Rightarrow$ $t^2 = 3$ $\Rightarrow$ $t = \pm\sqrt{3}$ $\Rightarrow$ $\dfrac{dy}{dx} = \dfrac{3(\pm\sqrt{3})^2 - 3}{2(\pm\sqrt{3})} = \dfrac{6}{2(\pm\sqrt{3})} = \pm\dfrac{3}{\sqrt{3}} = \pm\sqrt{3}$. When $t = \sqrt{3}$, an equation of the tangent line is $y - 0 = \sqrt{3}(x - 3)$ or $y = \sqrt{3}\,x - 3\sqrt{3}$. When $t = -\sqrt{3}$, an equation of the tangent line is $y - 0 = -\sqrt{3}(x - 3)$ or $y = -\sqrt{3}\,x + 3\sqrt{3}$.

(b) Horizontal tangent:

$dy/dx = 0$ $\Leftrightarrow$ $3t^2 - 3 = 0$ $\Leftrightarrow$ $3(t^2 - 1) = 0$ $\Leftrightarrow$

$t^2 = 1$ $\Leftrightarrow$ $t = \pm 1$.

$t = 1$ corresponds to the point

$(x, y) = (t^2, t^3 - 3t) = (1^2, 1^3 - 3 \cdot 1) = (1, -2)$ and

$t = -1$ to $(1, 2)$. Vertical tangent: dy/dx is

undefined $\Leftrightarrow$ $2t = 0$ $\Leftrightarrow$ $t = 0$.

The value $t = 0$ corresponds to the origin; that is, $(0, 0)$.

(c)

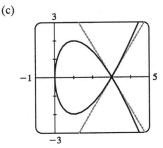

69. (a) Derive gives $g'(t) = \dfrac{45(t-2)^8}{(2t+1)^{10}}$ without simplifying. With either Maple or Mathematica, we first get

$g'(t) = 9\dfrac{(t-2)^8}{(2t+1)^9} - 18\dfrac{(t-2)^9}{(2t+1)^{10}}$, and the simplification command results in the above expression.

(b) Derive gives $y' = 2(x^3 - x + 1)^3(2x+1)^4(17x^3 + 6x^2 - 9x + 3)$ without simplifying.

With either Maple or Mathematica, we first get

$y' = 10(2x+1)^4(x^3 - x + 1)^4 + 4(2x+1)^5(x^3 - x + 1)^3(3x^2 - 1)$. If we use Mathematica's `Factor` or `Simplify`, or Maple's `factor`, we get the above expression, but Maple's `simplify` gives the polynomial expansion instead. For locating horizontal tangents, the factored form is the most helpful.

71. (a) $\dfrac{d}{dx}(\sin^n x \cos nx) = n\sin^{n-1} x \cos x \cos nx + \sin^n x(-n\sin nx)$ [Product Rule]

$= n\sin^{n-1} x(\cos nx \cos x - \sin nx \sin x)$ [factor out $n\sin^{n-1} x$]

$= n\sin^{n-1} x \cos(nx + x)$ [Addition Formula for cosine]

$= n\sin^{n-1} x \cos[(n+1)x]$ [factor out x]

(b) $\dfrac{d}{dx}(\cos^n x \cos nx) = n\cos^{n-1} x(-\sin x)\cos nx + \cos^n x(-n\sin nx)$ [Product Rule]

$= -n\cos^{n-1} x(\cos nx \sin x + \sin nx \cos x)$ [factor out $-n\cos^{n-1} x$]

$= -n\cos^{n-1} x \sin(nx + x)$ [Addition Formula for sine]

$= -n\cos^{n-1} x \sin[(n+1)x]$ [factor out x]

73. Since $\theta^\circ = \left(\dfrac{\pi}{180}\right)\theta$ rad, we have $\dfrac{d}{d\theta}(\sin\theta^\circ) = \dfrac{d}{d\theta}\left(\sin\dfrac{\pi}{180}\theta\right) = \dfrac{\pi}{180}\cos\dfrac{\pi}{180}\theta = \dfrac{\pi}{180}\cos\theta^\circ$.

75. $\dfrac{d^2y}{dx^2} = \dfrac{d}{dx}\left(\dfrac{dy}{dx}\right)$ [Leibniz notation for the second derivative]

$= \dfrac{d}{dx}\left(\dfrac{dy}{du}\dfrac{du}{dx}\right)$ [Chain Rule]

$= \dfrac{dy}{du} \cdot \dfrac{d}{dx}\left(\dfrac{du}{dx}\right) + \dfrac{du}{dx} \cdot \dfrac{d}{dx}\left(\dfrac{dy}{du}\right)$ [Product Rule]

$= \dfrac{dy}{du} \cdot \dfrac{d^2u}{dx^2} + \dfrac{du}{dx} \cdot \dfrac{d}{du}\left(\dfrac{dy}{du}\right) \cdot \dfrac{du}{dx}$ [dy/du is a function of u]

$= \dfrac{dy}{du}\dfrac{d^2u}{dx^2} + \dfrac{d^2y}{du^2}\left(\dfrac{du}{dx}\right)^2$

Or: Using function notation for $y = f(u)$ and $u = g(x)$, we have $y = f(g(x))$, so $y' = f'(g(x)) \cdot g'(x)$ [by the Chain Rule] $\Rightarrow$

$(y')' = [f'(g(x)) \cdot g'(x)]' = f'(g(x)) \cdot g''(x) + g'(x) \cdot f''(g(x)) \cdot g'(x) = f'(g(x)) \cdot g''(x) + f''(g(x)) \cdot [g'(x)]^2$.

◆3.6◆ Implicit Differentiation   •   •   •   •   •   •   •   •   •   •   •   •

1. (a) $\dfrac{d}{dx}(xy + 2x + 3x^2) = \dfrac{d}{dx}(4) \Rightarrow (x \cdot y' + y \cdot 1) + 2 + 6x = 0 \Rightarrow xy' = -y - 2 - 6x \Rightarrow$

$y' = \dfrac{-y - 2 - 6x}{x}$ or $y' = -6 - \dfrac{y+2}{x}$.

(b) $xy + 2x + 3x^2 = 4 \Rightarrow xy = 4 - 2x - 3x^2 \Rightarrow y = \dfrac{4 - 2x - 3x^2}{x} = \dfrac{4}{x} - 2 - 3x$, so $y' = -\dfrac{4}{x^2} - 3.$

(c) From part (a), $y' = \dfrac{-y - 2 - 6x}{x} = \dfrac{-(4/x - 2 - 3x) - 2 - 6x}{x} = \dfrac{-4/x - 3x}{x} = -\dfrac{4}{x^2} - 3.$

3. $\dfrac{d}{dx}\left(x^3 + x^2 y + 4y^2\right) = \dfrac{d}{dx}(6) \Rightarrow 3x^2 + \left(x^2 y' + y \cdot 2x\right) + 8yy' = 0 \Rightarrow x^2 y' + 8yy' = -3x^2 - 2xy$

$\Rightarrow (x^2 + 8y)y' = -3x^2 - 2xy \Rightarrow y' = -\dfrac{3x^2 + 2xy}{x^2 + 8y} = -\dfrac{x(3x + 2y)}{x^2 + 8y}$

5. $\dfrac{d}{dx}\left(x^2 y + xy^2\right) = \dfrac{d}{dx}(3x) \Rightarrow \left(x^2 y' + y \cdot 2x\right) + \left(x \cdot 2yy' + y^2 \cdot 1\right) = 3 \Rightarrow$

$x^2 y' + 2xyy' = 3 - 2xy - y^2 \Rightarrow y'\left(x^2 + 2xy\right) = 3 - 2xy - y^2 \Rightarrow y' = \dfrac{3 - 2xy - y^2}{x^2 + 2xy}$

7. $\sqrt{xy} = 1 + x^2 y \Rightarrow \tfrac{1}{2}(xy)^{-1/2}(xy' + y \cdot 1) = 0 + x^2 y' + y \cdot 2x \Rightarrow \dfrac{x}{2\sqrt{xy}}y' + \dfrac{y}{2\sqrt{xy}} = x^2 y' + 2xy$

$\Rightarrow y'\left(\dfrac{x}{2\sqrt{xy}} - x^2\right) = 2xy - \dfrac{y}{2\sqrt{xy}} \Rightarrow y'\left(\dfrac{x - 2x^2\sqrt{xy}}{2\sqrt{xy}}\right) = \dfrac{4xy\sqrt{xy} - y}{2\sqrt{xy}} \Rightarrow y' = \dfrac{4xy\sqrt{xy} - y}{x - 2x^2\sqrt{xy}}$

9. $\dfrac{d}{dx}(4\cos x \sin y) = \dfrac{d}{dx}(1) \Rightarrow 4[\cos x \cdot \cos y \cdot y' + \sin y \cdot (-\sin x)] = 0 \Rightarrow$

$y'(4\cos x \cos y) = 4\sin x \sin y \Rightarrow y' = \dfrac{4\sin x \sin y}{4\cos x \cos y} = \tan x \tan y$

11. $\cos(x - y) = xe^x \Rightarrow -\sin(x - y)(1 - y') = xe^x + e^x \Rightarrow 1 - y' = -\dfrac{(x + 1)e^x}{\sin(x - y)} \Rightarrow$

$y' = 1 + \dfrac{(x + 1)e^x}{\sin(x - y)}$

13. $\dfrac{d}{dx}\left(\dfrac{x^2}{16} - \dfrac{y^2}{9}\right) = \dfrac{d}{dx}(1) \Rightarrow \dfrac{x}{8} - \dfrac{2yy'}{9} = 0 \Rightarrow y' = \dfrac{9x}{16y}$. When $x = -5$ and $y = \tfrac{9}{4}$ we have

$y' = \dfrac{9(-5)}{16(9/4)} = -\dfrac{5}{4}$, so an equation of the tangent line is $y - \tfrac{9}{4} = -\tfrac{5}{4}(x + 5)$ or $y = -\tfrac{5}{4}x - 4.$

15. $y^2 = x^3(2 - x) = 2x^3 - x^4 \Rightarrow 2yy' = 6x^2 - 4x^3 \Rightarrow y' = \dfrac{3x^2 - 2x^3}{y}$. When $x = y = 1,$

$y' = \dfrac{3(1)^2 - 2(1)^3}{1} = 1$, so an equation of the tangent line is $y - 1 = 1(x - 1)$ or $y = x.$

17. $2\left(x^2 + y^2\right)^2 = 25\left(x^2 - y^2\right) \Rightarrow 4\left(x^2 + y^2\right)(2x + 2yy') = 25(2x - 2yy') \Rightarrow$

$4(x + yy')\left(x^2 + y^2\right) = 25(x - yy') \Rightarrow 4yy'\left(x^2 + y^2\right) + 25yy' = 25x - 4x\left(x^2 + y^2\right) \Rightarrow$

$y' = \dfrac{25x - 4x\left(x^2 + y^2\right)}{25y + 4y\left(x^2 + y^2\right)}$. When $x = 3$ and $y = 1$, $y' = \dfrac{75 - 120}{25 + 40} = -\dfrac{45}{65} = -\dfrac{9}{13}$, so an equation of the tangent

line is $y - 1 = -\tfrac{9}{13}(x - 3)$ or $y = -\tfrac{9}{13}x + \tfrac{40}{13}.$

19. (a) $y^2 = 5x^4 - x^2 \Rightarrow 2yy' = 5\left(4x^3\right) - 2x \Rightarrow y' = \dfrac{10x^3 - x}{y}.$ (b)

So at the point $(1, 2)$ we have $y' = \dfrac{10(1)^3 - 1}{2} = \dfrac{9}{2}$, and an

equation of the tangent line is $y - 2 = \tfrac{9}{2}(x - 1)$ or $y = \tfrac{9}{2}x - \tfrac{5}{2}.$

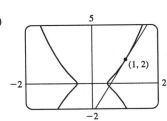

21. (a)

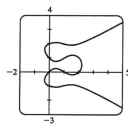

There are eight points with horizontal tangents:
four at $x \approx 1.57735$ and four at $x \approx 0.42265$.

(b) $y' = \dfrac{3x^2 - 6x + 2}{2(2y^3 - 3y^2 - y + 1)} \quad \Rightarrow \quad y' = -1$ at

$(0, 1)$ and $y' = \frac{1}{3}$ at $(0, 2)$.

Equations of the tangent lines are $y = -x + 1$

and $y = \frac{1}{3}x + 2$.

(c) $y' = 0 \quad \Rightarrow \quad 3x^2 - 6x + 2 = 0 \quad \Rightarrow$

$x = 1 \pm \frac{1}{3}\sqrt{3}$

(d) By multiplying the right side of the equation by $x - 3$, we
obtain the first graph.

By modifying the equation in other ways, we can generate
the other graphs.

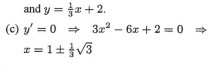

$y\left(y^2 - 1\right)(y - 2)$
$= x(x - 1)(x - 2)(x - 3)$

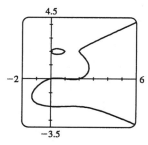

$y\left(y^2 - 4\right)(y - 2)$
$= x(x - 1)(x - 2)$

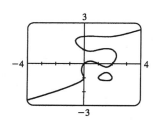

$y(y + 1)\left(y^2 - 1\right)(y - 2)$
$= x(x - 1)(x - 2)$

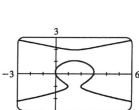

$(y + 1)\left(y^2 - 1\right)(y - 2)$
$= (x - 1)(x - 2)$

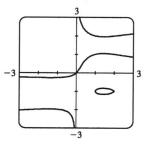

$x(y + 1)\left(y^2 - 1\right)(y - 2)$
$= y(x - 1)(x - 2)$

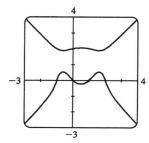

$y\left(y^2 + 1\right)(y - 2)$
$= x\left(x^2 - 1\right)(x - 2)$

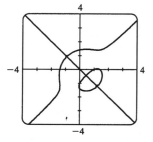

$y(y + 1)\left(y^2 - 2\right)$
$= x(x - 1)\left(x^2 - 2\right)$

23. From Exercise 17, a tangent to the lemniscate will be horizontal if $y' = 0 \Rightarrow 25x - 4x(x^2 + y^2) = 0 \Rightarrow$
$x[25 - 4(x^2 + y^2)] = 0 \Rightarrow x^2 + y^2 = \frac{25}{4}$ **(1)**. (Note that when x is 0, y is also 0, and there is no horizontal
tangent at the origin.) Substituting $\frac{25}{4}$ for $x^2 + y^2$ in the equation of the lemniscate, $2(x^2 + y^2)^2 = 25(x^2 - y^2)$,
we get $x^2 - y^2 = \frac{25}{8}$ **(2)**. Solving **(1)** and **(2)**, we have $x^2 = \frac{75}{16}$ and $y^2 = \frac{25}{16}$, so the four points are $\left(\pm\frac{5\sqrt{3}}{4}, \pm\frac{5}{4}\right)$.

25. (a) $x^4 + y^4 = 16 \Rightarrow 4x^3 + 4y^3 y' = 0 \Rightarrow y^3 y' = -x^3 \Rightarrow y' = -x^3/y^3$

(b) $y'' = -\dfrac{y^3(3x^2) - (x^3)(3y^2 y')}{(y^3)^2} = -\dfrac{3x^2 y^3 - 3x^3 y^2(-x^3/y^3)}{y^6} \cdot \dfrac{y}{y} = -\dfrac{3x^2 y^4 + 3x^6}{y^7}$

(c) $y'' = -\dfrac{3x^2(y^4 + x^4)}{y^7} = -\dfrac{3x^2(16)}{y^7} = -48\dfrac{x^2}{y^7}$

27. $y = \sin^{-1}(x^2) \Rightarrow y' = \dfrac{1}{\sqrt{1 - (x^2)^2}} \dfrac{d}{dx}(x^2) = \dfrac{2x}{\sqrt{1 - x^4}}$

29. $y = 2\sqrt{x}\tan^{-1}\sqrt{x} \Rightarrow$

$y' = 2\sqrt{x} \cdot \dfrac{d}{dx}\left(\tan^{-1}\sqrt{x}\right) + \tan^{-1}\sqrt{x} \cdot \dfrac{d}{dx}\left(2\sqrt{x}\right)$

$= 2\sqrt{x} \cdot \dfrac{1}{1 + (\sqrt{x})^2} \cdot \dfrac{1}{2\sqrt{x}} + \tan^{-1}\sqrt{x} \cdot \dfrac{2}{2\sqrt{x}} = \dfrac{1}{1 + x} + \dfrac{\tan^{-1}\sqrt{x}}{\sqrt{x}}$

31. $H(x) = (1 + x^2)\arctan x \Rightarrow H'(x) = (1 + x^2)\dfrac{1}{1 + x^2} + (\arctan x)(2x) = 1 + 2x\arctan x$

33. $y = \arcsin(\tan\theta) \Rightarrow y' = \dfrac{1}{\sqrt{1 - (\tan\theta)^2}} \cdot \dfrac{d}{d\theta}(\tan\theta) = \dfrac{\sec^2\theta}{\sqrt{1 - \tan^2\theta}}$

35. $f(x) = e^x - x^2\arctan x \Rightarrow$

$f'(x) = e^x - \left[x^2\left(\dfrac{1}{1 + x^2}\right) + (\arctan x)(2x)\right]$

$= e^x - \dfrac{x^2}{1 + x^2} - 2x\arctan x$

This is reasonable because the graphs show that f is increasing when f' is
positive, and f' is zero when f has a minimum.

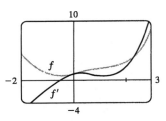

37. $2x^2 + y^2 = 3$ and $x = y^2$ intersect when $2x^2 + x - 3 = 0 \Leftrightarrow (2x + 3)(x - 1) = 0 \Leftrightarrow x = -\frac{3}{2}$ or 1, but
$-\frac{3}{2}$ is extraneous since $x = y^2$ is nonnegative. When $x = 1$, $1 = y^2 \Rightarrow y = \pm 1$, so there are two points of
intersection: $(1, \pm 1)$. $2x^2 + y^2 = 3 \Rightarrow 4x + 2yy' = 0 \Rightarrow y' = -2x/y$, and $x = y^2 \Rightarrow 1 = 2yy' \Rightarrow$
$y' = 1/(2y)$. At $(1, 1)$, the slopes are $m_1 = -2(1)/1 = -2$ and $m_2 = 1/(2 \cdot 1) = \frac{1}{2}$, so the curves are orthogonal
(since m_1 and m_2 are negative reciprocals of each other). By symmetry, the curves are also orthogonal at $(1, -1)$.

39.

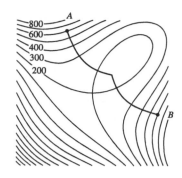

41. $x^2 + y^2 = r^2$ is a circle with center O and $ax + by = 0$ is a line through O.

$x^2 + y^2 = r^2 \;\Rightarrow\; 2x + 2yy' = 0 \;\Rightarrow\; y' = -x/y$, so the slope of the tangent line at $P_0\,(x_0, y_0)$ is $-x_0/y_0$. The slope of the line OP_0 is y_0/x_0, which is the negative reciprocal of $-x_0/y_0$. Hence, the curves are orthogonal, and the families of curves are orthogonal trajectories of each other.

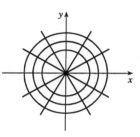

43. $y = cx^2 \;\Rightarrow\; y' = 2cx$ and $x^2 + 2y^2 = k \;\Rightarrow\; 2x + 4yy' = 0 \;\Rightarrow$

$2yy' = -x \;\Rightarrow\; y' = -\dfrac{x}{2(y)} = -\dfrac{x}{2(cx^2)} = -\dfrac{1}{2cx}$, so the curves are orthogonal.

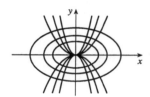

45. If the circle has radius r, its equation is $x^2 + y^2 = r^2 \;\Rightarrow\; 2x + 2yy' = 0 \;\Rightarrow\; y' = -\dfrac{x}{y}$, so the slope of the

tangent line at $P(x_0, y_0)$ is $-\dfrac{x_0}{y_0}$. The negative reciprocal of that slope is $\dfrac{-1}{-x_0/y_0} = \dfrac{y_0}{x_0}$, which is the slope of OP, so the tangent line at P is perpendicular to the radius OP.

47. To find the points at which the ellipse $x^2 - xy + y^2 = 3$ crosses the x-axis, let $y = 0$ and solve for x.

$y = 0 \;\Rightarrow\; x^2 - x(0) + 0^2 = 3 \;\Leftrightarrow\; x = \pm\sqrt{3}$. So the graph of the ellipse crosses the x-axis at the points $(\pm\sqrt{3}, 0)$. Using implicit differentiation to find y', we get $2x - xy' - y + 2yy' = 0 \;\Rightarrow\; y'(2y - x) = y - 2x$

$\Leftrightarrow\; y' = \dfrac{y - 2x}{2y - x}$. So y' at $(\sqrt{3}, 0)$ is $\dfrac{0 - 2\sqrt{3}}{2(0) - \sqrt{3}} = 2$ and y' at $(-\sqrt{3}, 0)$ is $\dfrac{0 + 2\sqrt{3}}{2(0) + \sqrt{3}} = 2$. Thus, the tangent lines at these points are parallel.

49. $x^2 y^2 + xy = 2 \;\Rightarrow\; x^2 \cdot 2yy' + y^2 \cdot 2x + x \cdot y' + y \cdot 1 = 0 \;\Leftrightarrow\; y'(2x^2 y + x) = -2xy^2 - y \;\Leftrightarrow$

$y' = -\dfrac{2xy^2 + y}{2x^2 y + x}$. So $-\dfrac{2xy^2 + y}{2x^2 y + x} = -1 \;\Leftrightarrow\; 2xy^2 + y = 2x^2 y + x \;\Leftrightarrow\; y(2xy + 1) = x(2xy + 1) \;\Leftrightarrow$

$y(2xy + 1) - x(2xy + 1) = 0 \;\Leftrightarrow\; (2xy + 1)(y - x) = 0 \;\Leftrightarrow\; xy = -\tfrac{1}{2}$ or $y = x$. But $xy = -\tfrac{1}{2} \;\Rightarrow$

$x^2 y^2 + xy = \tfrac{1}{4} - \tfrac{1}{2} \neq 2$, so we must have $x = y$. Then $x^2 y^2 + xy = 2 \;\Rightarrow\; x^4 + x^2 = 2 \;\Leftrightarrow$

$x^4 + x^2 - 2 = 0 \;\Leftrightarrow\; (x^2 + 2)(x^2 - 1) = 0$. So $x^2 = -2$, which is impossible, or $x^2 = 1 \;\Leftrightarrow\; x = \pm 1$.

Since $x = y$, the points on the curve where the tangent line has a slope of -1 are $(-1, -1)$ and $(1, 1)$.

51. (a) If $y = f^{-1}(x)$, then $f(y) = x$. Differentiating implicitly with respect to x and remembering that y is a function

of x, we get $f'(y)\dfrac{dy}{dx} = 1$, so $\dfrac{dy}{dx} = \dfrac{1}{f'(y)} \quad \Rightarrow \quad (f^{-1})'(x) = \dfrac{1}{f'(f^{-1}(x))}$.

(b) $f(4) = 5 \quad \Rightarrow \quad f^{-1}(5) = 4$. By part (a), $(f^{-1})'(5) = 1/f'(f^{-1}(5)) = 1/f'(4) = 1/\left(\frac{2}{3}\right) = \frac{3}{2}$.

53. (a) $y = J(x)$ and $xy'' + y' + xy = 0 \quad \Rightarrow \quad xJ''(x) + J'(x) + xJ(x) = 0$. If $x = 0$, we have $0 + J'(0) + 0 = 0$,

so $J'(0) = 0$.

(b) Differentiating $xy'' + y' + xy = 0$ implicitly, we get $xy''' + y'' \cdot 1 + y'' + xy' + y \cdot 1 = 0 \quad \Rightarrow$

$xy''' + 2y'' + xy' + y = 0$, so $xJ'''(x) + 2J''(x) + xJ'(x) + J(x) = 0$. If $x = 0$, we have

$0 + 2J''(0) + 0 + 1 \quad [J(0) = 1 \text{ is given}] \; = 0 \quad \Rightarrow \quad 2J''(0) = -1 \quad \Rightarrow \quad J''(0) = -\frac{1}{2}$.

3.7 Derivatives of Logarithmic Functions $\bullet$ $\bullet$ $\bullet$ $\bullet$ $\bullet$ $\bullet$ $\bullet$ $\bullet$

1. The differentiation formula for logarithmic functions, $\dfrac{d}{dx}(\log_a x) = \dfrac{1}{x \ln a}$, is simplest when $a = e$ because

$\ln e = 1$.

3. $f(\theta) = \ln(\cos\theta) \quad \Rightarrow \quad f'(\theta) = \dfrac{1}{\cos\theta}\dfrac{d}{d\theta}(\cos\theta) = \dfrac{-\sin\theta}{\cos\theta} = -\tan\theta$

5. $f(x) = \log_2(1 - 3x) \quad \Rightarrow \quad f'(x) = \dfrac{1}{(1 - 3x)\ln 2}\dfrac{d}{dx}(1 - 3x) = \dfrac{-3}{(1 - 3x)\ln 2}$ or $\dfrac{3}{(3x - 1)\ln 2}$

7. $f(x) = \sqrt[5]{\ln x} = (\ln x)^{1/5} \quad \Rightarrow \quad f'(x) = \frac{1}{5}(\ln x)^{-4/5}\dfrac{d}{dx}(\ln x) = \dfrac{1}{5(\ln x)^{4/5}}\cdot\dfrac{1}{x} = \dfrac{1}{5x\sqrt[5]{(\ln x)^4}}$

9. $f(x) = \sqrt{x}\ln x \quad \Rightarrow \quad f'(x) = \sqrt{x}\left(\dfrac{1}{x}\right) + (\ln x)\cdot\dfrac{1}{2\sqrt{x}} = \dfrac{1}{\sqrt{x}} + \dfrac{\ln x}{2\sqrt{x}} = \dfrac{2 + \ln x}{2\sqrt{x}}$

11. $F(t) = \ln\dfrac{(2t + 1)^3}{(3t - 1)^4} = \ln(2t + 1)^3 - \ln(3t - 1)^4 = 3\ln(2t + 1) - 4\ln(3t - 1) \quad \Rightarrow$

$F'(t) = 3\cdot\dfrac{1}{2t + 1}\cdot 2 - 4\cdot\dfrac{1}{3t - 1}\cdot 3 = \dfrac{6}{2t + 1} - \dfrac{12}{3t - 1}$, or combined, $\dfrac{-6(t + 3)}{(2t + 1)(3t - 1)}$.

13. $y = \dfrac{\ln x}{1 + x} \quad \Rightarrow \quad y' = \dfrac{(1 + x)(1/x) - (\ln x)(1)}{(1 + x)^2} = \dfrac{\dfrac{1 + x}{x} - \dfrac{x\ln x}{x}}{(1 + x)^2} = \dfrac{1 + x - x\ln x}{x(1 + x)^2}$

15. $y = \ln|x^3 - x^2| \quad \Rightarrow \quad y' = \dfrac{1}{x^3 - x^2}(3x^2 - 2x) = \dfrac{x(3x - 2)}{x^2(x - 1)} = \dfrac{3x - 2}{x(x - 1)}$

17. $y = \ln(e^{-x} + xe^{-x}) = \ln(e^{-x}(1 + x)) = \ln(e^{-x}) + \ln(1 + x) = -x + \ln(1 + x) \quad \Rightarrow$

$y' = -1 + \dfrac{1}{1 + x} = \dfrac{-1 - x + 1}{1 + x} = -\dfrac{x}{1 + x}$

19. $y = e^x\ln x \quad \Rightarrow \quad y' = e^x\cdot\dfrac{1}{x} + (\ln x)\cdot e^x = e^x\left(\dfrac{1}{x} + \ln x\right) \quad \Rightarrow$

$y'' = e^x\left(-\dfrac{1}{x^2} + \dfrac{1}{x}\right) + \left(\dfrac{1}{x} + \ln x\right)e^x = e^x\left(-\dfrac{1}{x^2} + \dfrac{1}{x} + \dfrac{1}{x} + \ln x\right) = e^x\left(\ln x + \dfrac{2}{x} - \dfrac{1}{x^2}\right)$

21. $f(x) = \dfrac{x}{1 - \ln(x-1)}$ $\Rightarrow$

$$f'(x) = \frac{[1 - \ln(x-1)] \cdot 1 - x \cdot \dfrac{-1}{x-1}}{[1 - \ln(x-1)]^2} = \frac{(x-1)[1 - \ln(x-1)] + x}{x-1}}{[1 - \ln(x-1)]^2} = \frac{x - 1 - (x-1)\ln(x-1) + x}{(x-1)[1 - \ln(x-1)]^2}$$

$$= \frac{2x - 1 - (x-1)\ln(x-1)}{(x-1)[1 - \ln(x-1)]^2}$$

$\text{Dom}(f) = \{x \mid x - 1 > 0 \quad \text{and} \quad 1 - \ln(x-1) \neq 0\} = \{x \mid x > 1 \quad \text{and} \quad \ln(x-1) \neq 1\}$

$= \{x \mid x > 1 \quad \text{and} \quad x - 1 \neq e^1\} = \{x \mid x > 1 \quad \text{and} \quad x \neq 1 + e\} = (1, 1+e) \cup (1+e, \infty)$

23. $y = \ln(x^2 - 3)$ $\Rightarrow$ $y' = \dfrac{1}{x^2 - 3} \cdot 2x = \dfrac{2x}{x^2 - 3}$.

$y'(2) = \dfrac{2(2)}{2^2 - 3} = 4$, so an equation of the tangent line at $(2, 0)$ is $y - 0 = 4(x - 2)$ or $y = 4x - 8$.

25. (a) The domain of $f(x) = x \ln x$ is $(0, \infty)$. $f'(x) = x(1/x) + (\ln x) \cdot 1 = 1 + \ln x$. So $f'(x) < 0$ when
$1 + \ln x < 0$ $\Leftrightarrow$ $\ln x < -1$ $\Leftrightarrow$ $x < e^{-1}$. Therefore, f is decreasing on $(0, 1/e)$.

(b) $f'(x) = 1 + \ln x$ $\Rightarrow$ $f''(x) = 1/x > 0$ for $x > 0$. So the curve is concave upward on $(0, \infty)$.

27. $y = (2x+1)^5 (x^4 - 3)^6$ $\Rightarrow$ $\ln y = \ln\left((2x+1)^5 (x^4 - 3)^6\right)$ $\Rightarrow$

$\ln y = 5 \ln(2x+1) + 6 \ln(x^4 - 3)$ $\Rightarrow$ $\dfrac{1}{y} y' = 5 \cdot \dfrac{1}{2x+1} \cdot 2 + 6 \cdot \dfrac{1}{x^4 - 3} \cdot 4x^3$ $\Rightarrow$

$y' = y\left(\dfrac{10}{2x+1} + \dfrac{24x^3}{x^4 - 3}\right) = (2x+1)^5 (x^4 - 3)^6 \left(\dfrac{10}{2x+1} + \dfrac{24x^3}{x^4 - 3}\right)$. [The answer could be simplified to

$y' = 2(2x+1)^4 (x^4 - 3)^5 (29x^4 + 12x^3 - 15)$, but this is unnecessary.]

29. $y = \dfrac{\sin^2 x \tan^4 x}{(x^2 + 1)^2}$ $\Rightarrow$ $\ln y = \ln(\sin^2 x \tan^4 x) - \ln(x^2 + 1)^2$ $\Rightarrow$

$\ln y = \ln(\sin x)^2 + \ln(\tan x)^4 - \ln(x^2 + 1)^2$ $\Rightarrow$ $\ln y = 2 \ln \sin x + 4 \ln \tan x - 2 \ln(x^2 + 1)$ $\Rightarrow$

$\dfrac{1}{y} y' = 2 \cdot \dfrac{1}{\sin x} \cdot \cos x + 4 \cdot \dfrac{1}{\tan x} \cdot \sec^2 x - 2 \cdot \dfrac{1}{x^2 + 1} \cdot 2x$ $\Rightarrow$

$y' = \dfrac{\sin^2 x \tan^4 x}{(x^2 + 1)^2}\left(2 \cot x + \dfrac{4 \sec^2 x}{\tan x} - \dfrac{4x}{x^2 + 1}\right)$

31. $y = x^x$ $\Rightarrow$ $\ln y = \ln x^x$ $\Rightarrow$ $\ln y = x \ln x$ $\Rightarrow$ $y'/y = x(1/x) + (\ln x) \cdot 1$ $\Rightarrow$
$y' = y(1 + \ln x)$ $\Rightarrow$ $y' = x^x(1 + \ln x)$

33. $y = x^{\sin x}$ $\Rightarrow$ $\ln y = \ln x^{\sin x}$ $\Rightarrow$ $\ln y = \sin x \ln x$ $\Rightarrow$ $\dfrac{y'}{y} = (\sin x) \cdot \dfrac{1}{x} + (\ln x)(\cos x)$ $\Rightarrow$

$y' = y\left(\dfrac{\sin x}{x} + \ln x \cos x\right)$ $\Rightarrow$ $y' = x^{\sin x}\left(\dfrac{\sin x}{x} + \ln x \cos x\right)$

35. $y = (\ln x)^x$ $\Rightarrow$ $\ln y = \ln(\ln x)^x$ $\Rightarrow$ $\ln y = x \ln \ln x$ $\Rightarrow$ $\dfrac{y'}{y} = x \cdot \dfrac{1}{\ln x} \cdot \dfrac{1}{x} + (\ln \ln x) \cdot 1$ $\Rightarrow$

$y' = y\left(\dfrac{x}{x \ln x} + \ln \ln x\right)$ $\Rightarrow$ $y' = (\ln x)^x \left(\dfrac{1}{\ln x} + \ln \ln x\right)$

37. $y = \ln(x^2 + y^2)$ $\Rightarrow$ $y' = \dfrac{1}{x^2 + y^2} \dfrac{d}{dx}(x^2 + y^2)$ $\Rightarrow$ $y' = \dfrac{2x + 2yy'}{x^2 + y^2}$ $\Rightarrow$ $x^2 y' + y^2 y' = 2x + 2yy'$

$\Rightarrow$ $x^2 y' + y^2 y' - 2yy' = 2x$ $\Rightarrow$ $(x^2 + y^2 - 2y)y' = 2x$ $\Rightarrow$ $y' = \dfrac{2x}{x^2 + y^2 - 2y}$

39. $f(x) = \ln(x-1) \Rightarrow f'(x) = 1/(x-1) = (x-1)^{-1} \Rightarrow f''(x) = -(x-1)^{-2} \Rightarrow$

$f'''(x) = 2(x-1)^{-3} \Rightarrow f^{(4)}(x) = -2 \cdot 3(x-1)^{-4} \Rightarrow \cdots \Rightarrow$

$f^{(n)}(x) = (-1)^{n-1} \cdot 2 \cdot 3 \cdot 4 \cdot \cdots \cdot (n-1)(x-1)^{-n} = (-1)^{n-1} \dfrac{(n-1)!}{(x-1)^n}$

41. If $f(x) = \ln(1+x)$, then $f'(x) = \dfrac{1}{1+x}$, so $f'(0) = 1$.

Thus, $\lim\limits_{x \to 0} \dfrac{\ln(1+x)}{x} = \lim\limits_{x \to 0} \dfrac{f(x)}{x} = \lim\limits_{x \to 0} \dfrac{f(x) - f(0)}{x-0} = f'(0) = 1$.

◆3.8 Linear Approximations and Differentials · · · · · · · ·

1. $f(x) = x^3 \Rightarrow f'(x) = 3x^2$, so $f(1) = 1$ and $f'(1) = 3$. With $a = 1$, $L(x) = f(a) + f'(a)(x-a)$
becomes $L(x) = f(1) + f'(1)(x-1) = 1 + 3(x-1) = 3x - 2$.

3. $f(x) = \cos x \Rightarrow f'(x) = -\sin x$, so $f\left(\frac{\pi}{2}\right) = 0$ and $f'\left(\frac{\pi}{2}\right) = -1$. Thus,
$L(x) = f\left(\frac{\pi}{2}\right) + f'\left(\frac{\pi}{2}\right)\left(x - \frac{\pi}{2}\right) = 0 - 1\left(x - \frac{\pi}{2}\right) = -x + \frac{\pi}{2}$.

5. $f(x) = \sqrt{1-x} \Rightarrow f'(x) = \dfrac{-1}{2\sqrt{1-x}}$, so $f(0) = 1$ and

$f'(0) = -\frac{1}{2}$. Therefore,

$$\sqrt{1-x} = f(x) \approx f(0) + f'(0)(x-0)$$

$$= 1 + \left(-\frac{1}{2}\right)(x-0) = 1 - \frac{1}{2}x$$

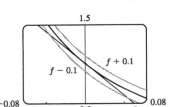

So $\sqrt{0.9} = \sqrt{1-0.1} \approx 1 - \frac{1}{2}(0.1) = 0.95$ and

$\sqrt{0.99} = \sqrt{1-0.01} \approx 1 - \frac{1}{2}(0.01) = 0.995$.

7. $f(x) = \sqrt{1+x} \Rightarrow f'(x) = \dfrac{1}{2\sqrt{1+x}}$, so $f(0) = 1$ and $f'(0) = \frac{1}{2}$.

Thus, $f(x) \approx f(0) + f'(0)(x-0) = 1 + \frac{1}{2}(x-0) = 1 + \frac{1}{2}x$.

We need $\sqrt{1+x} - 0.1 < 1 + \frac{1}{2}x < \sqrt{1+x} + 0.1$. By zooming in or

using an intersect feature, we see that this is true when $-0.69 < x < 1.09$.

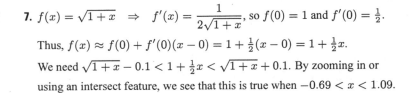

9. $f(x) = \dfrac{1}{(1+2x)^4} = (1+2x)^{-4} \Rightarrow$

$f'(x) = -4(1+2x)^{-5}(2) = \dfrac{-8}{(1+2x)^5}$, so $f(0) = 1$ and $f'(0) = -8$.

Thus, $f(x) \approx f(0) + f'(0)(x-0) = 1 + (-8)(x-0) = 1 - 8x$.

We need $1/(1+2x)^4 - 0.1 < 1 - 8x < 1/(1+2x)^4 + 0.1$, which is true

when $-0.045 < x < 0.055$.

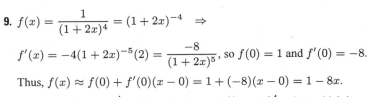

11. $y = f(x) = \sec x \Rightarrow f'(x) = \sec x \tan x$, so $f(0) = 1$ and $f'(0) = 1 \cdot 0 = 0$. The linear approximation of f
at 0 is $f(0) + f'(0)(x-0) = 1 + 0(x) = 1$. Since 0.08 is close to 0, approximating $\sec 0.08$ with 1 is reasonable.

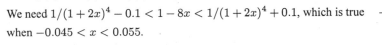

13. $y = f(x) = \ln x \implies f'(x) = 1/x$, so $f(1) = 0$ and $f'(1) = 1$. The linear approximation of f at 1 is
$f(1) + f'(1)(x - 1) = 0 + 1(x - 1) = x - 1$. Now $f(1.05) = \ln 1.05 \approx 1.05 - 1 = 0.05$, so the approximation
is reasonable.

15. (a) $f(x) = \sin x \implies f'(x) = \cos x$, so $f(0) = 0$ and $f'(0) = 1$.
Thus, $f(x) \approx f(0) + f'(0)(x - 0) = 0 + 1(x - 0) = x$.

(b)

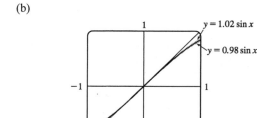

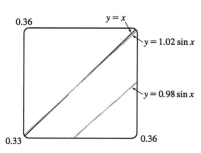

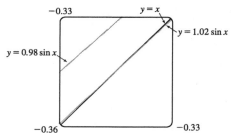

We want to know the values of x for which $y = x$ approximates $y = \sin x$ with less than a 2% difference; that
is, the values of x for which

$$\left| \frac{x - \sin x}{\sin x} \right| < 0.02 \quad \Leftrightarrow \quad -0.02 < \frac{x - \sin x}{\sin x} < 0.02 \quad \Leftrightarrow$$

$$\begin{cases} -0.02 \sin x < x - \sin x < 0.02 \sin x & \text{if } \sin x > 0 \\ -0.02 \sin x > x - \sin x > 0.02 \sin x & \text{if } \sin x < 0 \end{cases} \Leftrightarrow \begin{cases} 0.98 \sin x < x < 1.02 \sin x & \text{if } \sin x > 0 \\ 1.02 \sin x < x < 0.98 \sin x & \text{if } \sin x < 0 \end{cases}$$

In the first figure, we see that the graphs are very close to each other near $x = 0$. Changing the viewing rectangle
and using an intersect feature (see the second figure) we find that $y = x$ intersects $y = 1.02 \sin x$ at $x \approx 0.344$.
By symmetry, they also intersect at $x \approx -0.344$ (see the third figure.). Converting 0.344 radians to degrees, we
get $0.344 \left(\frac{180°}{\pi} \right) \approx 19.7° \approx 20°$, which verifies the statement.

17. (a) $y = e^{x/10} \implies dy = e^{x/10} \cdot \frac{1}{10} dx = \frac{1}{10} e^{x/10} dx$

(b) $x = 0$ and $dx = 0.1 \implies dy = \frac{1}{10} e^{0/10}(0.1) = 0.01$.
$\Delta y = f(x + \Delta x) - f(x) = e^{(x + \Delta x)/10} - e^x = e^{(0 + 0.1)/10} - e^0 = e^{1/100} - 1 \approx 0.0101$

19. (a) If x is the edge length, then $V = x^3 \Rightarrow dV = 3x^2\,dx$. When $x = 30$ and $dx = 0.1$,

$dV = 3(30)^2(0.1) = 270$, so the maximum possible error in computing the volume of the cube is about

270 cm^3. The relative error is calculated by dividing the change in V, ΔV, by V. We approximate ΔV with

dV.

Relative error $= \dfrac{\Delta V}{V} \approx \dfrac{dV}{V} = \dfrac{3x^2\,dx}{x^3} = 3\dfrac{dx}{x} = 3\left(\dfrac{0.1}{30}\right) = 0.01.$

Percentage error = relative error $\times\ 100\% = 0.01 \times 100\% = 1\%.$

(b) $S = 6x^2 \Rightarrow dS = 12x\,dx$. When $x = 30$ and $dx = 0.1$, $dS = 12(30)(0.1) = 36$, so the maximum

possible error in computing the surface area of the cube is about 36 cm^2.

Relative error $= \dfrac{\Delta S}{S} \approx \dfrac{dS}{S} = \dfrac{12x\,dx}{6x^2} = 2\dfrac{dx}{x} = 2\left(\dfrac{0.1}{30}\right) = 0.00\overline{6}.$

Percentage error = relative error $\times\ 100\% = 0.00\overline{6} \times 100\% = 0.\overline{6}\%.$

21. For a hemispherical dome, $V = \frac{2}{3}\pi r^3 \Rightarrow dV = 2\pi r^2\,dr$. When $r = \frac{1}{2}(50) = 25$ m and

$dr = 0.05$ cm $= 0.0005$ m, $dV = 2\pi(25)^2(0.0005) = \frac{5\pi}{8}$, so the amount of paint needed is about $\frac{5\pi}{8} \approx 2\text{ m}^3$.

Review

• CONCEPT CHECK •

1. (a) The Power Rule: If n is any real number, then $\dfrac{d}{dx}(x^n) = nx^{n-1}$. The derivative of a variable base raised to a

constant power is the power times the base raised to the power minus one.

(b) The Constant Multiple Rule: If c is a constant and f is a differentiable function, then $\dfrac{d}{dx}[cf(x)] = c\dfrac{d}{dx}f(x)$.

The derivative of a constant times a function is the constant times the derivative of the function.

(c) The Sum Rule: If f and g are both differentiable, then $\dfrac{d}{dx}[f(x) + g(x)] = \dfrac{d}{dx}f(x) + \dfrac{d}{dx}g(x)$. The derivative

of a sum of functions is the sum of the derivatives.

(d) The Difference Rule: If f and g are both differentiable, then $\dfrac{d}{dx}[f(x) - g(x)] = \dfrac{d}{dx}f(x) - \dfrac{d}{dx}g(x)$. The

derivative of a difference of functions is the difference of the derivatives.

(e) The Product Rule: If f and g are both differentiable, then $\dfrac{d}{dx}[f(x)g(x)] = f(x)\dfrac{d}{dx}g(x) + g(x)\dfrac{d}{dx}f(x)$.

The derivative of a product of two functions is the first function times the derivative of the second function plus

the second function times the derivative of the first function.

(f) The Quotient Rule: If f and g are both differentiable, then $\dfrac{d}{dx}\left[\dfrac{f(x)}{g(x)}\right] = \dfrac{g(x)\dfrac{d}{dx}f(x) - f(x)\dfrac{d}{dx}g(x)}{[g(x)]^2}$. The

derivative of a quotient of functions is the denominator times the derivative of the numerator minus the

numerator times the derivative of the denominator, all divided by the square of the denominator.

(g) The Chain Rule: If f and g are both differentiable and $F = f \circ g$ is the composite function defined by

$F(x) = f(g(x))$, then F is differentiable and F' is given by the product $F'(x) = f'(g(x))g'(x)$. The

derivative of a composite function is the derivative of the outer function evaluated at the inner function times the

derivative of the inner function.

2. (a) $y = x^n \implies y' = nx^{n-1}$

(b) $y = e^x \implies y' = e^x$

(c) $y = a^x \implies y' = a^x \ln a$

(d) $y = \ln x \implies y' = 1/x$

(e) $y = \log_a x \implies y' = 1/(x \ln a)$

(f) $y = \sin x \implies y' = \cos x$

(g) $y = \cos x \implies y' = -\sin x$

(h) $y = \tan x \implies y' = \sec^2 x$

(i) $y = \csc x \implies y' = -\csc x \cot x$

(j) $y = \sec x \implies y' = \sec x \tan x$

(k) $y = \cot x \implies y' = -\csc^2 x$

(l) $y = \sin^{-1} x \implies y' = 1/\sqrt{1 - x^2}$

(m) $y = \tan^{-1} x \implies y' = 1/(1 + x^2)$

3. (a) e is the number such that $\displaystyle\lim_{h \to 0} \frac{e^h - 1}{h} = 1$.

(b) $e = \displaystyle\lim_{x \to 0} (1 + x)^{1/x}$

(c) The differentiation formula for $y = a^x$ $[y' = a^x \ln a]$ is simplest when $a = e$ because $\ln e = 1$.

(d) The differentiation formula for $y = \log_a x$ $[y' = 1/(x \ln a)]$ is simplest when $a = e$ because $\ln e = 1$.

4. (a) Implicit differentiation consists of differentiating both sides of an equation involving x and y with respect to x, and then solving the resulting equation for y'.

(b) Logarithmic differentiation consists of taking natural logarithms of both sides of an equation $y = f(x)$, simplifying, differentiating implicitly with respect to x, and then solving the resulting equation for y'.

5. The linearization L of f at $x = a$ is $L(x) = f(a) + f'(a)(x - a)$.

▲ **TRUE–FALSE QUIZ** ▲

1. True. This is the Sum Rule.

3. True. This is the Chain Rule.

5. False. $\dfrac{d}{dx} f(\sqrt{x}) = \dfrac{f'(\sqrt{x})}{2\sqrt{x}}$ by the Chain Rule.

7. False. $\dfrac{d}{dx} 10^x = 10^x \ln 10$

9. True. $\dfrac{d}{dx}(\tan^2 x) = 2 \tan x \sec^2 x$, and $\dfrac{d}{dx}(\sec^2 x) = 2 \sec x (\sec x \tan x) = 2 \tan x \sec^2 x$.

11. True. $g(x) = x^5 \implies g'(x) = 5x^4 \implies g'(2) = 5(2)^4 = 80$, and by the definition of the derivative,

$$\lim_{x \to 2} \frac{g(x) - g(2)}{x - 2} = g'(2) = 80.$$

◆ **EXERCISES** ◆

1. $y = (x^4 - 3x^2 + 5)^3 \implies$

$y' = 3(x^4 - 3x^2 + 5)^2 \dfrac{d}{dx}(x^4 - 3x^2 + 5) = 3(x^4 - 3x^2 + 5)^2(4x^3 - 6x) = 6x(x^4 - 3x^2 + 5)^2(2x^2 - 3)$

3. $y = \sqrt{x} + \dfrac{1}{\sqrt[3]{x^4}} = x^{1/2} + x^{-4/3} \implies y' = \tfrac{1}{2}x^{-1/2} - \tfrac{4}{3}x^{-7/3} = \dfrac{1}{2\sqrt{x}} - \dfrac{4}{3\sqrt[3]{x^7}}$

5. $y = 2x\sqrt{x^2 + 1} \implies$

$y' = 2x \cdot \tfrac{1}{2}(x^2 + 1)^{-1/2}(2x) + \sqrt{x^2 + 1}\,(2) = \dfrac{2x^2}{\sqrt{x^2 + 1}} + 2\sqrt{x^2 + 1} = \dfrac{2x^2 + 2(x^2 + 1)}{\sqrt{x^2 + 1}} = \dfrac{2(2x^2 + 1)}{\sqrt{x^2 + 1}}$

7. $y = e^{\sin 2\theta} \;\Rightarrow\; y' = e^{\sin 2\theta}\dfrac{d}{d\theta}(\sin 2\theta) = e^{\sin 2\theta}(\cos 2\theta)(2) = 2\cos 2\theta\, e^{\sin 2\theta}$

9. $y = \dfrac{t}{1 - t^2} \;\Rightarrow\; y' = \dfrac{(1 - t^2)(1) - t(-2t)}{(1 - t^2)^2} = \dfrac{1 - t^2 + 2t^2}{(1 - t^2)^2} = \dfrac{t^2 + 1}{(1 - t^2)^2}$

11. $y = xe^{-1/x} \;\Rightarrow\; y' = xe^{-1/x}(1/x^2) + e^{-1/x}\cdot 1 = e^{-1/x}(1/x + 1)$

13. $\dfrac{d}{dx}(xy^4 + x^2 y) = \dfrac{d}{dx}(x + 3y) \;\Rightarrow\; x\cdot 4y^3 y' + y^4 \cdot 1 + x^2 \cdot y' + y\cdot 2x = 1 + 3y' \;\Rightarrow$

$y'(4xy^3 + x^2 - 3) = 1 - y^4 - 2xy \;\Rightarrow\; y' = \dfrac{1 - y^4 - 2xy}{4xy^3 + x^2 - 3}$

15. $y = \dfrac{\sec 2\theta}{1 + \tan 2\theta} \;\Rightarrow$

$y' = \dfrac{(1 + \tan 2\theta)(\sec 2\theta \tan 2\theta \cdot 2) - (\sec 2\theta)(\sec^2 2\theta \cdot 2)}{(1 + \tan 2\theta)^2} = \dfrac{2\sec 2\theta\left[(1 + \tan 2\theta)\tan 2\theta - \sec^2 2\theta\right]}{(1 + \tan 2\theta)^2}$

$= \dfrac{2\sec 2\theta\,(\tan 2\theta + \tan^2 2\theta - \sec^2 2\theta)}{(1 + \tan 2\theta)^2} = \dfrac{2\sec 2\theta\,(\tan 2\theta - 1)}{(1 + \tan 2\theta)^2}\;\left[1 + \tan^2 x = \sec^2 x\right]$

17. $y = e^{cx}(c\sin x - \cos x) \;\Rightarrow$

$y' = e^{cx}(c\cos x + \sin x) + ce^{cx}(c\sin x - \cos x)$

$= e^{cx}(c^2 \sin x - c\cos x + c\cos x + \sin x) = e^{cx}(c^2 \sin x + \sin x) = e^{cx}\sin x\,(c^2 + 1)$

19. $y = \log_5(1 + 2x) \;\Rightarrow\; y' = \dfrac{1}{(1 + 2x)\ln 5}\dfrac{d}{dx}(1 + 2x) = \dfrac{2}{(1 + 2x)\ln 5}$

21. $y = \ln\sin x - \tfrac{1}{2}\sin^2 x \;\Rightarrow\; y' = \dfrac{1}{\sin x}\cdot\cos x - \tfrac{1}{2}\cdot 2\sin x\cdot\cos x = \cot x - \sin x\cos x$

23. $y = x\tan^{-1}(4x) \;\Rightarrow\; y' = x\cdot\dfrac{1}{1 + (4x)^2}\cdot 4 + \tan^{-1}(4x)\cdot 1 = \dfrac{4x}{1 + 16x^2} + \tan^{-1}(4x)$

25. $y = \ln|\sec 5x + \tan 5x| \;\Rightarrow$

$y' = \dfrac{1}{\sec 5x + \tan 5x}(\sec 5x \tan 5x\cdot 5 + \sec^2 5x\cdot 5) = \dfrac{5\sec 5x\,(\tan 5x + \sec 5x)}{\sec 5x + \tan 5x} = 5\sec 5x$

27. $y = \cot(3x^2 + 5) \;\Rightarrow\; y' = -\csc^2(3x^2 + 5)(6x) = -6x\csc^2(3x^2 + 5)$

29. $y = \sin(\tan\sqrt{1 + x^3}) \;\Rightarrow\; y' = \cos(\tan\sqrt{1 + x^3})(\sec^2\sqrt{1 + x^3})\left[3x^2/(2\sqrt{1 + x^3})\right]$

31. $f(x) = 1/(2x - 1)^5 = (2x - 1)^{-5} \;\Rightarrow$

$f'(x) = -5(2x - 1)^{-6}(2) = -10(2x - 1)^{-6} \;\Rightarrow\; f''(x) = (-10)(-6)(2x - 1)^{-7}(2) = 120(2x - 1)^{-7}.$

$f''(0) = 120(-1)^{-7} = -120$

33. $f(x) = 2^x \;\Rightarrow\; f'(x) = 2^x\ln 2 \;\Rightarrow\; f''(x) = (2^x\ln 2)\ln 2 = 2^x(\ln 2)^2 \;\Rightarrow$

$f'''(x) = (2^x\ln 2)(\ln 2)^2 = 2^x(\ln 2)^3 \;\Rightarrow\;\cdots\;\Rightarrow\; f^{(n)}(x) = (2^x\ln 2)(\ln 2)^{n-1} = 2^x(\ln 2)^n$

35. (a) $f(x) = x\sqrt{5 - x} \;\Rightarrow$

$f'(x) = x\left[\tfrac{1}{2}(5 - x)^{-1/2}(-1)\right] + \sqrt{5 - x} = \dfrac{-x}{2\sqrt{5 - x}} + \sqrt{5 - x}\cdot\dfrac{2\sqrt{5 - x}}{2\sqrt{5 - x}}$

$= \dfrac{-x}{2\sqrt{5 - x}} + \dfrac{2(5 - x)}{2\sqrt{5 - x}} = \dfrac{-x + 10 - 2x}{2\sqrt{5 - x}} = \dfrac{10 - 3x}{2\sqrt{5 - x}}$

(b) At $(1, 2)$: $f'(1) = \tfrac{7}{4}$. So an equation of the tangent line is $y - 2 = \tfrac{7}{4}(x - 1)$ or $y = \tfrac{7}{4}x + \tfrac{1}{4}$.

At $(4, 4)$: $f'(4) = -\tfrac{2}{2} = -1$. So an equation of the tangent line is $y - 4 = -1(x - 4)$ or $y = -x + 8$.

(c)

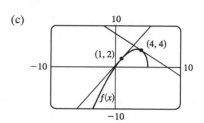

(d)

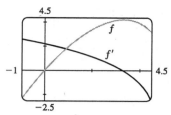

The graphs look reasonable, since f' is positive where f has tangents with positive slope, and f' is negative where f has tangents with negative slope.

37. $f(x) = xe^{\sin x}$ $\Rightarrow$ $f'(x) = x[e^{\sin x}(\cos x)] + e^{\sin x}(1) = e^{\sin x}(x \cos x + 1)$. As a check on our work, we notice from the graphs that $f'(x) > 0$ when f is increasing. Also, we see in the larger viewing rectangle a certain similarity in the graphs of f and f': the sizes of the oscillations of f and f' are linked.

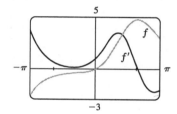

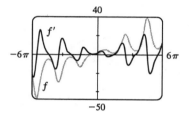

39. (a) $h(x) = f(x)g(x)$ $\Rightarrow$ $h'(x) = f(x)g'(x) + g(x)f'(x)$ $\Rightarrow$
$h'(2) = f(2)g'(2) + g(2)f'(2) = (3)(4) + (5)(-2) = 12 - 10 = 2$

(b) $F(x) = f(g(x))$ $\Rightarrow$ $F'(x) = f'(g(x))g'(x)$ $\Rightarrow$ $F'(2) = f'(g(2))g'(2) = f'(5)(4) = 11 \cdot 4 = 44$

41. $f(x) = x^2 g(x)$ $\Rightarrow$ $f'(x) = x^2 g'(x) + g(x)(2x) = x[xg'(x) + 2g(x)]$

43. $f(x) = [g(x)]^2$ $\Rightarrow$ $f'(x) = 2[g(x)]^1 \cdot g'(x) = 2g(x)g'(x)$

45. $f(x) = g(e^x)$ $\Rightarrow$ $f'(x) = g'(e^x)e^x$

47. $f(x) = \ln|g(x)|$ $\Rightarrow$ $f'(x) = \dfrac{1}{g(x)}g'(x) = \dfrac{g'(x)}{g(x)}$

49. $h(x) = \dfrac{f(x)g(x)}{f(x) + g(x)}$ $\Rightarrow$

$h'(x) = \dfrac{[f(x) + g(x)][f(x)g'(x) + g(x)f'(x)] - f(x)g(x)[f'(x) + g'(x)]}{[f(x) + g(x)]^2}$

$= \dfrac{[f(x)]^2 g'(x) + f(x)g(x)f'(x) + f(x)g(x)g'(x) + [g(x)]^2 f'(x) - f(x)g(x)f'(x) - f(x)g(x)g'(x)}{[f(x) + g(x)]^2}$

$= \dfrac{f'(x)[g(x)]^2 + g'(x)[f(x)]^2}{[f(x) + g(x)]^2}$

51. $y = [\ln(x + 4)]^2 \Rightarrow y' = 2[\ln(x + 4)]^1 \cdot \dfrac{1}{x + 4} \cdot 1 = 2\dfrac{\ln(x + 4)}{x + 4}$ and $y' = 0 \Leftrightarrow \ln(x + 4) = 0 \Leftrightarrow$

$x + 4 = e^0 \Rightarrow x + 4 = 1 \Leftrightarrow x = -3$, so the tangent is horizontal at the point $(-3, 0)$.

53. $x^2 + 2y^2 = 1 \Rightarrow 2x + 4yy' = 0 \Rightarrow y' = -x/(2y) = 1 \Leftrightarrow x = -2y$. Since the points lie on the ellipse,

we have $(-2y)^2 + 2y^2 = 1 \Rightarrow 6y^2 = 1 \Rightarrow y = \pm\frac{1}{\sqrt{6}}$. The points are $\left(-\frac{2}{\sqrt{6}}, \frac{1}{\sqrt{6}}\right)$ and $\left(\frac{2}{\sqrt{6}}, -\frac{1}{\sqrt{6}}\right)$.

55. $s(t) = Ae^{-ct}\cos(\omega t + \delta) \Rightarrow$

$v(t) = s'(t) = A\{e^{-ct}[-\omega\sin(\omega t + \delta)] + \cos(\omega t + \delta)(-ce^{-ct})\}$

$\qquad = -Ae^{-ct}[\omega\sin(\omega t + \delta) + c\cos(\omega t + \delta)] \Rightarrow$

$a(t) = v'(t) = -A\{e^{-ct}[\omega^2\cos(\omega t + \delta) - c\omega\sin(\omega t + \delta)]$

$\qquad\qquad\qquad + [\omega\sin(\omega t + \delta) + c\cos(\omega t + \delta)](-ce^{-ct})\}$

$\qquad = -Ae^{-ct}[\omega^2\cos(\omega t + \delta) - c\omega\sin(\omega t + \delta) - c\omega\sin(\omega t + \delta) - c^2\cos(\omega t + \delta)]$

$\qquad = -Ae^{-ct}[(\omega^2 - c^2)\cos(\omega t + \delta) - 2c\omega\sin(\omega t + \delta)]$

$\qquad = Ae^{-ct}[(c^2 - \omega^2)\cos(\omega t + \delta) + 2c\omega\sin(\omega t + \delta)]$

57. The linear density ρ is the rate of change of mass m with respect to length x. $m = x(1 + \sqrt{x}) = x + x^{3/2} \Rightarrow$

$\rho = dm/dx = 1 + \frac{3}{2}\sqrt{x}$, so the linear density when $x = 4$ is $1 + \frac{3}{2}\sqrt{4} = 4$ kg/m.

59. (a) $C(x) = 920 + 2x - 0.02x^2 + 0.00007x^3 \Rightarrow C'(x) = 2 - 0.04x + 0.00021x^2$

(b) $C'(100) = 2 - 4 + 2.1 = \$0.10/$unit. This value represents the rate at which costs are increasing as the hundredth unit is produced, and is the approximate cost of producing the 101st unit.

(c) The cost of producing the 101st item is $C(101) - C(100) = 990.10107 - 990 = \$0.10107$, slightly larger than $C'(100)$.

(d) $C''(x) = -0.04 + 0.00042x = 0 \Rightarrow x = \frac{0.04}{0.00042} \approx 95.24$ and C'' changes from negative to positive at this value of x. This is the value of x at which the marginal cost is minimized.

61. (a) $f(x) = \sqrt[3]{1 + 3x} = (1 + 3x)^{1/3} \Rightarrow f'(x) = (1 + 3x)^{-2/3}$, so the linearization of f at $a = 0$ is

$L(x) = f(0) + f'(0)(x - 0) = 1^{1/3} + 1^{-2/3}x = 1 + x$. Thus, $\sqrt[3]{1 + 3x} \approx 1 + x \Rightarrow$

$\sqrt[3]{1.03} = \sqrt[3]{1 + 3(0.01)} \approx 1 + (0.01) = 1.01$.

(b) The linear approximation is $\sqrt[3]{1 + 3x} \approx 1 + x$, so for the required accuracy we want $\sqrt[3]{1 + 3x} - 0.1 < 1 + x < \sqrt[3]{1 + 3x} + 0.1$. From the graph, it appears that this is true when $-0.23 < x < 0.40$.

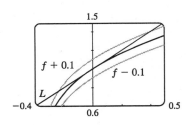

63. $\displaystyle\lim_{\theta \to \pi/3} \dfrac{\cos\theta - 0.5}{\theta - \pi/3} = \left[\dfrac{d}{d\theta}\cos\theta\right]_{\theta = \pi/3} = -\sin\dfrac{\pi}{3} = -\dfrac{\sqrt{3}}{2}$

65. $\displaystyle\lim_{x\to0}\frac{\sqrt{1+\tan x}-\sqrt{1+\sin x}}{x^3}=\lim_{x\to0}\frac{\left(\sqrt{1+\tan x}-\sqrt{1+\sin x}\right)\left(\sqrt{1+\tan x}+\sqrt{1+\sin x}\right)}{x^3\left(\sqrt{1+\tan x}+\sqrt{1+\sin x}\right)}$

$\displaystyle=\lim_{x\to0}\frac{(1+\tan x)-(1+\sin x)}{x^3\left(\sqrt{1+\tan x}+\sqrt{1+\sin x}\right)}=\lim_{x\to0}\frac{\sin x\left(1/\cos x-1\right)}{x^3\left(\sqrt{1+\tan x}+\sqrt{1+\sin x}\right)}\cdot\frac{\cos x}{\cos x}$

$\displaystyle=\lim_{x\to0}\frac{\sin x\left(1-\cos x\right)}{x^3\left(\sqrt{1+\tan x}+\sqrt{1+\sin x}\right)\cos x}\cdot\frac{1+\cos x}{1+\cos x}$

$\displaystyle=\lim_{x\to0}\frac{\sin x\cdot\sin^2 x}{x^3\left(\sqrt{1+\tan x}+\sqrt{1+\sin x}\right)\cos x\left(1+\cos x\right)}$

$\displaystyle=\left(\lim_{x\to0}\frac{\sin x}{x}\right)^3\lim_{x\to0}\frac{1}{\left(\sqrt{1+\tan x}+\sqrt{1+\sin x}\right)\cos x\left(1+\cos x\right)}$

$\displaystyle=1^3\cdot\frac{1}{\left(\sqrt{1}+\sqrt{1}\right)\cdot1\cdot(1+1)}=\frac{1}{4}$

Focus on Problem Solving

1. We must find a value x_0 such that the normal lines to the parabola $y = x^2$ at $x = \pm x_0$ intersect at a point one unit from the points $(\pm x_0, x_0^2)$. The normals to $y = x^2$ at $x = \pm x_0$ have slopes $-\dfrac{1}{\pm 2x_0}$ and pass through $(\pm x_0, x_0^2)$ respectively, so the normals have the equations $y - x_0^2 = -\dfrac{1}{2x_0}(x - x_0)$ and $y - x_0^2 = \dfrac{1}{2x_0}(x + x_0)$. The common y-intercept is $x_0^2 + \frac{1}{2}$. We want to find the value of x_0 for which the distance from $\left(0, x_0^2 + \frac{1}{2}\right)$ to $\left(x_0, x_0^2\right)$ equals 1. The square of the distance is $(x_0 - 0)^2 + \left[x_0^2 - \left(x_0^2 + \frac{1}{2}\right)\right]^2 = x_0^2 + \frac{1}{4} = 1 \iff x_0 = \pm\dfrac{\sqrt{3}}{2}$. For these values of x_0, the y-intercept is $x_0^2 + \frac{1}{2} = \frac{5}{4}$, so the center of the circle is at $\left(0, \frac{5}{4}\right)$.

Another solution: Let the center of the circle be $(0, a)$. Then the equation of the circle is $x^2 + (y - a)^2 = 1$. Solving with the equation of the parabola, $y = x^2$, we get $x^2 + \left(x^2 - a\right)^2 = 1 \iff x^2 + x^4 - 2ax^2 + a^2 = 1 \iff x^4 + (1 - 2a)x^2 + a^2 - 1 = 0$. The parabola and the circle will be tangent to each other when this quadratic equation in x^2 has equal roots; that is, when the discriminant is 0. Thus, $(1 - 2a)^2 - 4(a^2 - 1) = 0 \iff 1 - 4a + 4a^2 - 4a^2 + 4 = 0 \iff 4a = 5$, so $a = \frac{5}{4}$. The center of the circle is $\left(0, \frac{5}{4}\right)$.

3. (a) $f(x) = \sqrt{1 - \sqrt{2 - \sqrt{3 - x}}}$

$$D = \left\{ x \mid 3 - x \geq 0, 2 - \sqrt{3 - x} \geq 0, 1 - \sqrt{2 - \sqrt{3 - x}} \geq 0 \right\}$$

$$= \left\{ x \mid 3 \geq x, 2 \geq \sqrt{3 - x}, 1 \geq \sqrt{2 - \sqrt{3 - x}} \right\}$$

$$= \left\{ x \mid 3 \geq x,\ 4 \geq 3 - x, 1 \geq 2 - \sqrt{3 - x} \right\}$$

$$= \left\{ x \mid x \leq 3, x \geq -1, 1 \leq \sqrt{3 - x} \right\} = \left\{ x \mid x \leq 3, x \geq -1, 1 \leq 3 - x \right\}$$

$$= \left\{ x \mid x \leq 3, x \geq -1, x \leq 2 \right\} = \left\{ x \mid -1 \leq x \leq 2 \right\} = [-1, 2]$$

(b) $f(x) = \sqrt{1 - \sqrt{2 - \sqrt{3 - x}}} \implies$

$$f'(x) = \frac{1}{2\sqrt{1 - \sqrt{2 - \sqrt{3 - x}}}} \cdot \frac{d}{dx}\left(1 - \sqrt{2 - \sqrt{3 - x}}\right)$$

$$= \frac{1}{2\sqrt{1 - \sqrt{2 - \sqrt{3 - x}}}} \cdot \frac{-1}{2\sqrt{2 - \sqrt{3 - x}}} \frac{d}{dx}\left(2 - \sqrt{3 - x}\right)$$

$$= -\frac{1}{8\sqrt{1 - \sqrt{2 - \sqrt{3 - x}}}\,\sqrt{2 - \sqrt{3 - x}}\,\sqrt{3 - x}}$$

(c)

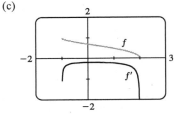

Note that f is always decreasing and f' is always negative.

5. We can assume without loss of generality that $\theta = 0$ at time $t = 0$, so that $\theta = 12\pi t$ rad. [The angular velocity of the wheel is 360 rpm $= 360 \cdot (2\pi \text{ rad})/(60 \text{ s}) = 12\pi$ rad/s.] Then the position of A as a function of time is

$$A = (40\cos\theta, 40\sin\theta) = (40\cos 12\pi t, 40\sin 12\pi t), \text{ so } \sin\alpha = \frac{y}{1.2 \text{ m}} = \frac{40\sin\theta}{120} = \frac{\sin\theta}{3} = \tfrac{1}{3}\sin 12\pi t.$$

(a) Differentiating the expression for $\sin\alpha$, we get $\cos\alpha \cdot \dfrac{d\alpha}{dt} = \tfrac{1}{3} \cdot 12\pi \cdot \cos 12\pi t = 4\pi\cos\theta$.

When $\theta = \frac{\pi}{3}$, we have $\sin\alpha = \tfrac{1}{3}\sin\theta = \frac{\sqrt{3}}{6}$, so $\cos\alpha = \sqrt{1 - \left(\frac{\sqrt{3}}{6}\right)^2} = \sqrt{\frac{11}{12}}$ and

$$\frac{d\alpha}{dt} = \frac{4\pi\cos\frac{\pi}{3}}{\cos\alpha} = \frac{2\pi}{\sqrt{11/12}} = \frac{4\pi\sqrt{3}}{\sqrt{11}} \approx 6.56 \text{ rad/s}.$$

(b) By the Law of Cosines, $|AP|^2 = |OA|^2 + |OP|^2 - 2|OA||OP|\cos\theta \Rightarrow$

$$120^2 = 40^2 + |OP|^2 - 2 \cdot 40|OP|\cos\theta \Rightarrow |OP|^2 - (80\cos\theta)|OP| - 12{,}800 = 0 \Rightarrow$$

$$|OP| = \tfrac{1}{2}\left(80\cos\theta \pm \sqrt{6400\cos^2\theta + 51{,}200}\right) = 40\cos\theta \pm 40\sqrt{\cos^2\theta + 8}$$

$$= 40\left(\cos\theta + \sqrt{8 + \cos^2\theta}\right) \text{ cm} \quad (\text{since } |OP| > 0)$$

As a check, note that $|OP| = 160$ cm when $\theta = 0$ and $|OP| = 80\sqrt{2}$ cm when $\theta = \frac{\pi}{2}$.

(c) By part (b), the x-coordinate of P is given by $x = 40\left(\cos\theta + \sqrt{8 + \cos^2\theta}\right)$, so

$$\frac{dx}{dt} = \frac{dx}{d\theta}\frac{d\theta}{dt} = 40\left(-\sin\theta - \frac{2\cos\theta\sin\theta}{2\sqrt{8 + \cos^2\theta}}\right) \cdot 12\pi = -480\pi\sin\theta\left(1 + \frac{\cos\theta}{\sqrt{8 + \cos^2\theta}}\right) \text{ cm/s}.$$

In particular, $dx/dt = 0$ cm/s when $\theta = 0$ and $dx/dt = -480\pi$ cm/s when $\theta = \frac{\pi}{2}$.

7. Consider the statement that $\dfrac{d^n}{dx^n}(e^{ax}\sin bx) = r^n e^{ax}\sin(bx + n\theta)$. For $n = 1$,

$$\frac{d}{dx}(e^{ax}\sin bx) = ae^{ax}\sin bx + be^{ax}\cos bx, \text{ and}$$

$$re^{ax}\sin(bx + \theta) = re^{ax}[\sin bx\cos\theta + \cos bx\sin\theta] = re^{ax}\left(\frac{a}{r}\sin bx + \frac{b}{r}\cos bx\right)$$

$$= ae^{ax}\sin bx + be^{ax}\cos bx$$

since $\tan\theta = \dfrac{b}{a} \Rightarrow \sin\theta = \dfrac{b}{r}$ and $\cos\theta = \dfrac{a}{r}$.

So the statement is true for $n = 1$. Assume it is true for $n = k$. Then

$$\frac{d^{k+1}}{dx^{k+1}}(e^{ax}\sin bx) = \frac{d}{dx}\left[r^k e^{ax}\sin(bx + k\theta)\right] = r^k ae^{ax}\sin(bx + k\theta) + r^k e^{ax}b\cos(bx + k\theta)$$

$$= r^k e^{ax}[a\sin(bx + k\theta) + b\cos(bx + k\theta)]$$

But

$$\sin\left[bx + (k+1)\theta\right] = \sin\left[(bx + k\theta) + \theta\right] = \sin(bx + k\theta)\cos\theta + \sin\theta\cos(bx + k\theta)$$

$$= \frac{a}{r}\sin(bx + k\theta) + \frac{b}{r}\cos(bx + k\theta)$$

Hence, $a\sin(bx + k\theta) + b\cos(bx + k\theta) = r\sin\left[bx + (k+1)\theta\right]$. So

$$\frac{d^{k+1}}{dx^{k+1}}\left(e^{ax}\sin bx\right) = r^k e^{ax}\left[a\sin(bx + k\theta) + b\cos(bx + k\theta)\right] = r^k e^{ax}\left[r\sin(bx + (k+1)\theta)\right]$$

$$= r^{k+1}e^{ax}\left[\sin(bx + (k+1)\theta)\right]$$

Therefore, the statement is true for all n by mathematical induction.

9. It seems from the figure that as P approaches the point $(0, 2)$ from the right, $x_T \to \infty$ and $y_T \to 2^+$. As P approaches the point $(3, 0)$ from the left, it appears that $x_T \to 3^+$ and $y_T \to \infty$. So we guess that $x_T \in (3, \infty)$ and $y_T \in (2, \infty)$. It is more difficult to estimate the range of values for x_N and y_N. We might perhaps guess that $x_N \in (0, 3)$, and $y_N \in (-\infty, 0)$ or $(-2, 0)$.

In order to actually solve the problem, we implicitly differentiate the equation of the ellipse to find the equation of the tangent line: $\dfrac{x^2}{9} + \dfrac{y^2}{4} = 1 \;\Rightarrow\; \dfrac{2x}{9} + \dfrac{2y}{4}y' = 0$, so $y' = -\dfrac{4}{9}\dfrac{x}{y}$. So at the point (x_0, y_0) on the ellipse, an equation of the tangent line is $y - y_0 = -\dfrac{4}{9}\dfrac{x_0}{y_0}(x - x_0)$ or $4x_0 x + 9y_0 y = 4x_0^2 + 9y_0^2$. This can be written as $\dfrac{x_0 x}{9} + \dfrac{y_0 y}{4} = \dfrac{x_0^2}{9} + \dfrac{y_0^2}{4} = 1$, because (x_0, y_0) lies on the ellipse. So an equation of the tangent line is $\dfrac{x_0 x}{9} + \dfrac{y_0 y}{4} = 1$.

Therefore, the x-intercept x_T for the tangent line is given by $\dfrac{x_0 x_T}{9} = 1 \;\Leftrightarrow\; x_T = \dfrac{9}{x_0}$, and the y-intercept y_T is given by $\dfrac{y_0 y_T}{4} = 1 \;\Leftrightarrow\; y_T = \dfrac{4}{y_0}$.

So as x_0 takes on all values in $(0, 3)$, x_T takes on all values in $(3, \infty)$, and as y_0 takes on all values in $(0, 2)$, y_T takes on all values in $(2, \infty)$.

At the point (x_0, y_0) on the ellipse, the slope of the normal line is $-\dfrac{1}{y'(x_0, y_0)} = \dfrac{9}{4}\dfrac{y_0}{x_0}$, and its equation is $y - y_0 = \dfrac{9}{4}\dfrac{y_0}{x_0}(x - x_0)$. So the x-intercept x_N for the normal line is given by $0 - y_0 = \dfrac{9}{4}\dfrac{y_0}{x_0}(x_N - x_0) \;\Rightarrow\;$ $x_N = -\dfrac{4x_0}{9} + x_0 = \dfrac{5x_0}{9}$, and the y-intercept y_N is given by $y_N - y_0 = \dfrac{9}{4}\dfrac{y_0}{x_0}(0 - x_0) \;\Rightarrow\;$ $y_N = -\dfrac{9y_0}{4} + y_0 = -\dfrac{5y_0}{4}$.

So as x_0 takes on all values in $(0, 3)$, x_N takes on all values in $\left(0, \frac{5}{3}\right)$, and as y_0 takes on all values in $(0, 2)$, y_N takes on all values in $\left(-\frac{5}{2}, 0\right)$.

11. If we divide $1 - x$ into x^n by long division, we find that $f(x) = \dfrac{x^n}{1-x} = -x^{n-1} - x^{n-2} - \cdots - x - 1 + \dfrac{1}{1-x}$.

This can also be seen by multiplying the last expression by $1 - x$ and canceling terms on the right-hand side. So we

let $g(x) = 1 + x + x^2 + \cdots + x^{n-1}$, so that $f(x) = \dfrac{1}{1-x} - g(x) \quad \Rightarrow \quad f^{(n)}(x) = \left(\dfrac{1}{1-x}\right)^{(n)} - g^{(n)}(x)$. But

g is a polynomial of degree $(n-1)$, so its nth derivative is 0, and therefore $f^{(n)}(x) = \left(\dfrac{1}{1-x}\right)^{(n)}$. Now

$\dfrac{d}{dx}(1-x)^{-1} = (-1)(1-x)^{-2}(-1) = (1-x)^{-2}$, $\dfrac{d^2}{dx^2}(1-x)^{-1} = (-2)(1-x)^{-3}(-1) = 2(1-x)^{-3}$,

$\dfrac{d^3}{dx^3}(1-x)^{-1} = (-3) \cdot 2(1-x)^{-4}(-1) = 3 \cdot 2(1-x)^{-4}$, $\dfrac{d^4}{dx^4}(1-x)^{-1} = 4 \cdot 3 \cdot 2(1-x)^{-5}$, and so on. So

after n differentiations, we will have $f^{(n)}(x) = \left(\dfrac{1}{1-x}\right)^{(n)} = \dfrac{n!}{(1-x)^{n+1}}$.

13. $y = \dfrac{x}{\sqrt{a^2-1}} - \dfrac{2}{\sqrt{a^2-1}} \arctan \dfrac{\sin x}{a + \sqrt{a^2-1} + \cos x}$. Let $k = a + \sqrt{a^2-1}$. Then

$$y' = \dfrac{1}{\sqrt{a^2-1}} - \dfrac{2}{\sqrt{a^2-1}} \cdot \dfrac{1}{1 + \sin^2 x/(k+\cos x)^2} \cdot \dfrac{\cos x(k+\cos x) + \sin^2 x}{(k+\cos x)^2}$$

$$= \dfrac{1}{\sqrt{a^2-1}} - \dfrac{2}{\sqrt{a^2-1}} \cdot \dfrac{k\cos x + \cos^2 x + \sin^2 x}{(k+\cos x)^2 + \sin^2 x} = \dfrac{1}{\sqrt{a^2-1}} - \dfrac{2}{\sqrt{a^2-1}} \cdot \dfrac{k\cos x + 1}{k^2 + 2k\cos x + 1}$$

$$= \dfrac{k^2 + 2k\cos x + 1 - 2k\cos x - 2}{\sqrt{a^2-1}\,(k^2 + 2k\cos x + 1)} = \dfrac{k^2 - 1}{\sqrt{a^2-1}\,(k^2 + 2k\cos x + 1)}$$

But $k^2 = 2a^2 + 2a\sqrt{a^2-1} - 1 = 2a(a + \sqrt{a^2-1}) - 1 = 2ak - 1$, so $k^2 + 1 = 2ak$, and

$k^2 - 1 = 2(ak - 1)$. So $y' = \dfrac{2(ak-1)}{\sqrt{a^2-1}\,(2ak + 2k\cos x)} = \dfrac{ak-1}{\sqrt{a^2-1}k(a+\cos x)}$. But

$ak - 1 = a^2 + a\sqrt{a^2-1} - 1 = k\sqrt{a^2-1}$, so $y' = 1/(a + \cos x)$.

15. (a)

If the two lines L_1 and L_2 have slopes m_1 and m_2 and angles of inclination ϕ_1 and ϕ_2, then $m_1 = \tan \phi_1$ and $m_2 = \tan \phi_2$. The triangle in the figure shows that $\phi_1 + \alpha + (180° - \phi_2) = 180°$ and so $\alpha = \phi_2 - \phi_1$. Therefore, using the identity for $\tan(x - y)$, we have

$$\tan \alpha = \tan(\phi_2 - \phi_1) = \dfrac{\tan \phi_2 - \tan \phi_1}{1 + \tan \phi_2 \tan \phi_1} \quad \text{and so}$$

$$\tan \alpha = \dfrac{m_2 - m_1}{1 + m_1 m_2}.$$

(b) (i) The parabolas intersect when $x^2 = (x-2)^2$ $\Rightarrow$ $x = 1$. If $y = x^2$, then $y' = 2x$, so the slope of the

tangent to $y = x^2$ at $(1, 1)$ is $m_1 = 2(1) = 2$. If $y = (x-2)^2$, then $y' = 2(x-2)$, so the slope of the

tangent to $y = (x-2)^2$ at $(1, 1)$ is $m_2 = 2(1-2) = -2$. Therefore,

$$\tan \alpha = \frac{m_2 - m_1}{1 + m_1 m_2} = \frac{-2-2}{1+2(-2)} = \frac{4}{3} \text{ and so } \alpha = \tan^{-1}\left(\frac{4}{3}\right) \approx 53°.$$

(ii) $x^2 - y^2 = 3$ and $x^2 - 4x + y^2 + 3 = 0$ intersect when $x^2 - 4x + (x^2 - 3) + 3 = 0$ $\Leftrightarrow$ $2x(x-2) = 0$

$\Rightarrow$ $x = 0$ or 2, but 0 is extraneous. If $x = 2$, then $y = \pm 1$. If $x^2 - y^2 = 3$ then $2x - 2yy' = 0$ $\Rightarrow$

$y' = x/y$ and $x^2 - 4x + y^2 + 3 = 0$ $\Rightarrow$ $2x - 4 + 2yy' = 0$ $\Rightarrow$ $y' = \dfrac{2-x}{y}$. At $(2, 1)$ the slopes are

$m_1 = 2$ and $m_2 = 0$, so $\tan \alpha = \frac{0-2}{1+2\cdot 0} = -2$ $\Rightarrow$ $\alpha \approx 117°$. At $(2, -1)$ the slopes are $m_1 = -2$ and

$m_2 = 0$, so $\tan \alpha = \frac{0-(-2)}{1+(-2)(0)} = 2$ $\Rightarrow$ $\alpha \approx 63°$.

17. Since $\angle ROQ = \angle OQP = \theta$, the triangle QOR is isosceles, so

$|QR| = |RO| = x$. By the Law of Cosines, $x^2 = x^2 + r^2 - 2rx \cos \theta$. Hence,

$2rx \cos \theta = r^2$, so $x = \dfrac{r^2}{2r \cos \theta} = \dfrac{r}{2 \cos \theta}$. Note that as $y \to 0^+$, $\theta \to 0^+$ (since

$\sin \theta = y/r$), and hence $x \to \dfrac{r}{2 \cos 0} = \dfrac{r}{2}$. Thus, as P is taken closer and closer

to the x-axis, the point R approaches the midpoint of the radius AO.

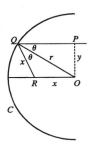

19. $y = x^4 - 2x^2 - x$ $\Rightarrow$ $y' = 4x^3 - 4x - 1$. The equation of the tangent line at $x = a$ is

$y - (a^4 - 2a^2 - a) = (4a^3 - 4a - 1)(x-a)$ or $y = (4a^3 - 4a - 1)x + (-3a^4 + 2a^2)$ and similarly for $x = b$.

So if at $x = a$ and $x = b$ we have the same tangent line, then $4a^3 - 4a - 1 = 4b^3 - 4b - 1$ and

$-3a^4 + 2a^2 = -3b^4 + 2b^2$. The first equation gives $a^3 - b^3 = a - b$ $\Rightarrow$ $(a-b)(a^2 + ab + b^2) = (a-b)$.

Assuming $a \neq b$, we have $1 = a^2 + ab + b^2$. The second equation gives $3(a^4 - b^4) = 2(a^2 - b^2)$ $\Rightarrow$

$3(a^2 - b^2)(a^2 + b^2) = 2(a^2 - b^2)$ which is true if $a = -b$. Substituting into $1 = a^2 + ab + b^2$ gives

$1 = a^2 - a^2 + a^2$ $\Rightarrow$ $a = \pm 1$ so that $a = 1$ and $b = -1$ or vice versa. Thus, the points $(1, -2)$ and $(-1, 0)$

have a common tangent line.

As long as there are only two such points, we are done. So we show that these are in fact the only two such

points. Suppose that $a^2 - b^2 \neq 0$. Then $3(a^2 - b^2)(a^2 + b^2) = 2(a^2 - b^2)$ gives $3(a^2 + b^2) = 2$ or $a^2 + b^2 = \frac{2}{3}$.

Thus, $ab = (a^2 + ab + b^2) - (a^2 + b^2) = 1 - \frac{2}{3} = \frac{1}{3}$, so $b = \dfrac{1}{3a}$. Hence, $a^2 + \dfrac{1}{9a^2} = \dfrac{2}{3}$, so $9a^4 + 1 = 6a^2$

$\Rightarrow$ $0 = 9a^4 - 6a^2 + 1 = (3a^2 - 1)^2$. So $3a^2 - 1 = 0$ $\Rightarrow$ $a^2 = \frac{1}{3}$ $\Rightarrow$ $b^2 = \dfrac{1}{9a^2} = \frac{1}{3} = a^2$, contradicting

our assumption that $a^2 \neq b^2$.

21.

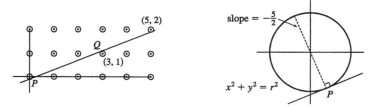

Because of the periodic nature of the lattice points, it suffices to consider the points in the 5×2 grid shown. We can

see that the minimum value of r occurs when there is a line with slope $\frac{2}{5}$ which touches the circle centered at $(3, 1)$

and the circles centered at $(0, 0)$ and $(5, 2)$. To find P, the point at which the line is tangent to the circle at $(0, 0)$,

we simultaneously solve $x^2 + y^2 = r^2$ and $y = -\frac{5}{2}x$ $\Rightarrow$ $x^2 + \frac{25}{4}x^2 = r^2$ $\Rightarrow$ $x^2 = \frac{4}{29}r^2$ $\Rightarrow$

$x = \frac{2}{\sqrt{29}}r$, $y = -\frac{5}{\sqrt{29}}r$. To find Q, we either use symmetry or solve $(x - 3)^2 + (y - 1)^2 = r^2$ and

$y - 1 = -\frac{5}{2}(x - 3)$. As above, we get $x = 3 - \frac{2}{\sqrt{29}}r$, $y = 1 + \frac{5}{\sqrt{29}}r$. Now the slope of the line PQ is $\frac{2}{5}$, so

$$m_{PQ} = \frac{1 + \frac{5}{\sqrt{29}}r - \left(-\frac{5}{\sqrt{29}}r\right)}{3 - \frac{2}{\sqrt{29}}r - \frac{2}{\sqrt{29}}r} = \frac{1 + \frac{10}{\sqrt{29}}r}{3 - \frac{4}{\sqrt{29}}r} = \frac{\sqrt{29} + 10r}{3\sqrt{29} - 4r} = \frac{2}{5}$$

$$\Rightarrow \qquad 5\sqrt{29} + 50r = 6\sqrt{29} - 8r$$

$$\Leftrightarrow \qquad 58r = \sqrt{29}$$

$$\Leftrightarrow \qquad r = \frac{\sqrt{29}}{58}$$

So the minimum value of r for which any line with slope $\frac{2}{5}$ intersects circles with radius r centered at the lattice

points on the plane is $r = \frac{\sqrt{29}}{58} \approx 0.093$.

4 Applications of Differentiation

 4.1 Related Rates • • • • • • • • • • • • • • • •

1. $V = x^3 \quad\Rightarrow\quad \dfrac{dV}{dt} = \dfrac{dV}{dx}\dfrac{dx}{dt} = 3x^2\dfrac{dx}{dt}$

3. $y = x^3 + 2x \quad\Rightarrow\quad \dfrac{dy}{dt} = \dfrac{dy}{dx}\dfrac{dx}{dt} = (3x^2 + 2)(5) = 5(3x^2 + 2)$. When $x = 2$, $\dfrac{dy}{dt} = 5(14) = 70$.

5. (a) Given: the rate of decrease of the surface area is $1 \text{ cm}^2/\text{min}$. If we let

(c)
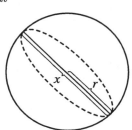

    t be time (in minutes) and S be the surface area (in cm^2), then we are

    given that $dS/dt = -1 \text{ cm}^2/\text{s}$.

  (b) Unknown: the rate of decrease of the diameter when the diameter is

    10 cm. If we let x be the diameter, then we want to find dx/dt when

    $x = 10$ cm.

  (d) If the radius is r and the diameter $x = 2r$, then $r = \frac{1}{2}x$ and $S = 4\pi r^2 = 4\pi\left(\frac{1}{2}x\right)^2 = \pi x^2 \quad\Rightarrow$

    $\dfrac{dS}{dt} = \dfrac{dS}{dx}\dfrac{dx}{dt} = 2\pi x\dfrac{dx}{dt}$.

  (e) $-1 = \dfrac{dS}{dt} = 2\pi x\dfrac{dx}{dt} \quad\Rightarrow\quad \dfrac{dx}{dt} = -\dfrac{1}{2\pi x}$. When $x = 10$, $\dfrac{dx}{dt} = -\dfrac{1}{20\pi}$. So the rate of decrease is

    $\frac{1}{20\pi}$ cm/min.

7. (a) Given: a plane flying horizontally at an altitude of 1 mi and a speed of 500 mi/h passes directly over a radar

    station. If we let t be time (in hours) and x be the horizontal distance traveled by the plane (in mi), then we are

    given that $dx/dt = 500$ mi/h.

  (b) Unknown: the rate at which the distance from the plane to the station is

(c)
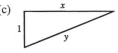

    increasing when it is 2 mi from the station. If we let y be the distance from

    the plane to the station, then we want to find dy/dt when $y = 2$ mi.

  (d) By the Pythagorean Theorem, $y^2 = x^2 + 1 \quad\Rightarrow\quad 2y(dy/dt) = 2x(dx/dt)$.

  (e) $\dfrac{dy}{dt} = \dfrac{x}{y}\dfrac{dx}{dt} = \dfrac{x}{y}(500)$. Since $y^2 = x^2 + 1$, when $y = 2$, $x = \sqrt{3}$, so $\dfrac{dy}{dt} = \dfrac{\sqrt{3}}{2}(500) = 250\sqrt{3} \approx 433$ mi/h.

9.

We are given that $\dfrac{dx}{dt} = 60$ mi/h and $\dfrac{dy}{dt} = 25$ mi/h. $z^2 = x^2 + y^2 \quad\Rightarrow$

$2z\dfrac{dz}{dt} = 2x\dfrac{dx}{dt} + 2y\dfrac{dy}{dt} \quad\Rightarrow\quad z\dfrac{dz}{dt} = x\dfrac{dx}{dt} + y\dfrac{dy}{dt} \quad\Rightarrow\quad \dfrac{dz}{dt} = \dfrac{1}{z}\left(x\dfrac{dx}{dt} + y\dfrac{dy}{dt}\right)$.

After 2 hours, $x = 2(60) = 120$ and $y = 2(25) = 50 \quad\Rightarrow\quad z = \sqrt{120^2 + 50^2} = 130$,

so $\dfrac{dz}{dt} = \dfrac{1}{z}\left(x\dfrac{dx}{dt} + y\dfrac{dy}{dt}\right) = \dfrac{120(60) + 50(25)}{130} = 65$ mi/h.

11.

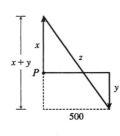

We are given that $\dfrac{dx}{dt} = 4$ ft/s and $\dfrac{dy}{dt} = 5$ ft/s. $z^2 = (x + y)^2 + 500^2$ $\Rightarrow$

$2z\dfrac{dz}{dt} = 2(x + y)\left(\dfrac{dx}{dt} + \dfrac{dy}{dt}\right)$. 15 minutes after the woman starts, we have

$x = (4 \text{ ft/s})(20 \text{ min})(60 \text{ s/min}) = 4800 \text{ ft}$ and $y = 5 \cdot 15 \cdot 60 = 4500$ $\Rightarrow$

$z = \sqrt{(4800 + 4500)^2 + 500^2} = \sqrt{86{,}740{,}000}$, so

$\dfrac{dz}{dt} = \dfrac{x + y}{z}\left(\dfrac{dx}{dt} + \dfrac{dy}{dt}\right) = \dfrac{4800 + 4500}{\sqrt{86{,}740{,}000}}(5 + 4) = \dfrac{837}{\sqrt{8674}} \approx 8.99$ ft/s.

13. $A = \frac{1}{2}bh$, where b is the base and h is the altitude. We are given that $\dfrac{dh}{dt} = 1$ cm/min and $\dfrac{dA}{dt} = 2$ cm^2/min.

Using the Product Rule, we have $\dfrac{dA}{dt} = \dfrac{1}{2}\left(b\dfrac{dh}{dt} + h\dfrac{db}{dt}\right)$. When $h = 10$ and $A = 100$, we have

$100 = \frac{1}{2}b(10)$ $\Rightarrow$ $\frac{1}{2}b = 10$ $\Rightarrow$ $b = 20$, so $2 = \frac{1}{2}\left(20 \cdot 1 + 10\dfrac{db}{dt}\right)$ $\Rightarrow$ $4 = 20 + 10\dfrac{db}{dt}$ $\Rightarrow$

$\dfrac{db}{dt} = \dfrac{4 - 20}{10} = -1.6$ cm/min.

15.

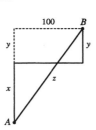

We are given that $\dfrac{dx}{dt} = 35$ km/h and $\dfrac{dy}{dt} = 25$ km/h. $z^2 = (x + y)^2 + 100^2$

$\Rightarrow$ $2z\dfrac{dz}{dt} = 2(x + y)\left(\dfrac{dx}{dt} + \dfrac{dy}{dt}\right)$. At 4:00 P.M., $x = 4(35) = 140$ and

$y = 4(25) = 100$ $\Rightarrow$ $z = \sqrt{(140 + 100)^2 + 100^2} = \sqrt{67{,}600} = 260$, so

$\dfrac{dz}{dt} = \dfrac{x + y}{z}\left(\dfrac{dx}{dt} + \dfrac{dy}{dt}\right) = \dfrac{140 + 100}{260}(35 + 25) = \dfrac{720}{13} \approx 55.4$ km/h.

17.

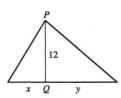

Using Q for the origin, we are given $\dfrac{dx}{dt} = -2$ ft/s and need to find $\dfrac{dy}{dt}$ when

$x = -5$. Using the Pythagorean Theorem twice, we have

$\sqrt{x^2 + 12^2} + \sqrt{y^2 + 12^2} = 39$, the total length of the rope. Differentiating

with respect to t, we get $\dfrac{x}{\sqrt{x^2 + 12^2}}\dfrac{dx}{dt} + \dfrac{y}{\sqrt{y^2 + 12^2}}\dfrac{dy}{dt} = 0$, so

$\dfrac{dy}{dt} = -\dfrac{x\sqrt{y^2 + 12^2}}{y\sqrt{x^2 + 12^2}}\dfrac{dx}{dt}$. Now when $x = -5$, $39 = \sqrt{(-5)^2 + 12^2} + \sqrt{y^2 + 12^2} = 13 + \sqrt{y^2 + 12^2}$ $\Leftrightarrow$

$\sqrt{y^2 + 12^2} = 26$, and $y = \sqrt{26^2 - 12^2} = \sqrt{532}$. So when $x = -5$,

$\dfrac{dy}{dt} = -\dfrac{(-5)(26)}{\sqrt{532}(13)}(-2) = -\dfrac{10}{\sqrt{133}} \approx -0.87$ ft/s. So cart B is moving towards Q at about 0.87 ft/s.

19.

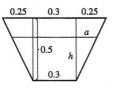

The figure is labeled in meters. The area A of a trapezoid is

$\frac{1}{2}(\text{base}_1 + \text{base}_2)(\text{height})$, and the volume V of the 10-meter-long trough is

$10A$. Thus, the volume of the trapezoid with height h is

$V = (10)\frac{1}{2}[0.3 + (0.3 + 2a)]h$. By similar triangles, $\dfrac{a}{h} = \dfrac{0.25}{0.5} = \dfrac{1}{2}$, so

$2a = h$ $\Rightarrow$ $V = 5(0.6 + h)h = 3h + 5h^2$. Now $\dfrac{dV}{dt} = \dfrac{dV}{dh}\dfrac{dh}{dt}$ $\Rightarrow$ $0.2 = (3 + 10h)\dfrac{dh}{dt}$ $\Rightarrow$

$\dfrac{dh}{dt} = \dfrac{0.2}{3 + 10h}$. When $h = 0.3$, $\dfrac{dh}{dt} = \dfrac{0.2}{3 + 10(0.3)} = \dfrac{0.2}{6}$ m/min $= \dfrac{10}{3}$ cm/min.

21.

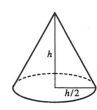

We are given that $\dfrac{dV}{dt} = 30$ ft^3/min. $V = \frac{1}{3}\pi r^2 h = \frac{1}{3}\pi\left(\dfrac{h}{2}\right)^2 h = \dfrac{\pi h^3}{12}$

$\Rightarrow \quad \dfrac{dV}{dt} = \dfrac{dV}{dh}\dfrac{dh}{dt} \quad \Rightarrow \quad 30 = \dfrac{\pi h^2}{4}\dfrac{dh}{dt} \quad \Rightarrow \quad \dfrac{dh}{dt} = \dfrac{120}{\pi h^2}.$ When

$h = 10$ ft, $\dfrac{dh}{dt} = \dfrac{120}{10^2\pi} = \dfrac{6}{5\pi} \approx 0.38$ ft/min.

23.

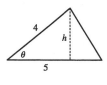

$A = \frac{1}{2}bh$, but $b = 5$ m and $\sin\theta = \dfrac{h}{4} \quad \Rightarrow \quad h = 4\sin\theta$, so

$A = \frac{1}{2}(5)(4\sin\theta) = 10\sin\theta$. We are given $\dfrac{d\theta}{dt} = 0.06$ rad/s, so

$\dfrac{dA}{dt} = \dfrac{dA}{d\theta}\dfrac{d\theta}{dt} = 10\cos\theta\dfrac{d\theta}{dt} = 0.6\cos\theta$. When $\theta = \frac{\pi}{3}$,

$\dfrac{dA}{dt} = 0.6\left(\cos\frac{\pi}{3}\right) = (0.6)\left(\frac{1}{2}\right) = 0.3$ m^2/s.

25. Differentiating both sides of $PV = C$ with respect to t and using the Product Rule gives us $P\dfrac{dV}{dt} + V\dfrac{dP}{dt} = 0$

$\Rightarrow \quad \dfrac{dV}{dt} = -\dfrac{V}{P}\dfrac{dP}{dt}$. When $V = 600$, $P = 150$ and $\dfrac{dP}{dt} = 20$, so we have $\dfrac{dV}{dt} = -\dfrac{600}{150}(20) = -80$. Thus, the

volume is decreasing at a rate of 80 cm^3/min.

27. With $R_1 = 80$ and $R_2 = 100$, $\dfrac{1}{R} = \dfrac{1}{R_1} + \dfrac{1}{R_2} = \dfrac{1}{80} + \dfrac{1}{100} = \dfrac{180}{8000} = \dfrac{9}{400}$, so $R = \dfrac{400}{9}$. Differentiating

$\dfrac{1}{R} = \dfrac{1}{R_1} + \dfrac{1}{R_2}$ with respect to t, we have $-\dfrac{1}{R^2}\dfrac{dR}{dt} = -\dfrac{1}{R_1^2}\dfrac{dR_1}{dt} - \dfrac{1}{R_2^2}\dfrac{dR_2}{dt} \quad \Rightarrow$

$\dfrac{dR}{dt} = R^2\left(\dfrac{1}{R_1^2}\dfrac{dR_1}{dt} + \dfrac{1}{R_2^2}\dfrac{dR_2}{dt}\right)$. When $R_1 = 80$ and $R_2 = 100$,

$\dfrac{dR}{dt} = \dfrac{400^2}{9^2}\left[\dfrac{1}{80^2}(0.3) + \dfrac{1}{100^2}(0.2)\right] = \dfrac{107}{810} \approx 0.132$ Ω/s.

29. (a)

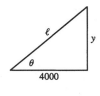

By the Pythagorean Theorem, $4000^2 + y^2 = \ell^2$. Differentiating with respect

to t, we obtain $2y\dfrac{dy}{dt} = 2\ell\dfrac{d\ell}{dt}$. We know that $\dfrac{dy}{dt} = 600$, so when $y = 3000$,

$\ell = \sqrt{4000^2 + 3000^2} = \sqrt{25{,}000{,}000} = 5000$ and

$\dfrac{d\ell}{dt} = \dfrac{y}{\ell}\dfrac{dy}{dt} = \dfrac{3000}{5000}(600) = \dfrac{1800}{5} = 360$ ft/s.

(b) Here $\tan\theta = \dfrac{y}{4000} \quad \Rightarrow \quad \dfrac{d}{dt}(\tan\theta) = \dfrac{d}{dt}\left(\dfrac{y}{4000}\right) \quad \Rightarrow \quad \sec^2\theta\dfrac{d\theta}{dt} = \dfrac{1}{4000}\dfrac{dy}{dt} \quad \Rightarrow \quad \dfrac{d\theta}{dt} = \dfrac{\cos^2\theta}{4000}\dfrac{dy}{dt}.$

When $y = 3000$, $\dfrac{dy}{dt} = 600$, $\ell = 5000$ and $\cos\theta = \dfrac{4000}{\ell} = \dfrac{4000}{5000} = \dfrac{4}{5}$, so

$\dfrac{d\theta}{dt} = \dfrac{(4/5)^2}{4000}(600) = 0.096$ rad/s.

31.

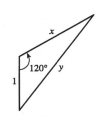

We are given that $\dfrac{dx}{dt} = 300$ km/h. By the Law of Cosines,

$$y^2 = x^2 + 1^2 - 2(1)(x)\cos 120° = x^2 + 1 - 2x\left(-\tfrac{1}{2}\right) = x^2 + x + 1, \text{ so}$$

$$2y\frac{dy}{dt} = 2x\frac{dx}{dt} + \frac{dx}{dt} \quad\Rightarrow\quad \frac{dy}{dt} = \frac{2x+1}{2y}\frac{dx}{dt}. \text{ After 1 minute,}$$

$$x = \tfrac{300}{60} = 5 \quad\Rightarrow\quad y = \sqrt{5^2 + 5 + 1} = \sqrt{31} \quad\Rightarrow$$

$$\frac{dy}{dt} = \frac{2(5)+1}{2\sqrt{31}}(300) = \frac{1650}{\sqrt{31}} \approx 296 \text{ km/h.}$$

33.

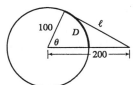

Let the distance between the runner and the friend be ℓ. Then by the Law of Cosines,

$$\ell^2 = 200^2 + 100^2 - 2 \cdot 200 \cdot 100 \cdot \cos\theta = 50{,}000 - 40{,}000\cos\theta \;(\star).$$

Differentiating implicitly with respect to t, we obtain

$$2\ell\frac{d\ell}{dt} = -40{,}000(-\sin\theta)\frac{d\theta}{dt}. \text{ Now if } D \text{ is the distance run when}$$

the angle is θ radians, then by the formula for the length of an arc on a circle, $s = r\theta$, we have $D = 100\theta$, so

$\theta = \dfrac{1}{100}D \;\Rightarrow\; \dfrac{d\theta}{dt} = \dfrac{1}{100}\dfrac{dD}{dt} = \dfrac{7}{100}$. To substitute into the expression for $\dfrac{d\ell}{dt}$, we must know $\sin\theta$ at the time

when $\ell = 200$, which we find from $(\star)$: $200^2 = 50{,}000 - 40{,}000\cos\theta \;\Leftrightarrow\; \cos\theta = \tfrac{1}{4} \;\Rightarrow$

$\sin\theta = \sqrt{1 - \left(\tfrac{1}{4}\right)^2} = \dfrac{\sqrt{15}}{4}$. Substituting, we get $2(200)\dfrac{d\ell}{dt} = 40{,}000\dfrac{\sqrt{15}}{4}\left(\tfrac{7}{100}\right) \;\Rightarrow$

$d\ell/dt = \dfrac{7\sqrt{15}}{4} \approx 6.78$ m/s. Whether the distance between them is increasing or decreasing depends on the direction in which the runner is running.

◆ 4.2 ◆ Maximum and Minimum Values • • • • • • • • • •

1. A function f has an **absolute minimum** at $x = c$ if $f(c)$ is the smallest function value on the entire domain of f, whereas f has a **local minimum** at c if $f(c)$ is the smallest function value when x is near c.

3. Absolute maximum at b; absolute minimum at d; local maxima at b and e; local minima at d and s; neither a maximum nor a minimum at a, c, r, and t.

5. Absolute maximum value is $f(4) = 4$; absolute minimum value is $f(7) = 0$; local maximum values are $f(4) = 4$ and $f(6) = 3$; local minimum values are $f(2) = 1$ and $f(5) = 2$.

7. The highest point must occur at $x = 0$ and the lowest point must occur at $x = 3$.

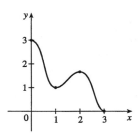

9. The figure has $f'(2) = 0$, so 2 is a critical number. There is an absolute maximum and an absolute minimum, but f has no *local* maximum or minimum.

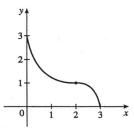

11. (a)

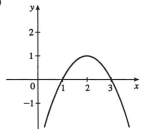

(b)

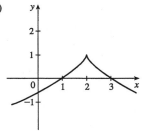

(c)

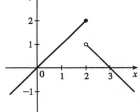

13. (a) *Note:* By the Extreme Value Theorem, f must *not* be continuous; because if it were, it would attain an absolute minimum.

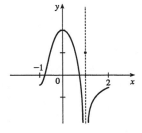

(b)

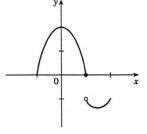

15. $f(x) = 8 - 3x$, $x \geq 1$. Absolute maximum
$f(1) = 5$; no local maximum. No absolute or
local minimum.

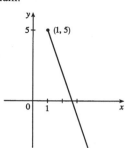

17. $f(x) = x^2$, $0 < x < 2$. No absolute or local
maximum or minimum value.

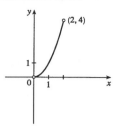

19. $f(\theta) = \sin \theta$, $-2\pi \leq \theta \leq 2\pi$. Absolute and
local maxima $f\left(-\frac{3\pi}{2}\right) = f\left(\frac{\pi}{2}\right) = 1$. Absolute
and local minima $f\left(-\frac{\pi}{2}\right) = f\left(\frac{3\pi}{2}\right) = -1$.

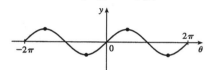

21. $f(x) = 1 - \sqrt{x}$. Absolute maximum $f(0) = 1$;
no local maximum. No absolute or local
minimum.

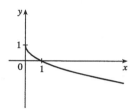

23. $f(x) = 5x^2 + 4x$ $\Rightarrow$ $f'(x) = 10x + 4$. $f'(x) = 0$ $\Rightarrow$ $x = -\frac{2}{5}$, so $-\frac{2}{5}$ is the only critical number.

25. $s(t) = 3t^4 + 4t^3 - 6t^2$ $\Rightarrow$ $s'(t) = 12t^3 + 12t^2 - 12t$. $s'(t) = 0$ $\Rightarrow$ $12t(t^2 + t - 1)$ $\Rightarrow$ $t = 0$ or
$t^2 + t - 1 = 0$. Using the quadratic formula to solve the latter equation gives us
$$t = \frac{-1 \pm \sqrt{1^2 - 4(1)(-1)}}{2(1)} = \frac{-1 \pm \sqrt{5}}{2} \approx 0.618, -1.618.$$ The three critical numbers are 0, $\dfrac{-1 \pm \sqrt{5}}{2}$.

27. $f(r) = \dfrac{r}{r^2 + 1}$ $\Rightarrow$ $f'(r) = \dfrac{(r^2 + 1)1 - r(2r)}{(r^2 + 1)^2} = \dfrac{-r^2 + 1}{(r^2 + 1)^2} = 0$ $\Leftrightarrow$ $r^2 = 1$ $\Leftrightarrow$ $r = \pm 1$, so these are
the critical numbers. Note that $f'(r)$ always exists since $r^2 + 1 \neq 0$.

29. $F(x) = x^{4/5}(x - 4)^2$ $\Rightarrow$
$F'(x) = x^{4/5} \cdot 2(x - 4) + (x - 4)^2 \cdot \frac{4}{5}x^{-1/5} = \frac{1}{5}x^{-1/5}(x - 4)\left[5 \cdot x \cdot 2 + (x - 4) \cdot 4\right]$
$= \dfrac{(x - 4)(14x - 16)}{5x^{1/5}} = \dfrac{2(x - 4)(7x - 8)}{5x^{1/5}} = 0$ when $x = 4$, $\frac{8}{7}$; and $F'(0)$ does not exist.
Critical numbers are 0, $\frac{8}{7}$, 4.

31. $f(\theta) = \sin^2(2\theta)$ $\Rightarrow$ $f'(\theta) = 2\sin(2\theta)\cos(2\theta)(2) = 2(2\sin 2\theta \cos 2\theta) = 2\left[\sin(2 \cdot 2\theta)\right]$ [double-angle
formula for the sine] $= 2\sin 4\theta = 0$ $\Leftrightarrow$ $\sin 4\theta = 0$ $\Leftrightarrow$ $4\theta = n\pi$, n an integer. So $\theta = n\pi/4$ are the critical
numbers.

33. $f(x) = x \ln x$ $\Rightarrow$ $f'(x) = x(1/x) + (\ln x) \cdot 1 = \ln x + 1$. $f'(x) = 0$ $\Leftrightarrow$ $\ln x = -1$ $\Leftrightarrow$
$x = e^{-1} = 1/e$. Therefore, the only critical number is $x = 1/e$.

35. $f(x) = 3x^2 - 12x + 5$, $[0, 3]$. $f'(x) = 6x - 12 = 0$ $\Leftrightarrow$ $x = 2$. Applying the Closed Interval Method, we find
that $f(0) = 5$, $f(2) = -7$, and $f(3) = -4$. So $f(0) = 5$ is the absolute maximum and $f(2) = -7$ is the absolute
minimum.

37. $f(x) = x^4 - 2x^2 + 3$, $[-2, 3]$. $f'(x) = 4x^3 - 4x = 4x(x^2 - 1) = 4x(x+1)(x-1) = 0 \Leftrightarrow x = -1, 0, 1$.
$f(-2) = 11$, $f(-1) = 2$, $f(0) = 3$, $f(1) = 2$, $f(3) = 66$. So $f(3) = 66$ is the absolute maximum and
$f(\pm 1) = 2$ is the absolute minimum.

39. $f(x) = x^2 + \dfrac{2}{x}$, $[\frac{1}{2}, 2]$. $f'(x) = 2x - \dfrac{2}{x^2} = 2\dfrac{x^3 - 1}{x^2} = 0 \Leftrightarrow x^3 - 1 = 0 \Leftrightarrow (x-1)(x^2 + x + 1) = 0$,
but $x^2 + x + 1 \neq 0$, so $x = 1$. The denominator is 0 at $x = 0$, but not in the desired interval. $f\left(\frac{1}{2}\right) = \frac{17}{4} = 4.25$,
$f(1) = 3$, $f(2) = 5$. So $f(2) = 5$ is the absolute maximum and $f(1) = 3$ is the absolute minimum.

41. $f(x) = \sin x + \cos x$, $\left[0, \frac{\pi}{3}\right]$. $f'(x) = \cos x - \sin x = 0 \Leftrightarrow \sin x = \cos x \Rightarrow \dfrac{\sin x}{\cos x} = 1 \Rightarrow \tan x = 1$
$\Rightarrow x = \frac{\pi}{4}$. $f(0) = 1$, $f\left(\frac{\pi}{4}\right) = \sqrt{2} \approx 1.41$, $f\left(\frac{\pi}{3}\right) = \frac{\sqrt{3}+1}{2} \approx 1.37$. So $f\left(\frac{\pi}{4}\right) = \sqrt{2}$ is the absolute maximum
and $f(0) = 1$ is the absolute minimum.

43. $f(x) = xe^{-x}$, $[0, 2]$. $f'(x) = x(-e^{-x}) + e^{-x} = e^{-x}(1 - x) = 0 \Leftrightarrow x = 1$. $f(0) = 0$,
$f(1) = e^{-1} = 1/e \approx 0.37$, $f(2) = 2/e^2 \approx 0.27$. So $f(1) = 1/e$ is the absolute maximum and $f(0) = 0$ is the
absolute minimum.

45.

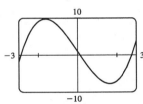

We see that $f'(x) = 0$ at about $x = -1.3$, 0.2, and 1.1. Since f'
exists everywhere, these are the only critical numbers.

47. (a)

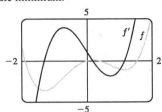

From the graph, it appears that the absolute maximum value is about
$f(-1.63) = 9.71$, and the absolute minimum value is about
$f(1.63) = -7.71$. These values make sense because the graph is
symmetric about the point $(0, 1)$. ($y = x^3 - 8x$ is symmetric about
the origin.)

(b) $f(x) = x^3 - 8x + 1 \Rightarrow f'(x) = 3x^2 - 8$. So $f'(x) = 0 \Rightarrow x = \pm\sqrt{\frac{8}{3}}$.
$f\left(\pm\sqrt{\frac{8}{3}}\right) = \left(\pm\sqrt{\frac{8}{3}}\right)^3 - 8\left(\pm\sqrt{\frac{8}{3}}\right) + 1 = \pm\frac{8}{3}\sqrt{\frac{8}{3}} \mp 8\sqrt{\frac{8}{3}} + 1$
$= -\frac{16}{3}\sqrt{\frac{8}{3}} + 1 = 1 - \frac{32\sqrt{6}}{9}$ (minimum) or $\frac{16}{3}\sqrt{\frac{8}{3}} + 1 = 1 + \frac{32\sqrt{6}}{9}$ (maximum)
(From the graph, we see that the extreme values do not occur at the endpoints.)

49. (a)

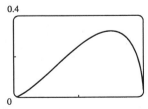

From the graph, it appears that the absolute maximum value is about
$f(0.75) = 0.32$, and the absolute minimum value is
$f(0) = f(1) = 0$; that is, at both endpoints.

(b) $f(x) = x\sqrt{x - x^2} \Rightarrow f'(x) = x \cdot \dfrac{1 - 2x}{2\sqrt{x - x^2}} + \sqrt{x - x^2} = \dfrac{(x - 2x^2) + (2x - 2x^2)}{2\sqrt{x - x^2}} = \dfrac{3x - 4x^2}{2\sqrt{x - x^2}}$.
So $f'(x) = 0 \Rightarrow 3x - 4x^2 = 0 \Rightarrow x(3 - 4x) = 0 \Rightarrow x = 0$ or $\frac{3}{4}$. $f(0) = f(1) = 0$ (minima), and
$f\left(\frac{3}{4}\right) = \frac{3}{4}\sqrt{\frac{3}{4} - \left(\frac{3}{4}\right)^2} = \frac{3\sqrt{3}}{16}$ (maximum).

51. The density is defined as $\rho = \dfrac{\text{mass}}{\text{volume}} = \dfrac{1000}{V(T)}$ (in g/cm^3). But a critical point of ρ will also be a critical point of

V [since $\dfrac{d\rho}{dT} = -1000V^{-2}\dfrac{dV}{dT}$ and V is never 0], and V is easier to differentiate than ρ.

$V(T) = 999.87 - 0.06426T + 0.0085043T^2 - 0.0000679T^3 \;\Rightarrow$

$V'(T) = -0.06426 + 0.0170086T - 0.0002037T^2$. Setting this equal to 0 and using the quadratic formula to find

T, we get $T = \dfrac{-0.0170086 \pm \sqrt{0.0170086^2 - 4 \cdot 0.0002037 \cdot 0.06426}}{2(-0.0002037)} \approx 3.9665\,°$ or $79.5318\,°$. Since we are

only interested in the region $0\,° \le T \le 30\,°$, we check the density ρ at the endpoints and at $3.9665\,°$:

$\rho(0) \approx \dfrac{1000}{999.87} \approx 1.00013$; $\rho(30) \approx \dfrac{1000}{1003.7641} \approx 0.99625$; $\rho(3.9665) \approx \dfrac{1000}{999.7447} \approx 1.000255$. So water has

its maximum density at about $3.9665\,°$C.

53. We apply the Closed Interval Method to the continuous function

$I(t) = 0.00009045t^5 + 0.001438t^4 - 0.06561t^3 + 0.4598t^2 - 0.6270t + 99.33$ on $[0, 10]$. Its derivative is

$I'(t) = 0.00045225t^4 + 0.005752t^3 - 0.19683t^2 + 0.9196t - 0.6270$. Since I' exists for all t, the only critical

numbers of I occur when $I'(t) = 0$. We use a root-finder on a computer algebra system (or a graphing device) to

find that $I'(t) = 0$ when $t \approx -29.7186$, 0.8231, 5.1309, or 11.0459, but only the second and third roots lie in the

interval $[0, 10]$. The values of I at these critical numbers are $I(0.8231) \approx 99.09$ and $I(5.1309) \approx 100.67$. The

values of I at the endpoints of the interval are $I(0) = 99.33$ and $I(10) \approx 96.86$. Comparing these four numbers,

we see that food was most expensive at $t \approx 5.1309$ (corresponding roughly to August, 1989) and cheapest at

$t = 10$ (midyear 1994).

55. (a) $v(r) = k(r_0 - r)r^2 = kr_0r^2 - kr^3 \;\Rightarrow$

$v'(r) = 2kr_0r - 3kr^2$. $v'(r) = 0 \;\Rightarrow$

$kr(2r_0 - 3r) = 0 \;\Rightarrow\; r = 0$ or $\frac{2}{3}r_0$ (but 0 is not in

the interval). Evaluating v at $\frac{1}{2}r_0$, $\frac{2}{3}r_0$, and r_0, we get

$v\left(\frac{1}{2}r_0\right) = \frac{1}{8}kr_0^3$, $v\left(\frac{2}{3}r_0\right) = \frac{4}{27}kr_0^3$, and $v(r_0) = 0$.

Since $\frac{4}{27} > \frac{1}{8}$, v attains its maximum value at

$r = \frac{2}{3}r_0$. This supports the statement in the text.

(b) From part (a), the maximum value of v is

$\frac{4}{27}kr_0^3$.

(c)

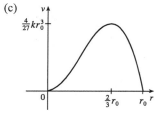

 4.3 **Derivatives and the Shapes of Curves** • • • • • • • •

1. $\dfrac{f(8) - f(0)}{8 - 0} = \dfrac{6 - 4}{8} = \dfrac{1}{4}$. The values of c which satisfy

$f'(c) = \frac{1}{4}$ seem to be about $c = 0.8$, 3.2, 4.4, and 6.1.

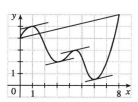

3. (a) Use the Increasing/Decreasing (I/D) Test.

(b) Use the Concavity Test.

(c) At any value of x where the concavity changes, we have an inflection point at $(x, f(x))$.

5. There is an inflection point at $x = 1$ because $f''(x)$ changes from negative to positive there, and so the graph of f changes from concave downward to concave upward. There is an inflection point at $x = 7$ because $f''(x)$ changes from positive to negative there, and so the graph of f changes from concave upward to concave downward.

7. (a) $f(x) = x^3 - 12x + 1 \implies f'(x) = 3x^2 - 12 = 3(x+2)(x-2)$. So $f'(x) > 0 \iff x > 2$ or $x < -2$ and $f'(x) < 0 \iff -2 < x < 2$. So f is increasing on $(-\infty, -2)$ and $(2, \infty)$ and decreasing on $(-2, 2)$.

(b) f changes from increasing to decreasing at $x = -2$ and from decreasing to increasing at $x = 2$. Thus, $f(-2) = 17$ is a local maximum and $f(2) = -15$ is a local minimum.

(c) $f''(x) = 6x$. $f''(x) > 0 \iff x > 0$ and $f''(x) < 0 \iff x < 0$. Thus, f is concave upward on $(0, \infty)$ and concave downward on $(-\infty, 0)$. There is an inflection point where the concavity changes, at $(0, f(0)) = (0, 1)$.

9. (a) $f(x) = x - 2\sin x$ on $(0, 3\pi) \implies f'(x) = 1 - 2\cos x$. $f'(x) > 0 \iff 1 - 2\cos x > 0 \iff \cos x < \frac{1}{2}$ $\iff \frac{\pi}{3} < x < \frac{5\pi}{3}$ or $\frac{7\pi}{3} < x < 3\pi$. $f'(x) < 0 \iff \cos x > \frac{1}{2} \iff 0 < x < \frac{\pi}{3}$ or $\frac{5\pi}{3} < x < \frac{7\pi}{3}$. So f is increasing on $\left(\frac{\pi}{3}, \frac{5\pi}{3}\right)$ and $\left(\frac{7\pi}{3}, 3\pi\right)$, and f is decreasing on $\left(0, \frac{\pi}{3}\right)$ and $\left(\frac{5\pi}{3}, \frac{7\pi}{3}\right)$.

(b) f changes from increasing to decreasing at $x = \frac{5\pi}{3}$, and from decreasing to increasing at $x = \frac{\pi}{3}$ and at $x = \frac{7\pi}{3}$. Thus, $f\left(\frac{5\pi}{3}\right) = \frac{5\pi}{3} + \sqrt{3} \approx 6.97$ is a local maximum and $f\left(\frac{\pi}{3}\right) = \frac{\pi}{3} - \sqrt{3} \approx -0.68$ and $f\left(\frac{7\pi}{3}\right) = \frac{7\pi}{3} - \sqrt{3} \approx 5.60$ are local minima.

(c) $f''(x) = 2\sin x > 0 \iff 0 < x < \pi$ and $2\pi < x < 3\pi$, $f''(x) < 0 \iff \pi < x < 2\pi$. Thus, f is concave upward on $(0, \pi)$ and $(2\pi, 3\pi)$, and f is concave downward on $(\pi, 2\pi)$. There are inflection points at (π, π) and $(2\pi, 2\pi)$.

11. (a) $y = f(x) = xe^x \implies f'(x) = xe^x + e^x = e^x(x+1)$. So $f'(x) > 0 \iff x + 1 > 0 \iff x > -1$. Thus, f is increasing on $(-1, \infty)$ and decreasing on $(-\infty, -1)$.

(b) f changes from decreasing to increasing at its only critical number, $x = -1$. Thus, $f(-1) = -e^{-1}$ is a local minimum.

(c) $f'(x) = e^x(x+1) \implies f''(x) = e^x(1) + (x+1)e^x = e^x(x+2)$. So $f''(x) > 0 \iff x + 2 > 0 \iff x > -2$. Thus, f is concave upward on $(-2, \infty)$ and concave downward on $(-\infty, -2)$. Since the concavity changes direction at $x = -2$, the point $\left(-2, -2e^{-2}\right)$ is an inflection point.

13. (a) $y = f(x) = \dfrac{\ln x}{\sqrt{x}}$. (Note that f is only defined for $x > 0$.)

$$f'(x) = \frac{\sqrt{x}\,(1/x) - \ln x\left(\frac{1}{2}x^{-1/2}\right)}{x} = \frac{\dfrac{1}{\sqrt{x}} - \dfrac{\ln x}{2\sqrt{x}}}{x} \cdot \frac{2\sqrt{x}}{2\sqrt{x}} = \frac{2 - \ln x}{2x^{3/2}} > 0 \iff$$

$2 - \ln x > 0 \iff \ln x < 2 \iff x < e^2$. Therefore f is increasing on $(0, e^2)$ and decreasing on (e^2, ∞).

(b) f changes from increasing to decreasing at $x = e^2$, so $f(e^2) = \dfrac{\ln e^2}{\sqrt{e^2}} = \dfrac{2}{e}$ is a local maximum.

(c) $f''(x) = \dfrac{2x^{3/2}(-1/x) - (2 - \ln x)\left(3x^{1/2}\right)}{(2x^{3/2})^2} = \dfrac{-2x^{1/2} + 3x^{1/2}(\ln x - 2)}{4x^3}$

$= \dfrac{x^{1/2}(-2 + 3\ln x - 6)}{4x^3} = \dfrac{3\ln x - 8}{4x^{5/2}}$

$f''(x) = 0 \iff \ln x = \frac{8}{3} \iff x = e^{8/3}$. $f''(x) > 0 \iff x > e^{8/3}$, so f is concave upward on $\left(e^{8/3}, \infty\right)$ and concave downward on $\left(0, e^{8/3}\right)$. There is an inflection point at $\left(e^{8/3}, \frac{8}{3}e^{-4/3}\right) \approx (14.39, 0.70)$.

15. $f(x) = x + \sqrt{1-x}$ $\Rightarrow$ $f'(x) = 1 + \frac{1}{2}(1-x)^{-1/2}(-1) = 1 - \dfrac{1}{2\sqrt{1-x}}$. Note that f is defined for

$1 - x \geq 0$; that is, for $x \leq 1$. $f'(x) = 0$ $\Rightarrow$ $2\sqrt{1-x} = 1$ $\Rightarrow$ $\sqrt{1-x} = \frac{1}{2}$ $\Rightarrow$ $1 - x = \frac{1}{4}$ $\Rightarrow$

$x = \frac{3}{4}$. f' does not exist at $x = 1$, but we can't have a local maximum or minimum at an endpoint.

   First Derivative Test: $f'(x) > 0$ $\Rightarrow$ $x < \frac{3}{4}$ and $f'(x) < 0$ $\Rightarrow$ $\frac{3}{4} < x < 1$. Since f' changes from

positive to negative at $x = \frac{3}{4}$, $f\left(\frac{3}{4}\right) = \frac{5}{4}$ is a local maximum.

   Second Derivative Test: $f''(x) = -\frac{1}{2}\left(-\frac{1}{2}\right)(1-x)^{-3/2}(-1) = -\dfrac{1}{4\left(\sqrt{1-x}\right)^3}$. $f''\left(\frac{3}{4}\right) = -2 < 0$ $\Rightarrow$

$f\left(\frac{3}{4}\right) = \frac{5}{4}$ is a local maximum.

   Preference: The First Derivative Test may be slightly easier to apply in this case.

17. (a) $f(x) = 2x^3 - 3x^2 - 12x$ $\Rightarrow$ $f'(x) = 6x^2 - 6x - 12 = 6\left(x^2 - x - 2\right) = 6(x-2)(x+1)$. $f'(x) > 0$

   $\Leftrightarrow$ $x < -1$ or $x > 2$ and $f'(x) < 0$ $\Leftrightarrow$ $-1 < x < 2$. So f is increasing on $(-\infty, -1)$ and $(2, \infty)$, and f

   is decreasing on $(-1, 2)$.

   (b) Since f changes from increasing to decreasing at $x = -1$, $f(-1) = 7$ is a local maximum value. Since f

   changes from decreasing to increasing at $x = 2$, $f(2) = -20$ is a local minimum value.

   (c) $f''(x) = 6(2x - 1)$ $\Rightarrow$ $f''(x) > 0$ on $\left(\frac{1}{2}, \infty\right)$ and $f''(x) < 0$ on   (d)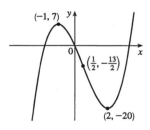

   $\left(-\infty, \frac{1}{2}\right)$. So f is concave upward on $\left(\frac{1}{2}, \infty\right)$ and concave

   downward on $\left(-\infty, \frac{1}{2}\right)$. There is a change in concavity at $x = \frac{1}{2}$,

   and we have an inflection point at $\left(\frac{1}{2}, -\frac{13}{2}\right)$.

19. (a) $h(x) = 3x^5 - 5x^3 + 3$ $\Rightarrow$ $h'(x) = 15x^4 - 15x^2 = 15x^2\left(x^2 - 1\right) = 0$ when $x = 0, \pm 1$. Since $15x^2$ is

   nonnegative, $h'(x) > 0$ $\Leftrightarrow$ $x^2 > 1$ $\Leftrightarrow$ $|x| > 1$ $\Leftrightarrow$ $x > 1$ or $x < -1$, so h is increasing on

   $(-\infty, -1)$ and $(1, \infty)$ and decreasing on $(-1, 1)$, with a horizontal tangent at $x = 0$.

   (b) Local maximum $h(-1) = 5$, local minimum $h(1) = 1$ (d)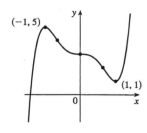

   (c) $h''(x) = 60x^3 - 30x = 30x\left(2x^2 - 1\right)$

   $\qquad = 60x\left(x + \frac{1}{\sqrt{2}}\right)\left(x - \frac{1}{\sqrt{2}}\right)$ $\Rightarrow$

   $h''(x) > 0$ when $x > \frac{1}{\sqrt{2}}$ or $-\frac{1}{\sqrt{2}} < x < 0$, so h is CU on

   $\left(-\frac{1}{\sqrt{2}}, 0\right)$ and $\left(\frac{1}{\sqrt{2}}, \infty\right)$ and CD on $\left(-\infty, -\frac{1}{\sqrt{2}}\right)$ and $\left(0, \frac{1}{\sqrt{2}}\right)$.

   Inflection points at $(0, 3)$ and $\left(\pm\frac{1}{\sqrt{2}}, 3 \mp \frac{7}{8}\sqrt{2}\right)$ [about

   $(-0.71, 4.24)$ and $(0.71, 1.76)$].

21. (a) $f(x) = x\sqrt{5-x}$ $\Rightarrow$

   $f'(x) = x \cdot \frac{1}{2}(5-x)^{-1/2}(-1) + (5-x)^{1/2} \cdot 1 = \frac{1}{2}(5-x)^{-1/2}[-x + 2(5-x)] = \dfrac{10 - 3x}{2\sqrt{5-x}} > 0$ $\Leftrightarrow$

   $x < \frac{10}{3}$, so f is increasing on $\left(-\infty, \frac{10}{3}\right)$ and decreasing on $\left(\frac{10}{3}, 5\right)$. [Note that the domain of f is $(-\infty, 5)$.]

(b) Local maximum $f\left(\frac{10}{3}\right) = \frac{10}{9}\sqrt{15} \approx 4.3$; no local minimum

(d)

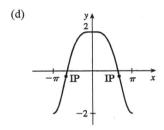

(c) $f''(x) = \dfrac{2(5-x)^{1/2}(-3) - (10-3x)\cdot 2\left(\frac{1}{2}\right)(5-x)^{-1/2}(-1)}{\left(2\sqrt{5-x}\right)^2}$

$= \dfrac{(5-x)^{-1/2}\left[-6(5-x) + (10-3x)\right]}{4(5-x)} = \dfrac{3x-20}{4(5-x)^{3/2}}$

$f''(x) < 0$ for $x < 5$, so f is CD on $(-\infty, 5)$. No IP

23. (a) $f(x) = 2\cos x + \sin^2 x$, $-\pi \le x \le \pi$. $f'(x) = -2\sin x + 2\sin x \cos x = 2\sin x\,(\cos x - 1)$. Since

$\cos x \le 1$, $\cos x - 1 \le 0$, so the sign of $f'(x)$ is the opposite of the sign of $\sin x$. Thus, $f'(x) > 0 \iff$

$\sin x < 0 \iff -\pi < x < 0$, so f is increasing on $(-\pi, 0)$ and decreasing on $(0, \pi)$.

(b) f changes from increasing to decreasing at $x = 0$, so $f(0) = 2$ is a local maximum. The absolute minimum

value of -2 occurs at $x = \pm\pi$ (the endpoints), but there is no local minimum.

(c) $f'(x) = -2\sin x + 2\sin x \cos x = -2\sin x + \sin 2x \quad\Rightarrow$

(d)

$f''(x) = -2\cos x + 2\cos 2x = 2\left(2\cos^2 x - \cos x - 1\right)$

$= 2(2\cos x + 1)(\cos x - 1) > 0 \iff$

$\cos x < -\frac{1}{2}$ $[\cos x - 1 \le 0] \iff x \in \left(-\pi, -\frac{2\pi}{3}\right)$ and

$\left(\frac{2\pi}{3}, \pi\right)$, so f is CU on these intervals and CD on $\left(-\frac{2\pi}{3}, \frac{2\pi}{3}\right)$. Note

that $f'' < 0$ on $\left(-\frac{2\pi}{3}, 0\right)$ and $\left(0, \frac{2\pi}{3}\right)$, so f is CD on these intervals

by the Concavity Test. In fact, since f' is decreasing on $\left(-\frac{2\pi}{3}, \frac{2\pi}{3}\right)$,

f is CD on $\left(-\frac{2\pi}{3}, \frac{2\pi}{3}\right)$. There are IP at $\left(\pm\frac{2\pi}{3}, -\frac{1}{4}\right)$.

25. (a) $\lim\limits_{x\to\pm\infty} \dfrac{1+x^2}{1-x^2} = \lim\limits_{x\to\pm\infty} \dfrac{(1/x^2)+1}{(1/x^2)-1} = -1$, so $y = -1$ is a HA. $\lim\limits_{x\to 1^-} \dfrac{1+x^2}{1-x^2} = \infty$, $\lim\limits_{x\to 1^+} \dfrac{1+x^2}{1-x^2} = -\infty$,

$\lim\limits_{x\to -1^-} \dfrac{1+x^2}{1-x^2} = -\infty$, $\lim\limits_{x\to -1^+} \dfrac{1+x^2}{1-x^2} = \infty$. So $x = 1$ and $x = -1$ are VA.

(b) $f(x) = \dfrac{1+x^2}{1-x^2} \Rightarrow f'(x) = \dfrac{(1-x^2)(2x) - (1+x^2)(-2x)}{(1-x^2)^2} = \dfrac{2x(1 - x^2 + 1 + x^2)}{(1-x^2)^2} = \dfrac{4x}{(1-x^2)^2} > 0$

$\iff x > 0$ $(x \ne 1)$, so f increases on $(0, 1)$, $(1, \infty)$ and decreases on $(-\infty, -1)$, $(-1, 0)$.

(c) $f(0) = 1$ is a local minimum.

(d) $f''(x) = \dfrac{(1-x^2)^2 \cdot 4 - 4x \cdot 2(1-x^2)(-2x)}{[(1-x^2)^2]^2}$

(e)

$= \dfrac{4(1-x^2)(1 - x^2 + 4x^2)}{(1-x^2)^4} = \dfrac{4(1+3x^2)}{(1-x^2)^3}$

Since the numerator is always positive, the sign of $f''(x)$ is the same

as the sign of $1 - x^2$. Thus, $f''(x) > 0 \iff 1 - x^2 > 0 \iff$

$x^2 < 1 \iff |x| < 1 \iff -1 < x < 1$, so f is CU on $(-1, 1)$

and CD on $(-\infty, -1)$ and $(1, \infty)$. There is no IP since $x = \pm 1$ are

not in the domain of f.

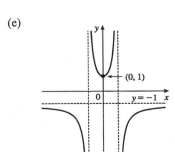

27. (a) $\lim\limits_{x \to \pm\infty} [x/(x^2 + 9)] = 0$, so $y = 0$ is a HA; no VA

(b) $f(x) = x/(x^2 + 9) \;\Rightarrow\; f'(x) = \dfrac{(x^2 + 9)(1) - x(2x)}{(x^2 + 9)^2} = \dfrac{9 - x^2}{(x^2 + 9)^2} > 0 \;\Leftrightarrow\; 9 - x^2 > 0 \;\Leftrightarrow\;$

$x^2 < 9 \;\Leftrightarrow\; |x| < 3 \;\Leftrightarrow\; -3 < x < 3$, so f is increasing on $(-3, 3)$ and decreasing on $(-\infty, -3)$
and $(3, \infty)$.

(c) Local minimum $f(-3) = -\frac{1}{6}$, local maximum $f(3) = \frac{1}{6}$

(d) $f''(x) = \dfrac{(x^2 + 9)^2(-2x) - (9 - x^2) \cdot 2(x^2 + 9)(2x)}{[(x^2 + 9)^2]^2} = \dfrac{(2x)(x^2 + 9)\left[-(x^2 + 9) - 2(9 - x^2)\right]}{(x^2 + 9)^4}$

$= \dfrac{2x(x^2 - 27)}{(x^2 + 9)^3} = 0 \;\Leftrightarrow\; x = 0, \pm\sqrt{27} = \pm 3\sqrt{3}$

$f''(x) > 0 \;\Leftrightarrow\; -3\sqrt{3} < x < 0$ or $x > 3\sqrt{3}$, so f is CU on

$(-3\sqrt{3}, 0)$ and $(3\sqrt{3}, \infty)$, and CD on $(-\infty, -3\sqrt{3})$ and

$(0, 3\sqrt{3})$. There are three inflection points: $(0, 0)$ and

$(\pm 3\sqrt{3}, \pm\frac{1}{12}\sqrt{3})$.

(e)

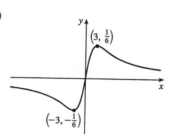

29. (a) $\lim\limits_{x \to \pm\infty} e^{-1/(x+1)} = 1$ since $-1/(x + 1) \to 0$, so $y = 1$ is a HA. $\lim\limits_{x \to -1+} e^{-1/(x+1)} = 0$ since

$-1/(x + 1) \to -\infty$, $\lim\limits_{x \to -1-} e^{-1/(x+1)} = \infty$ since $-1/(x + 1) \to \infty$, so $x = -1$ is a VA.

(b) $f(x) = e^{-1/(x+1)} \;\Rightarrow\; f'(x) = e^{-1/(x+1)}\left[-(-1)\dfrac{1}{(x+1)^2}\right]$ [Reciprocal Rule] $= e^{-1/(x+1)}/(x + 1)^2$

$\Rightarrow\; f'(x) > 0$ for all x except -1, so f is increasing on $(-\infty, -1)$ and $(-1, \infty)$.

(c) No local maximum or minimum

(d) $f''(x) = \dfrac{(x+1)^2 e^{-1/(x+1)}\left[1/(x+1)^2\right] - e^{-1/(x+1)}[2(x+1)]}{[(x+1)^2]^2}$

$= \dfrac{e^{-1/(x+1)}[1 - (2x+2)]}{(x+1)^4} = -\dfrac{e^{-1/(x+1)}(2x+1)}{(x+1)^4} \;\Rightarrow\;$

$f''(x) > 0 \;\Leftrightarrow\; 2x + 1 < 0 \;\Leftrightarrow\; x < -\frac{1}{2}$, so f is CU on $(-\infty, -1)$ and $(-1, -\frac{1}{2})$, and CD on $(-\frac{1}{2}, \infty)$.

f has an IP at $(-\frac{1}{2}, e^{-2})$.

(e)

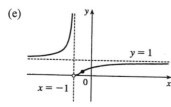

31. (a) From the graphs of

$f(x) = 3x^5 - 40x^3 + 30x^2$, it seems that f is concave upward on $(-2, 0.25)$ and $(2, \infty)$, and concave downward on $(-\infty, -2)$ and $(0.25, 2)$, with inflection points at about $(-2, 350)$, $(0.25, 1)$, and $(2, -100)$.

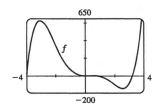
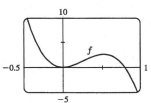

(b)

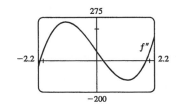

From the graph of $f''(x) = 60x^3 - 240x + 60$, it seems that f is CU on $(-2.1, 0.25)$ and $(1.9, \infty)$, and CD on $(-\infty, -2.1)$ and $(0.25, 1.9)$, with inflection points at about $(-2.1, 386)$, $(0.25, 1.3)$ and $(1.9, -87)$. (We have to check back on the graph of f to find the y-coordinates of the inflection points.)

33. (a)

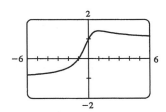

From the graph, we get us an estimate of $f(1) \approx 1.41$ as a local maximum, and no local minimum. $f(x) = \dfrac{x+1}{\sqrt{x^2+1}} \Rightarrow$

$f'(x) = \dfrac{1-x}{(x^2+1)^{3/2}}$. $f'(x) = 0 \Leftrightarrow x = 1$.

$f(1) = \dfrac{2}{\sqrt{2}} = \sqrt{2}$ is the exact value.

(b) From the graph in part (a), f increases most rapidly somewhere between $x = -\frac{1}{2}$ and $x = -\frac{1}{4}$. To find the exact value, we need to find the maximum value of f', which we can do by finding the critical numbers of f'.

$f''(x) = \dfrac{2x^2 - 3x - 1}{(x^2+1)^{5/2}} = 0 \Leftrightarrow x = \dfrac{3 \pm \sqrt{17}}{4}$. $x = \dfrac{3+\sqrt{17}}{4}$ corresponds to the *minimum* value of f'.

The maximum value of f' is at $\left(\dfrac{3-\sqrt{17}}{4}, \sqrt{\dfrac{7}{6} - \dfrac{\sqrt{17}}{6}} \right) \approx (-0.28, 0.69)$.

35. In Maple, we define f and then use the command
`plot(diff(diff(f,x),x),x=-3..3);`. In Mathematica, we define f and then use `Plot[Dt[Dt[f,x],x],{x,-3,3}]`. We see that $f'' > 0$ for $x > 0.1$ and $f'' < 0$ for $x < 0.1$. So f is concave up on $(0.1, \infty)$ and concave down on $(-\infty, 0.1)$.

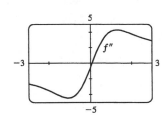

37. $y = -\dfrac{W}{24EI}x^4 + \dfrac{WL}{12EI}x^3 - \dfrac{WL^2}{24EI}x^2 = -\dfrac{W}{24EI}x^2\left(x^2 - 2Lx + L^2\right)$

$= \dfrac{-W}{24EI}x^2(x - L)^2 = cx^2(x - L)^2$

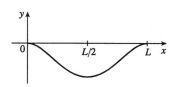

where $c = -\dfrac{W}{24EI}$ is a negative constant and $0 \le x \le L$. We sketch

$f(x) = cx^2(x - L)^2$ for $c = -1$. $f(0) = f(L) = 0$.

$f'(x) = cx^2[2(x - L)] + (x - L)^2(2cx) = 2cx(x - L)\left[x + (x - L)\right] = 2cx(x - L)(2x - L)$. So for

$0 < x < L$, $f'(x) > 0$ $\Leftrightarrow$ $x(x - L)(2x - L) < 0$ (since $c < 0$) $\Leftrightarrow$ $L/2 < x < L$ and $f'(x) < 0$ $\Leftrightarrow$

$0 < x < L/2$. So f is increasing on $(L/2, L)$ and decreasing on $(0, L/2)$, and there is a local and absolute

minimum at $(L/2, f(L/2)) = (L/2, cL^4/16)$.

$f'(x) = 2c\left[x(x - L)(2x - L)\right]$ $\Rightarrow$

$f''(x) = 2c\left[1(x - L)(2x - L) + x(1)(2x - L) + x(x - L)(2)\right] = 2c\left(6x^2 - 6Lx + L^2\right) = 0$ $\Leftrightarrow$

$x = \dfrac{6L \pm \sqrt{12L^2}}{12} = \tfrac{1}{2}L \pm \tfrac{\sqrt{3}}{6}L$, and these are the x-coordinates of the two inflection points.

39. From the graph, we estimate that the most rapid increase in the
number of VCRs occurs at about $t = 7$. To maximize the first
derivative, we need to determine the values for which the second

derivative is 0. $V(t) = \dfrac{75}{1 + 74e^{-0.6t}}$ $\Rightarrow$

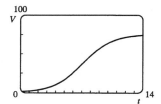

$V'(t) = -\dfrac{75\left[74e^{-0.6t}(-0.6)\right]}{\left(1 + 74e^{-0.6t}\right)^2} = \dfrac{3330e^{-0.6t}}{\left(1 + 74e^{-0.6t}\right)^2}$ $\Rightarrow$

$V''(t) = \dfrac{\left(1 + 74e^{-0.6t}\right)^2\left[3330e^{-0.6t}(-0.6)\right] - \left(3330e^{-0.6t}\right)2\left(1 + 74e^{-0.6t}\right)\left[74e^{-0.6t}(-0.6)\right]}{\left[(1 + 74e^{-0.6t})^2\right]^2}$

$= \dfrac{\left(1 + 74e^{-0.6t}\right)\left[3330e^{-0.6t}(-0.6)\right]\left[(1 + 74e^{-0.6t}) - 2(74e^{-0.6t})\right]}{(1 + 74e^{-0.6t})^4} = \dfrac{-1998e^{-0.6t}\left(1 - 74e^{-0.6t}\right)}{(1 + 74e^{-0.6t})^3}$

$V''(t) = 0$ $\Leftrightarrow$ $1 = 74e^{-0.6t}$ $\Leftrightarrow$ $e^{0.6t} = 74$ $\Leftrightarrow$ $0.6t = \ln 74$ $\Leftrightarrow$ $t = \tfrac{5}{3}\ln 74 \approx 7.173$ years, which
corresponds to early September 1987.

41. $f(x) = ax^3 + bx^2 + cx + d$ $\Rightarrow$ $f'(x) = 3ax^2 + 2bx + c$. We are
given that $f(1) = 0$ and $f(-2) = 3$, so $f(1) = a + b + c + d = 0$ and
$f(-2) = -8a + 4b - 2c + d = 3$. Also $f'(1) = 3a + 2b + c = 0$ and
$f'(-2) = 12a - 4b + c = 0$ by Fermat's Theorem. Solving these four

equations, we get $a = \tfrac{2}{9}$, $b = \tfrac{1}{3}$, $c = -\tfrac{4}{3}$, $d = \tfrac{7}{9}$, so the function is
$f(x) = \tfrac{1}{9}\left(2x^3 + 3x^2 - 12x + 7\right)$.

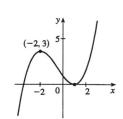

43. $f(x) = \tan x - x$ $\Rightarrow$ $f'(x) = \sec^2 x - 1 > 0$ for $0 < x < \tfrac{\pi}{2}$ since $\sec^2 x > 1$ for $0 < x < \tfrac{\pi}{2}$. So f is
increasing on $\left(0, \tfrac{\pi}{2}\right)$. Thus, $f(x) > f(0) = 0$ for $0 < x \le \tfrac{\pi}{2}$ $\Rightarrow$ $\tan x - x > 0$ $\Rightarrow$ $\tan x > x$ for
$0 < x < \tfrac{\pi}{2}$.

45. We are given that f is differentiable (and therefore continuous) everywhere. In particular, we can apply the Mean Value Theorem on the interval $[0, 4]$. There exists a number c in $(0, 4)$ such that $f(4) - f(0) = f'(c)(4 - 0)$, so $f(4) = f(0) + 4f'(c) = -3 + 4f'(c)$. We are given that $f'(x) \leq 5$ for all x, so in particular we know that $f'(c) \leq 5$. Multiplying both sides of this inequality by 4, we have $4f'(c) \leq 20$, so $f(4) = -3 + 4f'(c) \leq -3 + 20 = 17$. The largest possible value for $f(4)$ is 17.

47. Let $g(t)$ and $h(t)$ be the position functions of the two runners and let $f(t) = g(t) - h(t)$. By hypothesis, $f(0) = g(0) - h(0) = 0$ and $f(b) = g(b) - h(b) = 0$, where b is the finishing time. Then by the Mean Value

Theorem, there is a time c, with $0 < c < b$, such that $f'(c) = \dfrac{f(b) - f(0)}{b - 0}$. But $f(b) = f(0) = 0$, so $f'(c) = 0$.

Since $f'(c) = g'(c) - h'(c) = 0$, we have $g'(c) = h'(c)$. So at time c, both runners have the same velocity $g'(c) = h'(c)$.

49. Let the cubic function be $f(x) = ax^3 + bx^2 + cx + d$ $\Rightarrow$ $f'(x) = 3ax^2 + 2bx + c$ $\Rightarrow$ $f''(x) = 6ax + 2b$. So f is CU when $6ax + 2b > 0$ $\Leftrightarrow$ $x > -b/(3a)$, CD when $x < -b/(3a)$, and so the only point of inflection occurs when $x = -b/(3a)$. If the graph has three x-intercepts x_1, x_2 and x_3, then the expression for $f(x)$ must factor as $f(x) = a(x - x_1)(x - x_2)(x - x_3)$. Multiplying these factors together gives us $f(x) = a[x^3 - (x_1 + x_2 + x_3)x^2 + (x_1x_2 + x_1x_3 + x_2x_3)x - x_1x_2x_3]$. Equating the coefficients of the x^2-terms for the two forms of f gives us $b = -a(x_1 + x_2 + x_3)$. Hence, the x-coordinate of the point of inflection is $-\dfrac{b}{3a} = -\dfrac{-a(x_1 + x_2 + x_3)}{3a} = \dfrac{x_1 + x_2 + x_3}{3}$.

4.4 Graphing with Calculus *and* Calculators · · · · · · · ·

1. $f(x) = 4x^4 - 7x^2 + 4x + 6$ $\Rightarrow$ $f'(x) = 16x^3 - 14x + 4$ $\Rightarrow$ $f''(x) = 48x^2 - 14$

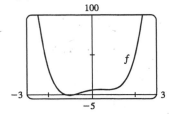

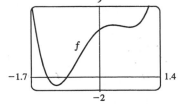

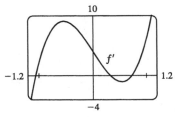

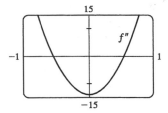

After finding suitable viewing rectangles (by ensuring that we have located all of the x-values where either $f' = 0$ or $f'' = 0$) we estimate from the graph of f' that f is increasing on $(-1.1, 0.3)$ and $(0.7, \infty)$ and decreasing on $(-\infty, -1.1)$ and $(0.3, 0.7)$, with a local maximum of $f(0.3) \approx 6.6$ and minima of $f(-1.1) \approx -1.1$ and $f(0.7) \approx 6.3$. We estimate from the graph of f'' that f is CU on $(-\infty, -0.5)$ and $(0.5, \infty)$ and CD on $(-0.5, 0.5)$, and that f has inflection points at about $(-0.5, 2.1)$ and $(0.5, 6.5)$.

3. $f(x) = \sqrt[3]{x^2 - 3x - 5}$ $\Rightarrow$ $f'(x) = \dfrac{1}{3}\dfrac{2x - 3}{(x^2 - 3x - 5)^{2/3}}$ $\Rightarrow$ $f''(x) = -\dfrac{2}{9}\dfrac{x^2 - 3x + 24}{(x^2 - 3x - 5)^{5/3}}$

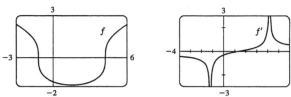

Note: With some CAS's, including Maple, it is necessary to define $f(x) = \dfrac{x^2 - 3x - 5}{|x^2 - 3x - 5|}|x^2 - 3x - 5|^{1/3}$, since
the CAS does not compute real cube roots of negative numbers. We estimate from the graph of f' that f is
increasing on $(1.5, \infty)$, and decreasing on $(-\infty, -1.5)$. f has no maximum. Minimum: $f(1.5) \approx -1.9$. From
the graph of f'', we estimate that f is CU on $(-1.2, 4.2)$ and CD on $(-\infty, -1.2)$ and $(4.2, \infty)$. IP at $(-1.2, 0)$
and $(4.2, 0)$.

5. $f(x) = \dfrac{x}{x^3 - x^2 - 4x + 1}$ $\Rightarrow$ $f'(x) = \dfrac{-2x^3 + x^2 + 1}{(x^3 - x^2 - 4x + 1)^2}$ $\Rightarrow$

$f''(x) = \dfrac{2(3x^5 - 3x^4 + 5x^3 - 6x^2 + 3x + 4)}{(x^3 - x^2 - 4x + 1)^3}$

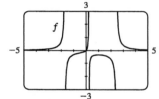
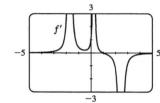
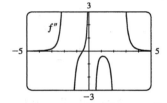

We estimate from the graph of f that $y = 0$ is a horizontal asymptote, and that there are vertical asymptotes at
$x = -1.7$, $x = 0.24$, and $x = 2.46$. From the graph of f', we estimate that f is increasing on $(-\infty, -1.7)$,
$(-1.7, 0.24)$, and $(0.24, 1)$, and that f is decreasing on $(1, 2.46)$ and $(2.46, \infty)$. There is a local maximum at
$f(1) = -\frac{1}{3}$. From the graph of f'', we estimate that f is CU on $(-\infty, -1.7)$, $(-0.506, 0.24)$, and $(2.46, \infty)$, and
that f is CD on $(-1.7, -0.506)$ and $(0.24, 2.46)$. There is an inflection point at $(-0.506, -0.192)$.

7. $f(x) = x^2 \sin x$, $-7 \le x \le 7$ $\Rightarrow$ $f'(x) = 2x \sin x + x^2 \cos x$ $\Rightarrow$ $f''(x) = 2 \sin x + 4x \cos x - x^2 \sin x$

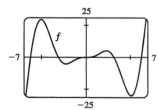
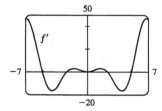
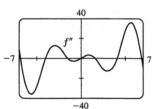

We estimate from the graph of f' that f is increasing on $(-7, -5.1)$, $(-2.3, 2.3)$, and $(5.1, 7)$ and decreasing on
$(-5.1, -2.3)$, and $(2.3, 5.1)$.
Local maxima: $f(-5.1) \approx 24.1$, $f(2.3) \approx 3.9$. Local minima: $f(-2.3) \approx -3.9$, $f(5.1) \approx -24.1$.
From the graph of f'', we estimate that f is CU on $(-7, -6.8)$, $(-4.0, -1.5)$, $(0, 1.5)$, and $(4.0, 6.8)$ and CD on
$(-6.8, -4.0)$, $(-1.5, 0)$, $(1.5, 4.0)$, and $(6.8, 7)$.
f has IP at $(-6.8, -24.4)$, $(-4.0, 12.0)$, $(-1.5, -2.3)$, $(0, 0)$, $(1.5, 2.3)$, $(4.0, -12.0)$ and $(6.8, 24.4)$.

9. $f(x) = 8x^3 - 3x^2 - 10 \;\Rightarrow\; f'(x) = 24x^2 - 6x \;\Rightarrow\; f''(x) = 48x - 6$

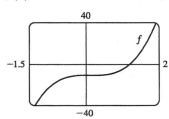
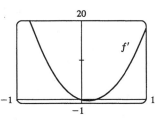
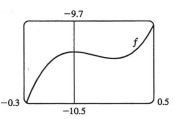

From the graphs, it appears that $f(x) = 8x^3 - 3x^2 - 10$ increases on $(-\infty, 0)$ and $(0.25, \infty)$ and decreases on $(0, 0.25)$; that f has a local maximum of $f(0) = -10.0$ and a local minimum of $f(0.25) \approx -10.1$; that f is CU on $(0.1, \infty)$ and CD on $(-\infty, 0.1)$; and that f has an IP at $(0.1, -10)$. To find the exact values, note that $f'(x) = 24x^2 - 6x = 6x(4x - 1)$, which is positive ($f$ is increasing) for $(-\infty, 0)$ and $\left(\frac{1}{4}, \infty\right)$, and negative ($f$ is decreasing) on $\left(0, \frac{1}{4}\right)$. By the FDT, f has a local maximum at $x = 0$: $f(0) = -10$; and f has a local minimum at $\frac{1}{4}$: $f\left(\frac{1}{4}\right) = \frac{1}{8} - \frac{3}{16} - 10 = -\frac{161}{16}$. $f''(x) = 48x - 6 = 6(8x - 1)$, which is positive ($f$ is CU) on $\left(\frac{1}{8}, \infty\right)$ and negative (f is CD) on $\left(-\infty, \frac{1}{8}\right)$. f has an IP at $\left(\frac{1}{8}, f\left(\frac{1}{8}\right)\right) = \left(\frac{1}{8}, -\frac{321}{32}\right)$.

11.

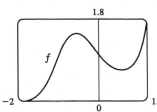
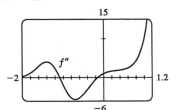

$f(x) = e^{x^3 - x} \to 0$ as $x \to -\infty$, and $f(x) \to \infty$ as $x \to \infty$. From the graph, it appears that f has a local minimum of about $f(0.58) = 0.68$, and a local maximum of about $f(-0.58) = 1.47$. To find the exact values, we calculate $f'(x) = (3x^2 - 1)e^{x^3 - x}$, which is 0 when $3x^2 - 1 = 0 \;\Leftrightarrow\; x = \pm\frac{1}{\sqrt{3}}$. The negative root corresponds to the local maximum $f\left(-\frac{1}{\sqrt{3}}\right) = e^{(-1/\sqrt{3})^3 - (-1/\sqrt{3})} = e^{2\sqrt{3}/9}$, and the positive root corresponds to the local minimum $f\left(\frac{1}{\sqrt{3}}\right) = e^{(1/\sqrt{3})^3 - (1/\sqrt{3})} = e^{-2\sqrt{3}/9}$. To estimate the inflection points, we calculate and graph $f''(x) = \frac{d}{dx}\left[(3x^2 - 1)e^{x^3 - x}\right] = (3x^2 - 1)e^{x^3 - x}(3x^2 - 1) + e^{x^3 - x}(6x) = e^{x^3 - x}\left(9x^4 - 6x^2 + 6x + 1\right)$.

From the graph, it appears that $f''(x)$ changes sign (and thus f has inflection points) at $x \approx -0.15$ and $x \approx -1.09$. From the graph of f, we see that these x-values correspond to inflection points at about $(-0.15, 1.15)$ and $(-1.09, 0.82)$.

13.

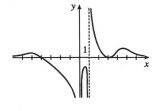

$f(x) = \dfrac{(x + 4)(x - 3)^2}{x^4(x - 1)}$ has VA at $x = 0$ and at $x = 1$ since

$$\lim_{x \to 0} f(x) = -\infty, \;\; \lim_{x \to 1^-} f(x) = -\infty \;\text{ and }\; \lim_{x \to 1^+} f(x) = \infty.$$

$$f(x) = \frac{\dfrac{x + 4}{x} \cdot \dfrac{(x - 3)^2}{x^2}}{\dfrac{x^4}{x^3} \cdot (x - 1)} \qquad \text{[dividing numerator and denominator by } x^3\text{]}$$

$$= \frac{(1 + 4/x)(1 - 3/x)^2}{x(x - 1)} \to 0 \text{ as } x \to \pm\infty, \text{ so } f \text{ is asymptotic}$$

to the x-axis. Since f is undefined at $x = 0$, it has no y-intercept. $f(x) = 0 \Rightarrow (x+4)(x-3)^2 = 0 \Rightarrow$ $x = -4$ or $x = 3$, so f has x-intercepts -4 and 3. Note, however, that the graph of f is only tangent to the x-axis and does not cross it at $x = 3$, since f is positive as $x \to 3^-$ and as $x \to 3^+$.

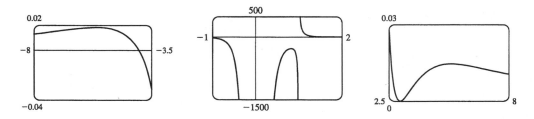

From these graphs, it appears that f has three maxima and one minimum. The maxima are approximately $f(-5.6) = 0.0182$, $f(0.82) = -281.5$ and $f(5.2) = 0.0145$ and we know (since the graph is tangent to the x-axis at $x = 3$) that the minimum is $f(3) = 0$.

15. $f(x) = \dfrac{x^2(x+1)^3}{(x-2)^2(x-4)^4} \Rightarrow f'(x) = -\dfrac{x(x+1)^2(x^3+18x^2-44x-16)}{(x-2)^3(x-4)^5}$ (from CAS).

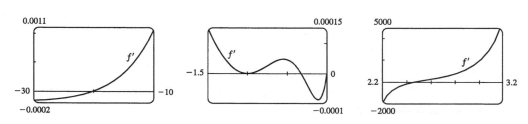

From the graphs of f', it seems that the critical points which indicate extrema occur at $x \approx -20$, -0.3, and 2.5, as estimated in Example 3. (There is another critical point at $x = -1$, but the sign of f' does not change there.) We differentiate again, obtaining $f''(x) = 2\dfrac{(x+1)(x^6+36x^5+6x^4-628x^3+684x^2+672x+64)}{(x-2)^4(x-4)^6}$.

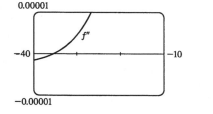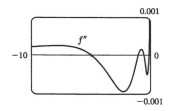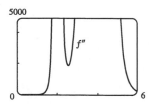

From the graphs of f'', it appears that f is CU on $(-35.3, -5.0)$, $(-1, -0.5)$, $(-0.1, 2)$, $(2, 4)$ and $(4, \infty)$ and CD on $(-\infty, -35.3)$, $(-5.0, -1)$ and $(-0.5, -0.1)$. We check back on the graphs of f to find the y-coordinates of the inflection points, and find that these points are approximately $(-35.3, -0.015)$, $(-5.0, -0.005)$, $(-1, 0)$, $(-0.5, 0.00001)$, and $(-0.1, 0.0000066)$.

17. $y = f(x) = \dfrac{\sin^2 x}{\sqrt{x^2+1}}$ with $0 \le x \le 3\pi$. From a CAS, $y' = \dfrac{\sin x \left[2(x^2+1)\cos x - x\sin x\right]}{(x^2+1)^{3/2}}$ and

$y'' = \dfrac{(4x^4+6x^2+5)\cos^2 x - 4x(x^2+1)\sin x \cos x - 2x^4 - 2x^2 - 3}{(x^2+1)^{5/2}}.$

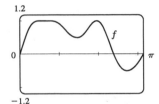

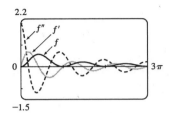

From the graph of f' and the formula for y', we determine that $y' = 0$ when $x = \pi$, 2π, 3π, or $x \approx 1.3$, 4.6, or 7.8. So f is increasing on $(0, 1.3)$, $(\pi, 4.6)$, and $(2\pi, 7.8)$. f is decreasing on $(1.3, \pi)$, $(4.6, 2\pi)$, and $(7.8, 3\pi)$. Local maxima: $f(1.3) \approx 0.6$, $f(4.6) \approx 0.21$, and $f(7.8) \approx 0.13$. Local minima: $f(\pi) = f(2\pi) = 0$. From the graph of f'', we see that $y'' = 0 \iff x \approx 0.6$, 2.1, 3.8, 5.4, 7.0, or 8.6. So f is CU on $(0, 0.6)$, $(2.1, 3.8)$, $(5.4, 7.0)$, and $(8.6, 3\pi)$. f is CD on $(0.6, 2.1)$, $(3.8, 5.4)$, and $(7.0, 8.6)$. There are IP at $(0.6, 0.25)$, $(2.1, 0.31)$, $(3.8, 0.10)$, $(5.4, 0.11)$, $(7.0, 0.061)$, and $(8.6, 0.065)$.

19.

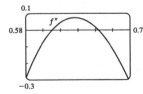

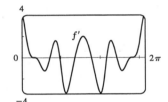

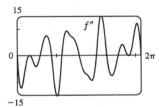

From the graph of $f(x) = \sin(x + \sin 3x)$ in the viewing rectangle $[0, \pi]$ by $[-1.2, 1.2]$, it looks like f has two maxima and two minima. If we calculate and graph $f'(x) = [\cos(x + \sin 3x)](1 + 3\cos 3x)$ on $[0, 2\pi]$, we see that the graph of f' appears to be almost tangent to the x-axis at about $x = 0.7$. The graph of $f'' = -[\sin(x + \sin 3x)](1 + 3\cos 3x)^2 + \cos(x + \sin 3x)(-9\sin 3x)$ is even more interesting near this x-value: it seems to just touch the x-axis.

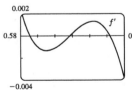

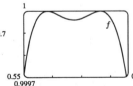

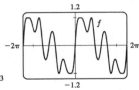

If we zoom in on this place on the graph of f'', we see that f'' actually does cross the axis twice near $x = 0.65$, indicating a change in concavity for a very short interval. If we look at the graph of f' on the same interval, we see that it changes sign three times near $x = 0.65$, indicating that what we had thought was a broad extremum at about $x = 0.7$ actually consists of three extrema (two maxima and a minimum). These maxima are roughly $f(0.59) = 1$ and $f(0.68) = 1$, and the minimum is roughly $f(0.64) = 0.99996$. There are also a maximum of about $f(1.96) = 1$ and minima of about $f(1.46) = 0.49$ and $f(2.73) = -0.51$. The points of inflection on $(0, \pi)$ are about $(0.61, 0.99998)$, $(0.66, 0.99998)$, $(1.17, 0.72)$, $(1.75, 0.77)$, and $(2.28, 0.34)$. On $(\pi, 2\pi)$, they are about $(4.01, -0.34)$, $(4.54, -0.77)$, $(5.11, -0.72)$, $(5.62, -0.99998)$, and $(5.67, -0.99998)$. There are also IP at $(0, 0)$ and $(\pi, 0)$. Note that the function is odd and periodic with period 2π, and it is also rotationally symmetric about all points of the form $((2n+1)\pi, 0)$, n an integer.

21.

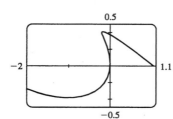 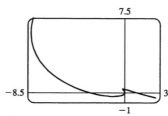

We graph the curve $x = t^4 - 2t^3 - 2t^2$, $y = t^3 - t$ in the viewing rectangle $[-2, 1.1]$ by $[-0.5, 0.5]$. This rectangle corresponds approximately to $t \in [-1, 0.8]$. We estimate that the curve has horizontal tangents at about $(-1, -0.4)$ and $(-0.17, 0.39)$ and vertical tangents at about $(0, 0)$ and $(-0.19, 0.37)$. We calculate

$\dfrac{dy}{dx} = \dfrac{dy/dt}{dx/dt} = \dfrac{3t^2 - 1}{4t^3 - 6t^2 - 4t}$. The horizontal tangents occur when $dy/dt = 3t^2 - 1 = 0 \iff t = \pm\frac{1}{\sqrt{3}}$, so

both horizontal tangents are shown in our graph. $t = \frac{1}{\sqrt{3}}$ corresponds to the point

$\left(\dfrac{-2\sqrt{3}-5}{9}, \dfrac{-2\sqrt{3}}{9}\right) \approx (-0.94, -0.38)$ and $t = -\frac{1}{\sqrt{3}}$ corresponds to $\left(\dfrac{2\sqrt{3}-5}{9}, \dfrac{2\sqrt{3}}{9}\right) \approx (-0.17, 0.38)$. The

vertical tangents occur when $dx/dt = 2t(2t^2 - 3t - 2) = 0 \iff 2t(2t+1)(t-2) = 0 \iff t = 0, -\frac{1}{2}$

or 2. It seems that we have missed one vertical tangent, and indeed if we plot the curve on the t-interval $[-1.2, 2.2]$ we see that there is another vertical tangent at $(-8, 6)$. The t-values and points at which there are vertical tangents are $t = 0$, $(0, 0)$; $t = -\frac{1}{2}$, $\left(-\frac{3}{16}, \frac{3}{8}\right)$; and $t = 2$, $(-8, 6)$.

23. $x = t^3 - ct$, $y = t^2$. For $c = 0$, there is a cusp at $(0, 0)$. For $c < 0$, there is a local minimum at $(0, 0)$. For $c > 0$, there is a loop whose size increases as c increases ($c = \frac{1}{2}$ and $c = 1$ are shown in the figure). The curve intersects itself on the y-axis; that is, when $x = 0 \iff t^3 - ct = 0 \iff t(t^2 - c) = 0 \iff t = 0, \pm\sqrt{c}$. Substituting $\pm\sqrt{c}$ for t gives us $y = c$, so the point of intersection is $(0, c)$.

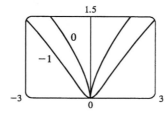 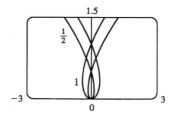

From the second figure, we see that the left- and rightmost points of the loop occur when there are vertical tangent lines. $dx/dt = 0 \implies 3t^2 - c = 0 \implies t = \pm\sqrt{c/3}$. The rightmost point occurs when $t = -\sqrt{c/3}$ and has coordinates $\left(\dfrac{2c\sqrt{3c}}{9}, \dfrac{c}{3}\right)$. The leftmost point occurs when $t = \sqrt{c/3}$ and has coordinates $\left(-\dfrac{2c\sqrt{3c}}{9}, \dfrac{c}{3}\right)$.

25. Note that $c = 0$ is a transitional value at which the graph consists of the x-axis. Also, we can see that if we substitute $-c$ for c, the function $f(x) = \dfrac{cx}{1 + c^2 x^2}$ will be reflected in the x-axis, so we investigate only positive values of c (except $c = -1$, as a demonstration of this reflective property). Also, f is an odd function. $\lim\limits_{x \to \pm\infty} f(x) = 0$, so $y = 0$ is a horizontal asymptote for all c. We calculate

$$f'(x) = \frac{(1 + c^2 x^2)c - cx(2c^2 x)}{(1 + c^2 x^2)^2} = -\frac{c(c^2 x^2 - 1)}{(1 + c^2 x^2)^2}. \; f'(x) = 0 \;\Leftrightarrow\; c^2 x^2 - 1 = 0 \;\Leftrightarrow\; x = \pm 1/c.$$ So there

is an absolute maximum of $f(1/c) = \frac{1}{2}$ and an absolute minimum of $f(-1/c) = -\frac{1}{2}$. These extrema have the same value regardless of c, but the maximum points move closer to the y-axis as c increases.

$$f''(x) = \frac{(-2c^3 x)(1 + c^2 x^2)^2 - (-c^3 x^2 + c)[2(1 + c^2 x^2)(2c^2 x)]}{(1 + c^2 x^2)^4}$$

$$= \frac{(-2c^3 x)(1 + c^2 x^2) + (c^3 x^2 - c)(4c^2 x)}{(1 + c^2 x^2)^3} = \frac{2c^3 x(c^2 x^2 - 3)}{(1 + c^2 x^2)^3}$$

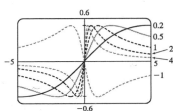

$f''(x) = 0 \;\Leftrightarrow\; x = 0$ or $\pm\sqrt{3}/c$, so there are inflection points at $(0, 0)$ and at $(\pm\sqrt{3}/c, \pm\sqrt{3}/4)$.

Again, the y-coordinate of the inflection points does not depend on c, but as c increases, both inflection points approach the y-axis.

27.

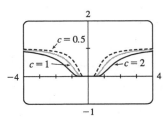

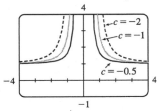

$c = 0$ is a transitional value — we get the graph of $y = 1$. For $c > 0$, we see that there is a HA at $y = 1$, and that the graph spreads out as c increases. At first glance there appears to be a minimum at $(0, 0)$, but $f(0)$ is undefined, so there is no minimum or maximum. For $c < 0$, we still have the HA at $y = 1$, but the range is $(1, \infty)$ rather than $(0, 1)$. We also have a VA at $x = 0$. $f(x) = e^{-c/x^2} \;\Rightarrow\; f'(x) = e^{-c/x^2}(-2c/x^3) \;\Rightarrow$

$f''(x) = \dfrac{2c(2c - 3x^2)}{x^6 e^{c/x^2}}$. $f'(x) \neq 0$ and $f'(x)$ exists for all $x \neq 0$ (and 0 is not in the domain of f), so there are no

maxima or minima. $f''(x) = 0 \;\Rightarrow\; x = \pm\sqrt{2c/3}$, so if $c > 0$, the inflection points spread out as c increases, and if $c < 0$, there are no IP. For $c > 0$, there are IP at $\left(\pm\sqrt{2c/3}, e^{-3/2}\right)$. Note that the y-coordinate of the IP is constant.

29. $f(x) = x^4 + cx^2 = x^2(x^2 + c)$. Note that f is an even function. For $c \geq 0$, the only x-intercept is the point $(0,0)$. We calculate $f'(x) = 4x^3 + 2cx = 4x(x^2 + \frac{1}{2}c)$ ⇒ $f''(x) = 12x^2 + 2c$. If $c \geq 0$, $x = 0$ is the only critical point and there is no inflection point. As we can see from the examples, there is no change in the basic shape of the graph for $c \geq 0$; it merely becomes steeper as c increases. For $c = 0$, the graph is the simple curve $y = x^4$. For $c < 0$, there are x-intercepts at 0 and at $\pm\sqrt{-c}$. Also, there is a maximum at $(0,0)$, and there are minima at $\left(\pm\sqrt{-\frac{1}{2}c}, -\frac{1}{4}c^2\right)$. As $c \to -\infty$, the x-coordinates of these minima get larger in absolute value, and the minimum points move downward. There are inflection points at $\left(\pm\sqrt{-\frac{1}{6}c}, -\frac{5}{36}c^2\right)$, which also move away from the origin as $c \to -\infty$.

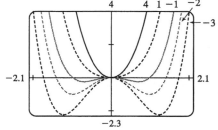

31. (a) $f(x) = cx^4 - 2x^2 + 1$. For $c = 0$, $f(x) = -2x^2 + 1$, a parabola whose vertex, $(0,1)$, is the absolute maximum. For $c > 0$, $f(x) = cx^4 - 2x^2 + 1$ opens upward with two minimum points. As $c \to 0$, the minimum points spread apart and move downward; they are below the x-axis for $0 < c < 1$ and above for $c > 1$. For $c < 0$, the graph opens downward, and has an absolute maximum at $x = 0$ and no local minimum.

(b) $f'(x) = 4cx^3 - 4x = 4cx(x^2 - 1/c)$ $(c \neq 0)$. If $c \leq 0$, 0 is the only critical number. $f''(x) = 12cx^2 - 4$, so $f''(0) = -4$ and there is a local maximum at $(0, f(0)) = (0,1)$, which lies on $y = 1 - x^2$. If $c > 0$, the critical numbers are 0 and $\pm 1/\sqrt{c}$. As before, there is a local maximum at $(0, f(0)) = (0,1)$, which lies on $y = 1 - x^2$. $f''(\pm 1/\sqrt{c}) = 12 - 4 = 8 > 0$, so there is a local minimum at $x = \pm 1/\sqrt{c}$. Here $f(\pm 1/\sqrt{c}) = c(1/c^2) - 2/c + 1 = -1/c + 1$. But $(\pm 1/\sqrt{c}, -1/c + 1)$ lies on $y = 1 - x^2$ since $1 - (\pm 1/\sqrt{c})^2 = 1 - 1/c$.

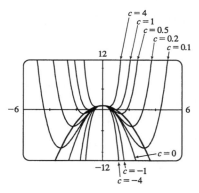

<h1>4.5 Indeterminate Forms and l'Hospital's Rule</h1>

Note: The use of l'Hospital's Rule is indicated by an H above the equal sign: $\overset{H}{=}$

1. (a) $\lim\limits_{x \to a} \dfrac{f(x)}{g(x)}$ is an indeterminate form of type $\dfrac{0}{0}$.

(b) $\lim\limits_{x \to a} \dfrac{f(x)}{p(x)} = 0$ because the numerator approaches 0 while the denominator becomes large.

(c) $\lim\limits_{x \to a} \dfrac{h(x)}{p(x)} = 0$ because the numerator approaches a finite number while the denominator becomes large.

(d) If $\lim\limits_{x \to a} p(x) = \infty$ and $f(x) \to 0$ through positive values, then $\lim\limits_{x \to a} \dfrac{p(x)}{f(x)} = \infty$. [For example, take $a = 0$,

$p(x) = 1/x^2$, and $f(x) = x^2$.] If $f(x) \to 0$ through negative values, then $\lim\limits_{x \to a} \dfrac{p(x)}{f(x)} = -\infty$. [For example,

take $a = 0$, $p(x) = 1/x^2$, and $f(x) = -x^2$.] If $f(x) \to 0$ through both positive and negative values, then the

limit might not exist. [For example, take $a = 0$, $p(x) = 1/x^2$, and $f(x) = x$.] It is not possible to evaluate

this limit.

(e) $\lim\limits_{x \to a} \dfrac{p(x)}{q(x)}$ is an indeterminate form of type $\dfrac{\infty}{\infty}$.

3. (a) When x is near a, $f(x)$ is near 0 and $p(x)$ is large, so $f(x) - p(x)$ is large negative. Thus,
$\lim\limits_{x \to a} [f(x) - p(x)] = -\infty$.

(b) $\lim\limits_{x \to a} [p(x) - q(x)]$ is an indeterminate form of type $\infty - \infty$.

(c) When x is near a, $p(x)$ and $q(x)$ are both large, so $p(x) + q(x)$ is large. Thus, $\lim\limits_{x \to a} [p(x) + q(x)] = \infty$.

5. We can simply factor the numerator to evaluate this limit.

$$\lim_{x \to -1} \frac{x^2 - 1}{x + 1} = \lim_{x \to -1} \frac{(x + 1)(x - 1)}{x + 1} = \lim_{x \to -1} (x - 1) = -2$$

7. $\lim\limits_{x \to 0} \dfrac{e^x - 1}{\sin x}$ is an indeterminate form of type $\dfrac{0}{0}$, so we'll apply l'Hospital's Rule.

$$\lim_{x \to 0} \frac{e^x - 1}{\sin x} \overset{H}{=} \lim_{x \to 0} \frac{e^x}{\cos x} = \frac{1}{1} = 1$$

9. This limit has the form $\dfrac{0}{0}$. $\lim\limits_{x \to 0} \dfrac{\tan px}{\tan qx} \overset{H}{=} \lim\limits_{x \to 0} \dfrac{p \sec^2 px}{q \sec^2 qx} = \dfrac{p(1)^2}{q(1)^2} = \dfrac{p}{q}$

11. $\lim\limits_{x \to 0^+} [(\ln x)/x] = -\infty$ since $\ln x \to -\infty$ as $x \to 0^+$ and dividing by small values of x just increases the

magnitude of the quotient $(\ln x)/x$. L'Hospital's Rule does not apply.

13. This limit has the form $\dfrac{0}{0}$. $\lim\limits_{t \to 0} \dfrac{5^t - 3^t}{t} \overset{H}{=} \lim\limits_{t \to 0} \dfrac{5^t \ln 5 - 3^t \ln 3}{1} = \ln 5 - \ln 3 = \ln \dfrac{5}{3}$

15. This limit has the form $\dfrac{0}{0}$. $\lim\limits_{x \to 0} \dfrac{e^x - 1 - x}{x^2} \overset{H}{=} \lim\limits_{x \to 0} \dfrac{e^x - 1}{2x} \overset{H}{=} \lim\limits_{x \to 0} \dfrac{e^x}{2} = \dfrac{1}{2}$

17. This limit has the form $\dfrac{0}{0}$. $\lim\limits_{x \to 0} \dfrac{\sin^{-1} x}{x} \overset{H}{=} \lim\limits_{x \to 0} \dfrac{1/\sqrt{1 - x^2}}{1} = \lim\limits_{x \to 0} \dfrac{1}{\sqrt{1 - x^2}} = \dfrac{1}{1} = 1$

19. This limit has the form $\dfrac{\infty}{\infty}$. $\lim\limits_{x \to \infty} \dfrac{x}{\ln(1 + 2e^x)} \overset{H}{=} \lim\limits_{x \to \infty} \dfrac{1}{\dfrac{1}{1 + 2e^x} \cdot 2e^x} = \lim\limits_{x \to \infty} \dfrac{1 + 2e^x}{2e^x} \overset{H}{=} \lim\limits_{x \to \infty} \dfrac{2e^x}{2e^x} = 1$

21. This limit has the form $0 \cdot (-\infty)$. We need to write this product as a quotient, but keep in mind that we will have to

differentiate both the numerator and the denominator. If we differentiate $\dfrac{1}{\ln x}$, we get a complicated expression that

results in a more difficult limit. Instead we write the quotient as $\dfrac{\ln x}{x^{-1/2}}$.

$$\lim_{x \to 0^+} \sqrt{x} \ln x = \lim_{x \to 0^+} \frac{\ln x}{x^{-1/2}} \overset{H}{=} \lim_{x \to 0^+} \frac{1/x}{-\frac{1}{2}x^{-3/2}} \cdot \frac{-2x^{3/2}}{-2x^{3/2}} = \lim_{x \to 0^+} (-2\sqrt{x}) = 0$$

23. $\lim\limits_{x\to\infty} e^{-x}\ln x = \lim\limits_{x\to\infty} \dfrac{\ln x}{e^x} \overset{\text{H}}{=} \lim\limits_{x\to\infty} \dfrac{1/x}{e^x} = \lim\limits_{x\to\infty} \dfrac{1}{xe^x} = 0$

25. $\lim\limits_{x\to\infty} x^3 e^{-x^2} = \lim\limits_{x\to\infty} \dfrac{x^3}{e^{x^2}} \overset{\text{H}}{=} \lim\limits_{x\to\infty} \dfrac{3x^2}{2xe^{x^2}} = \lim\limits_{x\to\infty} \dfrac{3x}{2e^{x^2}} \overset{\text{H}}{=} \lim\limits_{x\to\infty} \dfrac{3}{4xe^{x^2}} = 0$

27. $\lim\limits_{x\to 0}\left(\dfrac{1}{x} - \csc x\right) = \lim\limits_{x\to 0}\left(\dfrac{1}{x} - \dfrac{1}{\sin x}\right) = \lim\limits_{x\to 0} \dfrac{\sin x - x}{x\sin x}$

$\overset{\text{H}}{=} \lim\limits_{x\to 0} \dfrac{\cos x - 1}{x\cos x + \sin x} \overset{\text{H}}{=} \lim\limits_{x\to 0} \dfrac{-\sin x}{2\cos x - x\sin x} = \dfrac{0}{2} = 0$

29. As $x \to \infty$, $1/x \to 0$, and $e^{1/x} \to 1$. So the limit has the form $\infty - \infty$ and we will change the form to a product by factoring out x.

$$\lim\limits_{x\to\infty}\left(xe^{1/x} - x\right) = \lim\limits_{x\to\infty} x\left(e^{1/x} - 1\right) = \lim\limits_{x\to\infty} \dfrac{e^{1/x} - 1}{1/x} \overset{\text{H}}{=} \lim\limits_{x\to\infty} \dfrac{e^{1/x}(-1/x^2)}{-1/x^2} = \lim\limits_{x\to\infty} e^{1/x} = e^0 = 1$$

31. The limit $\lim\limits_{x\to 0^+} x^{\sin x}$ has the indeterminate form 0^0. We'll begin finding the given limit by letting $y = x^{\sin x}$ and taking the natural logarithm of both sides. Then we'll find the limit of $\ln y$ and lastly—don't forget—we must convert to an exponential form to find the given limit. $y = x^{\sin x} \Rightarrow \ln y = \sin x \ln x$, so

$$\lim\limits_{x\to 0^+} \ln y = \lim\limits_{x\to 0^+} \sin x \ln x = \lim\limits_{x\to 0^+} \dfrac{\ln x}{\csc x} \overset{\text{H}}{=} \lim\limits_{x\to 0^+} \dfrac{1/x}{-\csc x \cot x} = -\left(\lim\limits_{x\to 0^+} \dfrac{\sin x}{x}\right)\left(\lim\limits_{x\to 0^+} \tan x\right)$$

$$= -1 \cdot 0 = 0 \quad \Rightarrow \quad \lim\limits_{x\to 0^+} x^{\sin x} = \lim\limits_{x\to 0^+} e^{\ln y} = e^0 = 1$$

33. $y = (1 - 2x)^{1/x} \Rightarrow \ln y = \dfrac{1}{x}\ln(1 - 2x)$, so $\lim\limits_{x\to 0} \ln y = \lim\limits_{x\to 0} \dfrac{\ln(1 - 2x)}{x} \overset{\text{H}}{=} \lim\limits_{x\to 0} \dfrac{-2/(1 - 2x)}{1} = -2 \quad \Rightarrow$

$\lim\limits_{x\to 0}(1 - 2x)^{1/x} = \lim\limits_{x\to 0} e^{\ln y} = e^{-2}$.

35. As $x \to 0^+$, $\ln x \to -\infty$, and $-\ln x \to \infty$. So the given limit has the form ∞^0.

$y = (-\ln x)^x \Rightarrow \ln y = x\ln(-\ln x)$, so

$$\lim\limits_{x\to 0^+} \ln y = \lim\limits_{x\to 0^+} x\ln(-\ln x) = \lim\limits_{x\to 0^+} \dfrac{\ln(-\ln x)}{1/x} \overset{\text{H}}{=} \lim\limits_{x\to 0^+} \dfrac{[1/(-\ln x)](-1/x)}{-1/x^2} = \lim\limits_{x\to 0^+} \dfrac{-x}{\ln x} = 0 \quad \Rightarrow$$

$\lim\limits_{x\to 0^+} (-\ln x)^x = \lim\limits_{x\to 0^+} e^{\ln y} = e^0 = 1$.

37.

From the graph, it appears that $\lim\limits_{x\to\infty} x\left[\ln(x + 5) - \ln x\right] = 5$.

To prove this, we first note that

$$\ln(x + 5) - \ln x = \ln \dfrac{x + 5}{x} = \ln\left(1 + \dfrac{5}{x}\right) \to \ln 1 = 0 \text{ as } x \to \infty. \text{ Thus,}$$

$$\lim\limits_{x\to\infty} x\left[\ln(x + 5) - \ln x\right] = \lim\limits_{x\to\infty} \dfrac{\ln(x + 5) - \ln x}{1/x}$$

$$\overset{\text{H}}{=} \lim\limits_{x\to\infty} \dfrac{\dfrac{1}{x + 5} - \dfrac{1}{x}}{-1/x^2}$$

$$= \lim\limits_{x\to\infty}\left[\dfrac{x - (x + 5)}{x(x + 5)} \cdot \dfrac{-x^2}{1}\right] = \lim\limits_{x\to\infty} \dfrac{5x^2}{x^2 + 5x} = 5$$

39.

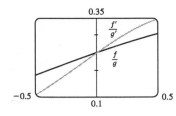

From the graph, it appears that

$$\lim_{x \to 0} \frac{f(x)}{g(x)} = \lim_{x \to 0} \frac{f'(x)}{g'(x)} = 0.25. \text{ We calculate}$$

$$\lim_{x \to 0} \frac{f(x)}{g(x)} = \lim_{x \to 0} \frac{e^x - 1}{x^3 + 4x} \overset{\text{H}}{=} \lim_{x \to 0} \frac{e^x}{3x^2 + 4} = \frac{1}{4}.$$

41. $\lim\limits_{x \to \infty} xe^{-x} = \lim\limits_{x \to \infty} (x/e^x) \overset{\text{H}}{=} \lim\limits_{x \to \infty} (1/e^x) = 0$, so $y = 0$ is a HA. $\lim\limits_{x \to -\infty} xe^{-x} = -\infty$. $f(x) = xe^{-x} \Rightarrow$

$f'(x) = x(-e^{-x}) + e^{-x} \cdot 1 = e^{-x}(1 - x) > 0 \iff 1 - x > 0 \iff x < 1$, so f is increasing on

$(-\infty, 1)$ and decreasing on $(1, \infty)$. By the FDT, $f(1) = 1/e$ is a local maximum.

$f''(x) = e^{-x}(-1) + (1 - x)(-e^{-x}) = e^{-x}(-1 - 1 + x) = e^{-x}(x - 2) > 0 \iff x - 2 > 0 \iff x > 2$,

so f is CU on $(2, \infty)$ and CD on $(-\infty, 2)$. IP is $(2, 2/e^2)$.

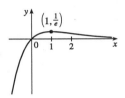

43. $\lim\limits_{x \to \infty} \frac{\ln x}{x} \overset{\text{H}}{=} \lim\limits_{x \to \infty} \frac{1/x}{1} = 0$, so $y = 0$ is a HA. Also $\lim\limits_{x \to 0^+} \frac{\ln x}{x} = -\infty$ since $\ln x \to -\infty$ and $x \to 0^+$, so $x = 0$

is a VA.

$f(x) = \dfrac{\ln x}{x} \Rightarrow f'(x) = \dfrac{x(1/x) - (\ln x)(1)}{x^2} = \dfrac{1 - \ln x}{x^2} = 0$ when $\ln x = 1 \iff x = e$. $f'(x) > 0 \iff$

$1 - \ln x > 0 \iff \ln x < 1 \iff 0 < x < e$. $f'(x) < 0 \iff x > e$. So f is increasing on $(0, e)$ and

decreasing on (e, ∞). By the FDT, $f(e) = 1/e$ is a local maximum.

$f''(x) = \dfrac{x^2(-1/x) - (1 - \ln x)(2x)}{(x^2)^2} = \dfrac{x(-1 - 2 + 2\ln x)}{x^4} = \dfrac{2\ln x - 3}{x^3}$, so $f''(x) > 0 \iff 2\ln x - 3 > 0$

$\iff \ln x > \dfrac{3}{2} \iff x > e^{3/2}$. $f''(x) < 0 \iff 0 < x < e^{3/2}$. So f is CU on $(e^{3/2}, \infty)$ and CD on

$(0, e^{3/2})$. There is an inflection point at $(e^{3/2}, \frac{3}{2}e^{-3/2})$.

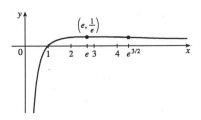

45. (a) $f(x) = x^2 \ln x$. The domain of f is $(0, \infty)$.

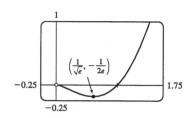

(b) $\lim\limits_{x \to 0^+} x^2 \ln x = \lim\limits_{x \to 0^+} \dfrac{\ln x}{1/x^2} \overset{\text{H}}{=} \lim\limits_{x \to 0^+} \dfrac{1/x}{-2/x^3} = \lim\limits_{x \to 0^+} \left(-\dfrac{x^2}{2}\right) = 0$. There is a hole at $(0,0)$.

(c) It appears that there is an IP at about $(0.2, -0.06)$ and a local minimum at $(0.6, -0.18)$. $f(x) = x^2 \ln x \Rightarrow$
$f'(x) = x^2(1/x) + (\ln x)(2x) = x(2 \ln x + 1) > 0 \iff \ln x > -\frac{1}{2} \iff x > e^{-1/2}$, so f is increasing on
$(1/\sqrt{e}, \infty)$, decreasing on $(0, 1/\sqrt{e})$. By the FDT, $f(1/\sqrt{e}) = -1/(2e)$ is a local minimum. This point is
approximately $(0.6065, -0.1839)$, which agrees with our estimate.
$f''(x) = x(2/x) + (2 \ln x + 1) = 2 \ln x + 3 > 0 \iff \ln x > -\frac{3}{2} \iff x > e^{-3/2}$, so f is CU on
$\left(e^{-3/2}, \infty\right)$ and CD on $\left(0, e^{-3/2}\right)$.
IP is $\left(e^{-3/2}, -3/(2e^3)\right) \approx (0.2231, -0.0747)$.

47. (a) $f(x) = x^{1/x}$

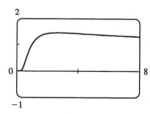

(b) Recall that $a^b = e^{b \ln a}$. $\lim\limits_{x \to 0^+} x^{1/x} = \lim\limits_{x \to 0^+} e^{(1/x) \ln x}$. As $x \to 0^+$, $\dfrac{\ln x}{x} \to -\infty$, so $x^{1/x} = e^{(1/x) \ln x} \to 0$.
This indicates that there is a hole at $(0,0)$. As $x \to \infty$, we have the indeterminate form ∞^0.

$\lim\limits_{x \to \infty} x^{1/x} = \lim\limits_{x \to \infty} e^{(1/x) \ln x}$, but $\lim\limits_{x \to \infty} \dfrac{\ln x}{x} \overset{\text{H}}{=} \lim\limits_{x \to \infty} \dfrac{1/x}{1} = 0$, so $\lim\limits_{x \to \infty} x^{1/x} = e^0 = 1$. This indicates that
$y = 1$ is a HA.

(c) Estimated maximum: $(2.72, 1.45)$. No estimated minimum. We use logarithmic differentiation to find any

critical numbers. $y = x^{1/x} \Rightarrow \ln y = \dfrac{1}{x} \ln x \Rightarrow \dfrac{y'}{y} = \dfrac{1}{x} \cdot \dfrac{1}{x} + (\ln x)\left(-\dfrac{1}{x^2}\right) \Rightarrow$

$y' = x^{1/x}\left(\dfrac{1 - \ln x}{x^2}\right) = 0 \Rightarrow \ln x = 1 \Rightarrow x = e$. For $0 < x < e$, $y' > 0$ and for $x > e$, $y' < 0$, so
$f(e) = e^{1/e}$ is a local maximum. This point is approximately $(2.7183, 1.4447)$, which agrees with our
estimate.

(d)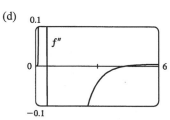

From the graph, we see that $f''(x) = 0$ at $x \approx 0.58$ and $x \approx 4.37$. Since f'' changes sign at these values, they are x-coordinates of inflection points.

49. If $c < 0$, then $\displaystyle\lim_{x \to -\infty} f(x) = \lim_{x \to -\infty} \frac{x}{e^{cx}} \overset{\text{H}}{=} \lim_{x \to -\infty} \frac{1}{ce^{cx}} = 0$, and $\displaystyle\lim_{x \to \infty} f(x) = \infty$.

If $c > 0$, then $\displaystyle\lim_{x \to -\infty} f(x) = -\infty$, and $\displaystyle\lim_{x \to \infty} f(x) \overset{\text{H}}{=} \lim_{x \to \infty} \frac{1}{ce^{cx}} = 0$.

If $c = 0$, then $f(x) = x$, so $\displaystyle\lim_{x \to \pm\infty} f(x) = \pm\infty$ respectively.

So we see that $c = 0$ is a transitional value. We now exclude the case $c = 0$, since we know how the function behaves in that case. To find the maxima and minima of f, we differentiate: $f(x) = xe^{-cx}$ $\Rightarrow$ $f'(x) = x(-ce^{-cx}) + e^{-cx} = (1 - cx)e^{-cx}$. This is 0 when $1 - cx = 0$ $\Leftrightarrow$ $x = 1/c$. If $c < 0$ then this represents a minimum of $f(1/c) = 1/(ce)$, since $f'(x)$ changes from negative to positive at $x = 1/c$; and if $c > 0$, it represents a maximum. As $|c|$ increases, the maximum or minimum gets closer to the origin. To find the inflection points, we differentiate again: $f'(x) = e^{-cx}(1 - cx)$ $\Rightarrow$ $f''(x) = e^{-cx}(-c) + (1 - cx)(-ce^{-cx}) = (cx - 2)ce^{-cx}$. This changes sign when $cx - 2 = 0$ $\Leftrightarrow$ $x = 2/c$. So as $|c|$ increases, the points of inflection get closer to the origin.

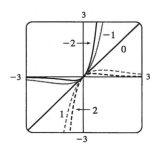

51. $\displaystyle\lim_{x \to \infty} \frac{e^x}{x^n} \overset{\text{H}}{=} \lim_{x \to \infty} \frac{e^x}{nx^{n-1}} \overset{\text{H}}{=} \lim_{x \to \infty} \frac{e^x}{n(n-1)x^{n-2}} \overset{\text{H}}{=} \cdots \overset{\text{H}}{=} \lim_{x \to \infty} \frac{e^x}{n!} = \infty$

53. First we will find $\displaystyle\lim_{n \to \infty} \left(1 + \frac{i}{n}\right)^{nt}$, which is of the form 1^∞. $y = \left(1 + \frac{i}{n}\right)^{nt}$ $\Rightarrow$ $\ln y = nt \ln\left(1 + \frac{i}{n}\right)$, so

$\displaystyle\lim_{n \to \infty} \ln y = \lim_{n \to \infty} nt \ln\left(1 + \frac{i}{n}\right) = t \lim_{n \to \infty} \frac{\ln(1 + i/n)}{1/n} \overset{\text{H}}{=} t \lim_{n \to \infty} \frac{(-i/n^2)}{(1 + i/n)(-1/n^2)} = t \lim_{n \to \infty} \frac{i}{1 + i/n} = ti$

$\Rightarrow$ $\displaystyle\lim_{n \to \infty} y = e^{it}$. Thus, as $n \to \infty$, $A = A_0\left(1 + \frac{i}{n}\right)^{nt} \to A_0 e^{it}$.

55. We see that both numerator and denominator approach 0, so we can use l'Hospital's Rule:

$$\lim_{x \to a} \frac{\sqrt{2a^3 x - x^4} - a\sqrt[3]{aax}}{a - \sqrt[4]{ax^3}} \overset{\text{H}}{=} \lim_{x \to a} \frac{\frac{1}{2}(2a^3 x - x^4)^{-1/2}(2a^3 - 4x^3) - a\left(\frac{1}{3}\right)(aax)^{-2/3}a^2}{-\frac{1}{4}(ax^3)^{-3/4}(3ax^2)}$$

$$= \frac{\frac{1}{2}(2a^3 a - a^4)^{-1/2}(2a^3 - 4a^3) - \frac{1}{3}a^3(a^2 a)^{-2/3}}{-\frac{1}{4}(aa^3)^{-3/4}(3aa^2)}$$

$$= \frac{(a^4)^{-1/2}(-a^3) - \frac{1}{3}a^3(a^3)^{-2/3}}{-\frac{3}{4}a^3(a^4)^{-3/4}} = \frac{-a - \frac{1}{3}a}{-\frac{3}{4}} = \frac{4}{3}\left(\frac{4}{3}a\right) = \frac{16}{9}a$$

57. Since $\lim\limits_{h \to 0}[f(x+h) - f(x-h)] = f(x) - f(x) = 0$ (f is differentiable and hence continuous) and $\lim\limits_{h \to 0} 2h = 0$,
we use l'Hospital's Rule:

$$\lim_{h \to 0} \frac{f(x+h) - f(x-h)}{2h} \stackrel{\text{H}}{=} \lim_{h \to 0} \frac{f'(x+h)(1) - f'(x-h)(-1)}{2} = \frac{f'(x) + f'(x)}{2} = \frac{2f'(x)}{2} = f'(x)$$

$\dfrac{f(x+h) - f(x-h)}{2h}$ is the slope of the secant line

between $(x-h, f(x-h))$ and $(x+h, f(x+h))$. As
$h \to 0$, this line gets closer to the tangent line and its slope
approaches $f'(x)$.

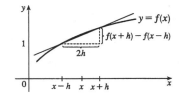

4.6 Optimization Problems • • • • • • • • • • • • •

1. (a)

First Number	Second Number	Product
1	22	22
2	21	42
3	20	60
4	19	76
5	18	90
6	17	102
7	16	112
8	15	120
9	14	126
10	13	130
11	12	132

We needn't consider pairs where the first number
is larger than the second, since we can just
interchange the numbers in such cases. The
answer appears to be 11 and 12, but we have
considered only integers in the table.

(b) Call the two numbers x and y. Then $x + y = 23$, so $y = 23 - x$. Call the product P. Then
$P = xy = x(23 - x) = 23x - x^2$, so we wish to maximize the function $P(x) = 23x - x^2$. Since
$P'(x) = 23 - 2x$, we see that $P'(x) = 0 \iff x = \frac{23}{2} = 11.5$. Thus, the maximum value of P is
$P(11.5) = (11.5)^2 = 132.25$ and it occurs when $x = y = 11.5$.
Or: Note that $P''(x) = -2 < 0$ for all x, so P is everywhere concave downward and the local maximum at
$x = 11.5$ must be an absolute maximum.

3. The two numbers are x and $\dfrac{100}{x}$, where $x > 0$. Minimize $f(x) = x + \dfrac{100}{x}$. $f'(x) = 1 - \dfrac{100}{x^2} = \dfrac{x^2 - 100}{x^2}$.
The critical number is $x = 10$. Since $f'(x) < 0$ for $0 < x < 10$ and $f'(x) > 0$ for $x > 10$, there is an absolute
minimum at $x = 10$. The numbers are 10 and 10.

5. If the rectangle has dimensions x and y, then its perimeter is $2x + 2y = 100$ m, so $y = 50 - x$. Thus, the area is $A = xy = x(50 - x)$. We wish to maximize the function $A(x) = x(50 - x) = 50x - x^2$, where $0 < x < 50$. Since $A'(x) = 50 - 2x = -2(x - 25)$, $A'(x) > 0$ for $0 < x < 25$ and $A'(x) < 0$ for $25 < x < 50$. Thus, A has an absolute maximum at $x = 25$, and $A(25) = 25^2 = 625$ m². The dimensions of the rectangle that maximize its area are $x = y = 25$ m. (The rectangle is a square.)

7. (a)

  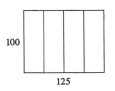

The areas of the three figures are 12,500, 12,500, and 9000 ft². There appears to be a maximum area of at least 12,500 ft².

(b) Let x denote the length of each of two sides and three dividers. Let y denote the length of the other two sides.

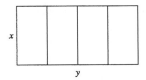

(c) Area $A = $ length $\times$ width $= y \cdot x$

(d) Length of fencing $= 750$ ⇒ $5x + 2y = 750$

(e) $5x + 2y = 750$ ⇒ $y = 375 - \frac{5}{2}x$ ⇒ $A(x) = \left(375 - \frac{5}{2}x\right)x = 375x - \frac{5}{2}x^2$

(f) $A'(x) = 375 - 5x = 0$ ⇒ $x = 75$. Since $A''(x) = -5 < 0$ there is an absolute maximum when $x = 75$. Then $y = \frac{375}{2} = 187.5$. The largest area is $75\left(\frac{375}{2}\right) = 14{,}062.5$ ft². These values of x and y are between the values in the first and second figures in part (a). Our original estimate was low.

9. Let b be the length of the base of the box and h the height. The surface area is $1200 = b^2 + 4hb$ ⇒ $h = (1200 - b^2)/(4b)$. The volume is $V = b^2h = b^2(1200 - b^2)/4b = 300b - b^3/4$ ⇒ $V'(b) = 300 - \frac{3}{4}b^2$. $V'(b) = 0$ ⇒ $300 = \frac{3}{4}b^2$ ⇒ $b^2 = 400$ ⇒ $b = \sqrt{400} = 20$. Since $V'(b) > 0$ for $0 < b < 20$ and $V'(b) < 0$ for $b > 20$, there is an absolute maximum when $b = 20$ by the First Derivative Test for Absolute Extreme Values (see page 310). If $b = 20$, then $h = (1200 - 20^2)/(4 \cdot 20) = 10$, so the largest possible volume is $b^2h = (20)^2(10) = 4000$ cm³.

11. (a) Let the rectangle have sides x and y and area A, so $A = xy$ or $y = A/x$. The problem is to minimize the perimeter $= 2x + 2y = 2x + 2A/x = P(x)$. Now $P'(x) = 2 - 2A/x^2 = 2(x^2 - A)/x^2$. So the critical number is $x = \sqrt{A}$. Since $P'(x) < 0$ for $0 < x < \sqrt{A}$ and $P'(x) > 0$ for $x > \sqrt{A}$, there is an absolute minimum at $x = \sqrt{A}$. The sides of the rectangle are $\sqrt{A}$ and $A/\sqrt{A} = \sqrt{A}$, so the rectangle is a square.

(b) Let p be the perimeter and x and y the lengths of the sides, so $p = 2x + 2y$ ⇒ $2y = p - 2x$ ⇒ $y = \frac{1}{2}p - x$. The area is $A(x) = x\left(\frac{1}{2}p - x\right) = \frac{1}{2}px - x^2$. Now $A'(x) = 0$ ⇒ $\frac{1}{2}p - 2x = 0$ ⇒ $2x = \frac{1}{2}p$ ⇒ $x = \frac{1}{4}p$. Since $A''(x) = -2 < 0$, there is an absolute maximum for A when $x = \frac{1}{4}p$ by the Second Derivative Test. The sides of the rectangle are $\frac{1}{4}p$ and $\frac{1}{2}p - \frac{1}{4}p = \frac{1}{4}p$, so the rectangle is a square.

13. The distance from a point (x, y) on the line $y = 4x + 7$ to the origin is $\sqrt{(x - 0)^2 + (y - 0)^2} = \sqrt{x^2 + y^2}$.
However, it is easier to work with the *square* of the distance; that is,

$$D(x) = \left(\sqrt{x^2 + y^2}\right)^2 = x^2 + y^2 = x^2 + (4x + 7)^2.$$ Because the distance is positive, its minimum value will

occur at the same point as the minimum value of D.

$D'(x) = 2x + 2(4x + 7)(4) = 34x + 56$, so $D'(x) = 0 \iff x = -\frac{28}{17}$.

$D''(x) = 34 > 0$, so D is concave upward for all x. Thus, D has an absolute minimum at $x = -\frac{28}{17}$. The point

closest to the origin is $(x, y) = \left(-\frac{28}{17}, 4\left(-\frac{28}{17}\right) + 7\right) = \left(-\frac{28}{17}, \frac{7}{17}\right)$.

15.

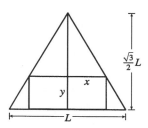

The height h of the equilateral triangle with sides of length L is $\frac{\sqrt{3}}{2} L$,

since $h^2 + (L/2)^2 = L^2 \Rightarrow h^2 = L^2 - \frac{1}{4}L^2 = \frac{3}{4}L^2 \Rightarrow$

$h = \frac{\sqrt{3}}{2} L$. Using similar triangles, $\dfrac{\frac{\sqrt{3}}{2}L - y}{x} = \dfrac{\frac{\sqrt{3}}{2}L}{L/2} = \sqrt{3} \Rightarrow$

$\sqrt{3}\, x = \dfrac{\sqrt{3}}{2}L - y \Rightarrow y = \dfrac{\sqrt{3}}{2}L - \sqrt{3}\, x \Rightarrow y = \dfrac{\sqrt{3}}{2}(L - 2x).$

The area of the inscribed rectangle is $A(x) = (2x)y = \sqrt{3}\, x(L - 2x) = \sqrt{3}\, Lx - 2\sqrt{3}\, x^2$, where $0 \le x \le L/2$.
Now $0 = A'(x) = \sqrt{3}\, L - 4\sqrt{3}\, x \Rightarrow x = \sqrt{3}\, L/(4\sqrt{3}) = L/4$. Since $A(0) = A(L/2) = 0$, the maximum

occurs when $x = L/4$, and $y = \frac{\sqrt{3}}{2}L - \frac{\sqrt{3}}{4}L = \frac{\sqrt{3}}{4}L$, so the dimensions are $L/2$ and $\frac{\sqrt{3}}{4}L$.

17.

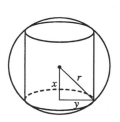

The cylinder has surface area

2(area of the base) + (lateral surface area)

$= 2\pi(\text{radius})^2 + 2\pi(\text{radius})(\text{height}) = 2\pi y^2 + 2\pi y(2x).$

Now $x^2 + y^2 = r^2 \Rightarrow y^2 = r^2 - x^2 \Rightarrow y = \sqrt{r^2 - x^2}$, so the

surface area is

$$S(x) = 2\pi(r^2 - x^2) + 4\pi x \sqrt{r^2 - x^2}, \quad 0 \le x \le r$$

$$= 2\pi r^2 - 2\pi x^2 + 4\pi\left(x\sqrt{r^2 - x^2}\right)$$

Thus, $S'(x) = 0 - 4\pi x + 4\pi\left[x \cdot \frac{1}{2}(r^2 - x^2)^{-1/2}(-2x) + (r^2 - x^2)^{1/2} \cdot 1\right]$

$= 4\pi\left[-x - \dfrac{x^2}{\sqrt{r^2 - x^2}} + \sqrt{r^2 - x^2}\right] = 4\pi \cdot \dfrac{-x\sqrt{r^2 - x^2} - x^2 + r^2 - x^2}{\sqrt{r^2 - x^2}}$

$S'(x) = 0 \Rightarrow x\sqrt{r^2 - x^2} = r^2 - 2x^2 \quad (\star) \Rightarrow \left(x\sqrt{r^2 - x^2}\right)^2 = (r^2 - 2x^2)^2 \Rightarrow$

$x^2(r^2 - x^2) = r^4 - 4r^2x^2 + 4x^4 \Rightarrow r^2x^2 - x^4 = r^4 - 4r^2x^2 + 4x^4 \Rightarrow 5x^4 - 5r^2x^2 + r^4 = 0$. This is a

quadratic equation in x^2. By the quadratic formula, $x^2 = \frac{5 \pm \sqrt{5}}{10}r^2$, but we reject the root with the $+$ sign since it

doesn't satisfy $(\star)$. [The right side is negative and the left side is positive.] So $x = \sqrt{\frac{5 - \sqrt{5}}{10}}\, r$. Since

$S(0) = S(r) = 0$, the maximum surface area occurs at the critical number and

$x^2 = \frac{5 - \sqrt{5}}{10}r^2 \Rightarrow y^2 = r^2 - \frac{5 - \sqrt{5}}{10}r^2 = \frac{5 + \sqrt{5}}{10}r^2 \Rightarrow$ the surface area is

$2\pi\left(\frac{5 + \sqrt{5}}{10}\right)r^2 + 4\pi\sqrt{\frac{5 - \sqrt{5}}{10}}\sqrt{\frac{5 + \sqrt{5}}{10}}r^2 = \pi r^2\left[2 \cdot \frac{5 + \sqrt{5}}{10} + 4\frac{\sqrt{(5 - \sqrt{5})(5 + \sqrt{5})}}{10}\right] = \pi r^2\left[\frac{5 + \sqrt{5}}{5} + \frac{2\sqrt{20}}{5}\right] =$

$\pi r^2\left[\frac{5 + \sqrt{5} + 2 \cdot 2\sqrt{5}}{5}\right] = \pi r^2\left[\frac{5 + 5\sqrt{5}}{5}\right] = \pi r^2(1 + \sqrt{5}).$

19.

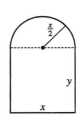

Perimeter $= 30 \Rightarrow 2y + x + \pi\left(\dfrac{x}{2}\right) = 30 \Rightarrow$

$y = \dfrac{1}{2}\left(30 - x - \dfrac{\pi x}{2}\right) = 15 - \dfrac{x}{2} - \dfrac{\pi x}{4}$. The area is the area of the

rectangle plus the area of the semicircle, or $xy + \dfrac{1}{2}\pi\left(\dfrac{x}{2}\right)^2$, so

$$A(x) = x\left(15 - \dfrac{x}{2} - \dfrac{\pi x}{4}\right) + \dfrac{1}{8}\pi x^2 = 15x - \dfrac{1}{2}x^2 - \dfrac{\pi}{8}x^2.$$

$A'(x) = 15 - \left(1 + \dfrac{\pi}{4}\right)x = 0 \Rightarrow x = \dfrac{15}{1 + \pi/4} = \dfrac{60}{4 + \pi}$. $A''(x) = -\left(1 + \dfrac{\pi}{4}\right) < 0$, so this gives a

maximum. The dimensions are $x = \dfrac{60}{4 + \pi}$ ft and $y = 15 - \dfrac{30}{4 + \pi} - \dfrac{15\pi}{4 + \pi} = \dfrac{60 + 15\pi - 30 - 15\pi}{4 + \pi} = \dfrac{30}{4 + \pi}$ ft,

so the height of the rectangle is half the base.

21.

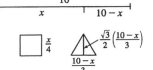

Let x be the length of the wire used for the square. The total area is

$$A(x) = \left(\dfrac{x}{4}\right)^2 + \dfrac{1}{2}\left(\dfrac{10 - x}{3}\right)\dfrac{\sqrt{3}}{2}\left(\dfrac{10 - x}{3}\right)$$

$$= \dfrac{1}{16}x^2 + \dfrac{\sqrt{3}}{36}(10 - x)^2, \quad 0 \le x \le 10$$

$A'(x) = \dfrac{1}{8}x - \dfrac{\sqrt{3}}{18}(10 - x) = 0 \Leftrightarrow \dfrac{9}{72}x + \dfrac{4\sqrt{3}}{72}x - \dfrac{40\sqrt{3}}{72} = 0 \Leftrightarrow x = \dfrac{40\sqrt{3}}{9 + 4\sqrt{3}}$. Now

$A(0) = \left(\dfrac{\sqrt{3}}{36}\right)100 \approx 4.81$, $A(10) = \dfrac{100}{16} = 6.25$ and $A\left(\dfrac{40\sqrt{3}}{9 + 4\sqrt{3}}\right) \approx 2.72$, so

(a) The maximum area occurs when $x = 10$ m, and all the wire is used for the square.

(b) The minimum area occurs when $x = \dfrac{40\sqrt{3}}{9 + 4\sqrt{3}} \approx 4.35$ m.

23.

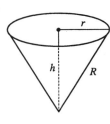

$h^2 + r^2 = R^2 \Rightarrow V = \dfrac{\pi}{3}r^2 h = \dfrac{\pi}{3}(R^2 - h^2)h = \dfrac{\pi}{3}(R^2 h - h^3)$.

$V'(h) = \dfrac{\pi}{3}(R^2 - 3h^2) = 0$ when $h = \dfrac{1}{\sqrt{3}}R$. This gives an absolute

maximum, since $V'(h) > 0$ for $0 < h < \dfrac{1}{\sqrt{3}}R$ and $V'(h) < 0$ for

$h > \dfrac{1}{\sqrt{3}}R$. The maximum volume is

$$V\left(\dfrac{1}{\sqrt{3}}R\right) = \dfrac{\pi}{3}\left(\dfrac{1}{\sqrt{3}}R^3 - \dfrac{1}{3\sqrt{3}}R^3\right) = \dfrac{2}{9\sqrt{3}}\pi R^3.$$

25. $S = 6sh - \dfrac{3}{2}s^2 \cot\theta + 3s^2\dfrac{\sqrt{3}}{2}\csc\theta$

(a) $\dfrac{dS}{d\theta} = \dfrac{3}{2}s^2 \csc^2\theta - 3s^2\dfrac{\sqrt{3}}{2}\csc\theta\cot\theta$ or $\dfrac{3}{2}s^2\csc\theta\left(\csc\theta - \sqrt{3}\cot\theta\right)$.

(b) $\dfrac{dS}{d\theta} = 0$ when $\csc\theta - \sqrt{3}\cot\theta = 0 \Rightarrow \dfrac{1}{\sin\theta} - \sqrt{3}\dfrac{\cos\theta}{\sin\theta} = 0 \Rightarrow \cos\theta = \dfrac{1}{\sqrt{3}}$. The First Derivative

Test shows that the minimum surface area occurs when $\theta = \cos^{-1}\left(\dfrac{1}{\sqrt{3}}\right) \approx 55°$.

(c)

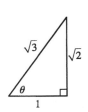

If $\cos\theta = \dfrac{1}{\sqrt{3}}$, then $\cot\theta = \dfrac{1}{\sqrt{2}}$ and $\csc\theta = \dfrac{\sqrt{3}}{\sqrt{2}}$, so the surface area is

$$S = 6sh - \dfrac{3}{2}s^2\dfrac{1}{\sqrt{2}} + 3s^2\dfrac{\sqrt{3}}{2}\dfrac{\sqrt{3}}{\sqrt{2}} = 6sh - \dfrac{3}{2\sqrt{2}}s^2 + \dfrac{9}{2\sqrt{2}}s^2$$

$$= 6sh + \dfrac{6}{2\sqrt{2}}s^2 = 6s\left(h + \dfrac{1}{2\sqrt{2}}s\right)$$

27.

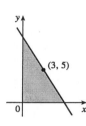

The total illumination is $I(x) = \dfrac{3k}{x^2} + \dfrac{k}{(10-x)^2}$, $0 < x < 10$. Then

$$I'(x) = \frac{-6k}{x^3} + \frac{2k}{(10-x)^3} = 0 \quad \Rightarrow \quad 6k(10-x)^3 = 2kx^3 \quad \Rightarrow$$

$$3(10-x)^3 = x^3 \quad \Rightarrow \quad \sqrt[3]{3}\,(10-x) = x \quad \Rightarrow \quad 10\sqrt[3]{3} - \sqrt[3]{3}\,x = x$$

$$\Rightarrow \quad 10\sqrt[3]{3} = x + \sqrt[3]{3}\,x \quad \Rightarrow \quad 10\sqrt[3]{3} = (1+\sqrt[3]{3})x \quad \Rightarrow$$

$$x = \frac{10\sqrt[3]{3}}{1+\sqrt[3]{3}} \approx 5.9 \text{ ft. This gives a minimum since } I''(x) > 0 \text{ for}$$

$0 < x < 10$.

29.

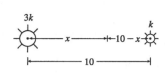

The line with slope m (where $m < 0$) through $(3, 5)$ has equation $y - 5 = m(x - 3)$ or $y = mx + (5 - 3m)$. The y-intercept is $5 - 3m$ and the x-intercept is $-5/m + 3$. So the triangle has area $A(m) = \frac{1}{2}(5 - 3m)(-5/m + 3) = 15 - 25/(2m) - \frac{9}{2}m$.

Now $A'(m) = \dfrac{25}{2m^2} - \dfrac{9}{2} = 0 \iff m^2 = \frac{25}{9} \quad \Rightarrow \quad m = -\frac{5}{3}$ (since $m < 0$).

$A''(m) = -\dfrac{25}{m^3} > 0$, so there is an absolute minimum when $m = -\frac{5}{3}$. Thus, an equation of the line is $y - 5 = -\frac{5}{3}(x - 3)$ or $y = -\frac{5}{3}x + 10$.

31. Note that $|AD| = |AP| + |PD| \quad \Rightarrow \quad 5 = x + |PD| \quad \Rightarrow \quad |PD| = 5 - x$. Using the Pythagorean Theorem for $\triangle PDB$ and $\triangle PDC$ gives us

$$L(x) = |AP| + |BP| + |CP| = x + \sqrt{(5-x)^2 + 2^2} + \sqrt{(5-x)^2 + 3^2}$$

$$= x + \sqrt{x^2 - 10x + 29} + \sqrt{x^2 - 10x + 34}$$

$$\Rightarrow \quad L'(x) = 1 + \frac{x-5}{\sqrt{x^2 - 10x + 29}} + \frac{x-5}{\sqrt{x^2 - 10x + 34}}.$$

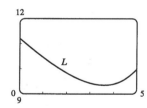

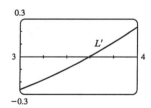

From the graphs of L and L', it seems that the minimum value of L is about $L(3.59) = 9.35$ m.

33.

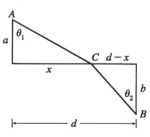

The total time is

$$T(x) = (\text{time from } A \text{ to } C) + (\text{time from } C \text{ to } B)$$

$$= \frac{\sqrt{a^2 + x^2}}{v_1} + \frac{\sqrt{b^2 + (d-x)^2}}{v_2}, \quad 0 < x < d$$

$$T'(x) = \frac{x}{v_1\sqrt{a^2 + x^2}} - \frac{d-x}{v_2\sqrt{b^2 + (d-x)^2}} = \frac{\sin\theta_1}{v_1} - \frac{\sin\theta_2}{v_2}.$$

The minimum occurs when $T'(x) = 0 \quad \Rightarrow \quad \dfrac{\sin\theta_1}{v_1} = \dfrac{\sin\theta_2}{v_2}$.

35.

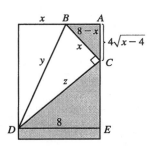

$y^2 = x^2 + z^2$, but triangles CDE and BCA are similar, so

$z/8 = x/(4\sqrt{x-4}) \Rightarrow z = 2x/\sqrt{x-4}$. Thus, we minimize

$f(x) = y^2 = x^2 + 4x^2/(x-4) = x^3/(x-4), \quad 4 < x \le 8.$

$f'(x) = \dfrac{(x-4)(3x^2) - x^3}{(x-4)^2} = \dfrac{x^2[3(x-4) - x]}{(x-4)^2} = \dfrac{2x^2(x-6)}{(x-4)^2} = 0$

when $x = 6$. $f'(x) < 0$ when $x < 6$, $f'(x) > 0$ when $x > 6$, so the

minimum occurs when $x = 6$ in.

37.

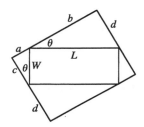

In the small triangle with sides a and c and hypotenuse W, $\sin\theta = \dfrac{a}{W}$

and $\cos\theta = \dfrac{c}{W}$. In the triangle with sides b and d and hypotenuse L,

$\sin\theta = \dfrac{d}{L}$ and $\cos\theta = \dfrac{b}{L}$. Thus, $a = W\sin\theta$, $c = W\cos\theta$, $d = L\sin\theta$,

and $b = L\cos\theta$, so the area of the circumscribed rectangle is

$$\begin{aligned}
A(\theta) &= (a+b)(c+d) = (W\sin\theta + L\cos\theta)(W\cos\theta + L\sin\theta)\\
&= W^2\sin\theta\,\cos\theta + WL\sin^2\theta + LW\cos^2\theta + L^2\sin\theta\,\cos\theta\\
&= LW\sin^2\theta + LW\cos^2\theta + (L^2 + W^2)\sin\theta\,\cos\theta\\
&= LW(\sin^2\theta + \cos^2\theta) + (L^2 + W^2)\cdot\tfrac{1}{2}\cdot 2\sin\theta\,\cos\theta\\
&= LW + \tfrac{1}{2}(L^2 + W^2)\sin 2\theta, \quad 0 \le \theta \le \tfrac{\pi}{2}
\end{aligned}$$

This expression shows, without calculus, that the maximum value of $A(\theta)$ occurs when $\sin 2\theta = 1 \Leftrightarrow 2\theta = \tfrac{\pi}{2}$
$\Rightarrow \theta = \tfrac{\pi}{4}$. So the maximum area is $A(\tfrac{\pi}{4}) = LW + \tfrac{1}{2}(L^2 + W^2) = \tfrac{1}{2}(L^2 + 2LW + W^2) = \tfrac{1}{2}(L + W)^2$.

39.

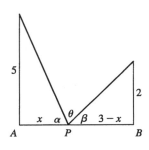

From the figure, $\tan\alpha = \dfrac{5}{x}$ and $\tan\beta = \dfrac{2}{3-x}$. Since

$\alpha + \beta + \theta = 180° = \pi$, $\theta = \pi - \tan^{-1}\left(\dfrac{5}{x}\right) - \tan^{-1}\left(\dfrac{2}{3-x}\right) \Rightarrow$

$\dfrac{d\theta}{dx} = -\dfrac{1}{1 + \left(\dfrac{5}{x}\right)^2}\left(-\dfrac{5}{x^2}\right) - \dfrac{1}{1 + \left(\dfrac{2}{3-x}\right)^2}\left[\dfrac{2}{(3-x)^2}\right]$

$= \dfrac{x^2}{x^2 + 25}\cdot\dfrac{5}{x^2} - \dfrac{(3-x)^2}{(3-x)^2 + 4}\cdot\dfrac{2}{(3-x)^2}$. Now

$\dfrac{d\theta}{dx} = 0 \Rightarrow \dfrac{5}{x^2 + 25} = \dfrac{2}{x^2 - 6x + 13} \Rightarrow 2x^2 + 50 = 5x^2 - 30x + 65 \Rightarrow$

$3x^2 - 30x + 15 = 0 \Rightarrow x^2 - 10x + 5 = 0 \Rightarrow x = 5 \pm 2\sqrt{5}$. We reject the root with the $+$ sign, since it is

larger than 3. $d\theta/dx > 0$ for $x < 5 - 2\sqrt{5}$ and $d\theta/dx < 0$ for $x > 5 - 2\sqrt{5}$, so θ is maximized when

$|AP| = x = 5 - 2\sqrt{5} \approx 0.53$.

41. (a)

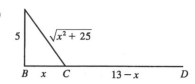

If k = energy/km over land, then energy/km over water = $1.4k$. So the total energy is
$$E = 1.4k\sqrt{25 + x^2} + k(13 - x), \, 0 \le x \le 13, \text{ and so}$$
$$\frac{dE}{dx} = \frac{1.4kx}{(25 + x^2)^{1/2}} - k.$$

Set $\frac{dE}{dx} = 0$: $1.4kx = k(25 + x^2)^{1/2}$ $\Rightarrow$ $1.96x^2 = x^2 + 25$ $\Rightarrow$ $0.96x^2 = 25$ $\Rightarrow$ $x = \frac{5}{\sqrt{0.96}} \approx 5.1$.
Testing against the value of E at the endpoints: $E(0) = 1.4k(5) + 13k = 20k$, $E(5.1) \approx 17.9k$, $E(13) \approx 19.5k$. Thus, to minimize energy, the bird should fly to a point about 5.1 km from B.

(b) If W/L is large, the bird would fly to a point C that is closer to B than to D to minimize the energy used flying over water. If W/L is small, the bird would fly to a point C that is closer to D than to B to minimize the distance of the flight. $E = W\sqrt{25 + x^2} + L(13 - x)$ $\Rightarrow$ $\frac{dE}{dx} = \frac{Wx}{\sqrt{25 + x^2}} - L = 0$ when

$\frac{W}{L} = \frac{\sqrt{25 + x^2}}{x}$. By the same sort of argument as in part (a), this ratio will give the minimal expenditure of energy if the bird heads for the point x km from B.

(c) For flight direct to D, $x = 13$, so from part (b), $W/L = \frac{\sqrt{25 + 13^2}}{13} \approx 1.07$. There is no value of W/L for which the bird should fly directly to B. But note that $\lim\limits_{x \to 0^+} (W/L) = \infty$, so if the point at which E is a minimum is close to B, then W/L is large.

(d) Assuming that the birds instinctively choose the path that minimizes the energy expenditure, we can use the equation for $dE/dx = 0$ from part (a) with $1.4k = c$, $x = 4$, and $k = 1$: $c(4) = 1 \cdot (25 + 4^2)^{1/2}$ $\Rightarrow$ $c = \sqrt{41}/4 \approx 1.6$.

43. (a) Distance = rate × time, so time = distance/rate. $T_1 = \frac{D}{c_1}$,
$$T_2 = \frac{2\,|PR|}{c_1} + \frac{|RS|}{c_2} = \frac{2h\sec\theta}{c_1} + \frac{D - 2h\tan\theta}{c_2}, \, T_3 = \frac{2\sqrt{h^2 + D^2/4}}{c_1} = \frac{\sqrt{4h^2 + D^2}}{c_1}.$$

(b) $\frac{dT_2}{d\theta} = \frac{2h}{c_1} \cdot \sec\theta\tan\theta - \frac{2h}{c_2}\sec^2\theta = 0$ when $2h\sec\theta\left(\frac{1}{c_1}\tan\theta - \frac{1}{c_2}\sec\theta\right) = 0$ $\Rightarrow$

$\frac{1}{c_1}\frac{\sin\theta}{\cos\theta} - \frac{1}{c_2}\frac{1}{\cos\theta} = 0$ $\Rightarrow$ $\frac{\sin\theta}{c_1\cos\theta} = \frac{1}{c_2\cos\theta}$ $\Rightarrow$ $\sin\theta = \frac{c_1}{c_2}$. The First Derivative Test shows that this gives a minimum.

(c) Using part (a) with $D = 1$ and $T_1 = 0.26$, we have $T_1 = \frac{D}{c_1}$ $\Rightarrow$

$c_1 = \frac{1}{0.26} \approx 3.85$ km/s. $T_3 = \frac{\sqrt{4h^2 + D^2}}{c_1}$ $\Rightarrow$ $4h^2 + D^2 = T_3^2 c_1^2$ $\Rightarrow$

$h = \frac{1}{2}\sqrt{T_3^2 c_1^2 - D^2} = \frac{1}{2}\sqrt{(0.34)^2(1/0.26)^2 - 1^2} \approx 0.42$ km. To find c_2, we use $\sin\theta = \frac{c_1}{c_2}$ from part (b)

and $T_2 = \frac{2h\sec\theta}{c_1} + \frac{D - 2h\tan\theta}{c_2}$ from part (a). From the figure,

$\sin \theta = \dfrac{c_1}{c_2} \Rightarrow \sec \theta = \dfrac{c_2}{\sqrt{c_2^2 - c_1^2}}$ and $\tan \theta = \dfrac{c_1}{\sqrt{c_2^2 - c_1^2}}$, so

$T_2 = \dfrac{2hc_2}{c_1\sqrt{c_2^2 - c_1^2}} + \dfrac{D\sqrt{c_2^2 - c_1^2} - 2hc_1}{c_2\sqrt{c_2^2 - c_1^2}}$.

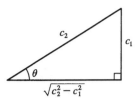

Using the values for T_2 [given as 0.32], h, c_1, and D, we can graph

$Y_1 = T_2$ and $Y_2 = \dfrac{2hc_2}{c_1\sqrt{c_2^2 - c_1^2}} + \dfrac{D\sqrt{c_2^2 - c_1^2} - 2hc_1}{c_2\sqrt{c_2^2 - c_1^2}}$ and find their intersection points. Doing so gives us

$c_2 \approx 4.10$ and 7.66, but if $c_2 = 4.10$, then $\theta = \arcsin(c_1/c_2) \approx 69.6°$, which implies that point S is to the left of point R in the diagram. So $c_2 = 7.66$ km/s.

4.7 Applications to Economics • • • • • • • • • • •

1. (a) $C(0)$ represents the fixed costs of production, such as rent, utilities, machinery etc., which are incurred even when nothing is produced.

(b) The inflection point is the point at which $C''(x)$ changes from negative to positive; that is, the marginal cost $C'(x)$ changes from decreasing to increasing. Thus, the marginal cost is minimized at the inflection point.

(c) The marginal cost function is $C'(x)$. We graph it as in Example 1 in Section 2.8.

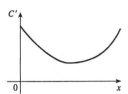

3. $c(x) = 21.4 - 0.002x$ and $c(x) = C(x)/x \Rightarrow C(x) = 21.4x - 0.002x^2$. $C'(x) = 21.4 - 0.004x$ and $C'(1000) = 17.4$. This means that the cost of producing the 1001st unit is about $17.40.

5. (a) The cost function is $C(x) = 40{,}000 + 300x + x^2$, so the cost at a production level of 1000 is $C(1000) = \$1{,}340{,}000$. The average cost function is $c(x) = \dfrac{C(x)}{x} = \dfrac{40{,}000}{x} + 300 + x$ and $c(1000) = \$1340/$unit. The marginal cost function is $C'(x) = 300 + 2x$ and $C'(1000) = \$2300/$unit.

(b) See the box preceding Example 1. We must have $C'(x) = c(x) \Leftrightarrow 300 + 2x = \dfrac{40{,}000}{x} + 300 + x \Leftrightarrow$

$x = \dfrac{40{,}000}{x} \Rightarrow x^2 = 40{,}000 \Rightarrow x = \sqrt{40{,}000} = 200$. This gives a minimum value of the average cost function $c(x)$ since $c''(x) = \dfrac{80{,}000}{x^3} > 0$.

(c) The minimum average cost is $c(200) = \$700/$unit.

7. (a) $C(x) = 3700 + 5x - 0.04x^2 + 0.0003x^3 \Rightarrow C'(x) = 5 - 0.08x + 0.0009x^2$ (marginal cost).

$c(x) = \dfrac{C(x)}{x} = \dfrac{3700}{x} + 5 - 0.04x + 0.0003x^2$ (average cost).

(b)

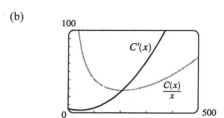

(c) $c'(x) = -\dfrac{3700}{x^2} - 0.04 + 0.0006x = 0 \implies$

$3700 + 0.04x^2 - 0.0006x^3 = 0 \implies$

$x_1 \approx 208.51. \quad c(x_1) \approx \$27.45/\text{unit}.$

The graphs intersect at $(208.51, 27.45)$, so the
production level that minimizes average cost
is about 209 units.

(d) The marginal cost is given by $C'(x)$, so to find its minimum value we'll find the derivative of C'; that is, C''.
$C''(x) = -0.08 + 0.0018x = 0 \implies x_1 = \frac{800}{18} = 44.4\overline{4}.$ $C'(x_1) = \$3.22/\text{unit}.$
$C'''(x) = 0.0018 > 0$ for all x, so this is the minimum marginal cost. C''' is the second derivative of C'.

9. $C(x) = 680 + 4x + 0.01x^2$, $p(x) = 12 - x/500$. Then $R(x) = xp(x) = 12x - x^2/500$. If the profit is
maximum, then $R'(x) = C'(x)$ [See the box preceding Example 2.] $\iff 12 - x/250 = 4 + 0.02x \iff$
$8 = 0.024x \iff x = 8/0.024 = \frac{1000}{3}.$ The profit is maximized if $P''(x) < 0$, but since
$P''(x) = R''(x) - C''(x)$, we can just check the condition $R''(x) < C''(x)$. Now
$R''(x) = -\frac{1}{250} < 0.02 = C''(x)$, so $x = \frac{1000}{3}$ gives a maximum.

11. $C(x) = 0.001x^3 - 0.3x^2 + 6x + 900$. The marginal cost is $C'(x) = 0.003x^2 - 0.6x + 6$.
$C'(x)$ is increasing when $C''(x) > 0 \iff 0.006x - 0.6 > 0 \iff x > 0.6/0.006 = 100$. So $C'(x)$ starts to
increase when $x = 100$.

13. (a) $C(x) = 1200 + 12x - 0.1x^2 + 0.0005x^3.$ $R(x) = xp(x) = 29x - 0.00021x^2.$

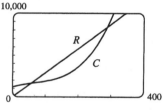

Since the profit is maximized when $R'(x) = C'(x)$, we examine the
curves R and C in the figure, looking for x-values at which the slopes of
the tangent lines are equal. It appears that $x = 200$ is a good estimate.

(b) $R'(x) = C'(x) \implies 29 - 0.00042x = 12 - 0.2x + 0.0015x^2 \implies 0.0015x^2 - 0.19958x - 17 = 0 \implies$
$x \approx 192.06$ (for $x > 0$). As in Exercise 9, $R''(x) < C''(x) \implies -0.00042 < -0.2 + 0.003x \iff$
$0.003x > 0.19958 \iff x > 66.5.$ Our value of 192 is in this range, so we have a maximum profit when we
produce 192 yards of fabric.

15. (a) We are given that the demand function p is linear and $p(27{,}000) = 10$, $p(33{,}000) = 8$, so the slope is
$\frac{10-8}{27{,}000 - 33{,}000} = -\frac{1}{3000}$ and an equation of the line is $y - 10 = \left(-\frac{1}{3000}\right)(x - 27{,}000) \implies$
$y = p(x) = -\frac{1}{3000}x + 19 = 19 - (x/3000).$

(b) The revenue is $R(x) = xp(x) = 19x - (x^2/3000) \implies R'(x) = 19 - (x/1500) = 0$ when $x = 28{,}500.$
Since $R''(x) = -1/1500 < 0$, the maximum revenue occurs when $x = 28{,}500 \implies$ the price is
$p(28{,}500) = \$9.50.$

17. (a) As in Example 3, we see that the demand function p is linear. We are given that $p(1000) = 450$ and deduce that
$p(1100) = 440$, since a $10 reduction in price increases sales by 100 per week. The slope for p is
$\frac{440 - 450}{1100 - 1000} = -\frac{1}{10}$, so an equation is $p - 450 = -\frac{1}{10}(x - 1000)$ or $p(x) = -\frac{1}{10}x + 550$.

(b) $R(x) = xp(x) = -\frac{1}{10}x^2 + 550x$. $R'(x) = -\frac{1}{5}x + 550 = 0$ when $x = 5(550) = 2750$.
$p(2750) = 275$, so the rebate should be $450 - 275 = \$175$.

(c) $C(x) = 68{,}000 + 150x \quad \Rightarrow$
$P(x) = R(x) - C(x) = -\frac{1}{10}x^2 + 550x - 68{,}000 - 150x = -\frac{1}{10}x^2 + 400x - 68{,}000$,
$P'(x) = -\frac{1}{5}x + 400 = 0$ when $x = 2000$. $p(2000) = 350$. Therefore, the rebate to maximize profits should
be $450 - 350 = \$100$.

 4.8 **Newton's Method** · · · · · · · · · · · · · · · · · ·

1.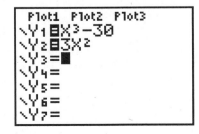

The tangent line at $x = 1$ intersects the x-axis at $x \approx 2.3$, so $x_2 \approx 2.3$.
The tangent line at $x = 2.3$ intersects the x-axis at $x \approx 3$, so $x_3 \approx 3.0$.

3. Since $x_1 = 3$ and $y = 5x - 4$ is tangent to $y = f(x)$ at $x = 3$, we simply need to find where the tangent line
intersects the x-axis. $y = 0 \quad \Rightarrow \quad 5x_2 - 4 = 0 \quad \Rightarrow \quad x_2 = \frac{4}{5}$.

5. $f(x) = x^4 - 20 \quad \Rightarrow \quad f'(x) = 4x^3$, so $x_{n+1} = x_n - \dfrac{f(x_n)}{f'(x_n)} = x_n - \dfrac{x_n^4 - 20}{4x_n^3}$.

Now $x_1 = 2 \quad \Rightarrow \quad x_2 = 2 - \dfrac{2^4 - 20}{4(2)^3} = 2.125 \quad \Rightarrow \quad x_3 = 2.125 - \dfrac{(2.125)^4 - 20}{4(2.125)^3} \approx 2.1148$.

7. To approximate $x = \sqrt[3]{30}$ (so that $x^3 = 30$), we can take $f(x) = x^3 - 30$. So $f'(x) = 3x^2$, and thus,
$x_{n+1} = x_n - \dfrac{x_n^3 - 30}{3x_n^2}$. Since $\sqrt[3]{27} = 3$ and 27 is close to 30, we'll use $x_1 = 3$. We need to find approximations
until they agree to eight decimal places. $x_1 = 3 \quad \Rightarrow \quad x_2 \approx 3.11111111$, $x_3 \approx 3.10723734$,
$x_4 \approx 3.10723251 \approx x_5$. So $\sqrt[3]{30} \approx 3.10723251$, to eight decimal places.
Here is a quick and easy method for finding the iterations for Newton's method on a programmable calculator. (The
screens shown are from the TI-83 Plus, but the method is similar on other calculators.) Assign $f(x) = x^3 - 30$ to
Y$_1$, and $f'(x) = 3x^2$ to Y$_2$. Now store $x_1 = 3$ in X and then enter X $-$ Y$_1$/Y$_2 \to$ X to get $x_2 = 3.\overline{1}$. By
successively pressing the ENTER key, you get the approximations x_3, x_4, ...

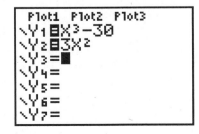

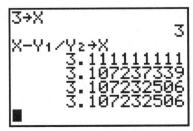

9.

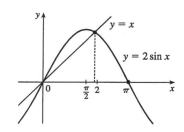

From the graph it appears that there is a root near 2, so we take $x_1 = 2$.
Write the equation as $f(x) = 2 \sin x - x = 0$. Then $f'(x) = 2 \cos x - 1$,
so $x_{n+1} = x_n - \dfrac{2 \sin x_n - x_n}{2 \cos x_n - 1}$. Now $x_1 = 2$, $x_2 \approx 1.900996$,
$x_3 \approx 1.895512$, $x_4 \approx 1.895494 \approx x_5$. So the root is 1.895494, to six
decimal places.

11.

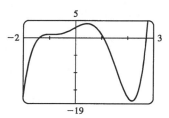

$f(x) = x^5 - x^4 - 5x^3 - x^2 + 4x + 3 \quad \Rightarrow$
$f'(x) = 5x^4 - 4x^3 - 15x^2 - 2x + 4 \quad \Rightarrow$
$x_{n+1} = x_n - \dfrac{x_n^5 - x_n^4 - 5x_n^3 - x_n^2 + 4x_n + 3}{5x_n^4 - 4x_n^3 - 15x_n^2 - 2x_n + 4}$. From the graph of f,
there appear to be roots near -1.4, 1.1, and 2.7.

$x_1 = -1.4$	$x_1 = 1.1$	$x_1 = 2.7$
$x_2 \approx -1.39210970$	$x_2 \approx 1.07780402$	$x_2 \approx 2.72046250$
$x_3 \approx -1.39194698$	$x_3 \approx 1.07739442$	$x_3 \approx 2.71987870$
$x_4 \approx -1.39194691 \approx x_5$	$x_4 \approx 1.07739428 \approx x_5$	$x_4 \approx 2.71987822 \approx x_5$

To eight decimal places, the roots of the equation are -1.39194691, 1.07739428, and 2.71987822.

13.

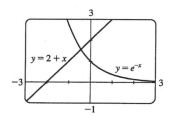

Solving $e^{-x} = 2 + x$ is the same as solving $f(x) = e^{-x} - x - 2 = 0$.
$f'(x) = -e^{-x} - 1 \quad \Rightarrow \quad x_{n+1} = x_n - \dfrac{e^{-x_n} - x_n - 2}{-e^{-x_n} - 1}$. From the
graph of $y = e^{-x}$ and $y = 2 + x$, there appears to be a root near
$x = -0.5$. Now $x_1 = -0.5 \quad \Rightarrow \quad x_2 \approx -0.44385167$,
$x_3 \approx -0.44285470$, $x_4 \approx -0.44285440 \approx x_5$. To eight decimal places,
the root of the equation is -0.44285440.

15.

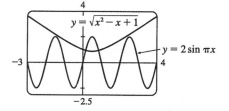

From the graph, we see that there are roots of this equation
near 0.2 and 0.8. $f(x) = \sqrt{x^2 - x + 1} - 2 \sin \pi x \quad \Rightarrow$
$f'(x) = \dfrac{2x - 1}{2\sqrt{x^2 - x + 1}} - 2\pi \cos \pi x$, so
$x_{n+1} = x_n - \dfrac{\sqrt{x_n^2 - x_n + 1} - 2 \sin \pi x_n}{\dfrac{2x_n - 1}{2\sqrt{x_n^2 - x_n + 1}} - 2\pi \cos \pi x_n}$.

Taking $x_1 = 0.2$, we get $x_2 \approx 0.15212015$, $x_3 \approx 0.15438067$, $x_4 \approx 0.15438500 \approx x_5$. Taking $x_1 = 0.8$, we get
$x_2 \approx 0.84787985$, $x_3 \approx 0.84561933$, $x_4 \approx 0.84561500 \approx x_5$. To eight decimal places, the roots of the equation
are 0.15438500 and 0.84561500.

17.

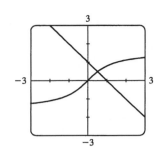

From the graph of $y = \tan^{-1} x$ and $y = 1 - x$, there appears to be a point of intersection near $x = 0.5$. Solving $\tan^{-1} x = 1 - x$ is the same as solving $f(x) = \tan^{-1} x + x - 1 = 0$. $f(x) = \tan^{-1} x + x - 1 \Rightarrow$

$$f'(x) = \frac{1}{1 + x^2} + 1, \text{ so } x_{n+1} = x_n - \frac{\tan^{-1} x_n + x_n - 1}{1/(1 + x_n^2) + 1}.$$

Now $x_1 = 0.5 \Rightarrow x_2 \approx 0.52019577$, $x_3 \approx 0.52026899 \approx x_4$.
To eight decimal places, the root of the equation is 0.52026899.

19. (a) $f(x) = x^2 - a \Rightarrow f'(x) = 2x$, so Newton's method gives

$$x_{n+1} = x_n - \frac{x_n^2 - a}{2x_n} = x_n - \frac{1}{2}x_n + \frac{a}{2x_n} = \frac{1}{2}x_n + \frac{a}{2x_n} = \frac{1}{2}\left(x_n + \frac{a}{x_n}\right).$$

(b) Using (a) with $a = 1000$ and $x_1 = \sqrt{900} = 30$, we get $x_2 \approx 31.666667$, $x_3 \approx 31.622807$, and $x_4 \approx 31.622777 \approx x_5$. So $\sqrt{1000} \approx 31.622777$.

21. $f(x) = x^3 - 3x + 6 \Rightarrow f'(x) = 3x^2 - 3$. If $x_1 = 1$, then $f'(x_1) = 0$ and the tangent line used for approximating x_2 is horizontal. Attempting to find x_2 results in trying to divide by zero.

23. For $f(x) = x^{1/3}$, $f'(x) = \frac{1}{3}x^{-2/3}$ and

$$x_{n+1} = x_n - \frac{f(x_n)}{f'(x_n)} = x_n - \frac{x_n^{1/3}}{\frac{1}{3}x_n^{-2/3}} = x_n - 3x_n = -2x_n.$$

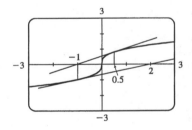

Therefore, each successive approximation becomes twice as large as the previous one in absolute value, so the sequence of approximations fails to converge to the root, which is 0. In the figure, we have $x_1 = 0.5$, $x_2 = -2(0.5) = -1$, and $x_3 = -2(-1) = 2$.

25.

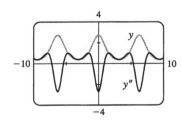

From the figure, we see that $y = f(x) = e^{\cos x}$ is periodic with period 2π. To find the x-coordinates of the IP, we only need to approximate the zeros of y'' on $[0, \pi]$. $f'(x) = -e^{\cos x} \sin x \Rightarrow$
$f''(x) = e^{\cos x}(\sin^2 x - \cos x)$. Since $e^{\cos x} \neq 0$, we will use Newton's method with $g(x) = \sin^2 x - \cos x$, $g'(x) = 2\sin x \cos x + \sin x$, and $x_1 = 1$. $x_2 \approx 0.904173$, $x_3 \approx 0.904557 \approx x_4$. Thus, $(0.904557, 1.855277)$ is the IP.

27.

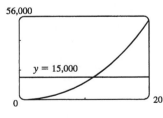

The volume of the silo, in terms of its radius, is
$$V(r) = \pi r^2 (30) + \frac{1}{2}\left(\frac{4}{3}\pi r^3\right) = 30\pi r^2 + \frac{2}{3}\pi r^3.$$
From a graph of V, we see that $V(r) = 15{,}000$ at $r \approx 11$ ft. Now we use Newton's method to solve the equation $V(r) - 15{,}000 = 0$.
$$\frac{dV}{dr} = 60\pi r + 2\pi r^2, \text{ so } r_{n+1} = r_n - \frac{30\pi r_n^2 + \frac{2}{3}\pi r_n^3 - 15{,}000}{60\pi r_n + 2\pi r_n^2}. \text{ Taking}$$
$r_1 = 11$, we get $r_2 \approx 11.2853$, $r_3 \approx 11.2807 \approx r_4$. So in order for the silo to hold $15{,}000$ ft^3 of grain, its radius must be about 11.2807 ft.

29. In this case, $A = 18{,}000$, $R = 375$, and $n = 5(12) = 60$. So the formula $A = \dfrac{R}{i}\left[1 - (1+i)^{-n}\right]$ becomes

$18{,}000 = \dfrac{375}{x}\left[1 - (1+x)^{-60}\right] \;\Leftrightarrow\; 48x = 1 - (1+x)^{-60}$ [multiply each term by $(1+x)^{60}$] $\Leftrightarrow$

$48x(1+x)^{60} - (1+x)^{60} + 1 = 0$. Let the LHS be called $f(x)$, so that

$$f'(x) = 48x(60)(1+x)^{59} + 48(1+x)^{60} - 60(1+x)^{59}$$

$$= 12(1+x)^{59}\left[4x(60) + 4(1+x) - 5\right] = 12(1+x)^{59}(244x - 1)$$

$x_{n+1} = x_n - \dfrac{48x_n(1+x_n)^{60} - (1+x_n)^{60} + 1}{12(1+x_n)^{59}(244x_n - 1)}$. An interest rate of 1%/month seems like a reasonable estimate

for $x = i$. So let $x_1 = 1\% = 0.01$, and we get $x_2 \approx 0.0082202$, $x_3 \approx 0.0076802$, $x_4 \approx 0.0076291$,

$x_5 \approx 0.0076286 \approx x_6$. Thus, the dealer is charging a monthly interest rate of 0.76286% (or 9.55%/year,

compounded monthly).

4.9 Antiderivatives • • • • • • • • • • • • • • •

1. $f(x) = 6x^2 - 8x + 3 \;\Rightarrow\; F(x) = 6\dfrac{x^{2+1}}{2+1} - 8\dfrac{x^{1+1}}{1+1} + 3x + C = 2x^3 - 4x^2 + 3x + C$

Check: $F'(x) = 2\cdot 3x^2 - 4\cdot 2x + 3 + 0 = 6x^2 - 8x + 3 = f(x)$

3. $f(x) = 5x^{1/4} - 7x^{3/4} \;\Rightarrow\; F(x) = 5\dfrac{x^{1/4+1}}{\frac{1}{4}+1} - 7\dfrac{x^{3/4+1}}{\frac{3}{4}+1} + C = 5\dfrac{x^{5/4}}{5/4} - 7\dfrac{x^{7/4}}{7/4} + C = 4x^{5/4} - 4x^{7/4} + C$

Check: $F'(x) = 4\cdot\frac{5}{4}x^{1/4} - 4\cdot\frac{7}{4}x^{3/4} = 5x^{1/4} - 7x^{3/4} = f(x)$

5. $f(x) = \dfrac{10}{x^9} = 10x^{-9}$ has domain $(-\infty, 0) \cup (0, \infty)$, so $F(x) = \begin{cases} \dfrac{10x^{-8}}{-8} + C_1 = -\dfrac{5}{4x^8} + C_1 & \text{if } x < 0 \\ -\dfrac{5}{4x^8} + C_2 & \text{if } x > 0 \end{cases}$

See Example 1(b) for a similar exercise.

7. $g(t) = \dfrac{t^3 + 2t^2}{\sqrt{t}} = t^{5/2} + 2t^{3/2} \;\Rightarrow\; G(t) = \dfrac{t^{7/2}}{7/2} + \dfrac{2t^{5/2}}{5/2} + C = \frac{2}{7}t^{7/2} + \frac{4}{5}t^{5/2} + C$

Note that g has domain $(0, \infty)$.

9. $f(t) = 3\cos t - 4\sin t \;\Rightarrow\; F(t) = 3(\sin t) - 4(-\cos t) + C = 3\sin t + 4\cos t + C$

11. $f(x) = 2x + 5\left(1 - x^2\right)^{-1/2} = 2x + \dfrac{5}{\sqrt{1 - x^2}} \;\Rightarrow\; F(x) = x^2 + 5\sin^{-1} x + C$

13. $f(x) = 5x^4 - 2x^5 \;\Rightarrow\; F(x) = 5\cdot\dfrac{x^5}{5} - 2\cdot\dfrac{x^6}{6} + C = x^5 - \frac{1}{3}x^6 + C.$

$F(0) = 4 \;\Rightarrow\; 0^5 - \frac{1}{3}\cdot 0^6 + C = 4 \;\Rightarrow\; C = 4$, so

$F(x) = x^5 - \frac{1}{3}x^6 + 4$. The graph confirms our answer since $f(x) = 0$
when F has a local maximum, f is positive when F is increasing, and f is
negative when F is decreasing.

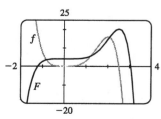

15. $f''(x) = 6x + 12x^2 \;\Rightarrow\; f'(x) = 6\cdot\dfrac{x^2}{2} + 12\cdot\dfrac{x^3}{3} + C = 3x^2 + 4x^3 + C \;\Rightarrow$

$f(x) = 3\cdot\dfrac{x^3}{3} + 4\cdot\dfrac{x^4}{4} + Cx + D = x^3 + x^4 + Cx + D$ [C and D are just arbitrary constants]

17. $f''(x) = 1 + x^{4/5} \Rightarrow f'(x) = x + \frac{5}{9}x^{9/5} + C \Rightarrow$

$f(x) = \frac{1}{2}x^2 + \frac{5}{9} \cdot \frac{5}{14}x^{14/5} + Cx + D = \frac{1}{2}x^2 + \frac{25}{126}x^{14/5} + Cx + D$

19. $f'(x) = 3\cos x + 5\sin x \Rightarrow f(x) = 3\sin x - 5\cos x + C.$

$f(0) = 4 \Rightarrow -5 + C = 4 \Rightarrow C = 9,$ so $f(x) = 3\sin x - 5\cos x + 9.$

21. $f''(x) = x \Rightarrow f'(x) = \frac{1}{2}x^2 + C.$ $f'(0) = 2 \Rightarrow C = 2,$ so $f'(x) = \frac{1}{2}x^2 + 2 \Rightarrow$

$f(x) = \frac{1}{6}x^3 + 2x + D.$ $f(0) = -3 \Rightarrow D = -3,$ so $f(x) = \frac{1}{6}x^3 + 2x - 3.$

23. $f''(x) = x^{-2}, x > 0 \Rightarrow f'(x) = -1/x + C \Rightarrow f(x) = -\ln|x| + Cx + D = -\ln x + Cx + D$ (since

$x > 0$). $f(1) = 0 \Rightarrow C + D = 0$ and $f(2) = 0 \Rightarrow -\ln 2 + 2C + D = 0 \Rightarrow$

$-\ln 2 + 2C - C$ (since $D = -C$) $= 0 \Rightarrow -\ln 2 + C = 0 \Rightarrow C = \ln 2$ and $D = -\ln 2.$ So

$f(x) = -\ln x + (\ln 2)x - \ln 2.$

25. Given $f'(x) = 2x + 1,$ we have $f(x) = x^2 + x + C.$ Since f passes through $(1, 6),$

$f(1) = 6 \Rightarrow 1^2 + 1 + C = 6 \Rightarrow C = 4.$ Therefore, $f(x) = x^2 + x + 4$ and $f(2) = 2^2 + 2 + 4 = 10.$

27.

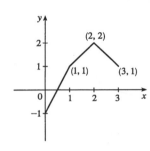

$f'(x) = \begin{cases} 2 & \text{if } 0 \le x < 1 \\ 1 & \text{if } 1 < x < 2 \\ -1 & \text{if } 2 < x \le 3 \end{cases} \Rightarrow f(x) = \begin{cases} 2x + C & \text{if } 0 \le x < 1 \\ x + D & \text{if } 1 < x < 2 \\ -x + E & \text{if } 2 < x \le 3 \end{cases}$

$f(0) = -1 \Rightarrow 2(0) + C = -1 \Rightarrow C = -1.$ Starting at the point

$(0, -1)$ and moving to the right on a line with slope 2 gets us to the point

$(1, 1).$ The slope for $1 < x < 2$ is 1, so we get to the point $(2, 2).$ Here we

have used the fact that f is continuous. We can include the point $x = 1$ on

either the first or the second part of $f.$ The line connecting $(1, 1)$ to $(2, 2)$ is $y = x,$ so $D = 0.$ The slope for

$2 < x \le 3$ is $-1,$ so we get to $(3, 1).$ $f(3) = 1 \Rightarrow -3 + E = 1 \Rightarrow E = 4.$ Thus,

$$f(x) = \begin{cases} 2x - 1 & \text{if } 0 \le x \le 1 \\ x & \text{if } 1 < x < 2 \\ -x + 4 & \text{if } 2 \le x \le 3 \end{cases}$$

Note that $f'(x)$ does not exist at $x = 1$ or at $x = 2.$

29.

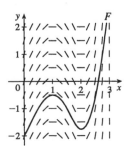

31.

x	$f(x)$
0	1
0.5	0.959
1.0	0.841
1.5	0.665
2.0	0.455
2.5	0.239
3.0	0.047

x	$f(x)$
3.5	-0.100
4.0	-0.189
4.5	-0.217
5.0	-0.192
5.5	-0.128
6.0	-0.047

We compute slopes [values of $f(x) = (\sin x)/x$ for $0 < x < 2\pi$] as in the table [$\lim_{x \to 0+} f(x) = 1$] and draw a direction field as in Example 5. Then we use the direction field to graph F starting at $(0, 0)$.

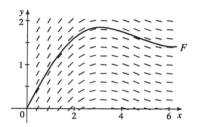

33.

Remember that the given table values of f are the slopes of F at any x. For example, at $x = 1.4$, the slope of F is $f(1.4) = 0$.

35. $v(t) = s'(t) = \sin t - \cos t \Rightarrow s(t) = -\cos t - \sin t + C.$ $s(0) = -1 + C$ and $s(0) = 0 \Rightarrow$
$-1 + C = 0 \Rightarrow C = 1$, so $s(t) = -\cos t - \sin t + 1.$

37. (a) We first observe that since the stone is dropped 450 m above the ground, $v(0) = 0$ and $s(0) = 450$.
$v'(t) = a(t) = -9.8 \Rightarrow v(t) = -9.8t + C.$ Now $v(0) = 0 \Rightarrow C = 0$, so $v(t) = -9.8t \Rightarrow$
$s(t) = -4.9t^2 + D.$ Last, $s(0) = 450 \Rightarrow D = 450 \Rightarrow s(t) = 450 - 4.9t^2.$

(b) The stone reaches the ground when $s(t) = 0.$ $450 - 4.9t^2 = 0 \Rightarrow t^2 = 450/4.9 \Rightarrow$
$t_1 = \sqrt{450/4.9} \approx 9.58$ s.

(c) The velocity with which the stone strikes the ground is $v(t_1) = -9.8\sqrt{450/4.9} \approx -93.9$ m/s.

(d) This is just reworking parts (a) and (b) with $v(0) = -5.$ Using $v(t) = -9.8t + C,$ $v(0) = -5 \Rightarrow$
$0 + C = -5 \Rightarrow v(t) = -9.8t - 5.$ So $s(t) = -4.9t^2 - 5t + D$ and $s(0) = 450 \Rightarrow D = 450 \Rightarrow$
$s(t) = -4.9t^2 - 5t + 450.$ Solving $s(t) = 0$ by using the quadratic formula gives us
$t = (5 \pm \sqrt{8845})/(-9.8) \Rightarrow t_1 \approx 9.09$ s.

39. By Exercise 38 with $a = -9.8,$ $s(t) = -4.9t^2 + v_0 t + s_0$ and $v(t) = s'(t) = -9.8t + v_0.$ So
$[v(t)]^2 = (-9.8t + v_0)^2 = (9.8)^2 t^2 - 19.6v_0 t + v_0^2 = v_0^2 + 96.04t^2 - 19.6v_0 t = v_0^2 - 19.6(-4.9t^2 + v_0 t).$ But
$-4.9t^2 + v_0 t$ is just $s(t)$ without the s_0 term; that is, $s(t) - s_0.$ Thus, $[v(t)]^2 = v_0^2 - 19.6[s(t) - s_0].$

41. Marginal cost $= 1.92 - 0.002x = C'(x) \Rightarrow C(x) = 1.92x - 0.001x^2 + K.$ But
$C(1) = 1.92 - 0.001 + K = 562 \Rightarrow K = 560.081.$ Therefore, $C(x) = 1.92x - 0.001x^2 + 560.081 \Rightarrow$
$C(100) = 742.081,$ so the cost of producing 100 items is \$742.08.

43. Using Exercise 38 with $a = -32,$ $v_0 = 0,$ and $s_0 = h$ (the height of the cliff), we know that the height at time t is
$s(t) = -16t^2 + h.$ $v(t) = s'(t) = -32t$ and $v(t) = -120 \Rightarrow -32t = -120 \Rightarrow t = 3.75,$ so
$0 = s(3.75) = -16(3.75)^2 + h \Rightarrow h = 16(3.75)^2 = 225$ ft.

45. $a(t) = k$, the initial velocity is 30 mi/h $= 30 \cdot \frac{5280}{3600} = 44$ ft/s, and the final velocity (after 5 seconds) is

50 mi/h $= 50 \cdot \frac{5280}{3600} = \frac{220}{3}$ ft/s. So $v(t) = kt + C$ and $v(0) = 44 \;\Rightarrow\; C = 44$. Thus, $v(t) = kt + 44 \;\Rightarrow\;$

$v(5) = 5k + 44$. But $v(5) = \frac{220}{3}$, so $5k + 44 = \frac{220}{3} \;\Rightarrow\; 5k = \frac{88}{3} \;\Rightarrow\; k = \frac{88}{15} \approx 5.87$ ft/s^2.

47. (a) The Mean Value Theorem says that there exists a number c in the interval (x_1, x_2) such that

$$H'(c) = \frac{H(x_2) - H(x_1)}{x_2 - x_1}. \text{ Since } H = G - F \text{ and } G \text{ and } F \text{ are antiderivatives of } f,$$

$H'(c) = G'(c) - F'(c) = f(c) - f(c) = 0$. So now $\dfrac{H(x_2) - H(x_1)}{x_2 - x_1} = 0 \;\Rightarrow\; H(x_2) - H(x_1) = 0$

$(x_2 \neq x_1) \;\Rightarrow\; H(x_2) = H(x_1)$. Since this is true for any $x_1 < x_2$ in I, H must be a constant function.

(b) We have $H = G - F$ and $H(x) = C$, so $C = G - F \;\Rightarrow\; G(x) = F(x) + C$. Thus, any antiderivative G can be expressed as $F(x) + C$.

49. (a) First note that 90 mi/h $= 90 \times \frac{5280}{3600}$ ft/s $= 132$ ft/s. Then $a(t) = 4$ ft/s$^2 \;\Rightarrow\; v(t) = 4t + C$, but $v(0) = 0$

$\Rightarrow C = 0$. Now $4t = 132$ when $t = \frac{132}{4} = 33$ s, so it takes 33 s to reach 132 ft/s. Therefore, taking

$s(0) = 0$, we have $s(t) = 2t^2$, $0 \le t \le 33$. So $s(33) = 2178$ ft. 15 minutes $= 15(60) = 900$ s, so for

$33 < t \le 933$ we have $v(t) = 132$ ft/s $\;\Rightarrow\; s(933) = 132(900) + 2178 = 120{,}978$ ft $= 22.9125$ mi.

(b) As in part (a), the train accelerates for 33 s and travels 2178 ft while doing so. Similarly, it decelerates for 33 s and travels 2178 ft at the end of its trip. During the remaining $900 - 66 = 834$ s it travels at 132 ft/s, so the distance traveled is $132 \cdot 834 = 110{,}088$ ft. Thus, the total distance is

$2178 + 110{,}088 + 2178 = 114{,}444$ ft $= 21.675$ mi.

(c) 45 mi $= 45(5280) = 237{,}600$ ft. Subtract $2(2178)$ to take care of the speeding up and slowing down, and we have 233,244 ft at 132 ft/s for a trip of $233{,}244/132 = 1767$ s at 90 mi/h. The total time is

$1767 + 2(33) = 1833$ s $= 30$ min 33 s $= 30.55$ min.

(d) $37.5(60) = 2250$ s. $2250 - 2(33) = 2184$ s at maximum speed. $2184(132) + 2(2178) = 292{,}644$ total feet or $292{,}644/5280 = 55.425$ mi.

4 Review

━━━━━━━━━━━━━━━━ • CONCEPT CHECK • ━━━━━━━━━━━━━━━━

1. A function f has an **absolute maximum** at $x = c$ if $f(c)$ is the largest function value on the entire domain of f, whereas f has a **local maximum** at c if $f(c)$ is the largest function value when x is near c. See Figure 4 in Section 4.2.

2. (a) See Theorem 4.2.3.

(b) See the Closed Interval Method before Example 6 in Section 4.2.

3. (a) See Theorem 4.2.4.

(b) See Definition 4.2.5.

4. See the Mean Value Theorem in Section 4.3. Geometrical interpretation — there is some point P on the graph of a function f [on the interval (a, b)] where the tangent line is parallel to the secant line that connects $(a, f(a))$ and $(b, f(b))$.

5. (a) See the I/D Test before Example 2 in Section 4.3.

(b) See the Concavity Test before Example 4 in Section 4.3.

6. (a) See the First Derivative Test after Example 2 in Section 4.3.

(b) See the Second Derivative Test before Example 4 in Section 4.3.

(c) See the note before Example 5 in Section 4.3.

7. (a) See l'Hospital's Rule and the three notes that follow it in Section 4.5.

(b) Write fg as $\dfrac{f}{1/g}$ or $\dfrac{g}{1/f}$.

(c) Convert the difference into a quotient using a common denominator, rationalizing, factoring, or some other method.

(d) Convert the power to a product by taking the natural logarithm of both sides of $y = f^g$ or by writing f^g as $e^{g \ln f}$.

8. Without calculus you could get misleading graphs that fail to show the most interesting features of a function. See the first paragraph in Section 4.4.

9. (a) See Figure 3 in Section 4.8.

(b) $x_2 = x_1 - \dfrac{f(x_1)}{f'(x_1)}$

(c) $x_{n+1} = x_n - \dfrac{f(x_n)}{f'(x_n)}$

(d) Newton's method is likely to fail or to work very slowly when $f'(x_1)$ is close to 0.

10. (a) See the definition at the beginning of Section 4.9.

(b) If F_1 and F_2 are both antiderivatives of f on an interval I, then they differ by a constant.

▲ TRUE–FALSE QUIZ ▲

1. False. For example, take $f(x) = x^3$, then $f'(x) = 3x^2$ and $f'(0) = 0$, but $f(0) = 0$ is not a maximum or minimum; $(0, 0)$ is an inflection point.

3. False. For example, $f(x) = x$ is continuous on $(0, 1)$ but attains neither a maximum nor a minimum value on $(0, 1)$. Don't confuse this with f being continuous on the *closed* interval $[a, b]$, which would make the statement true.

5. True. This is an example of part (b) of the I/D Test.

7. False. $f'(x) = g'(x) \implies f(x) = g(x) + C$. For example, if $f(x) = x + 2$ and $g(x) = x + 1$, then $f'(x) = g'(x) = 1$, but $f(x) \neq g(x)$.

9. True. The graph of one such function is sketched.

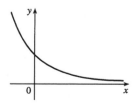

11. True. By the Mean Value Theorem, there exists a number c in $(0, 1)$ such that
$$f(1) - f(0) = f'(c)(1 - 0) = f'(c).$$ Since $f'(c)$ is nonzero, $f(1) - f(0) \neq 0$, so $f(1) \neq f(0)$.

13. False. $\displaystyle\lim_{x \to 0} \frac{x}{e^x} = \frac{\displaystyle\lim_{x \to 0} x}{\displaystyle\lim_{x \to 0} e^x} = \frac{0}{1} = 0$, not 1.

◆ **EXERCISES** ◆

1. $f(x) = 10 + 27x - x^3$, $0 \le x \le 4$. $f'(x) = 27 - 3x^2 = -3(x^2 - 9) = -3(x + 3)(x - 3) = 0$ only when $x = 3$ (since -3 is not in the domain). $f'(x) > 0$ for $x < 3$ and $f'(x) < 0$ for $x > 3$, so $f(3) = 64$ is a local maximum value. Checking the endpoints, we find $f(0) = 10$ and $f(4) = 54$. Thus, $f(0) = 10$ is the absolute minimum value and $f(3) = 64$ is the absolute maximum value.

3. $f(x) = \dfrac{x}{x^2 + x + 1}$, $-2 \le x \le 0$. $f'(x) = \dfrac{(x^2 + x + 1)(1) - x(2x + 1)}{(x^2 + x + 1)^2} = \dfrac{1 - x^2}{(x^2 + x + 1)^2} = 0 \Leftrightarrow x = -1$ (since 1 is not in the domain). $f'(x) < 0$ for $-2 < x < -1$ and $f'(x) > 0$ for $-1 < x < 0$, so $f(-1) = -1$ is a local and absolute minimum value. $f(-2) = -\frac{2}{3}$ and $f(0) = 0$, so $f(0) = 0$ is an absolute maximum value.

5. (a) $f(x) = 2 - 2x - x^3$ is a polynomial, so there is no asymptote.

(b) $f'(x) = -2 - 3x^2 = -1(3x^2 + 2) < 0$, so f is decreasing on $\mathbb{R}$.

(c) No local extrema

(d) $f''(x) = -6x < 0$ on $(0, \infty)$ and $f''(x) > 0$ on $(-\infty, 0)$, so f is CD on $(0, \infty)$ and CU on $(-\infty, 0)$. There is an IP at $(0, 2)$.

(e)

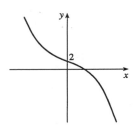

7. (a) $f(x) = x + \sqrt{1 - x}$ has no asymptote.

(b) $f'(x) = 1 - 1/(2\sqrt{1 - x}) = 0 \Leftrightarrow 2\sqrt{1 - x} = 1 \Leftrightarrow 1 - x = \frac{1}{4} \Leftrightarrow x = \frac{3}{4}$ and $f'(x) > 0 \Leftrightarrow x < \frac{3}{4}$, so f is increasing on $(-\infty, \frac{3}{4})$ and decreasing on $(\frac{3}{4}, 1)$.

(c) $f(\frac{3}{4}) = \frac{3}{4} + \sqrt{1 - \frac{3}{4}} = \frac{3}{4} + \sqrt{\frac{1}{4}} = \frac{3}{4} + \frac{1}{2} = \frac{5}{4}$ is a local maximum.

(d) $f''(x) = -\dfrac{1}{4(1 - x)^{3/2}} < 0$ on the domain of f, so f is CD on $(-\infty, 1)$. No IP.

(e)

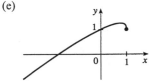

9. (a) $y = f(x) = \sin^2 x - 2\cos x$ has no asymptote.

(b) $y' = 2\sin x \cos x + 2\sin x = 2\sin x (\cos x + 1)$. $y' = 0 \Leftrightarrow \sin x = 0$ or $\cos x = -1 \Leftrightarrow x = n\pi$ or $x = (2n + 1)\pi$. $y' > 0$ when $\sin x > 0$, since $\cos x + 1 \ge 0$ for all x. Therefore, $y' > 0$ (and so f is increasing) on $(2n\pi, (2n + 1)\pi)$; $y' < 0$ (and so f is decreasing) on $((2n - 1)\pi, 2n\pi)$ or equivalently, $((2n + 1)\pi, (2n + 2)\pi)$.

(c) Local maxima are $f((2n + 1)\pi) = 2$; local minima are $f(2n\pi) = -2$.

(d) $y' = \sin 2x + 2\sin x \Rightarrow$

$$y'' = 2\cos 2x + 2\cos x = 2(2\cos^2 x - 1) + 2\cos x = 4\cos^2 x + 2\cos x - 2$$

$$= 2(2\cos^2 x + \cos x - 1) = 2(2\cos x - 1)(\cos x + 1)$$

$y'' = 0 \Leftrightarrow \cos x = \frac{1}{2}$ or $-1 \Leftrightarrow x = 2n\pi \pm \frac{\pi}{3}$ or $x = (2n + 1)\pi$. $y'' > 0$ (and so f is CU) on $(2n\pi - \frac{\pi}{3}, 2n\pi + \frac{\pi}{3})$; $y'' \le 0$ (and so f is CD) on $(2n\pi + \frac{\pi}{3}, 2n\pi + \frac{5\pi}{3})$. There are inflection points at $(2n\pi \pm \frac{\pi}{3}, -\frac{1}{4})$.

(e)

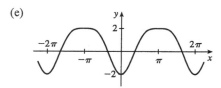

11. (a) $\lim\limits_{x \to \pm\infty} \left(e^x + e^{-3x}\right) = \infty$, no asymptote.

(b) $y = f(x) = e^x + e^{-3x}$ $\Rightarrow$

$f'(x) = e^x - 3e^{-3x} = e^{-3x}\left(e^{4x} - 3\right) > 0$ $\Leftrightarrow$ $e^{4x} > 3$ $\Leftrightarrow$

$4x > \ln 3$ $\Leftrightarrow$ $x > \frac{1}{4}\ln 3$, so f is increasing on $\left(\frac{1}{4}\ln 3, \infty\right)$

and decreasing on $\left(-\infty, \frac{1}{4}\ln 3\right)$.

(c) $f\left(\frac{1}{4}\ln 3\right) = 3^{1/4} + 3^{-3/4} \approx 1.75$ is a local and absolute

minimum.

(d) $f''(x) = e^x + 9e^{-3x} > 0$, so f is CU on $(-\infty, \infty)$. No IP.

(e)

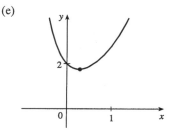

13. $f(x) = \dfrac{x^2 - 1}{x^3}$ $\Rightarrow$ $f'(x) = \dfrac{x^3(2x) - (x^2 - 1)3x^2}{x^6} = \dfrac{3 - x^2}{x^4}$ $\Rightarrow$

$f''(x) = \dfrac{x^4(-2x) - (3 - x^2)4x^3}{x^8} = \dfrac{2x^2 - 12}{x^5}$

Estimates: From the graphs of f' and f'', it appears that f is increasing on

$(-1.73, 0)$ and $(0, 1.73)$ and decreasing on $(-\infty, -1.73)$ and $(1.73, \infty)$;

f has a local maximum of about $f(1.73) = 0.38$ and a local minimum of

about $f(-1.7) = -0.38$; f is CU on $(-2.45, 0)$ and $(2.45, \infty)$, and CD

on $(-\infty, -2.45)$ and $(0, 2.45)$; and f has inflection points at about

$(-2.45, -0.34)$ and $(2.45, 0.34)$.

Exact: Now $f'(x) = \dfrac{3 - x^2}{x^4}$ is positive for $0 < x^2 < 3$, that is, f is

increasing on $\left(-\sqrt{3}, 0\right)$ and $\left(0, \sqrt{3}\right)$; and $f'(x)$ is negative (and so f is

decreasing) on $\left(-\infty, -\sqrt{3}\right)$ and $\left(\sqrt{3}, \infty\right)$. $f'(x) = 0$ when $x = \pm\sqrt{3}$.

f' goes from positive to negative at $x = \sqrt{3}$, so f has a local maximum of

$f\left(\sqrt{3}\right) = \dfrac{\left(\sqrt{3}\right)^2 - 1}{\left(\sqrt{3}\right)^3} = \dfrac{2\sqrt{3}}{9}$; and since f is odd, we know that maxima on

the interval $(0, \infty)$ correspond to minima on $(-\infty, 0)$, so f has a local

minimum of $f\left(-\sqrt{3}\right) = -\dfrac{2\sqrt{3}}{9}$. Also, $f''(x) = \dfrac{2x^2 - 12}{x^5}$ is positive (so

f is CU) on $\left(-\sqrt{6}, 0\right)$ and $\left(\sqrt{6}, \infty\right)$, and negative (so f is CD) on

$\left(-\infty, -\sqrt{6}\right)$ and $\left(0, \sqrt{6}\right)$. There are IP at $\left(\sqrt{6}, \frac{5\sqrt{6}}{36}\right)$ and

$\left(-\sqrt{6}, -\frac{5\sqrt{6}}{36}\right)$.

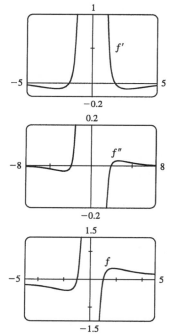

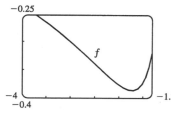

15. $f(x) = 3x^6 - 5x^5 + x^4 - 5x^3 - 2x^2 + 2 \quad \Rightarrow \quad f'(x) = 18x^5 - 25x^4 + 4x^3 - 15x^2 - 4x \quad \Rightarrow$
$f''(x) = 90x^4 - 100x^3 + 12x^2 - 30x - 4$

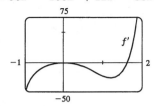 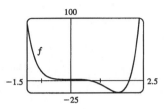 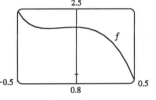

From the graphs of f' and f'', it appears that f is increasing on $(-0.23, 0)$ and $(1.62, \infty)$ and decreasing on $(-\infty, -0.23)$ and $(0, 1.62)$; f has a local maximum of about $f(0) = 2$ and local minima

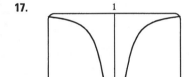 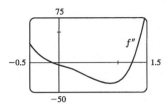

of about $f(-0.23) = 1.96$ and $f(1.62) = -19.2$; f is CU on $(-\infty, -0.12)$ and $(1.24, \infty)$ and CD on $(-0.12, 1.24)$; and f has inflection points at about $(-0.12, 1.98)$ and $(1.2, -12.1)$.

17.

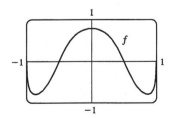

From the graph, we estimate the points of inflection to be about $(\pm 0.82, 0.22)$. $f(x) = e^{-1/x^2} \quad \Rightarrow \quad f'(x) = 2x^{-3}e^{-1/x^2} \quad \Rightarrow$
$f''(x) = 2\left[x^{-3}\left(2x^{-3}\right)e^{-1/x^2} + e^{-1/x^2}\left(-3x^{-4}\right)\right]$
$\qquad = 2x^{-6}e^{-1/x^2}\left(2 - 3x^2\right).$

This is 0 when $2 - 3x^2 = 0 \quad \Leftrightarrow \quad x = \pm\sqrt{\frac{2}{3}}$, so the inflection points are $\left(\pm\sqrt{\frac{2}{3}}, e^{-3/2}\right)$.

19. $f(x) = \arctan(\cos(3\arcsin x))$. We use a CAS to compute f' and f'', and to graph f, f', and f'':

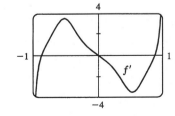 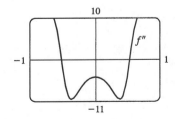

From the graph of f', it appears that the only maximum occurs at $x = 0$ and there are minima at $x = \pm 0.87$. From the graph of f'', it appears that there are inflection points at $x = \pm 0.52$.

21. The family of functions $f(x) = \ln(\sin x + C)$ all have the same period and all have maximum values at $x = \frac{\pi}{2} + 2\pi n$. Since the domain of ln is $(0, \infty)$, f has a graph only if $\sin x + C > 0$ somewhere. Since $-1 \leq \sin x \leq 1$, this happens if $C > -1$, that is, f has no graph if $C \leq -1$. Similarly, if $C > 1$, then $\sin x + C > 0$ and f is continuous on $(-\infty, \infty)$. As C increases, the graph of f is shifted vertically upward and flattens out.

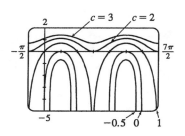

If $-1 < C \leq 1$, f is defined where $\sin x + C > 0 \quad \Leftrightarrow \quad \sin x > -C \quad \Leftrightarrow \quad \sin^{-1}(-C) < x < \pi - \sin^{-1}(-C)$. Since the period is 2π, the domain of f is $\left(2n\pi + \sin^{-1}(-C), (2n+1)\pi - \sin^{-1}(-C)\right)$, n an integer.

23. For $(1, 6)$ to be on the curve $y = x^3 + ax^2 + bx + 1$, we have that $6 = 1 + a + b + 1 \quad \Rightarrow \quad b = 4 - a$. Now $y' = 3x^2 + 2ax + b$ and $y'' = 6x + 2a$. Also, for $(1, 6)$ to be an inflection point it must be true that $y''(1) = 6(1) + 2a = 0 \quad \Rightarrow \quad a = -3 \quad \Rightarrow \quad b = 4 - (-3) = 7$.

25. $\displaystyle\lim_{x \to \pi} \frac{\sin x}{x^2 - \pi^2} \overset{H}{=} \lim_{x \to \pi} \frac{\cos x}{2x} = -\frac{1}{2\pi}$

27. $\displaystyle\lim_{x \to \infty} \frac{\ln(\ln x)}{\ln x} \overset{H}{=} \lim_{x \to \infty} \frac{(1/\ln x)(1/x)}{1/x} = \lim_{x \to \infty} \frac{1}{\ln x} = 0$

29. $\displaystyle\lim_{x \to 0} \frac{\ln(1-x) + x + \frac{1}{2}x^2}{x^3} \overset{H}{=} \lim_{x \to 0} \frac{-\dfrac{1}{1-x} + 1 + x}{3x^2} \overset{H}{=} \lim_{x \to 0} \frac{-\dfrac{1}{(1-x)^2} + 1}{6x} \overset{H}{=} \lim_{x \to 0} \frac{-\dfrac{2}{(1-x)^3}}{6} = -\frac{2}{6} = -\frac{1}{3}$

31. $\displaystyle\lim_{x \to 0} \left(\csc^2 x - x^{-2}\right) = \lim_{x \to 0} \left[\frac{1}{\sin^2 x} - \frac{1}{x^2}\right] = \lim_{x \to 0} \frac{x^2 - \sin^2 x}{x^2 \sin^2 x} \overset{H}{=} \lim_{x \to 0} \frac{2x - \sin 2x}{2x \sin^2 x + x^2 \sin 2x}$

$\overset{H}{=} \displaystyle\lim_{x \to 0} \frac{2 - 2\cos 2x}{2\sin^2 x + 4x \sin 2x + 2x^2 \cos 2x} \overset{H}{=} \lim_{x \to 0} \frac{4 \sin 2x}{6 \sin 2x + 12x \cos 2x - 4x^2 \sin 2x}$

$\overset{H}{=} \displaystyle\lim_{x \to 0} \frac{8 \cos 2x}{24 \cos 2x - 32x \sin 2x - 8x^2 \cos 2x} = \frac{8}{24} = \frac{1}{3}$

33. We are given $d\theta/dt = -0.25$ rad/h.

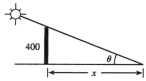

$x = 400 \cot \theta \quad \Rightarrow \quad \dfrac{dx}{dt} = -400 \csc^2 \theta \, \dfrac{d\theta}{dt}$. When $\theta = \frac{\pi}{6}$,

$\dfrac{dx}{dt} = -400(2)^2(-0.25) = 400$ ft/h.

35. Given $dh/dt = 5$ and $dx/dt = 15$, find dz/dt. $z^2 = x^2 + h^2 \quad \Rightarrow$

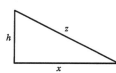

$2z \dfrac{dz}{dt} = 2x \dfrac{dx}{dt} + 2h \dfrac{dh}{dt} \quad \Rightarrow \quad \dfrac{dz}{dt} = \dfrac{1}{z}(15x + 5h)$. When $t = 3$,

$h = 45 + 3(5) = 60$ and $x = 15(3) = 45 \Rightarrow z = 75$, so

$\dfrac{dz}{dt} = \frac{1}{75}[15(45) + 5(60)] = 13$ ft/s.

37. Call the two integers x and y. Then $x + 4y = 1000$, so $x = 1000 - 4y$. Their product is $P = xy = (1000 - 4y)y$, so our problem is to maximize the function $P(y) = 1000y - 4y^2$, where $0 < y < 250$ and y is an integer. $P'(y) = 1000 - 8y$, so $P'(y) = 0 \quad \Leftrightarrow \quad y = 125$. $P''(y) = -8 < 0$, so $P(125) = 62{,}500$ is an absolute maximum. Since the optimal y turned out to be an integer, we have found the desired pair of numbers, namely $x = 1000 - 4(125) = 500$ and $y = 125$.

39.

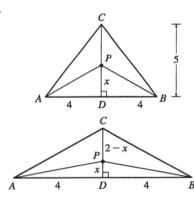

By similar triangles, $\dfrac{y}{x} = \dfrac{r}{\sqrt{x^2 - 2rx}}$, so the area of the triangle is

$$A(x) = \tfrac{1}{2}(2y)x = xy = \frac{rx^2}{\sqrt{x^2 - 2rx}} \quad \Rightarrow$$

$$A'(x) = \frac{2rx\sqrt{x^2 - 2rx} - rx^2(x - r)/\sqrt{x^2 - 2rx}}{x^2 - 2rx}$$

$$= \frac{rx^2\,(x - 3r)}{(x^2 - 2rx)^{3/2}} = 0 \text{ when } x = 3r.$$

$A'(x) < 0$ when $2r < x < 3r$, $A'(x) > 0$ when $x > 3r$. So $x = 3r$ gives a minimum and $A(3r) = r(9r^2)/(\sqrt{3}\,r) = 3\sqrt{3}\,r^2$.

41.

We minimize

$$L(x) = |PA| + |PB| + |PC| = 2\sqrt{x^2 + 16} + (5 - x),$$

$0 \le x \le 5$. $L'(x) = 2x/\sqrt{x^2 + 16} - 1 = 0 \quad \Leftrightarrow$

$$2x = \sqrt{x^2 + 16} \quad \Leftrightarrow \quad 4x^2 = x^2 + 16 \quad \Leftrightarrow \quad x = \tfrac{4}{\sqrt{3}}.$$

$L(0) = 13$, $L\left(\tfrac{4}{\sqrt{3}}\right) \approx 11.9$, $L(5) \approx 12.8$, so the minimum occurs when $x = \tfrac{4}{\sqrt{3}} \approx 2.3$.

If $|CD| = 2$, $L(x)$ changes from $(5 - x)$ to $(2 - x)$ with $0 \le x \le 2$. But we still get $L'(x) = 0 \quad \Leftrightarrow \quad x = \tfrac{4}{\sqrt{3}}$, which isn't in the interval $[0, 2]$. Now $L(0) = 10$ and $L(2) = 2\sqrt{20} = 4\sqrt{5} \approx 8.9$. The minimum occurs when $P = C$.

43. $v = K\sqrt{\dfrac{L}{C} + \dfrac{C}{L}} \quad \Rightarrow \quad \dfrac{dv}{dL} = \dfrac{K}{2\sqrt{(L/C) + (C/L)}}\left(\dfrac{1}{C} - \dfrac{C}{L^2}\right) = 0 \quad \Leftrightarrow \quad \dfrac{1}{C} = \dfrac{C}{L^2} \quad \Leftrightarrow \quad L^2 = C^2 \quad \Leftrightarrow$

$L = C$. This gives the minimum velocity since $v' < 0$ for $0 < L < C$ and $v' > 0$ for $L > C$.

45. Let x denote the number of \$1 decreases in ticket price. Then the ticket price is $\$12 - \$1(x)$, and the average attendance is $11{,}000 + 1000(x)$. Now the revenue per game is

$$R(x) = (\text{price per person}) \times (\text{number of people per game})$$

$$= (12 - x)(11{,}000 + 1000x) = -1000x^2 + 1000x + 132{,}000$$

for $0 \le x \le 4$ (since the seating capacity is 15,000) $\Rightarrow$ $R'(x) = -2000x + 1000 = 0 \quad \Leftrightarrow \quad x = 0.5$. This is a maximum since $R''(x) = -2000 < 0$ for all x. Now we must check the value of $R(x) = (12 - x)(11{,}000 + 1000x)$ at $x = 0.5$ and at the endpoints of the domain to see which value of x gives the maximum value of R. $R(0) = (12)(11{,}000) = 132{,}000$, $R(0.5) = (11.5)(11{,}500) = 132{,}250$, and $R(4) = (8)(15{,}000) = 120{,}000$. Thus, the maximum revenue of \$132,250 per game occurs when the average attendance is 11,500 and the ticket price is \$11.50.

47. $f(x) = x^6 + 2x^2 - 8x + 3$ $\Rightarrow$ $f'(x) = 6x^5 + 4x - 8$. We want to find the minimum of f, so we examine the graph of f' looking for values at which f' changes from negative to positive.

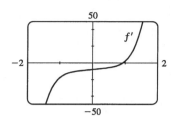

From the graph, we see that this occurs at $x \approx 1$. So we will use Newton's method with $g(x) = f'(x) = 6x^5 + 4x - 8$, $g'(x) = 30x^4 + 4$, and

$$x_1 = 1. \ x_{n+1} = x_n - \frac{6x_n^5 + 4x_n - 8}{30x_n^4 + 4} \text{ gives us } x_2 \approx 0.941176,$$

$x_3 \approx 0.934068$, $x_4 \approx 0.933975 \approx x_5$. Thus, $f(x_5) \approx -2.063421$ is the absolute minimum value of f.

49. $f(x) = e^x - (2/\sqrt{x}) = e^x - 2x^{-1/2}$ $\Rightarrow$

$$F(x) = e^x - 2\frac{x^{-1/2+1}}{-1/2+1} + C = e^x - 2\frac{x^{1/2}}{1/2} + C = e^x - 4\sqrt{x} + C$$

51. $f'(x) = 2/(1+x^2)$ $\Rightarrow$ $f(x) = 2\arctan x + C$. $f(0) = 2\arctan 0 + C = 0 + C = C$ and $f(0) = -1$ $\Rightarrow$ $C = -1$. Therefore, $f(x) = 2\arctan x - 1$.

53. $f''(x) = x^3 + x$ $\Rightarrow$ $f'(x) = \frac{1}{4}x^4 + \frac{1}{2}x^2 + C$. Now $f'(0) = C$ and $f'(0) = 1$ $\Rightarrow$ $f'(x) = \frac{1}{4}x^4 + \frac{1}{2}x^2 + 1$ $\Rightarrow$ $f(x) = \frac{1}{20}x^5 + \frac{1}{6}x^3 + x + D$. Now $f(0) = D$ and $f(0) = -1$ $\Rightarrow$ $f(x) = \frac{1}{20}x^5 + \frac{1}{6}x^3 + x - 1$.

55. (a) Since f is 0 just to the left of the y-axis, we must have a minimum of F at the same place since we are increasing through $(0,0)$ on F. There must be a local maximum to the left of $x = -3$, since f changes from positive to negative there.

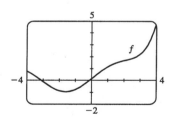

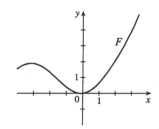

(b) $f(x) = 0.1e^x + \sin x$ $\Rightarrow$ $F(x) = 0.1e^x - \cos x + C$.
$F(0) = 0$ $\Rightarrow$ $0.1 - 1 + C = 0$ $\Rightarrow$ $C = 0.9$, so
$F(x) = 0.1e^x - \cos x + 0.9$.

(c)

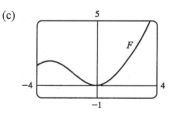

57. Choosing the positive direction to be upward, we have $a(t) = -9.8 \Rightarrow v(t) = -9.8t + v_0$, but $v(0) = 0 = v_0$
$\Rightarrow v(t) = -9.8t = s'(t) \Rightarrow s(t) = -4.9t^2 + s_0$, but $s(0) = s_0 = 500 \Rightarrow s(t) = -4.9t^2 + 500$. When
$s = 0$, $-4.9t^2 + 500 = 0 \Rightarrow t_1 = \sqrt{\frac{500}{4.9}} \approx 10.1 \Rightarrow v(t_1) = -9.8\sqrt{\frac{500}{4.9}} \approx -98.995$ m/s. Since the
canister has been designed to withstand an impact velocity of 100 m/s, the canister will *not burst*.

59. (a)

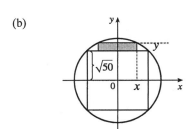

The cross-sectional area of the rectangular beam is
$A = 2x \cdot 2y = 4xy = 4x\sqrt{100 - x^2}$, $0 \le x \le 10$, so
$$\frac{dA}{dx} = 4x\left(\tfrac{1}{2}\right)\left(100 - x^2\right)^{-1/2}(-2x) + \left(100 - x^2\right)^{1/2} \cdot 4$$
$$= \frac{-4x^2}{\left(100 - x^2\right)^{1/2}} + 4\left(100 - x^2\right)^{1/2}$$
$$= \frac{4\left[-x^2 + \left(100 - x^2\right)\right]}{\left(100 - x^2\right)^{1/2}}.$$

$\frac{dA}{dx} = 0$ when $-x^2 + \left(100 - x^2\right) = 0 \Rightarrow x^2 = 50 \Rightarrow x = \sqrt{50} \approx 7.07 \Rightarrow$
$y = \sqrt{100 - \left(\sqrt{50}\right)^2} = \sqrt{50}$. Since $A(0) = A(10) = 0$, the rectangle of maximum area is a square.

(b)

The cross-sectional area of each rectangular plank (shaded in the figure) is
$A = 2x\left(y - \sqrt{50}\right) = 2x\left[\sqrt{100 - x^2} - \sqrt{50}\right]$, $0 \le x \le \sqrt{50}$, so
$$\frac{dA}{dx} = 2\left(\sqrt{100 - x^2} - \sqrt{50}\right) + 2x\left(\tfrac{1}{2}\right)\left(100 - x^2\right)^{-1/2}(-2x)$$
$$= 2\left(100 - x^2\right)^{1/2} - 2\sqrt{50} - \frac{2x^2}{\left(100 - x^2\right)^{1/2}}$$

Set $\frac{dA}{dx} = 0$: $\left(100 - x^2\right) - \sqrt{50}\left(100 - x^2\right)^{1/2} - x^2 = 0 \Rightarrow 100 - 2x^2 = \sqrt{50}\left(100 - x^2\right)^{1/2} \Rightarrow$
$10{,}000 - 400x^2 + 4x^4 = 50\left(100 - x^2\right) \Rightarrow 4x^4 - 350x^2 + 5000 = 0 \Rightarrow 2x^4 - 175x^2 + 2500 = 0$
$\Rightarrow x^2 = \frac{175 \pm \sqrt{10{,}625}}{4} \approx 69.52$ or $17.98 \Rightarrow x \approx 8.34$ or 4.24.
But $8.34 > \sqrt{50}$, so $x_1 \approx 4.24 \Rightarrow y - \sqrt{50} = \sqrt{100 - x_1^2} - \sqrt{50} \approx 1.99$. Each plank should have
dimensions about $8\frac{1}{2}$ inches by 2 inches.

(c) From the figure in part (a), the width is $2x$ and the depth is $2y$, so the strength is
$S = k(2x)(2y)^2 = 8kxy^2 = 8kx\left(100 - x^2\right) = 800kx - 8kx^3$, $0 \le x \le 10$. $dS/dx = 800k - 24kx^2 = 0$
when $24kx^2 = 800k \Rightarrow x^2 = \frac{100}{3} \Rightarrow x = \frac{10}{\sqrt{3}} \Rightarrow y = \sqrt{\frac{200}{3}} = \frac{10\sqrt{2}}{\sqrt{3}} = \sqrt{2}\,x$. Since
$S(0) = S(10) = 0$, the maximum strength occurs when $x = \frac{10}{\sqrt{3}}$. The dimensions should be
$\frac{20}{\sqrt{3}} \approx 11.55$ inches by $\frac{20\sqrt{2}}{\sqrt{3}} \approx 16.33$ inches.

61. (a) $I = \dfrac{k \cos\theta}{d^2} = \dfrac{k(h/d)}{d^2} = k\dfrac{h}{d^3} = k\dfrac{h}{\left(\sqrt{40^2 + h^2}\right)^3} = k\dfrac{h}{(1600 + h^2)^{3/2}} \quad \Rightarrow$

$\dfrac{dI}{dh} = k\dfrac{(1600 + h^2)^{3/2} - h\frac{3}{2}(1600 + h^2)^{1/2} \cdot 2h}{\left[(1600 + h^2)^{3/2}\right]^2} = \dfrac{k(1600 + h^2)^{1/2}(1600 + h^2 - 3h^2)}{(1600 + h^2)^3}$

$= \dfrac{k(1600 - 2h^2)}{(1600 + h^2)^{5/2}}$

Set $dI/dh = 0$: $1600 - 2h^2 = 0 \Rightarrow h^2 = 800 \Rightarrow h = \sqrt{800} = 20\sqrt{2}$. By the First Derivative Test, I has a local maximum at $h = 20\sqrt{2} \approx 28$ ft.

(b)

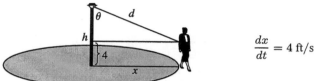

$\dfrac{dx}{dt} = 4$ ft/s

$I = \dfrac{k\cos\theta}{d^2} = \dfrac{k[(h-4)/d]}{d^2} = \dfrac{k(h-4)}{d^3} = \dfrac{k(h-4)}{[(h-4)^2 + x^2]^{3/2}} = k(h-4)\left[(h-4)^2 + x^2\right]^{-3/2}$

$\dfrac{dI}{dt} = \dfrac{dI}{dx} \cdot \dfrac{dx}{dt} = k(h-4)\left(-\tfrac{3}{2}\right)\left[(h-4)^2 + x^2\right]^{-5/2} \cdot 2x \cdot \dfrac{dx}{dt}$

$= k(h-4)(-3x)\left[(h-4)^2 + x^2\right]^{-5/2} \cdot 4 = \dfrac{-12xk(h-4)}{[(h-4)^2 + x^2]^{5/2}}$

$\dfrac{dI}{dt}\bigg|_{x=40} = -\dfrac{480k(h-4)}{[(h-4)^2 + 1600]^{5/2}}$

Focus on Problem Solving

1. Let $y = f(x) = e^{-x^2}$. The area of the rectangle under the curve from $-x$ to x is $A(x) = 2xe^{-x^2}$ where $x \geq 0$.

We maximize $A(x)$: $A'(x) = 2e^{-x^2} - 4x^2e^{-x^2} = 2e^{-x^2}(1 - 2x^2) = 0 \Rightarrow x = \frac{1}{\sqrt{2}}$. This gives a maximum

since $A'(x) > 0$ for $0 \leq x < \frac{1}{\sqrt{2}}$ and $A'(x) < 0$ for $x > \frac{1}{\sqrt{2}}$. We next determine the points of inflection of $f(x)$.

Notice that $f'(x) = -2xe^{-x^2} = -A(x)$. So $f''(x) = -A'(x)$ and hence, $f''(x) < 0$ for $-\frac{1}{\sqrt{2}} < x < \frac{1}{\sqrt{2}}$ and

$f''(x) > 0$ for $x < -\frac{1}{\sqrt{2}}$ and $x > \frac{1}{\sqrt{2}}$. So $f(x)$ changes concavity at $x = \pm\frac{1}{\sqrt{2}}$, and the two vertices of the

rectangle of largest area are at the inflection points.

3. First, we recognize some symmetry in the inequality: $\dfrac{e^{x+y}}{xy} \geq e^2 \Leftrightarrow \dfrac{e^x}{x} \cdot \dfrac{e^y}{y} \geq e \cdot e$. This suggests that we

need to show that $\dfrac{e^x}{x} \geq e$ for $x > 0$. If we can do this, then the inequality $\dfrac{e^y}{y} \geq e$ is true, and the given inequality

follows. $f(x) = \dfrac{e^x}{x} \Rightarrow f'(x) = \dfrac{xe^x - e^x}{x^2} = \dfrac{e^x(x - 1)}{x^2} = 0 \Rightarrow x = 1$. By the First Derivative Test, we

have a minimum of $f(1) = e$, so $e^x/x \geq e$ for all x.

5. First we show that $x(1 - x) \leq \frac{1}{4}$ for all x. Let $f(x) = x(1 - x) = x - x^2$. Then $f'(x) = 1 - 2x$. This is 0 when

$x = \frac{1}{2}$ and $f'(x) > 0$ for $x < \frac{1}{2}$, $f'(x) < 0$ for $x > \frac{1}{2}$, so the absolute maximum of f is $f(\frac{1}{2}) = \frac{1}{4}$. Thus,

$x(1 - x) \leq \frac{1}{4}$ for all x.

Now suppose that the given assertion is false, that is, $a(1 - b) > \frac{1}{4}$ and $b(1 - a) > \frac{1}{4}$. Multiply these inequalities:

$a(1 - b)b(1 - a) > \frac{1}{16} \Rightarrow [a(1 - a)][b(1 - b)] > \frac{1}{16}$. But we know that $a(1 - a) \leq \frac{1}{4}$ and $b(1 - b) \leq \frac{1}{4} \Rightarrow$

$[a(1 - a)][b(1 - b)] \leq \frac{1}{16}$. Thus, we have a contradiction, so the given assertion is proved.

7. Differentiating $x^2 + xy + y^2 = 12$ implicitly with respect to x gives $2x + y + x\dfrac{dy}{dx} + 2y\dfrac{dy}{dx} = 0$, so

$\dfrac{dy}{dx} = -\dfrac{2x + y}{x + 2y}$. At a highest or lowest point, $\dfrac{dy}{dx} = 0 \Leftrightarrow y = -2x$. Substituting $-2x$ for y in the original

equation gives $x^2 + x(-2x) + (-2x)^2 = 12$, so $3x^2 = 12$ and $x = \pm 2$. If $x = 2$, then $y = -2x = -4$, and if

$x = -2$ then $y = 4$. Thus, the highest and lowest points are $(-2, 4)$ and $(2, -4)$.

9. $f(x) = \dfrac{1}{1 + |x|} + \dfrac{1}{1 + |x - 2|}$

$$= \begin{cases} \dfrac{1}{1 - x} + \dfrac{1}{1 - (x - 2)} & \text{if } x < 0 \\[2mm] \dfrac{1}{1 + x} + \dfrac{1}{1 - (x - 2)} & \text{if } 0 \leq x < 2 \\[2mm] \dfrac{1}{1 + x} + \dfrac{1}{1 + (x - 2)} & \text{if } x \geq 2 \end{cases} \Rightarrow f'(x) = \begin{cases} \dfrac{1}{(1 - x)^2} + \dfrac{1}{(3 - x)^2} & \text{if } x < 0 \\[2mm] \dfrac{-1}{(1 + x)^2} + \dfrac{1}{(3 - x)^2} & \text{if } 0 < x < 2 \\[2mm] \dfrac{-1}{(1 + x)^2} - \dfrac{1}{(x - 1)^2} & \text{if } x > 2 \end{cases}$$

We see that $f'(x) > 0$ for $x < 0$ and $f'(x) < 0$ for $x > 2$. For $0 < x < 2$, we have

$$f'(x) = \frac{1}{(3-x)^2} - \frac{1}{(x+1)^2} = \frac{(x^2 + 2x + 1) - (x^2 - 6x + 9)}{(3-x)^2(x+1)^2} = \frac{8(x-1)}{(3-x)^2(x+1)^2}, \text{ so } f'(x) < 0 \text{ for}$$

$0 < x < 1$, $f'(1) = 0$ and $f'(x) > 0$ for $1 < x < 2$. We have shown that $f'(x) > 0$ for $x < 0$; $f'(x) < 0$ for $0 < x < 1$; $f'(x) > 0$ for $1 < x < 2$; and $f'(x) < 0$ for $x > 2$. Therefore, by the First Derivative Test, the local maxima of f are at $x = 0$ and $x = 2$, where f takes the value $\frac{4}{3}$. Therefore, $\frac{4}{3}$ is the absolute maximum value of f.

11. We first show that $\dfrac{x}{1+x^2} < \tan^{-1} x$ for $x > 0$. Let $f(x) = \tan^{-1} x - \dfrac{x}{1+x^2}$. Then

$$f'(x) = \frac{1}{1+x^2} - \frac{1(1+x^2) - x(2x)}{(1+x^2)^2} = \frac{(1+x^2) - (1-x^2)}{(1+x^2)^2} = \frac{2x^2}{(1+x^2)^2} > 0 \text{ for } x > 0. \text{ So } f(x) \text{ is}$$

increasing on $(0, \infty)$. Hence, $0 < x \implies 0 = f(0) < f(x) = \tan^{-1} x - \dfrac{x}{1+x^2}$. So $\dfrac{x}{1+x^2} < \tan^{-1} x$

for $0 < x$. We next show that $\tan^{-1} x < x$ for $x > 0$. Let $h(x) = x - \tan^{-1} x$. Then

$$h'(x) = 1 - \frac{1}{1+x^2} = \frac{x^2}{1+x^2} > 0. \text{ Hence, } h(x) \text{ is increasing on } (0, \infty). \text{ So for } 0 < x,$$

$0 = h(0) < h(x) = x - \tan^{-1} x$. Hence, $\tan^{-1} x < x$ for $x > 0$, and we conclude that $\dfrac{x}{1+x^2} < \tan^{-1} x < x$

for $x > 0$.

13. $A = (x_1, x_1^2)$ and $B = (x_2, x_2^2)$, where x_1 and x_2 are the solutions of the quadratic equation $x^2 = mx + b$. Let $P = (x, x^2)$ and set $A_1 = (x_1, 0)$, $B_1 = (x_2, 0)$, and $P_1 = (x, 0)$. Let $f(x)$ denote the area of triangle PAB. Then $f(x)$ can be expressed in terms of the areas of three trapezoids as follows:

$$f(x) = \text{area}(A_1 ABB_1) - \text{area}(A_1 APP_1) - \text{area}(B_1 BPP_1)$$
$$= \tfrac{1}{2}(x_1^2 + x_2^2)(x_2 - x_1) - \tfrac{1}{2}(x_1^2 + x^2)(x - x_1) - \tfrac{1}{2}(x^2 + x_2^2)(x_2 - x)$$

After expanding and canceling terms, we get

$f(x) = \tfrac{1}{2}(x_2 x_1^2 - x_1 x_2^2 - x x_1^2 + x_1 x^2 - x_2 x^2 + x x_2^2) = \tfrac{1}{2}[x_1^2(x_2 - x) + x_2^2(x - x_1) + x^2(x_1 - x_2)]$

$f'(x) = \tfrac{1}{2}[-x_1^2 + x_2^2 + 2x(x_1 - x_2)]$. $f''(x) = \tfrac{1}{2}[2(x_1 - x_2)] = x_1 - x_2 < 0$ since $x_2 > x_1$.

$f'(x) = 0 \implies 2x(x_1 - x_2) = x_1^2 - x_2^2 \implies x_P = \tfrac{1}{2}(x_1 + x_2)$.

$$f(x_P) = \tfrac{1}{2}\left(x_1^2\left[\tfrac{1}{2}(x_2 - x_1)\right] + x_2^2\left[\tfrac{1}{2}(x_2 - x_1)\right] + \tfrac{1}{4}(x_1 + x_2)^2(x_1 - x_2)\right)$$
$$= \tfrac{1}{2}\left[\tfrac{1}{2}(x_2 - x_1)(x_1^2 + x_2^2) - \tfrac{1}{4}(x_2 - x_1)(x_1 + x_2)^2\right]$$
$$= \tfrac{1}{8}(x_2 - x_1)\left[2(x_1^2 + x_2^2) - (x_1^2 + 2x_1 x_2 + x_2^2)\right]$$
$$= \tfrac{1}{8}(x_2 - x_1)(x_1^2 - 2x_1 x_2 + x_2^2) = \tfrac{1}{8}(x_2 - x_1)(x_1 - x_2)^2 = \tfrac{1}{8}(x_2 - x_1)(x_2 - x_1)^2$$
$$= \tfrac{1}{8}(x_2 - x_1)^3$$

To put this in terms of m and b, we solve the system $y = x_1^2$ and $y = mx_1 + b$, giving us $x_1^2 - mx_1 - b = 0 \implies$ $x_1 = \tfrac{1}{2}(m - \sqrt{m^2 + 4b})$. Similarly, $x_2 = \tfrac{1}{2}(m + \sqrt{m^2 + 4b})$. The area is then

$\tfrac{1}{8}(x_2 - x_1)^3 = \tfrac{1}{8}(\sqrt{m^2 + 4b})^3$, and is attained at the point $P(x_P, x_P^2) = P(\tfrac{1}{2}m, \tfrac{1}{4}m^2)$.

Note: Another way to get an expression for $f(x)$ is to use the formula for an area of a triangle in terms of the coordinates of the vertices: $f(x) = \tfrac{1}{2}[(x_2 x_1^2 - x_1 x_2^2) + (x_1 x^2 - x x_1^2) + (x x_2^2 - x_2 x^2)]$.

15. (a) $A = \frac{1}{2}bh$ with $\sin\theta = h/c$, so $A = \frac{1}{2}bc\sin\theta$. But A is a

constant, so differentiating this equation with respect to t, we

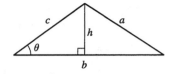

get $\dfrac{dA}{dt} = 0 = \dfrac{1}{2}\left[bc\cos\theta\,\dfrac{d\theta}{dt} + b\,\dfrac{dc}{dt}\sin\theta + \dfrac{db}{dt}c\sin\theta\right]$

$\Rightarrow \quad bc\cos\theta\,\dfrac{d\theta}{dt} = -\sin\theta\left[b\,\dfrac{dc}{dt} + c\,\dfrac{db}{dt}\right] \quad \Rightarrow \quad \dfrac{d\theta}{dt} = -\tan\theta\left[\dfrac{1}{c}\dfrac{dc}{dt} + \dfrac{1}{b}\dfrac{db}{dt}\right].$

(b) We use the Law of Cosines to get the length of side a in terms of those of b and c, and then

we differentiate implicitly with respect to t: $a^2 = b^2 + c^2 - 2bc\cos\theta \quad \Rightarrow$

$2a\,\dfrac{da}{dt} = 2b\,\dfrac{db}{dt} + 2c\,\dfrac{dc}{dt} - 2\left[bc(-\sin\theta)\dfrac{d\theta}{dt} + b\,\dfrac{dc}{dt}\cos\theta + \dfrac{db}{dt}c\cos\theta\right] \quad \Rightarrow$

$\dfrac{da}{dt} = \dfrac{1}{a}\left(b\,\dfrac{db}{dt} + c\,\dfrac{dc}{dt} + bc\sin\theta\,\dfrac{d\theta}{dt} - b\,\dfrac{dc}{dt}\cos\theta - c\,\dfrac{db}{dt}\cos\theta\right).$ Now we substitute our value of a from the

Law of Cosines and the value of $d\theta/dt$ from part (a), and simplify (primes signify differentiation by t):

$$\dfrac{da}{dt} = \dfrac{bb' + cc' + bc\sin\theta\left[-\tan\theta(c'/c + b'/b)\right] - (bc' + cb')(\cos\theta)}{\sqrt{b^2 + c^2 - 2bc\cos\theta}}$$

$$= \dfrac{bb' + cc' - [\sin^2\theta(bc' + cb') + \cos^2\theta(bc' + cb')]/\cos\theta}{\sqrt{b^2 + c^2 - 2bc\cos\theta}} = \dfrac{bb' + cc' - (bc' + cb')\sec\theta}{\sqrt{b^2 + c^2 - 2bc\cos\theta}}$$

17. (a) Let $y = |AD|$, $x = |AB|$, and $1/x = |AC|$, so that $|AB| \cdot |AC| = 1$.

We compute the area $\mathcal{A}$ of $\triangle ABC$ in two ways. First,

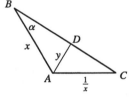

$\mathcal{A} = \frac{1}{2}|AB||AC|\sin\frac{2\pi}{3} = \frac{1}{2} \cdot 1 \cdot \frac{\sqrt{3}}{2} = \frac{\sqrt{3}}{4}.$ Second,

$\mathcal{A} = (\text{area of } \triangle ABD) + (\text{area of } \triangle ACD)$

$\quad = \frac{1}{2}|AB||AD|\sin\frac{\pi}{3} + \frac{1}{2}|AD||AC|\sin\frac{\pi}{3}$

$\quad = \frac{1}{2}xy\frac{\sqrt{3}}{2} + \frac{1}{2}y(1/x)\frac{\sqrt{3}}{2} = \frac{\sqrt{3}}{4}y(x + 1/x)$

Equating the two expressions for the area, we get $\frac{\sqrt{3}}{4}y\left(x + \dfrac{1}{x}\right) = \frac{\sqrt{3}}{4} \quad \Leftrightarrow \quad y = \dfrac{1}{x + 1/x} = \dfrac{x}{x^2 + 1}, x > 0.$

Another method: Use the Law of Sines on the triangles ABD and ABC. In $\triangle ABD$, we have

$\angle A + \angle B + \angle D = 180° \quad \Leftrightarrow \quad 60° + \alpha + \angle D = 180° \quad \Leftrightarrow \quad \angle D = 120° - \alpha.$ Thus,

$\dfrac{x}{y} = \dfrac{\sin(120° - \alpha)}{\sin\alpha} = \dfrac{\sin 120° \cos\alpha - \cos 120° \sin\alpha}{\sin\alpha} = \dfrac{\frac{\sqrt{3}}{2}\cos\alpha + \frac{1}{2}\sin\alpha}{\sin\alpha} \quad \Rightarrow \quad \dfrac{x}{y} = \dfrac{\sqrt{3}}{2}\cot\alpha + \frac{1}{2},$

and by a similar argument with $\triangle ABC$, $\frac{\sqrt{3}}{2}\cot\alpha = x^2 + \frac{1}{2}$. Eliminating $\cot\alpha$ gives $\dfrac{x}{y} = \left(x^2 + \frac{1}{2}\right) + \frac{1}{2} \quad \Rightarrow$

$y = \dfrac{x}{x^2 + 1}, x > 0.$

(b) We differentiate our expression for y with respect to x to find the maximum:

$\dfrac{dy}{dx} = \dfrac{(x^2 + 1) - x(2x)}{(x^2 + 1)^2} = \dfrac{1 - x^2}{(x^2 + 1)^2} = 0$ when $x = 1$. This indicates a maximum by the First Derivative

Test, since $y'(x) > 0$ for $0 < x < 1$ and $y'(x) < 0$ for $x > 1$, so the maximum value of y is $y(1) = \frac{1}{2}$.

19. Let $s_A(t)$ and $s_B(t)$ be the position functions for cars A and B and let $f(t) = s_A(t) - s_B(t)$. Since A passed B twice (B passed A once), there must be three values of t such that $f(t) = 0$. Let these times be denoted $t_1, t_2,$ and t_3. By the Mean Value Theorem, we know that for some number c_1 in (t_1, t_2), $f'(c_1) = \dfrac{f(t_2) - f(t_1)}{t_2 - t_1}$, but $f(t_2) - f(t_1) = 0$, so $f'(c_1) = 0$ $\Leftrightarrow$ $s_A'(c_1) - s_B'(c_1) = 0$ $\Leftrightarrow$ $v_A(c_1) - v_B(c_1) = 0$ $\Rightarrow$ $v_A(c_1) = v_B(c_1)$. By a similar argument, there exists some number c_2 in (t_2, t_3) such that $v_A(c_2) = v_B(c_2)$. Now let $g(t) = v_A(t) - v_B(t)$ and apply the Mean Value Theorem on $[c_1, c_2]$ with $c_1 < c < c_2$.

$g'(c) = \dfrac{g(c_2) - g(c_1)}{c_2 - c_1}$, but $g(c_2) = g(c_1) = 0$, so $g'(c) = 0$ $\Rightarrow$ $v_A'(c) - v_B'(c) = 0$ $\Rightarrow$ $a_A(c) - a_B(c) = 0$ $\Rightarrow$ $a_A(c) = a_B(c)$; that is, A and B had equal accelerations at $t = c$.

21.

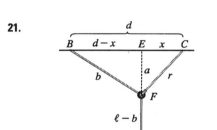

Let $a = |EF|$ and $b = |BF|$ as shown in the figure. Since $\ell = |BF| + |FD|$, $|FD| = \ell - b$. Now

$$|ED| = |EF| + |FD| = a + \ell - b$$
$$= \sqrt{r^2 - x^2} + \ell - \sqrt{(d-x)^2 + a^2}$$
$$= \sqrt{r^2 - x^2} + \ell - \sqrt{(d-x)^2 + \left(\sqrt{r^2 - x^2}\right)^2}$$
$$= \sqrt{r^2 - x^2} + \ell - \sqrt{d^2 - 2dx + x^2 + r^2 - x^2}$$

Let $f(x) = \sqrt{r^2 - x^2} + \ell - \sqrt{d^2 + r^2 - 2dx}$.

$f'(x) = \frac{1}{2}(r^2 - x^2)^{-1/2}(-2x) - \frac{1}{2}(d^2 + r^2 - 2dx)^{-1/2}(-2d) = \dfrac{-x}{\sqrt{r^2 - x^2}} + \dfrac{d}{\sqrt{d^2 + r^2 - 2dx}}$.

$f'(x) = 0$ $\Rightarrow$ $\dfrac{x}{\sqrt{r^2 - x^2}} = \dfrac{d}{\sqrt{d^2 + r^2 - 2dx}}$ $\Rightarrow$ $\dfrac{x^2}{r^2 - x^2} = \dfrac{d^2}{d^2 + r^2 - 2dx}$ $\Rightarrow$

$d^2x^2 + r^2x^2 - 2dx^3 = d^2r^2 - d^2x^2$ $\Rightarrow$ $0 = 2dx^3 - 2d^2x^2 - r^2x^2 + d^2r^2$ $\Rightarrow$

$0 = 2dx^2(x - d) - r^2(x^2 - d^2)$ $\Rightarrow$ $0 = 2dx^2(x - d) - r^2(x + d)(x - d)$ $\Rightarrow$

$0 = (x - d)[2dx^2 - r^2(x + d)]$

But $d > r > x$, so $x \neq d$. Thus, we solve $2dx^2 - r^2x - dr^2 = 0$ for x:

$x = \dfrac{-(-r^2) \pm \sqrt{(-r^2)^2 - 4(2d)(-dr^2)}}{2(2d)} = \dfrac{r^2 \pm \sqrt{r^4 + 8d^2r^2}}{4d}$. Because $\sqrt{r^4 + 8d^2r^2} > r^2$, the "negative" can be discarded. Thus,

$$x = \dfrac{r^2 + \sqrt{r^2}\sqrt{r^2 + 8d^2}}{4d} = \dfrac{r^2 + r\sqrt{r^2 + 8d^2}}{4d} \quad (r > 0)$$
$$= \dfrac{r}{4d}\left(r + \sqrt{r^2 + 8d^2}\right)$$

The maximum value of $|ED|$ occurs at this value of x.

23.

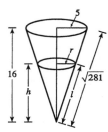

By similar triangles, $\dfrac{r}{5} = \dfrac{h}{16} \;\Rightarrow\; r = \dfrac{5h}{16}$. The volume of the cone is

$$V = \tfrac{1}{3}\pi r^2 h = \tfrac{1}{3}\pi \left(\frac{5h}{16}\right)^2 h = \frac{25\pi}{768}h^3, \text{ so } \frac{dV}{dt} = \frac{25\pi}{256}h^2\,\frac{dh}{dt}. \text{ Now the rate of}$$

change of the volume is also equal to the difference of what is being added

($2 \text{ cm}^3/\text{min}$) and what is oozing out ($k\pi rl$, where πrl is the area of the cone and k

is a proportionality constant). Thus, $\dfrac{dV}{dt} = 2 - k\pi rl$.

Equating the two expressions for $\dfrac{dV}{dt}$ and substituting $h = 10$, $\dfrac{dh}{dt} = -0.3$, $r = \dfrac{5(10)}{16} = \dfrac{25}{8}$, and $\dfrac{l}{\sqrt{281}} = \dfrac{10}{16}$

$\Leftrightarrow\; l = \tfrac{5}{8}\sqrt{281}$, we get $\frac{25\pi}{256}(10)^2(-0.3) = 2 - k\pi \frac{25}{8}\cdot\frac{5}{8}\sqrt{281} \;\Leftrightarrow\; \dfrac{125k\pi\sqrt{281}}{64} = 2 + \dfrac{750\pi}{256}$. Solving for k

gives us $k = \dfrac{256 + 375\pi}{250\pi\sqrt{281}}$. To maintain a certain height, the rate of oozing, $k\pi rl$, must equal the rate of the liquid

being poured in; that is, $\dfrac{dV}{dt} = 0$. $k\pi rl = \dfrac{256 + 375\pi}{250\pi\sqrt{281}}\cdot\pi\cdot\dfrac{25}{8}\cdot\dfrac{5\sqrt{281}}{8} = \dfrac{256 + 375\pi}{128} \approx 11.204 \text{ cm}^3/\text{min}.$

5 Integrals

 Areas and Distances • • • • • • • • • • • • •

1. (a) Since f is *increasing*, we can obtain a *lower* estimate by using *left* endpoints. We are instructed to use five rectangles, so $n = 5$.

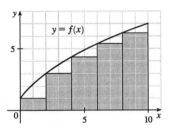

$$L_5 = \sum_{i=1}^{5} f(x_{i-1})\,\Delta x \quad [\Delta x = \tfrac{b-a}{n} = \tfrac{10-0}{5} = 2]$$

$$= f(x_0) \cdot 2 + f(x_1) \cdot 2 + f(x_2) \cdot 2 + f(x_3) \cdot 2 + f(x_4) \cdot 2$$

$$= 2\,[f(0) + f(2) + f(4) + f(6) + f(8)]$$

$$\approx 2(1 + 3 + 4.3 + 5.4 + 6.3) = 2(20) = 40$$

Since f is *increasing*, we can obtain an *upper* estimate by using *right* endpoints.

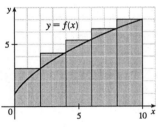

$$R_5 = \sum_{i=1}^{5} f(x_i)\,\Delta x$$

$$= 2\,[f(x_1) + f(x_2) + f(x_3) + f(x_4) + f(x_5)]$$

$$= 2\,[f(2) + f(4) + f(6) + f(8) + f(10)]$$

$$\approx 2(3 + 4.3 + 5.4 + 6.3 + 7) = 2(26) = 52$$

Comparing R_5 to L_5, we see that we have added the area of the rightmost rectangle, $f(10) \cdot 2$, to the sum and subtracted the area of the leftmost rectangle, $f(0) \cdot 2$, from the sum.

(b) $L_{10} = \sum_{i=1}^{10} f(x_{i-1})\,\Delta x \quad [\Delta x = \tfrac{10-0}{10} = 1]$

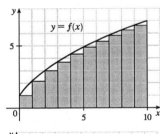

$$= 1\,[f(x_0) + f(x_1) + \cdots + f(x_9)]$$

$$= f(0) + f(1) + \cdots + f(9)$$

$$\approx 1 + 2.1 + 3 + 3.7 + 4.3 + 4.9 + 5.4 + 5.8 + 6.3 + 6.7$$

$$= 43.2$$

$$R_{10} = \sum_{i=1}^{10} f(x_i)\,\Delta x = f(1) + f(2) + \cdots + f(10)$$

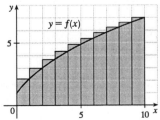

$$= L_{10} + 1 \cdot f(10) - 1 \cdot f(0) \quad \begin{bmatrix} \text{add rightmost rectangle,} \\ \text{subtract leftmost} \end{bmatrix}$$

$$= 43.2 + 7 - 1 = 49.2$$

3. (a) $R_4 = \sum_{i=1}^{4} f(x_i)\,\Delta x \quad [\Delta x = \frac{5-1}{4} = 1]$

$ = f(x_1)\cdot 1 + f(x_2)\cdot 1 + f(x_3)\cdot 1 + f(x_4)\cdot 1$

$ = f(2) + f(3) + f(4) + f(5)$

$ = \frac{1}{2} + \frac{1}{3} + \frac{1}{4} + \frac{1}{5} = \frac{77}{60} = 1.28\overline{3}$

Since f is *decreasing* on $[1, 5]$, an *underestimate* is obtained by using the *right* endpoint approximation, R_4.

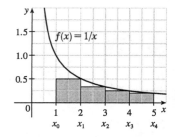

(b) $L_4 = \sum_{i=1}^{4} f(x_{i-1})\,\Delta x$

$ = f(1) + f(2) + f(3) + f(4)$

$ = 1 + \frac{1}{2} + \frac{1}{3} + \frac{1}{4} = \frac{25}{12} = 2.08\overline{3}$

L_4 is an overestimate. Alternatively, we could just add the area of the leftmost rectangle and subtract the area of the rightmost; that is,

$L_4 = R_4 + f(1)\cdot 1 - f(5)\cdot 1.$

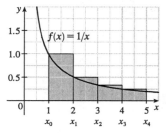

5. (a) $f(x) = 1 + x^2$ and $\Delta x = \dfrac{2-(-1)}{3} = 1 \;\Rightarrow$

$R_3 = 1\cdot f(0) + 1\cdot f(1) + 1\cdot f(2) = 1\cdot 1 + 1\cdot 2 + 1\cdot 5 = 8.$

$\Delta x = \dfrac{2-(-1)}{6} = 0.5 \;\Rightarrow$

$R_6 = 0.5[f(-0.5) + f(0) + f(0.5) + f(1) + f(1.5) + f(2)]$

$ = 0.5(1.25 + 1 + 1.25 + 2 + 3.25 + 5)$

$ = 0.5(13.75) = 6.875$

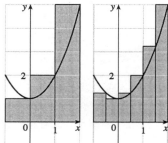

(b) $L_3 = 1\cdot f(-1) + 1\cdot f(0) + 1\cdot f(1) = 1\cdot 2 + 1\cdot 1 + 1\cdot 2 = 5$

$L_6 = 0.5[f(-1) + f(-0.5) + f(0) + f(0.5) + f(1) + f(1.5)]$

$ = 0.5(2 + 1.25 + 1 + 1.25 + 2 + 3.25)$

$ = 0.5(10.75) = 5.375$

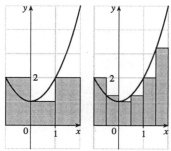

(c) $M_3 = 1\cdot f(-0.5) + 1\cdot f(0.5) + 1\cdot f(1.5)$

$ = 1\cdot 1.25 + 1\cdot 1.25 + 1\cdot 3.25 = 5.75$

$M_6 = 0.5[f(-0.75) + f(-0.25) + f(0.25)$

$ + f(0.75) + f(1.25) + f(1.75)]$

$ = 0.5(1.5625 + 1.0625 + 1.0625 + 1.5625 + 2.5625 + 4.0625)$

$ = 0.5(11.875) = 5.9375$

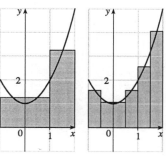

(d) M_6 appears to be the best estimate.

7. Here is one possible algorithm (ordered sequence of operations) for calculating the sums:

1 Let SUM = 0, X_MIN = 0, X_MAX = π, N = 10 (or 30 or 50, depending on which sum we are calculating), DELTA_X = (X_MAX − X_MIN)/N, and RIGHT_ENDPOINT = X_MIN + DELTA_X.

2 Repeat steps 2a, 2b in sequence until RIGHT_ENDPOINT > X_MAX.

2a Add sin (RIGHT_ENDPOINT) to SUM.

2b Add DELTA_X to RIGHT_ENDPOINT.

At the end of this procedure, (DELTA_X) · (SUM) is equal to the answer we are looking for. We find that

$$R_{10} = \frac{\pi}{10} \sum_{i=1}^{10} \sin\left(\frac{i\pi}{10}\right) \approx 1.9835, \; R_{30} = \frac{\pi}{30} \sum_{i=1}^{30} \sin\left(\frac{i\pi}{30}\right) \approx 1.9982, \text{ and } R_{50} = \frac{\pi}{50} \sum_{i=1}^{50} \sin\left(\frac{i\pi}{50}\right) \approx 1.9993.$$

It appears that the exact area is 2.

Shown below is program SUMRIGHT and its output from a TI-83 Plus calculator. To generalize the program, we have input (rather than assign) values for Xmin, Xmax, and N. Also, the function, sin x, is assigned to Y_1, enabling us to evaluate any right sum merely by changing Y_1 and running the program.

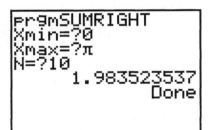

9. In Maple, we have to perform a number of steps before getting a numerical answer. After loading the student package [command: `with(student);`] we use the command `left_sum:=leftsum(x^(1/2),x=1..4,10 [or 30, or 50]);` which gives us the expression in summation notation. To get a numerical approximation to the sum, we use `evalf(left_sum);`. Mathematica does not have a special command for these sums, so we must type them in manually. For example, the first left sum is given by `(3/10)*Sum[Sqrt[1+3(i-1)/10],{i,1,10}]`, and we use the N command on the resulting output to get a numerical approximation.

In Derive, we use the `LEFT_RIEMANN` command to get the left sums, but must define the right sums ourselves. (We can define a new function using `LEFT_RIEMANN` with k ranging from 1 to n instead of from 0 to $n - 1$.)

(a) With $f(x) = \sqrt{x}$, $1 \leq x \leq 4$, the left sums are of the form $L_n = \frac{3}{n} \sum_{i=1}^{n} \sqrt{1 + \frac{3(i-1)}{n}}$. Specifically,

$L_{10} \approx 4.5148$, $L_{30} \approx 4.6165$, and $L_{50} \approx 4.6366$. The right sums are of the form $R_n = \frac{3}{n} \sum_{i=1}^{n} \sqrt{1 + \frac{3i}{n}}$.

Specifically, $R_{10} \approx 4.8148$, $R_{30} \approx 4.7165$, and $R_{50} \approx 4.6966$.

(b) In Maple, we use the `leftbox` and `rightbox` commands (with the same arguments as `leftsum` and `rightsum` above) to generate the graphs.

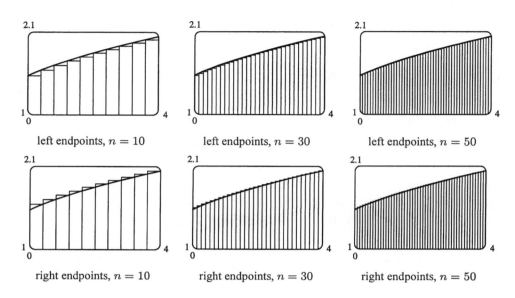

left endpoints, $n = 10$ left endpoints, $n = 30$ left endpoints, $n = 50$

right endpoints, $n = 10$ right endpoints, $n = 30$ right endpoints, $n = 50$

(c) We know that since $\sqrt{x}$ is an increasing function on $(1, 4)$, all of the left sums are smaller than the actual area, and all of the right sums are larger than the actual area. Since the left sum with $n = 50$ is about $4.637 > 4.6$ and the right sum with $n = 50$ is about $4.697 < 4.7$, we conclude that $4.6 < L_{50} <$ exact area $< R_{50} < 4.7$, so the exact area is between 4.6 and 4.7.

11. Since v is an increasing function, L_6 will give us a lower estimate and R_6 will give us an upper estimate.

$$L_6 = (0 \text{ ft/s})(0.5 \text{ s}) + (6.2)(0.5) + (10.8)(0.5) + (14.9)(0.5) + (18.1)(0.5) + (19.4)(0.5)$$
$$= 0.5(69.4) = 34.7 \text{ ft}$$
$$R_6 = 0.5(6.2 + 10.8 + 14.9 + 18.1 + 19.4 + 20.2) = 0.5(89.6) = 44.8 \text{ ft}$$

13. For a decreasing function, using left endpoints gives us an overestimate and using right endpoints results in an underestimate. We will use M_6 to get an estimate. $\Delta t = 1$, so

$$M_6 = 1[v(0.5) + v(1.5) + v(2.5) + v(3.5) + v(4.5) + v(5.5)]$$
$$\approx 55 + 40 + 28 + 18 + 10 + 4 = 155 \text{ ft}$$

For a very rough check on the above calculation, we can draw a line from $(0, 70)$ to $(6, 0)$ and calculate the area of the triangle: $\frac{1}{2}(70)(6) = 210$. This is clearly an overestimate, so our midpoint estimate of 155 is reasonable.

15. $f(x) = \sqrt[4]{x}$, $1 \le x \le 16$. $\Delta x = (16 - 1)/n = 15/n$ and $x_i = 1 + i \Delta x = 1 + 15i/n$.

$$A = \lim_{n \to \infty} R_n = \lim_{n \to \infty} \sum_{i=1}^{n} f(x_i) \Delta x = \lim_{n \to \infty} \sum_{i=1}^{n} \sqrt[4]{1 + \frac{15i}{n}} \cdot \frac{15}{n}.$$

17. $\lim\limits_{n\to\infty}\sum\limits_{i=1}^{n}\dfrac{\pi}{4n}\tan\dfrac{i\pi}{4n}$ can be interpreted as the area of the region lying under the graph of $y=\tan x$ on the interval

$\left[0,\frac{\pi}{4}\right]$, since for $y=\tan x$ on $\left[0,\frac{\pi}{4}\right]$ with $\Delta x=\dfrac{\pi/4-0}{n}=\dfrac{\pi}{4n}$, $x_i=0+i\,\Delta x=\dfrac{i\pi}{4n}$, and $x_i^{*}=x_i$, the

expression for the area is $A=\lim\limits_{n\to\infty}\sum\limits_{i=1}^{n}f(x_i^{*})\,\Delta x=\lim\limits_{n\to\infty}\sum\limits_{i=1}^{n}\tan\left(\dfrac{i\pi}{4n}\right)\dfrac{\pi}{4n}$. Note that this answer is not unique,

since the expression for the area is the same for the function $y=\tan(x+k\pi)$ on the interval $\left[k\pi,\,k\pi+\frac{\pi}{4}\right]$, where k is any integer.

19. (a) $y=f(x)=x^5.$ $\Delta x=\dfrac{2-0}{n}=\dfrac{2}{n}$ and $x_i=0+i\,\Delta x=\dfrac{2i}{n}$.

$$A=\lim_{n\to\infty}R_n=\lim_{n\to\infty}\sum_{i=1}^{n}f(x_i)\,\Delta x=\lim_{n\to\infty}\sum_{i=1}^{n}\left(\dfrac{2i}{n}\right)^5\cdot\dfrac{2}{n}=\lim_{n\to\infty}\sum_{i=1}^{n}\dfrac{32i^5}{n^5}\cdot\dfrac{2}{n}=\lim_{n\to\infty}\dfrac{64}{n^6}\sum_{i=1}^{n}i^5.$$

(b) $\sum\limits_{i=1}^{n}i^5=\dfrac{n^2(n+1)^2(2n^2+2n-1)}{12}$

(c) $\lim\limits_{n\to\infty}\dfrac{64}{n^6}\cdot\dfrac{n^2(n+1)^2(2n^2+2n-1)}{12}=\dfrac{64}{12}\lim\limits_{n\to\infty}\dfrac{(n^2+2n+1)(2n^2+2n-1)}{n^2\cdot n^2}$

$$=\dfrac{16}{3}\lim_{n\to\infty}\left(1+\dfrac{2}{n}+\dfrac{1}{n^2}\right)\left(2+\dfrac{2}{n}-\dfrac{1}{n^2}\right)=\tfrac{16}{3}\cdot1\cdot2=\tfrac{32}{3}$$

21. $y=f(x)=\cos x.$ $\Delta x=\dfrac{b-0}{n}=\dfrac{b}{n}$ and $x_i=0+i\,\Delta x=\dfrac{bi}{n}$.

$$A=\lim_{n\to\infty}R_n=\lim_{n\to\infty}\sum_{i=1}^{n}f(x_i)\,\Delta x=\lim_{n\to\infty}\sum_{i=1}^{n}\cos\left(\dfrac{bi}{n}\right)\cdot\dfrac{b}{n}=\lim_{n\to\infty}\left[\dfrac{b\sin\left(b\left(\dfrac{1}{2n}+1\right)\right)}{2n\sin\left(\dfrac{b}{2n}\right)}-\dfrac{b}{2n}\right]=\sin b$$

If $b=\frac{\pi}{2}$, then $A=\sin\frac{\pi}{2}=1$.

5.2 The Definite Integral • • • • • • • • • • • •

1. $R_4=\sum\limits_{i=1}^{4}f(x_i)\,\Delta x$ $[x_i^{*}=x_i$ is a right endpoint and $\Delta x=0.5]$

$=0.5\left[f(0.5)+f(1)+f(1.5)+f(2)\right]$ $[f(x)=2-x^2]$

$=0.5\left[1.75+1+(-0.25)+(-2)\right]$

$=0.5(0.5)=0.25$

The Riemann sum represents the sum of the areas of the two rectangles above the x-axis minus the sum of the areas of the two rectangles below the x-axis; that is, the *net area* of the rectangles with respect to the x-axis.

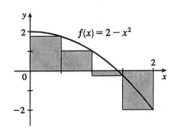

3. $M_5 = \sum\limits_{i=1}^{5} f(\overline{x}_i)\,\Delta x$ $[x_i^* = \overline{x}_i = \frac{1}{2}(x_{i-1} + x_i)$ is a midpoint and $\Delta x = 1]$

$\qquad = 1\,[f(1.5) + f(2.5) + f(3.5)$

$\qquad\qquad\qquad + f(4.5) + f(5.5)]$ $[f(x) = \sqrt{x} - 2]$

$\qquad \approx -0.856759$

The Riemann sum represents the sum of the areas of the two rectangles
above the x-axis minus the sum of the areas of the three rectangles below
the x-axis.

$f(x) = \sqrt{x} - 2$

5. $\Delta x = (b - a)/n = (8 - 0)/4 = 8/4 = 2.$

(a) Using the right endpoints to approximate $\int_0^8 f(x)\,dx$, we have

$\sum\limits_{i=1}^{4} f(x_i)\,\Delta x = 2[f(2) + f(4) + f(6) + f(8)] \approx 2[1 + 2 + (-2) + 1] = 4.$

(b) Using the left endpoints to approximate $\int_0^8 f(x)\,dx$, we have

$\sum\limits_{i=1}^{4} f(x_{i-1})\,\Delta x = 2[f(0) + f(2) + f(4) + f(6)] \approx 2[2 + 1 + 2 + (-2)] = 6.$

(c) Using the midpoint of each subinterval to approximate $\int_0^8 f(x)\,dx$, we have

$\sum\limits_{i=1}^{4} f(\overline{x}_i)\,\Delta x = 2[f(1) + f(3) + f(5) + f(7)] \approx 2[3 + 2 + 1 + (-1)] = 10.$

7. Since f is increasing, $L_5 \le \int_0^{25} f(x)\,dx \le R_5.$

$\qquad$ Lower estimate $= L_5 = \sum\limits_{i=1}^{5} f(x_{i-1})\,\Delta x = 5[f(0) + f(5) + f(10) + f(15) + f(20)]$

$\qquad\qquad\qquad\qquad\qquad = 5(-42 - 37 - 25 - 6 + 15) = 5(-95) = -475$

$\qquad$ Upper estimate $= R_5 = \sum\limits_{i=1}^{5} f(x_i)\,\Delta x = 5[f(5) + f(10) + f(15) + f(20) + f(25)]$

$\qquad\qquad\qquad\qquad\qquad = 5(-37 - 25 - 6 + 15 + 36) = 5(-17) = -85$

9. $\Delta x = (10 - 0)/5 = 2$, so the endpoints are 0, 2, 4, 6, 8, and 10, and the midpoints are 1, 3, 5, 7, and 9.
The Midpoint Rule gives

$\int_0^{10} \sin\sqrt{x}\,dx \approx \sum\limits_{i=1}^{5} f(\overline{x}_i)\,\Delta x = 2\left(\sin\sqrt{1} + \sin\sqrt{3} + \sin\sqrt{5} + \sin\sqrt{7} + \sin\sqrt{9}\right) \approx 6.4643.$

11. $\Delta x = (2 - 1)/10 = 0.1$, so the endpoints are $1.0, 1.1, \ldots, 2.0$ and the midpoints are $1.05, 1.15, \ldots, 1.95$.
The Midpoint Rule gives

$\int_1^2 \sqrt{1 + x^2}\,dx \approx \sum\limits_{i=1}^{10} f(\overline{x}_i)\,\Delta x = 0.1\left[\sqrt{1 + (1.05)^2} + \sqrt{1 + (1.15)^2} + \cdots + \sqrt{1 + (1.95)^2}\right] \approx 1.8100.$

13. In Maple, we use the command `with(student);` to load the sum and box commands, then
`m:=middlesum(sqrt(1+x^2),x=1..2,10);` which gives us the sum in summation notation, then
`M:=evalf(m);` which gives $M_{10} \approx 1.81001414$, confirming the result of Exercise 11. The command
`middlebox(sqrt(1+x^2),x=1..2,10);` generates the graph. Repeating for $n = 20$ and $n = 30$ gives
$M_{20} \approx 1.81007263$ and $M_{30} \approx 1.81008347$.

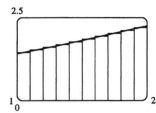 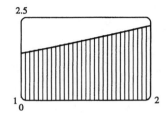 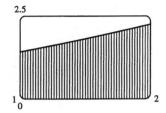

15. We'll create the table of values to approximate $\int_0^\pi \sin x \, dx$ by using the
program in the solution to Exercise 5.1.7 with $Y_1 = \sin x$, Xmin $= 0$,
Xmax $= \pi$, and $n = 5, 10, 50$, and 100.

The values of R_n appear to be approaching 2.

n	R_n
5	1.933766
10	1.983524
50	1.999342
100	1.999836

17. On $[0, \pi]$, $\displaystyle\lim_{n\to\infty} \sum_{i=1}^{n} x_i \sin x_i \, \Delta x = \int_0^\pi x \sin x \, dx$.

19. On $[0, 1]$, $\displaystyle\lim_{n\to\infty} \sum_{i=1}^{n} \left[2(x_i^*)^2 - 5x_i^*\right] \Delta x = \int_0^1 \left(2x^2 - 5x\right) dx$.

21. Note that $\Delta x = \dfrac{5 - (-1)}{n} = \dfrac{6}{n}$ and $x_i = -1 + i\,\Delta x = -1 + \dfrac{6i}{n}$.

$$\int_{-1}^{5} (1 + 3x)\, dx = \lim_{n\to\infty} \sum_{i=1}^{n} f(x_i)\,\Delta x = \lim_{n\to\infty} \sum_{i=1}^{n} \left[1 + 3\left(-1 + \frac{6i}{n}\right)\right] \frac{6}{n}$$

$$= \lim_{n\to\infty} \frac{6}{n} \sum_{i=1}^{n} \left[-2 + \frac{18i}{n}\right] = \lim_{n\to\infty} \frac{6}{n} \left[\sum_{i=1}^{n}(-2) + \sum_{i=1}^{n} \frac{18i}{n}\right]$$

$$= \lim_{n\to\infty} \frac{6}{n} \left[-2n + \frac{18}{n} \sum_{i=1}^{n} i\right] = \lim_{n\to\infty} \frac{6}{n} \left[-2n + \frac{18}{n} \cdot \frac{n(n+1)}{2}\right]$$

$$= \lim_{n\to\infty} \left[-12 + \frac{108}{n^2} \cdot \frac{n(n+1)}{2}\right] = \lim_{n\to\infty} \left[-12 + 54\frac{n+1}{n}\right]$$

$$= \lim_{n\to\infty} \left[-12 + 54\left(1 + \frac{1}{n}\right)\right] = -12 + 54 \cdot 1 = 42$$

23. Note that $\Delta x = \dfrac{2-0}{n} = \dfrac{2}{n}$ and $x_i = 0 + i\,\Delta x = \dfrac{2i}{n}$.

$$\int_0^2 (2 - x^2)\,dx = \lim_{n\to\infty} \sum_{i=1}^n f(x_i)\,\Delta x = \lim_{n\to\infty} \sum_{i=1}^n \left(2 - \frac{4i^2}{n^2}\right)\left(\frac{2}{n}\right)$$

$$= \lim_{n\to\infty} \frac{2}{n}\left[\sum_{i=1}^n 2 - \frac{4}{n^2}\sum_{i=1}^n i^2\right] = \lim_{n\to\infty} \frac{2}{n}\left(2n - \frac{4}{n^2}\sum_{i=1}^n i^2\right)$$

$$= \lim_{n\to\infty}\left[4 - \frac{8}{n^3}\cdot\frac{n(n+1)(2n+1)}{6}\right] = \lim_{n\to\infty}\left(4 - \frac{4}{3}\cdot\frac{n+1}{n}\cdot\frac{2n+1}{n}\right)$$

$$= \lim_{n\to\infty}\left[4 - \frac{4}{3}\left(1 + \frac{1}{n}\right)\left(2 + \frac{1}{n}\right)\right] = 4 - \tfrac{4}{3}\cdot 1\cdot 2 = \tfrac{4}{3}$$

25. Note that $\Delta x = \dfrac{2-1}{n} = \dfrac{1}{n}$ and $x_i = 1 + i\,\Delta x = 1 + i(1/n) = 1 + i/n$.

$$\int_1^2 x^3\,dx = \lim_{n\to\infty}\sum_{i=1}^n f(x_i)\,\Delta x = \lim_{n\to\infty}\sum_{i=1}^n \left(1 + \frac{i}{n}\right)^3\left(\frac{1}{n}\right) = \lim_{n\to\infty}\frac{1}{n}\sum_{i=1}^n \left(\frac{n+i}{n}\right)^3$$

$$= \lim_{n\to\infty}\frac{1}{n^4}\sum_{i=1}^n (n^3 + 3n^2 i + 3ni^2 + i^3) = \lim_{n\to\infty}\frac{1}{n^4}\left[\sum_{i=1}^n n^3 + \sum_{i=1}^n 3n^2 i + \sum_{i=1}^n 3ni^2 + \sum_{i=1}^n i^3\right]$$

$$= \lim_{n\to\infty}\frac{1}{n^4}\left[n\cdot n^3 + 3n^2\sum_{i=1}^n i + 3n\sum_{i=1}^n i^2 + \sum_{i=1}^n i^3\right]$$

$$= \lim_{n\to\infty}\left[1 + \frac{3}{n^2}\cdot\frac{n(n+1)}{2} + \frac{3}{n^3}\cdot\frac{n(n+1)(2n+1)}{6} + \frac{1}{n^4}\cdot\frac{n^2(n+1)^2}{4}\right]$$

$$= \lim_{n\to\infty}\left[1 + \frac{3}{2}\cdot\frac{n+1}{n} + \frac{1}{2}\cdot\frac{n+1}{n}\cdot\frac{2n+1}{n} + \frac{1}{4}\cdot\frac{(n+1)^2}{n^2}\right]$$

$$= \lim_{n\to\infty}\left[1 + \frac{3}{2}\left(1 + \frac{1}{n}\right) + \frac{1}{2}\left(1 + \frac{1}{n}\right)\left(2 + \frac{1}{n}\right) + \frac{1}{4}\left(1 + \frac{1}{n}\right)^2\right] = 1 + \frac{3}{2} + \frac{1}{2}\cdot 2 + \frac{1}{4} = 3.75$$

27. $\Delta x = (\pi - 0)/n = \pi/n$ and $x_i^* = x_i = \pi i/n$.

$$\int_0^\pi \sin 5x\,dx = \lim_{n\to\infty}\sum_{i=1}^n (\sin 5x_i)\left(\frac{\pi}{n}\right) = \lim_{n\to\infty}\sum_{i=1}^n \left(\sin\frac{5\pi i}{n}\right)\frac{\pi}{n} \overset{\text{CAS}}{=} \pi\lim_{n\to\infty}\frac{1}{n}\cot\left(\frac{5\pi}{2n}\right) \overset{\text{CAS}}{=} \pi\left(\frac{2}{5\pi}\right) = \frac{2}{5}$$

29. (a) Think of $\int_0^2 f(x)\,dx$ as the area of a trapezoid with bases 1 and 3 and height 2. The area of a trapezoid is $A = \frac{1}{2}(b + B)h$, so $\int_0^2 f(x)\,dx = \frac{1}{2}(1 + 3)2 = 4$.

(b) $\int_0^5 f(x)\,dx = \int_0^2 f(x)\,dx + \int_2^3 f(x)\,dx + \int_3^5 f(x)\,dx$

$$\qquad\qquad\text{trapezoid}\qquad\text{rectangle}\qquad\text{triangle}$$

$$= \tfrac{1}{2}(1+3)2 + \quad 3\cdot 1 \quad + \quad \tfrac{1}{2}\cdot 2\cdot 3 \quad = 4 + 3 + 3 = 10$$

(c) $\int_5^7 f(x)\,dx$ is the negative of the area of the triangle with base 2 and height 3. $\int_5^7 f(x)\,dx = -\frac{1}{2}\cdot 2\cdot 3 = -3$.

(d) As in parts (a) and (c), $\int_7^9 f(x)\,dx = -\frac{1}{2}(3 + 2)2 = -5$. Thus,

$$\int_0^9 f(x)\,dx = \int_0^5 f(x)\,dx + \int_5^7 f(x)\,dx + \int_7^9 f(x)\,dx = 10 - 3 - 5 = 2.$$

31. $\int_1^3 (1 + 2x)\, dx$ can be interpreted as the area under the graph of $f(x) = 1 + 2x$

between $x = 1$ and $x = 3$. This is equal to the area of the rectangle plus the

area of the triangle, so $\int_1^3 (1 + 2x)\, dx = A = 2 \cdot 3 + \frac{1}{2} \cdot 2 \cdot 4 = 10$.

Or: Use the formula for the area of a trapezoid: $A = \frac{1}{2}(3 + 7)(2) = 10$.

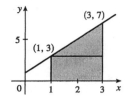

33. $\int_{-3}^0 \left(1 + \sqrt{9 - x^2}\right) dx$ can be interpreted as the area under the graph of

$f(x) = 1 + \sqrt{9 - x^2}$ between $x = -3$ and $x = 0$. This is equal to one-quarter

the area of the circle with radius 3, plus the area of the rectangle, so

$\int_{-3}^0 \left(1 + \sqrt{9 - x^2}\right) dx = \frac{1}{4}\pi \cdot 3^2 + 1 \cdot 3 = 3 + \frac{9}{4}\pi$.

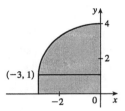

35. $\int_{-2}^2 (1 - |x|)\, dx$ can be interpreted as the area of the middle triangle minus the

areas of the outside triangles, so $\int_{-2}^2 (1 - |x|)\, dx = \frac{1}{2} \cdot 2 \cdot 1 - 2 \cdot \frac{1}{2} \cdot 1 \cdot 1 = 0$.

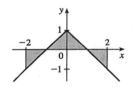

37. $\int_9^4 \sqrt{t}\, dt = -\int_4^9 \sqrt{t}\, dt$ (because we reversed the limits of integration)

$\qquad = -\int_4^9 \sqrt{x}\, dx$ (we can use any letter without changing the value of the integral)

$\qquad = -\frac{38}{3}$

39. $\int_1^3 f(x)\, dx + \int_3^6 f(x)\, dx + \int_6^{12} f(x)\, dx = \int_1^6 f(x)\, dx + \int_6^{12} f(x)\, dx = \int_1^{12} f(x)\, dx$

41. $\int_2^5 f(x)\, dx + \int_5^8 f(x)\, dx = \int_2^8 f(x)\, dx \quad \Rightarrow \quad \int_2^5 f(x)\, dx + 2.5 = 1.7 \quad \Rightarrow \quad \int_2^5 f(x)\, dx = 1.7 - 2.5 = -0.8$

43. $\int_0^1 (5 - 6x^2)\, dx = \int_0^1 5\, dx - 6\int_0^1 x^2\, dx = 5(1 - 0) - 6\left(\frac{1}{3}\right) = 5 - 2 = 3$

45. $\int_1^3 e^{x+2}\, dx = \int_1^3 e^x \cdot e^2\, dx = e^2 \int_1^3 e^x\, dx = e^2(e^3 - e) = e^5 - e^3$

47. On $[1, 3]$, $\ln 1 \le \ln x \le \ln 3 \quad \Rightarrow \quad 0(3 - 1) \le \int_1^3 \ln x\, dx \le (\ln 3)(3 - 1) \quad \Rightarrow \quad 0 \le \int_1^3 \ln x\, dx \le 2\ln 3$.

49. $\displaystyle\lim_{n\to\infty} \sum_{i=1}^n \frac{i^4}{n^5} = \lim_{n\to\infty} \sum_{i=1}^n \frac{i^4}{n^4} \cdot \frac{1}{n} = \lim_{n\to\infty} \sum_{i=1}^n \left(\frac{i}{n}\right)^4 \frac{1}{n}$. At this point, we need to recognize the limit as being of

the form $\displaystyle\lim_{n\to\infty} \sum_{i=1}^n f(x_i)\, \Delta x$, where $\Delta x = (1 - 0)/n = 1/n$, $x_i = 0 + i\,\Delta x = i/n$, and $f(x) = x^4$. Thus, the

definite integral is $\int_0^1 x^4\, dx$.

5.3 Evaluating Definite Integrals • • • • • • • • • • •

1. If $w'(t)$ is the rate of change of weight in pounds per year, then $w(t)$ represents the weight in pounds of the child at

age t. We know from the Total Change Theorem that $\int_5^{10} w'(t)\, dt = w(10) - w(5)$, so the integral represents the

increase in the child's weight (in pounds) between the ages of 5 and 10.

3. Since $r(t)$ is the rate at which oil leaks, we can write $r(t) = -V'(t)$, where $V(t)$ is the volume of oil at time t. [Note that the minus sign is needed because V is decreasing, so $V'(t)$ is negative, but $r(t)$ is positive.] Thus, by the Total Change Theorem, $\int_0^{120} r(t)\,dt = -\int_0^{120} V'(t)\,dt = -[V(120) - V(0)] = V(0) - V(120)$, which is the number of gallons of oil that leaked from the tank in the first two hours (120 minutes).

5. By the Total Change Theorem, $\int_{1000}^{5000} R'(x)\,dx = R(5000) - R(1000)$, so it represents the increase in revenue when production is increased from 1000 units to 5000 units.

7. In general, the unit of measurement for $\int_a^b f(x)\,dx$ is the product of the unit for $f(x)$ and the unit for x. Since $f(x)$ is measured in newtons and x is measured in meters, the units for $\int_0^{100} f(x)\,dx$ are newton-meters. (A newton-meter is abbreviated N-m and is called a joule.)

9. $\displaystyle\int_{-1}^{3} x^5\,dx = \left[\frac{x^6}{6}\right]_{-1}^{3} = \frac{3^6}{6} - \frac{(-1)^6}{6} = \frac{729-1}{6} = \frac{364}{3}$

11. $\int_2^8 (4x+3)\,dx = \left[\frac{4}{2}x^2 + 3x\right]_2^8 = (2\cdot 8^2 + 3\cdot 8) - (2\cdot 2^2 + 3\cdot 2) = 152 - 14 = 138$

13. $\displaystyle\int_0^4 \sqrt{x}\,dx = \int_0^4 x^{1/2}\,dx = \left[\frac{x^{3/2}}{3/2}\right]_0^4 = \left[\frac{2x^{3/2}}{3}\right]_0^4 = \frac{2(4)^{3/2}}{3} - 0 = \frac{2\cdot 8}{3} = \frac{16}{3}$

15. $\int_{-1}^0 (2x - e^x)\,dx = \left[x^2 - e^x\right]_{-1}^0 = (0-1) - (1 - e^{-1}) = -2 + 1/e$

17. $\displaystyle\int_1^2 \frac{3}{t^4}\,dt = 3\int_1^2 t^{-4}\,dt = 3\left[\frac{t^{-3}}{-3}\right]_1^2 = \frac{3}{-3}\left[\frac{1}{t^3}\right]_1^2 = -1\left(\frac{1}{8} - 1\right) = \frac{7}{8}$

19. $\displaystyle\int_1^2 \frac{x^2+1}{\sqrt{x}}\,dx = \int_1^2 \left(x^{3/2} + x^{-1/2}\right)dx = \left[\frac{x^{5/2}}{5/2} + \frac{x^{1/2}}{1/2}\right]_1^2 = \left[\frac{2}{5}x^{5/2} + 2x^{1/2}\right]_1^2$
$$= \left(\tfrac{2}{5}\cdot 4\sqrt{2} + 2\sqrt{2}\right) - \left(\tfrac{2}{5} + 2\right) = \tfrac{18}{5}\sqrt{2} - \tfrac{12}{5}$$

21. $\int_{\pi/4}^{\pi/3} \sin t\,dt = \left[-\cos t\right]_{\pi/4}^{\pi/3} = \left(-\cos\frac{\pi}{3}\right) - \left(-\cos\frac{\pi}{4}\right) = -\frac{1}{2} + \frac{\sqrt{2}}{2} = \frac{\sqrt{2}-1}{2}$

23. $\displaystyle\int_0^1 u(\sqrt{u} + \sqrt[3]{u})\,du = \int_0^1 u\left(u^{1/2} + u^{1/3}\right)du = \int_0^1 \left(u^{3/2} + u^{4/3}\right)du = \left[\frac{u^{5/2}}{5/2} + \frac{u^{7/3}}{7/3}\right]_0^1$
$$= \left[\tfrac{2}{5}u^{5/2} + \tfrac{3}{7}u^{7/3}\right]_0^1 = \tfrac{2}{5} + \tfrac{3}{7} = \tfrac{29}{35}$$

25. $\int_{\pi/6}^{\pi/3} \csc^2\theta\,d\theta = \left[-\cot\theta\right]_{\pi/6}^{\pi/3} = \left(-\cot\frac{\pi}{3}\right) - \left(-\cot\frac{\pi}{6}\right) = -\frac{1}{3}\sqrt{3} + \sqrt{3} = \frac{2}{3}\sqrt{3}$

27. $\displaystyle\int_1^9 \frac{1}{2x}\,dx = \frac{1}{2}\int_1^9 \frac{1}{x}\,dx = \frac{1}{2}\left[\ln|x|\right]_1^9 = \frac{1}{2}(\ln 9 - \ln 1) = \frac{1}{2}\ln 9 - 0 = \ln 9^{1/2} = \ln 3$

29. $\displaystyle\int_8^9 2^t\,dt = \left[\frac{1}{\ln 2}2^t\right]_8^9 = \frac{1}{\ln 2}(2^9 - 2^8) = \frac{1}{\ln 2}\cdot 2^8(2^1 - 1) = \frac{2^8}{\ln 2}, \text{ or } \frac{256}{\ln 2}$

31. $\displaystyle\int_1^{\sqrt{3}} \frac{6}{1+x^2}\,dx = 6\left[\tan^{-1}x\right]_1^{\sqrt{3}} = 6\left(\tan^{-1}\sqrt{3} - \tan^{-1}1\right) = 6\left(\frac{\pi}{3} - \frac{\pi}{4}\right) = 6\left(\frac{\pi}{12}\right) = \frac{\pi}{2}$

33. $\displaystyle\int_0^{\pi/4} \frac{1+\cos^2\theta}{\cos^2\theta}\,d\theta = \int_0^{\pi/4} \left(\frac{1}{\cos^2\theta} + \frac{\cos^2\theta}{\cos^2\theta}\right)d\theta = \int_0^{\pi/4}(\sec^2\theta + 1)\,d\theta$
$$= \left[\tan\theta + \theta\right]_0^{\pi/4} = \left(\tan\frac{\pi}{4} + \frac{\pi}{4}\right) - (0+0) = 1 + \frac{\pi}{4}$$

35. It appears that the area under the graph is about $\frac{2}{3}$ of the area of the

viewing rectangle, or about $\frac{2}{3}\pi \approx 2.1$. The actual area is

$\int_0^\pi \sin x \, dx = [-\cos x]_0^\pi = (-\cos \pi) - (-\cos 0) = -(-1) + 1 = 2.$

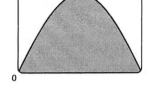

37. The graph shows that $y = x + x^2 - x^4$ has x-intercepts at $x = 0$ and at

$x \approx 1.32$. So the area of the region that lies under the curve and above the

x-axis is about

$\int_0^{1.32} (x + x^2 - x^4) \, dx = \left[\frac{1}{2}x^2 + \frac{1}{3}x^3 - \frac{1}{5}x^5\right]_0^{1.32}$

$= \left[\frac{1}{2}(1.32)^2 + \frac{1}{3}(1.32)^3 - \frac{1}{5}(1.32)^5\right] - 0$

≈ 0.84

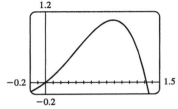

39. $\int_{-1}^2 x^3 \, dx = \left[\frac{1}{4}x^4\right]_{-1}^2 = 4 - \frac{1}{4} = \frac{15}{4} = 3.75$

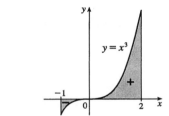

41. $\dfrac{d}{dx}\left[\sqrt{x^2 + 1} + C\right] = \dfrac{d}{dx}\left[(x^2 + 1)^{1/2} + C\right] = \frac{1}{2}(x^2 + 1)^{-1/2} \cdot 2x = \dfrac{x}{\sqrt{x^2 + 1}}$

43. $\int x\sqrt{x} \, dx = \int x^{3/2} \, dx = \frac{2}{5}x^{5/2} + C.$

The members of the family in the figure correspond to $C = 5, 3, 0,$

$-2,$ and $-4.$

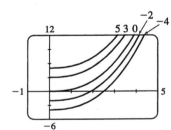

45. $\int (1 - t)(2 + t^2) \, dt = \int (2 - 2t + t^2 - t^3) \, dt = 2t - 2\dfrac{t^2}{2} + \dfrac{t^3}{3} - \dfrac{t^4}{4} + C = 2t - t^2 + \frac{1}{3}t^3 - \frac{1}{4}t^4 + C$

47. $\displaystyle\int \dfrac{\sin x}{1 - \sin^2 x} \, dx = \int \dfrac{\sin x}{\cos^2 x} \, dx = \int \dfrac{1}{\cos x} \cdot \dfrac{\sin x}{\cos x} \, dx = \int \sec x \tan x \, dx = \sec x + C$

49. $A = \int_0^2 (2y - y^2) \, dy = \left[y^2 - \frac{1}{3}y^3\right]_0^2 = \left(4 - \frac{8}{3}\right) - 0 = \frac{4}{3}$

51. (a) displacement $= \int_0^3 (3t - 5) \, dt = \left[\frac{3}{2}t^2 - 5t\right]_0^3 = \frac{27}{2} - 15 = -\frac{3}{2}$ m

(b) distance traveled $= \int_0^3 |3t - 5| \, dt = \int_0^{5/3} (5 - 3t) \, dt + \int_{5/3}^3 (3t - 5) \, dt$

$= \left[5t - \frac{3}{2}t^2\right]_0^{5/3} + \left[\frac{3}{2}t^2 - 5t\right]_{5/3}^3 = \frac{25}{3} - \frac{3}{2} \cdot \frac{25}{9} + \frac{27}{2} - 15 - \left(\frac{3}{2} \cdot \frac{25}{9} - \frac{25}{3}\right) = \frac{41}{6}$ m

53. (a) $v'(t) = a(t) = t + 4 \implies v(t) = \frac{1}{2}t^2 + 4t + C \implies v(0) = C = 5 \implies v(t) = \frac{1}{2}t^2 + 4t + 5$ m/s

(b) distance traveled $= \int_0^{10} |v(t)| \, dt = \int_0^{10} \left|\frac{1}{2}t^2 + 4t + 5\right| dt = \int_0^{10} \left(\frac{1}{2}t^2 + 4t + 5\right) dt$

$= \left[\frac{1}{6}t^3 + 2t^2 + 5t\right]_0^{10} = \frac{500}{3} + 200 + 50 = 416\frac{2}{3}$ m

55. Since $m'(x) = \rho(x)$, $m = \int_0^4 \rho(x) \, dx = \int_0^4 (9 + 2\sqrt{x}) \, dx = \left[9x + \frac{4}{3}x^{3/2}\right]_0^4 = 36 + \frac{32}{3} - 0 = \frac{140}{3} = 46\frac{2}{3}$ kg.

57. Let s be the position of the car. We know from Equation 2 that $s(100) - s(0) = \int_0^{100} v(t) \, dt$. We use the Midpoint Rule for $0 \le t \le 100$ with $n = 5$. Note that the length of each of the five time intervals is 20 seconds $= \frac{20}{3600}$ hour $= \frac{1}{180}$ hour. So the distance traveled is

$$\int_0^{100} v(t) \, dt \approx \frac{1}{180}[v(10) + v(30) + v(50) + v(70) + v(90)]$$

$$= \frac{1}{180}(38 + 58 + 51 + 53 + 47)$$

$$= \frac{247}{180} \approx 1.4 \text{ miles}$$

59. From Equation 2, the increase in cost if the production level is raised from 2000 yards to 4000 yards is $C(4000) - C(2000) = \int_{2000}^{4000} C'(x) \, dx$.

$$\int_{2000}^{4000} C'(x) \, dx = \int_{2000}^{4000} (3 - 0.01x + 0.000006x^2) \, dx$$

$$= \left[3x - 0.005x^2 + 0.000002x^3\right]_{2000}^{4000} = 60{,}000 - 2{,}000 = \$58{,}000$$

61. (a) We can find the area between the Lorenz curve and the line $y = x$ by subtracting the area under $y = L(x)$ from the area under $y = x$. Thus,

$$\text{coefficient of inequality} = \frac{\text{area between Lorenz curve and line } y = x}{\text{area under line } y = x} = \frac{\int_0^1 [x - L(x)] \, dx}{\int_0^1 x \, dx}$$

$$= \frac{\int_0^1 [x - L(x)] \, dx}{[x^2/2]_0^1} = \frac{\int_0^1 [x - L(x)] \, dx}{1/2} = 2 \int_0^1 [x - L(x)] \, dx$$

(b) $L(x) = \frac{5}{12}x^2 + \frac{7}{12}x \implies L(50\%) = L\left(\frac{1}{2}\right) = \frac{5}{48} + \frac{7}{24} = \frac{19}{48} = 0.39583$, so the bottom 50% of the households receive at most about 40% of the income. Using the result in part (a),

$$\text{coefficient of inequality} = 2\int_0^1 [x - L(x)] \, dx = 2\int_0^1 \left(x - \frac{5}{12}x^2 - \frac{7}{12}x\right) dx$$

$$= 2\int_0^1 \left(\frac{5}{12}x - \frac{5}{12}x^2\right) dx = 2\int_0^1 \frac{5}{12}(x - x^2) \, dx$$

$$= \frac{5}{6}\left[\frac{1}{2}x^2 - \frac{1}{3}x^3\right]_0^1 = \frac{5}{6}\left(\frac{1}{2} - \frac{1}{3}\right) = \frac{5}{6}\left(\frac{1}{6}\right) = \frac{5}{36}$$

63. The second derivative is the derivative of the first derivative, so we'll apply the Total Change Theorem with $F = h'$. $\int_1^2 h''(u) \, du = \int_1^2 (h')'(u) \, du = h'(2) - h'(1) = 5 - 2 = 3$. The other information is unnecessary.

5.4 **The Fundamental Theorem of Calculus** • • • • • • •

1. The precise version of this statement is given by the Fundamental Theorem of Calculus. See the statement of this theorem and the paragraph that follows it on page 384.

3. (a) $g(x) = \int_0^x f(t)\,dt$.

(d)

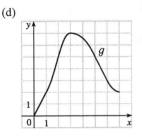

$g(0) = \int_0^0 f(t)\,dt = 0$, $g(1) = \int_0^1 f(t)\,dt = 1 \cdot 2 = 2$ [rectangle],

$g(2) = \int_0^2 f(t)\,dt = \int_0^1 f(t)\,dt + \int_1^2 f(t)\,dt = g(1) + \int_1^2 f(t)\,dt$

$= 2 + 1 \cdot 2 + \frac{1}{2} \cdot 1 \cdot 2 = 5$ [rectangle plus triangle],

$g(3) = \int_0^3 f(t)\,dt = g(2) + \int_2^3 f(t)\,dt = 5 + \frac{1}{2} \cdot 1 \cdot 4 = 7$,

$g(6) = g(3) + \int_3^6 f(t)\,dt$ [the integral is negative since f lies under the x-axis]

$= 7 + \left[-\left(\frac{1}{2} \cdot 2 \cdot 2 + 1 \cdot 2\right)\right] = 7 - 4 = 3$

(b) g is increasing on $(0, 3)$ because as x increases from 0 to 3, we keep adding more area.

(c) g has a maximum value when we start subtracting area; that is, at $x = 3$.

5.

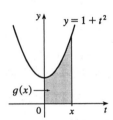

(a) By FTC1, $g(x) = \int_0^x (1 + t^2)\,dt \;\Rightarrow\; g'(x) = f(x) = 1 + x^2$.

(b) By FTC2, $g(x) = \int_0^x (1 + t^2)\,dt = \left[t + \frac{1}{3}t^3\right]_0^x = \left(x + \frac{1}{3}x^3\right) - 0$

$\Rightarrow\; g'(x) = 1 + x^2$.

7. $f(t) = \sqrt{1 + 2t}$ and $g(x) = \int_0^x \sqrt{1 + 2t}\,dt$, so by FTC1, $g'(x) = f(x) = \sqrt{1 + 2x}$.

9. $f(t) = t^2 \sin t$ and $g(y) = \int_2^y t^2 \sin t\,dt$, so by FTC1, $g'(y) = f(y) = y^2 \sin y$.

11. Let $u = \dfrac{1}{x}$. Then $\dfrac{du}{dx} = -\dfrac{1}{x^2}$. Also, $\dfrac{dh}{dx} = \dfrac{dh}{du}\dfrac{du}{dx}$, so

$h'(x) = \dfrac{d}{dx}\displaystyle\int_2^{1/x} \arctan t\,dt = \dfrac{d}{du}\displaystyle\int_2^u \arctan t\,dt \cdot \dfrac{du}{dx} = \arctan u\,\dfrac{du}{dx} = -\dfrac{\arctan(1/x)}{x^2}$.

13. Let $u = \sqrt{x}$. Then $\dfrac{du}{dx} = \dfrac{1}{2\sqrt{x}}$. Also, $\dfrac{dy}{dx} = \dfrac{dy}{du}\dfrac{du}{dx}$, so

$y' = \dfrac{d}{dx}\displaystyle\int_3^{\sqrt{x}} \dfrac{\cos t}{t}\,dt = \dfrac{d}{du}\displaystyle\int_3^u \dfrac{\cos t}{t}\,dt \cdot \dfrac{du}{dx} = \dfrac{\cos u}{u} \cdot \dfrac{1}{2\sqrt{x}} = \dfrac{\cos \sqrt{x}}{\sqrt{x}} \cdot \dfrac{1}{2\sqrt{x}} = \dfrac{\cos \sqrt{x}}{2x}$.

15. $g(x) = \displaystyle\int_{2x}^{3x} \dfrac{u^2 - 1}{u^2 + 1}\,du = \displaystyle\int_{2x}^0 \dfrac{u^2 - 1}{u^2 + 1}\,du + \displaystyle\int_0^{3x} \dfrac{u^2 - 1}{u^2 + 1}\,du = -\displaystyle\int_0^{2x} \dfrac{u^2 - 1}{u^2 + 1}\,du + \displaystyle\int_0^{3x} \dfrac{u^2 - 1}{u^2 + 1}\,du \;\Rightarrow$

$g'(x) = -\dfrac{(2x)^2 - 1}{(2x)^2 + 1} \cdot \dfrac{d}{dx}(2x) + \dfrac{(3x)^2 - 1}{(3x)^2 + 1} \cdot \dfrac{d}{dx}(3x) = -2 \cdot \dfrac{4x^2 - 1}{4x^2 + 1} + 3 \cdot \dfrac{9x^2 - 1}{9x^2 + 1}$

17. $F(x) = \displaystyle\int_1^x f(t)\,dt \;\Rightarrow\; F'(x) = f(x) = \displaystyle\int_1^{x^2} \dfrac{\sqrt{1 + u^4}}{u}\,du \left[\text{since } f(t) = \displaystyle\int_1^{t^2} \dfrac{\sqrt{1 + u^4}}{u}\,du\right] \;\Rightarrow$

$F''(x) = f'(x) = \dfrac{\sqrt{1 + (x^2)^4}}{x^2} \cdot \dfrac{d}{dx}(x^2) = \dfrac{\sqrt{1 + x^8}}{x^2} \cdot 2x = \dfrac{2\sqrt{1 + x^8}}{x}$. So $F''(2) = \sqrt{1 + 2^8} = \sqrt{257}$.

19. (a) By FTC1, $g'(x) = f(x)$. So $g'(x) = f(x) = 0$ at $x = 1, 3, 5, 7$, and 9. g has local maxima at $x = 1$ and 5 (since $f = g'$ changes from positive to negative there) and local minima at $x = 3$ and 7. There is no local maximum or minimum at $x = 9$, since f is not defined for $x > 9$.

(b) We can see from the graph that $\left|\int_0^1 f\, dt\right| < \left|\int_1^3 f\, dt\right| < \left|\int_3^5 f\, dt\right| < \left|\int_5^7 f\, dt\right| < \left|\int_7^9 f\, dt\right|$.
So $g(1) = \left|\int_0^1 f\, dt\right|$, $g(5) = \int_0^5 f\, dt = g(1) - \left|\int_1^3 f\, dt\right| + \left|\int_3^5 f\, dt\right|$, and
$g(9) = \int_0^9 f\, dt = g(5) - \left|\int_5^7 f\, dt\right| + \left|\int_7^9 f\, dt\right|$. Thus, $g(1) < g(5) < g(9)$, and so the absolute maximum of $g(x)$ occurs at $x = 9$.

(c) g is concave downward on those intervals where $g'' < 0$. But $g'(x) = f(x)$, so $g''(x) = f'(x)$, which is negative on (approximately) $\left(\frac{1}{2}, 2\right)$, $(4, 6)$ and $(8, 9)$. So g is concave downward on these intervals.

(d)
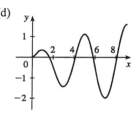

21. (a) The Fresnel Function $S(x) = \int_0^x \sin\left(\frac{\pi}{2}t^2\right) dt$ has local maximum values where $0 = S'(x) = \sin\left(\frac{\pi}{2}x^2\right)$ and S' changes from positive to negative. For $x > 0$, this happens when $\frac{\pi}{2}x^2 = (2n-1)\pi$ [odd multiples of π] $\Leftrightarrow$ $x^2 = 2(2n-1)$ $\Leftrightarrow$ $x = \sqrt{4n-2}$, n any positive integer. For $x < 0$, S' changes from positive to negative where $\frac{\pi}{2}x^2 = 2n\pi$ [even multiples of π] $\Leftrightarrow$ $x^2 = 4n$ $\Leftrightarrow$ $x = -2\sqrt{n}$. S' does not change sign at $x = 0$.

(b) S is concave upward on those intervals where $S''(x) > 0$. Differentiating our expression for $S'(x)$, we get $S''(x) = \cos\left(\frac{\pi}{2}x^2\right)\left(2\frac{\pi}{2}x\right) = \pi x \cos\left(\frac{\pi}{2}x^2\right)$. For $x > 0$, $S''(x) > 0$ where $\cos\left(\frac{\pi}{2}x^2\right) > 0$ $\Leftrightarrow$ $0 < \frac{\pi}{2}x^2 < \frac{\pi}{2}$ or $\left(2n - \frac{1}{2}\right)\pi < \frac{\pi}{2}x^2 < \left(2n + \frac{1}{2}\right)\pi$, n any integer $\Leftrightarrow$ $0 < x < 1$ or $\sqrt{4n-1} < x < \sqrt{4n+1}$, n any positive integer. For $x < 0$, $S''(x) > 0$ where $\cos\left(\frac{\pi}{2}x^2\right) < 0$ $\Leftrightarrow$ $\left(2n - \frac{3}{2}\right)\pi < \frac{\pi}{2}x^2 < \left(2n - \frac{1}{2}\right)\pi$, n any integer $\Leftrightarrow$ $4n - 3 < x^2 < 4n - 1$ $\Leftrightarrow$ $\sqrt{4n-3} < |x| < \sqrt{4n-1}$ $\Rightarrow$ $\sqrt{4n-3} < -x < \sqrt{4n-1}$ $\Rightarrow$ $-\sqrt{4n-3} > x > -\sqrt{4n-1}$, so the intervals of upward concavity for $x < 0$ are $\left(-\sqrt{4n-1}, -\sqrt{4n-3}\right)$, n any positive integer. To summarize: S is concave upward on the intervals $(0, 1)$, $(-\sqrt{3}, -1)$, $(\sqrt{3}, \sqrt{5})$, $(-\sqrt{7}, -\sqrt{5})$, $(\sqrt{7}, 3)$, ...

(c) In Maple, we use `plot({int(sin(Pi*t^2/2),t=0..x),0.2},x=0..2);`. Note that Maple recognizes the Fresnel function, calling it `FresnelS(x)`. In Mathematica, we use `Plot[{Integrate[Sin[Pi*t^2/2],{t,0,x}],0.2},{x,0,2}]`. In Derive, we load the utility file FRESNEL and plot FRESNEL_SIN(x). From the graphs, we see that $\int_0^x \sin\left(\frac{\pi}{2}t^2\right) dt = 0.2$ at $x \approx 0.74$.

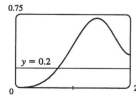

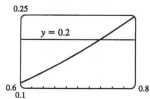

23. By FTC2, $\int_1^x f'(t)\, dt = f(x) - f(1)$ $\Rightarrow$ $f(x) = f(1) + \int_1^x f'(t)\, dt = f(1) + \int_1^x \left(2^t/t\right) dt$. This integral cannot be expressed in a simpler form. Since we want $f(1) = 0$, we have $f(x) = \int_1^x \left(2^t/t\right) dt$.

25. Using FTC1, we differentiate both sides of $6 + \int_a^x \frac{f(t)}{t^2}\,dt = 2\sqrt{x}$ to get $\frac{f(x)}{x^2} = 2\frac{1}{2\sqrt{x}}$ $\Rightarrow$ $f(x) = x^{3/2}$.

To find a, we substitute $x = a$ in the original equation to obtain $6 + \int_a^a \frac{f(t)}{t^2}\,dt = 2\sqrt{a}$ $\Rightarrow$ $6 + 0 = 2\sqrt{a}$ $\Rightarrow$

$3 = \sqrt{a}$ $\Rightarrow$ $a = 9$.

27. (a) Let $F(t) = \int_0^t f(s)\,ds$. Then, by FTC1, $F'(t) = f(t) =$ rate of depreciation, so $F(t)$ represents the loss in value over the interval $[0, t]$.

(b) $C(t) = \frac{1}{t}\left[A + \int_0^t f(s)\,ds\right] = \frac{A + F(t)}{t}$ represents the average expenditure per unit of t during the interval

$[0, t]$, assuming that there has been only one overhaul during that time period. The company wants to minimize average expenditure.

(c) $C(t) = \frac{1}{t}\left[A + \int_0^t f(s)\,ds\right]$. Using FTC1, we have $C'(t) = -\frac{1}{t^2}\left[A + \int_0^t f(s)\,ds\right] + \frac{1}{t}f(t)$.

$C'(t) = 0$ $\Rightarrow$ $t\,f(t) = A + \int_0^t f(s)\,ds$ $\Rightarrow$ $f(t) = \frac{1}{t}\left[A + \int_0^t f(s)\,ds\right] = C(t)$.

5.5 The Substitution Rule • • • • • • • • • • • • • •

1. Let $u = 3x$. Then $du = 3\,dx$, so $dx = \frac{1}{3}du$. Thus,

$\int \cos 3x\,dx = \int \cos u\left(\frac{1}{3}\,du\right) = \frac{1}{3}\int \cos u\,du = \frac{1}{3}\sin u + C = \frac{1}{3}\sin 3x + C$. Don't forget that it is often very easy to check an indefinite integration by differentiating your answer. In this case,

$\frac{d}{dx}\left(\frac{1}{3}\sin 3x + C\right) = \frac{1}{3}(\cos 3x) \cdot 3 = \cos 3x$, the desired result.

3. Let $u = x^3 + 1$. Then $du = 3x^2\,dx$, so $x^2\,dx = \frac{1}{3}du$. Thus,

$$\int x^2\sqrt{x^3 + 1}\,dx = \int \sqrt{u}\left(\frac{1}{3}\,du\right) = \frac{1}{3}\frac{u^{3/2}}{3/2} + C = \frac{1}{3}\cdot\frac{2}{3}u^{3/2} + C = \frac{2}{9}\left(x^3 + 1\right)^{3/2} + C$$

5. Let $u = 1 + 2x$. Then $du = 2\,dx$, so

$$\int \frac{4}{(1 + 2x)^3}\,dx = 4\int u^{-3}\left(\frac{1}{2}\,du\right) = 2\frac{u^{-2}}{-2} + C = -\frac{1}{u^2} + C = -\frac{1}{(1 + 2x)^2} + C$$

7. Let $u = x^2 + 3$. Then $du = 2x\,dx$, so $\int 2x\left(x^2 + 3\right)^4\,dx = \int u^4\,du = \frac{1}{5}u^5 + C = \frac{1}{5}\left(x^2 + 3\right)^5 + C$.

9. Let $u = \ln x$. Then $du = \frac{dx}{x}$, so $\int \frac{(\ln x)^2}{x}\,dx = \int u^2\,du = \frac{1}{3}u^3 + C = \frac{1}{3}(\ln x)^3 + C$.

11. Let $u = x - 1$. Then $du = dx$, so $\int \sqrt{x - 1}\,dx = \int u^{1/2}\,du = \frac{2}{3}u^{3/2} + C = \frac{2}{3}(x - 1)^{3/2} + C$.

13. Let $u = 5 - 3x$. Then $du = -3\,dx$, so $\int \frac{dx}{5 - 3x} = -\frac{1}{3}\int \frac{1}{u}\,du = -\frac{1}{3}\ln|u| + C = -\frac{1}{3}\ln|5 - 3x| + C$.

15. Let $u = 1 + x + 2x^2$. Then $du = (1 + 4x)dx$, so

$$\int \frac{1 + 4x}{\sqrt{1 + x + 2x^2}}\,dx = \int \frac{du}{\sqrt{u}} = \int u^{-1/2}\,du = \frac{u^{1/2}}{1/2} + C = 2\sqrt{1 + x + 2x^2} + C$$

17. Let $u = t + 1$. Then $du = dt$, so $\int \frac{2}{(t + 1)^6}\,dt = 2\int u^{-6}\,du = -\frac{2}{5}u^{-5} + C = -\frac{2}{5(t + 1)^5} + C$.

19. Let $u = 3\theta$. Then $du = 3\,d\theta$, so $\int \sin 3\theta\,d\theta = \int \sin u\left(\frac{1}{3}du\right) = \frac{1}{3}(-\cos u) + C = -\frac{1}{3}\cos 3\theta + C$.

21. Let $u = 1 + e^x$. Then $du = e^x\,dx$, so $\int e^x\sqrt{1 + e^x}\,dx = \int \sqrt{u}\,du = \frac{2}{3}u^{3/2} + C = \frac{2}{3}(1 + e^x)^{3/2} + C$.

Or: Let $u = \sqrt{1 + e^x}$. Then $u^2 = 1 + e^x$ and $2u\,du = e^x\,dx$, so

$$\int e^x\sqrt{1 + e^x}\,dx = \int u \cdot 2u\,du = \frac{2}{3}u^3 + C = \frac{2}{3}(1 + e^x)^{3/2} + C$$

23. Let $u = \cos x$. Then $du = -\sin x\,dx$, so $\int \cos^4 x \sin x\,dx = \int u^4(-du) = -\frac{1}{5}u^5 + C = -\frac{1}{5}\cos^5 x + C$.

25. Let $u = \cot x$. Then $du = -\csc^2 x\,dx$, so

$$\int \sqrt{\cot x}\,\csc^2 x\,dx = \int \sqrt{u}\,(-du) = -\frac{u^{3/2}}{3/2} + C = -\frac{2}{3}(\cot x)^{3/2} + C$$

27. $\int \dfrac{e^x + 1}{e^x}\,dx = \int (1 + e^{-x})\,dx = x - e^{-x} + C$ [Substitute $u = -x$.]

29. Let $u = \sec x$. Then $du = \sec x \tan x\,dx$, so

$$\int \sec^3 x \tan x\,dx = \int \sec^2 x\,(\sec x \tan x)\,dx = \int u^2\,du = \frac{1}{3}u^3 + C = \frac{1}{3}\sec^3 x + C$$

31. Let $u = 1 + x^2$. Then $du = 2x\,dx$, so

$$\int \frac{1 + x}{1 + x^2}\,dx = \int \frac{1}{1 + x^2}\,dx + \int \frac{x}{1 + x^2}\,dx = \tan^{-1} x + \int \frac{\frac{1}{2}du}{u} = \tan^{-1} x + \frac{1}{2}\ln|u| + C$$
$$= \tan^{-1} x + \frac{1}{2}\ln|1 + x^2| + C = \tan^{-1} x + \frac{1}{2}\ln(1 + x^2) + C \quad [\text{since } 1 + x^2 > 0].$$

In Exercises 33–35, let $f(x)$ denote the integrand and $F(x)$ its antiderivative (with $C = 0$).

33. $f(x) = \dfrac{3x - 1}{(3x^2 - 2x + 1)^4}$.

$u = 3x^2 - 2x + 1 \quad\Rightarrow\quad du = (6x - 2)\,dx = 2(3x - 1)\,dx$, so

$$\int \frac{3x - 1}{(3x^2 - 2x + 1)^4}\,dx = \int \frac{1}{u^4}\left(\frac{1}{2}du\right) = \frac{1}{2}\int u^{-4}\,du$$

$$= -\frac{1}{6}u^{-3} + C = -\frac{1}{6(3x^2 - 2x + 1)^3} + C$$

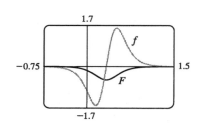

Notice that at $x = \frac{1}{3}$, f changes from negative to positive, and F has a local minimum.

35. $f(x) = \sin^3 x \cos x$. $u = \sin x \quad\Rightarrow\quad du = \cos x\,dx$, so

$$\int \sin^3 x \cos x\,dx = \int u^3\,du = \frac{1}{4}u^4 + C = \frac{1}{4}\sin^4 x + C$$

Note that at $x = \frac{\pi}{2}$, f changes from positive to negative and F has a local maximum. Also, both f and F are periodic with period π, so at $x = 0$ and at $x = \pi$, f changes from negative to positive and F has local minima.

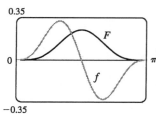

37. Let $u = x - 1$, so $du = dx$. When $x = 0$, $u = -1$; when $x = 2$, $u = 1$. Thus, $\int_0^2 (x - 1)^{25}\,dx = \int_{-1}^1 u^{25}\,du = 0$ by Theorem 6(b), since $f(u) = u^{25}$ is an odd function.

39. Let $u = 1 + 2x^3$, so $du = 6x^2\,dx$. When $x = 0$, $u = 1$; when $x = 1$, $u = 3$. Thus,

$$\int_0^1 x^2\left(1 + 2x^3\right)^5 dx = \int_1^3 u^5\left(\frac{1}{6}du\right) = \frac{1}{6}\left[\frac{1}{6}u^6\right]_1^3 = \frac{1}{36}(3^6 - 1^6) = \frac{1}{36}(729 - 1) = \frac{728}{36} = \frac{182}{9}$$

41. Let $u = \pi t$, so $du = \pi\, dt$. When $t = 0$, $u = 0$; when $t = 1$, $u = \pi$. Thus,

$$\int_0^1 \cos \pi t\, dt = \int_0^\pi \cos u \left(\tfrac{1}{\pi}\, du\right) = \tfrac{1}{\pi}\left[\sin u\right]_0^\pi = \tfrac{1}{\pi}\left(0 - 0\right) = 0$$

43. Let $u = \sqrt{x}$, so $du = \dfrac{1}{2\sqrt{x}}\, dx$. When $x = 1$, $u = 1$; when $x = 4$, $u = 2$. Thus,

$$\int_1^4 \frac{e^{\sqrt{x}}}{\sqrt{x}}\, dx = \int_1^2 e^u(2\, du) = 2\left[e^u\right]_1^2 = 2\left(e^2 - e\right)$$

45. Let $u = x - 1$, so $u + 1 = x$ and $du = dx$. When $x = 1$, $u = 0$; when $x = 2$, $u = 1$. Thus,

$$\int_1^2 x\sqrt{x-1}\, dx = \int_0^1 (u+1)\sqrt{u}\, du = \int_0^1 \left(u^{3/2} + u^{1/2}\right) du = \left[\tfrac{2}{5}u^{5/2} + \tfrac{2}{3}u^{3/2}\right]_0^1 = \tfrac{2}{5} + \tfrac{2}{3} = \tfrac{16}{15}$$

47. Let $u = 1 + 2x$, so $du = 2\, dx$. When $x = 0$, $u = 1$; when $x = 13$, $u = 27$. Thus,

$$\int_0^{13} \frac{dx}{\sqrt[3]{(1+2x)^2}} = \int_1^{27} u^{-2/3}\left(\tfrac{1}{2}\, du\right) = \left[\tfrac{1}{2}\cdot 3u^{1/3}\right]_1^{27} = \tfrac{3}{2}(3 - 1) = 3$$

49. $\displaystyle\int_{-\pi/6}^{\pi/6} \tan^3 \theta\, d\theta = 0$ by Theorem 6(b), since $f(\theta) = \tan^3 \theta$ is an odd function.

51. Let $u = \ln x$, so $du = \dfrac{dx}{x}$. When $x = e$, $u = 1$; when $x = e^4$; $u = 4$. Thus,

$$\int_e^{e^4} \frac{dx}{x\sqrt{\ln x}} = \int_1^4 u^{-1/2}\, du = 2\left[u^{1/2}\right]_1^4 = 2(2 - 1) = 2$$

53. From the graph, it appears that the area under the curve is about

$1 + \left(\text{a little more than } \tfrac{1}{2}\cdot 1 \cdot 0.7\right)$, or about 1.4. The exact area is given by

$A = \int_0^1 \sqrt{2x+1}\, dx$. Let $u = 2x + 1$, so $du = 2\, dx$. The limits change to

$2 \cdot 0 + 1 = 1$ and $2 \cdot 1 + 1 = 3$, and

$A = \int_1^3 \sqrt{u}\left(\tfrac{1}{2}\, du\right) = \tfrac{1}{2}\left[\tfrac{2}{3}u^{3/2}\right]_1^3 = \tfrac{1}{3}\left(3\sqrt{3} - 1\right) = \sqrt{3} - \tfrac{1}{3} \approx 1.399$.

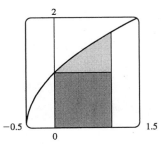

55. First write the integral as a sum of two integrals:

$\int_{-2}^2 (x+3)\sqrt{4-x^2}\, dx = \int_{-2}^2 x\sqrt{4-x^2}\, dx + \int_{-2}^2 3\sqrt{4-x^2}\, dx$. The first integral is 0 by Theorem 6(b), since

$f(x) = x\sqrt{4-x^2}$ is an odd function and we are integrating from $x = -2$ to $x = 2$. The second integral we

interpret as three times the area of a semicircle with radius 2, so the original integral is equal to

$0 + 3 \cdot \tfrac{1}{2}\left(\pi \cdot 2^2\right) = 6\pi$.

57. First Figure   Let $u = \sqrt{x}$, so $du = \dfrac{1}{2\sqrt{x}}\, dx$ and $dx = 2\sqrt{x}\, du = 2u\, du$. When $x = 0$, $u = 0$; when

$x = 1$, $u = 1$. Thus, $A_1 = \displaystyle\int_0^1 e^{\sqrt{x}}\, dx = \int_0^1 e^u(2u\, du) = 2\int_0^1 ue^u\, du$.

Second Figure $A_2 = \int_0^1 2xe^x\, dx = 2\int_0^1 ue^u\, du$.

Third Figure   Let $u = \sin x$, so $du = \cos x\, dx$. When $x = 0$, $u = 0$; when $x = \tfrac{\pi}{2}$, $u = 1$. Thus,

$$A_3 = \int_0^{\pi/2} e^{\sin x}\sin 2x\, dx = \int_0^{\pi/2} e^{\sin x}(2\sin x\,\cos x)\, dx = \int_0^1 e^u(2u\, du) = 2\int_0^1 ue^u\, du.$$

Since $A_1 = A_2 = A_3$, all three areas are equal.

59. The volume of inhaled air in the lungs at time t is

$$V(t) = \int_0^t f(u)\,du = \int_0^t \tfrac{1}{2} \sin\left(\tfrac{2\pi}{5}u\right)du = \int_0^{2\pi t/5} \tfrac{1}{2} \sin v\left(\tfrac{5}{2\pi}\,dv\right) \quad [\text{substitute } v = \tfrac{2\pi}{5}u,\ dv = \tfrac{2\pi}{5}\,du]$$

$$= \tfrac{5}{4\pi}\left[-\cos v\right]_0^{2\pi t/5} = \tfrac{5}{4\pi}\left[-\cos\left(\tfrac{2\pi}{5}t\right) + 1\right] = \tfrac{5}{4\pi}\left[1 - \cos\left(\tfrac{2\pi}{5}t\right)\right] \text{ liters}$$

61. Let $u = 2x$. Then $du = 2\,dx$, so $\int_0^2 f(2x)\,dx = \int_0^4 f(u)\left(\tfrac{1}{2}\,du\right) = \tfrac{1}{2}\int_0^4 f(u)\,du = \tfrac{1}{2}(10) = 5$.

63. Let $u = -x$. Then $du = -dx$, so

$$\int_a^b f(-x)\,dx = \int_{-a}^{-b} f(u)(-du) = \int_{-b}^{-a} f(u)\,du = \int_{-b}^{-a} f(x)\,dx. \text{ From}$$

the diagram, we see that the equality follows from the fact that we are
reflecting the graph of f, and the limits of integration, about the y-axis.

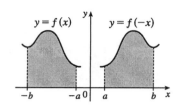

65. Let $u = 1 - x$. Then $x = 1 - u$ and $dx = -du$, so

$$\int_0^1 x^a(1-x)^b\,dx = \int_1^0 (1-u)^a\,u^b(-du) = \int_0^1 u^b(1-u)^a\,du = \int_0^1 x^b(1-x)^a\,dx$$

◆ 5.6 Integration by Parts · · · · · · · · · · · · · ·

1. Let $u = \ln x$, $dv = x\,dx$ $\Rightarrow$ $du = dx/x$, $v = \tfrac{1}{2}x^2$. Then by Equation 2, $\int u\,dv = uv - \int v\,du$,

$$\int x \ln x\,dx = \tfrac{1}{2}x^2 \ln x - \int \tfrac{1}{2}x^2(dx/x) = \tfrac{1}{2}x^2 \ln x - \tfrac{1}{2}\int x\,dx = \tfrac{1}{2}x^2 \ln x - \tfrac{1}{2}\cdot\tfrac{1}{2}x^2 + C$$

$$= \tfrac{1}{2}x^2 \ln x - \tfrac{1}{4}x^2 + C$$

Note: A mnemonic device which is helpful for selecting u when using integration by parts is the LIATE principle of precedence for u:

<u>L</u>ogarithmic

<u>I</u>nverse trigonometric

<u>A</u>lgebraic

<u>T</u>rigonometric

<u>E</u>xponential

If the integrand has several factors, then we try to choose among them a u which appears as high as possible on the list. For example, in Exercise 3 the integrand is xe^{2x}, which is the product of an algebraic function (x) and an exponential function (e^{2x}). Since <u>A</u>lgebraic appears before <u>E</u>xponential, we choose $u = x$. Sometimes the integration turns out to be similar regardless of the selection of u and dv, but it is advisable to refer to LIATE when in doubt.

3. Let $u = x$, $dv = e^{2x}\,dx$ $\Rightarrow$ $du = dx$, $v = \tfrac{1}{2}e^{2x}$. Then by Equation 2,

$\int xe^{2x}\,dx = \tfrac{1}{2}xe^{2x} - \int \tfrac{1}{2}e^{2x}\,dx = \tfrac{1}{2}xe^{2x} - \tfrac{1}{4}e^{2x} + C$.

5. Let $u = x$, $dv = \sin 4x\,dx$ $\Rightarrow$ $du = dx$, $v = -\tfrac{1}{4}\cos 4x$. Then by Equation 2,

$\int x \sin 4x\,dx = -\tfrac{1}{4}x \cos 4x - \int \left(-\tfrac{1}{4}\cos 4x\right)dx = -\tfrac{1}{4}x \cos 4x + \tfrac{1}{4}\left(\tfrac{1}{4}\sin 4x\right) + C$

$= -\tfrac{1}{4}x \cos 4x + \tfrac{1}{16}\sin 4x + C$.

7. First let $u = x^2$, $dv = \cos 3x\, dx$ $\Rightarrow$ $du = 2x\, dx$, $v = \frac{1}{3}\sin 3x$. Then by Equation 2, the original integral

$I = \int x^2 \cos 3x\, dx = \frac{1}{3}x^2 \sin 3x - \frac{2}{3}\int x \sin 3x\, dx$. To evaluate the last integral, we next let $U = x$,

$dV = \sin 3x\, dx$ $\Rightarrow$ $dU = dx$, $V = -\frac{1}{3}\cos 3x$. So

$\int x \sin 3x\, dx = -\frac{1}{3}x\cos 3x + \frac{1}{3}\int \cos 3x\, dx = -\frac{1}{3}x\cos 3x + \frac{1}{9}\sin 3x + C_1$. Substituting for $\int x \sin 3x\, dx$,

we get $I = \frac{1}{3}x^2 \sin 3x - \frac{2}{3}\left(-\frac{1}{3}x\cos 3x + \frac{1}{9}\sin 3x + C_1\right) = \frac{1}{3}x^2 \sin 3x + \frac{2}{9}x\cos 3x - \frac{2}{27}\sin 3x + C$,

where $C = -\frac{2}{3}C_1$.

9. First let $u = (\ln x)^2$, $dv = dx$ $\Rightarrow$ $du = 2\ln x \cdot \frac{1}{x}\, dx$, $v = x$. Then by Equation 2,

$I = \int (\ln x)^2\, dx = x(\ln x)^2 - 2\int x \ln x \cdot \frac{1}{x}\, dx = x(\ln x)^2 - 2\int \ln x\, dx$. Next let $U = \ln x$, $dV = dx$ $\Rightarrow$

$dU = 1/x\, dx$, $V = x$ to get $\int \ln x\, dx = x\ln x - \int x \cdot (1/x)\, dx = x\ln x - \int dx = x\ln x - x + C_1$. Thus,

$I = x(\ln x)^2 - 2(x\ln x - x + C_1) = x(\ln x)^2 - 2x\ln x + 2x + C$, where $C = -2C_1$.

11. Let $u = \ln r$, $dv = r^3 dr$ $\Rightarrow$ $du = (1/r)\, dr$, $v = \frac{1}{4}r^4$. Then

$$\int r^3 \ln r\, dr = \frac{1}{4}r^4 \ln r - \int \frac{1}{4}r^4 (1/r)\, dr = \frac{1}{4}r^4 \ln r - \frac{1}{4}\int r^3 dr = \frac{1}{4}r^4 \ln r - \frac{1}{4}\left(\frac{1}{4}r^4\right) + C$$
$$= \frac{1}{4}r^4 \ln r - \frac{1}{16}r^4 + C, \text{ or } \frac{1}{16}r^4(4\ln r - 1) + C.$$

13. First let $u = \sin 3\theta$, $dv = e^{2\theta}\, d\theta$ $\Rightarrow$ $du = 3\cos 3\theta\, d\theta$, $v = \frac{1}{2}e^{2\theta}$.

Then $I = \int e^{2\theta} \sin 3\theta\, d\theta = \frac{1}{2}e^{2\theta}\sin 3\theta - \frac{3}{2}\int e^{2\theta}\cos 3\theta\, d\theta$.

Next let $U = \cos 3\theta$, $dV = e^{2\theta}\, d\theta$ $\Rightarrow$ $dU = -3\sin 3\theta\, d\theta$, $V = \frac{1}{2}e^{2\theta}$ to get

$\int e^{2\theta}\cos 3\theta\, d\theta = \frac{1}{2}e^{2\theta}\cos 3\theta + \frac{3}{2}\int e^{2\theta}\sin 3\theta\, d\theta$. Substituting in the previous formula gives

$I = \frac{1}{2}e^{2\theta}\sin 3\theta - \frac{3}{4}e^{2\theta}\cos 3\theta - \frac{9}{4}\int e^{2\theta}\sin 3\theta\, d\theta = \frac{1}{2}e^{2\theta}\sin 3\theta - \frac{3}{4}e^{2\theta}\cos 3\theta - \frac{9}{4}I$ $\Rightarrow$

$\frac{13}{4}I = \frac{1}{2}e^{2\theta}\sin 3\theta - \frac{3}{4}e^{2\theta}\cos 3\theta + C_1$. Hence, $I = \frac{1}{13}e^{2\theta}(2\sin 3\theta - 3\cos 3\theta) + C$, where $C = \frac{4}{13}C_1$.

15. Let $u = t$, $dv = e^{-t}\, dt$ $\Rightarrow$ $du = dt$, $v = -e^{-t}$. By Formula 6,

$\int_0^1 te^{-t}\, dt = \left[-te^{-t}\right]_0^1 + \int_0^1 e^{-t}\, dt = -1/e + \left[-e^{-t}\right]_0^1 = -1/e - 1/e + 1 = 1 - 2/e$.

17. Let $u = x$, $dv = \cos 2x\, dx$ $\Rightarrow$ $du = dx$, $v = \frac{1}{2}\sin 2x\, dx$. By Formula 6,

$\int_0^{\pi/2} x \cos 2x\, dx = \left[\frac{1}{2}x\sin 2x\right]_0^{\pi/2} - \frac{1}{2}\int_0^{\pi/2}\sin 2x\, dx = 0 + \left[\frac{1}{4}\cos 2x\right]_0^{\pi/2} = \frac{1}{4}(-1-1) = -\frac{1}{2}$.

19. Let $u = \sin^{-1} x$, $dv = dx$ $\Rightarrow$ $du = \dfrac{dx}{\sqrt{1-x^2}}$, $v = x$. By Formula 6,

$I = \int_0^{1/2} \sin^{-1} x\, dx = \left[x\sin^{-1} x\right]_0^{1/2} - \int_0^{1/2} \dfrac{x\, dx}{\sqrt{1-x^2}} = \dfrac{1}{2}\cdot\dfrac{\pi}{6} - \int_0^{1/2}\dfrac{x\, dx}{\sqrt{1-x^2}}$. To evaluate the last

integral, let $t = 1 - x^2$, so $dt = -2x\, dx$ and $x\, dx = -\frac{1}{2}dt$. When $x = 0$, $t = 1$; when $x = \frac{1}{2}$, $t = \frac{3}{4}$. So

$\displaystyle\int_0^{1/2}\dfrac{x\, dx}{\sqrt{1-x^2}} = \int_1^{3/4}\dfrac{1}{\sqrt{t}}\left(-\dfrac{1}{2}dt\right) = \dfrac{1}{2}\int_{3/4}^1 t^{-1/2}dt = \dfrac{1}{2}\left[2t^{1/2}\right]_{3/4}^1 = \sqrt{1} - \sqrt{\dfrac{3}{4}} = 1 - \dfrac{\sqrt{3}}{2}$.

Thus, $I = \dfrac{\pi}{12} - \left(1 - \dfrac{\sqrt{3}}{2}\right) = \dfrac{\pi}{12} - 1 + \dfrac{\sqrt{3}}{2} = \dfrac{1}{12}\left(\pi - 12 + 6\sqrt{3}\right)$.

21. $I = \int_1^4 \ln\sqrt{x}\, dx = \int_1^4 \ln x^{1/2}dx = \int_1^4 \frac{1}{2}\ln x\, dx = \frac{1}{2}\int_1^4 \ln x\, dx = \frac{1}{2}\left[x\ln x - x\right]_1^4$ as in Example 2. So

$I = \frac{1}{2}\left[(4\ln 4 - 4) - (0 - 1)\right] = 2\ln 4 - \frac{3}{2}$.

23. Let $u = \ln(\sin\theta)$, $dv = \cos\theta\, d\theta$ $\Rightarrow$ $du = (1/\sin\theta)\cos\theta\, d\theta$, $v = \sin\theta$. Then

$\int_{\pi/6}^{\pi/2}\cos\theta\,\ln(\sin\theta)\, d\theta = \left[\sin\theta\,\ln(\sin\theta)\right]_{\pi/6}^{\pi/2} - \int_{\pi/6}^{\pi/2}\cos\theta\, d\theta = (1\cdot\ln 1) - \left(\frac{1}{2}\cdot\ln\frac{1}{2}\right) - \left[\sin\theta\right]_{\pi/6}^{\pi/2}$

$= 0 - \frac{1}{2}\ln 2^{-1} - \left(1 - \frac{1}{2}\right) = \frac{1}{2}\ln 2 - \frac{1}{2}$, or $\frac{1}{2}(\ln 2 - 1)$.

25. Let $w = \sqrt{x}$, so that $x = w^2$ and $dx = 2w\,dw$. Thus, $\int \sin \sqrt{x}\,dx = \int 2w \sin w\,dw$. Now use parts with $u = 2w$, $dv = \sin w\,dw$, $du = 2\,dw$, $v = -\cos w$ to get

$$\int 2w \sin w\,dw = -2w \cos w + \int 2 \cos w\,dw = -2w \cos w + 2 \sin w + C$$
$$= -2\sqrt{x} \cos \sqrt{x} + 2 \sin \sqrt{x} + C = 2\left(\sin \sqrt{x} - \sqrt{x} \cos \sqrt{x}\right) + C$$

27. Let $x = \theta^2$, so that $dx = 2\theta\,d\theta$. Thus, $\int_{\sqrt{\pi/2}}^{\sqrt{\pi}} \theta^3 \cos(\theta^2)\,d\theta = \int_{\sqrt{\pi/2}}^{\sqrt{\pi}} \theta^2 \cos(\theta^2) \cdot \frac{1}{2}(2\theta\,d\theta) = \frac{1}{2}\int_{\pi/2}^{\pi} x \cos x\,dx$.

Now use parts with $u = x$, $dv = \cos x\,dx$, $du = dx$, $v = \sin x$ to get

$$\frac{1}{2}\int_{\pi/2}^{\pi} x \cos x\,dx = \frac{1}{2}\left([x \sin x]_{\pi/2}^{\pi} - \int_{\pi/2}^{\pi} \sin x\,dx\right) = \frac{1}{2}[x \sin x + \cos x]_{\pi/2}^{\pi}$$
$$= \frac{1}{2}(\pi \sin \pi + \cos \pi) - \frac{1}{2}\left(\frac{\pi}{2} \sin \frac{\pi}{2} + \cos \frac{\pi}{2}\right) = \frac{1}{2}(\pi \cdot 0 - 1) - \frac{1}{2}\left(\frac{\pi}{2} \cdot 1 + 0\right) = -\frac{1}{2} - \frac{\pi}{4}$$

In Exercises 29–31, let $f(x)$ denote the integrand and $F(x)$ its antiderivative (with $C = 0$).

29. Let $u = x$, $dv = \cos \pi x\,dx$ $\Rightarrow$ $du = dx$, $v = (\sin \pi x)/\pi$. Then

$$\int x \cos \pi x\,dx = x \cdot \frac{\sin \pi x}{\pi} - \int \frac{\sin \pi x}{\pi}\,dx = \frac{x \sin \pi x}{\pi} + \frac{\cos \pi x}{\pi^2} + C.$$

We see from the graph that this is reasonable, since F has extreme values where f is 0.

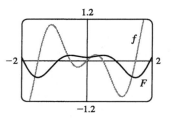

31. Let $u = 2x + 3$, $dv = e^x\,dx$ $\Rightarrow$ $du = 2\,dx$, $v = e^x$. Then
$\int (2x + 3)e^x\,dx = (2x + 3)e^x - 2\int e^x\,dx = (2x + 3)e^x - 2e^x + C = (2x + 1)e^x + C$. We see from the graph that this is reasonable, since F has a minimum where f changes from negative to positive.

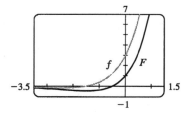

33. (a) Take $n = 2$ in Example 6 to get $\int \sin^2 x\,dx = -\frac{1}{2} \cos x \sin x + \frac{1}{2}\int 1\,dx = \frac{x}{2} - \frac{\sin 2x}{4} + C$.

(b) $\int \sin^4 x\,dx = -\frac{1}{4} \cos x \sin^3 x + \frac{3}{4}\int \sin^2 x\,dx = -\frac{1}{4} \cos x \sin^3 x + \frac{3}{8}x - \frac{3}{16} \sin 2x + C$.

35. (a) From Example 6, $\int \sin^n x\,dx = -\frac{1}{n} \cos x \sin^{n-1} x + \frac{n-1}{n}\int \sin^{n-2} x\,dx$. Using (6),

$$\int_0^{\pi/2} \sin^n x\,dx = \left[-\frac{\cos x \sin^{n-1} x}{n}\right]_0^{\pi/2} + \frac{n-1}{n}\int_0^{\pi/2} \sin^{n-2} x\,dx$$
$$= (0 - 0) + \frac{n-1}{n}\int_0^{\pi/2} \sin^{n-2} x\,dx = \frac{n-1}{n}\int_0^{\pi/2} \sin^{n-2} x\,dx$$

(b) Using $n = 3$ in part (a), we have $\int_0^{\pi/2} \sin^3 x\,dx = \frac{2}{3}\int_0^{\pi/2} \sin x\,dx = \left[-\frac{2}{3} \cos x\right]_0^{\pi/2} = \frac{2}{3}$.

Using $n = 5$ in part (a), we have $\int_0^{\pi/2} \sin^5 x\,dx = \frac{4}{5}\int_0^{\pi/2} \sin^3 x\,dx = \frac{4}{5} \cdot \frac{2}{3} = \frac{8}{15}$.

(c) The formula holds for $n = 1$ (that is, $2n + 1 = 3$) by (b). Assume it holds for some $k \geq 1$. Then

$$\int_0^{\pi/2} \sin^{2k+1} x \, dx = \frac{2 \cdot 4 \cdot 6 \cdots (2k)}{3 \cdot 5 \cdot 7 \cdots (2k+1)}. \text{ By Example 6,}$$

$$\int_0^{\pi/2} \sin^{2k+3} x \, dx = \frac{2k+2}{2k+3} \int_0^{\pi/2} \sin^{2k+1} x \, dx = \frac{2k+2}{2k+3} \cdot \frac{2 \cdot 4 \cdot 6 \cdots (2k)}{3 \cdot 5 \cdot 7 \cdots (2k+1)}$$

$$= \frac{2 \cdot 4 \cdot 6 \cdots (2k)[2\,(k+1)]}{3 \cdot 5 \cdot 7 \cdots (2k+1)[2\,(k+1)+1]},$$

so the formula holds for $n = k + 1$. By induction, the formula holds for all $n \geq 1$.

37. Let $u = (\ln x)^n$, $dv = dx$ $\Rightarrow$ $du = n(\ln x)^{n-1}(dx/x)$, $v = x$. By Equation 2,
$\int (\ln x)^n \, dx = x(\ln x)^n - \int nx(\ln x)^{n-1}(dx/x) = x(\ln x)^n - n \int (\ln x)^{n-1} \, dx.$

39. Take $n = 3$ in Exercise 37 to get
$\int (\ln x)^3 \, dx = x\,(\ln x)^3 - 3 \int (\ln x)^2 \, dx = x(\ln x)^3 - 3x(\ln x)^2 + 6x \ln x - 6x + C$ (by Exercise 9).
Or: Instead of using Exercise 9, apply Exercise 37 again with $n = 2$.

41. Since $v(t) > 0$ for all t, the desired distance is $s(t) = \int_0^t v(w)dw = \int_0^t w^2 e^{-w} \, dw$.
First let $u = w^2$, $dv = e^{-w} \, dw$ $\Rightarrow$ $du = 2w \, dw$, $v = -e^{-w}$. Then $s(t) = \left[-w^2 e^{-w}\right]_0^t + 2 \int_0^t we^{-w} \, dw$.
Next let $U = w$, $dV = e^{-w} \, dw$ $\Rightarrow$ $dU = dw$, $V = -e^{-w}$. Then

$$s(t) = -t^2 e^{-t} + 2\left(\left[-we^{-w}\right]_0^t + \int_0^t e^{-w} \, dw\right) = -t^2 e^{-t} + 2\left(-te^{-t} + 0 + \left[-e^{-w}\right]_0^t\right)$$

$$= -t^2 e^{-t} + 2\left(-te^{-t} - e^{-t} + 1\right) = -t^2 e^{-t} - 2te^{-t} - 2e^{-t} + 2$$

$$= 2 - e^{-t}\left(t^2 + 2t + 2\right) \text{ meters}$$

43. Take $g(x) = x$ and $g'(x) = 1$ in Equation 1.

45. Suppose $f(0) = g(0) = 0$ and let $u = f(x)$, $dv = g''(x) \, dx$ $\Rightarrow$ $du = f'(x) \, dx$, $v = g'(x)$.
Then $\int_0^a f(x)g''(x) \, dx = [f(x)g'(x)]_0^a - \int_0^a f'(x)g'(x) \, dx = f(a)g'(a) - \int_0^a f'(x)g'(x) \, dx.$
Now let $U = f'(x)$, $dV = g'(x) \, dx$ $\Rightarrow$ $dU = f''(x) \, dx$ and $V = g(x)$, so
$\int_0^a f'(x)g'(x) \, dx = [f'(x)g(x)]_0^a - \int_0^a f''(x)g(x) \, dx = f'(a)g(a) - \int_0^a f''(x)g(x) \, dx.$ Combining the two
results, we get $\int_0^a f(x)g''(x) \, dx = f(a)g'(a) - f'(a)g(a) + \int_0^a f''(x)g(x) \, dx.$

◆5.7◆ Additional Techniques of Integration · · · · · · · · · ·

The symbols $\overset{s}{=}$ and $\overset{c}{=}$ indicate the use of the substitutions $\{u = \sin x, \ du = \cos x \, dx\}$ and $\{u = \cos x, \ du = -\sin x \, dx\}$, respectively.

1. $\int \sin^3 x \cos^2 x \, dx = \int \sin^2 x \cos^2 x \sin x \, dx = \int (1 - \cos^2 x)\cos^2 x \sin x \, dx \overset{c}{=} \int (1 - u^2)u^2(-du)$
$= \int (u^2 - 1)u^2 \, du = \int (u^4 - u^2)du = \frac{1}{5}u^5 - \frac{1}{3}u^3 + C = \frac{1}{5}\cos^5 x - \frac{1}{3}\cos^3 x + C$

3. $\int_{\pi/2}^{3\pi/4} \sin^5 x \cos^3 x \, dx = \int_{\pi/2}^{3\pi/4} \sin^5 x \cos^2 x \cos x \, dx = \int_{\pi/2}^{3\pi/4} \sin^5 x \left(1 - \sin^2 x\right) \cos x \, dx$

$$\overset{s}{=} \int_1^{\sqrt{2}/2} u^5 \left(1 - u^2\right) du = \int_1^{\sqrt{2}/2} \left(u^5 - u^7\right) du = \left[\frac{1}{6} u^6 - \frac{1}{8} u^8\right]_1^{\sqrt{2}/2}$$

$$= \left(\frac{1/8}{6} - \frac{1/16}{8}\right) - \left(\frac{1}{6} - \frac{1}{8}\right) = -\frac{11}{384}$$

5. $\int \cos^4 t \, dt = \int \left(\cos^2 t\right)^2 dt = \int \left[\frac{1}{2}(1 + \cos 2t)\right]^2 dt = \frac{1}{4} \int \left(1 + 2\cos 2t + \cos^2 2t\right) dt$

$$= \frac{1}{4} t + \frac{1}{4} \sin 2t + \frac{1}{4} \int \frac{1}{2}(1 + \cos 4t) \, dt = \frac{1}{4} \left[t + \sin 2t + \frac{1}{2} t + \frac{1}{8} \sin 4t\right] + C$$

$$= \frac{1}{4} \left(\frac{3}{2} t + \sin 2t + \frac{1}{8} \sin 4t\right) + C = \frac{3}{8} t + \frac{1}{4} \sin 2t + \frac{1}{32} \sin 4t + C$$

7. Let $u = \sec x$. Then $du = \sec x \tan x \, dx$, so

$$\int \tan^3 x \sec x \, dx = \int \left(\tan^2 x\right)(\tan x \sec x) \, dx = \int \left(\sec^2 x - 1\right)(\sec x \tan x \, dx)$$

$$= \int \left(u^2 - 1\right) du = \frac{1}{3} u^3 - u + C = \frac{1}{3} \sec^3 x - \sec x + C$$

9. $x = 3 \sin \theta$, where $-\pi/2 \le \theta \le \pi/2$. Then $dx = 3 \cos \theta \, d\theta$ and

$\sqrt{9 - x^2} = \sqrt{9 - 9 \sin^2 \theta} = \sqrt{9 \cos^2 \theta} = 3 \, |\cos \theta| = 3 \cos \theta$. (Note that $\cos \theta \ge 0$ because
$-\pi/2 \le \theta \le \pi/2$.) Thus, substitution gives

$$\int \frac{\sqrt{9 - x^2}}{x^2} \, dx = \int \frac{3 \cos \theta}{9 \sin^2 \theta} 3 \cos \theta \, d\theta = \int \frac{\cos^2 \theta}{\sin^2 \theta} \, d\theta = \int \cot^2 \theta \, d\theta$$

$$= \int \left(\csc^2 \theta - 1\right) d\theta = -\cot \theta - \theta + C$$

Since this is an indefinite integral, we must return to the original
variable x. This can be done either by using trigonometric identities to
express $\cot \theta$ in terms of $\sin \theta = x/3$ or by drawing a diagram, as
shown, where θ is interpreted as an angle of a right triangle.

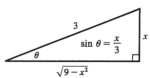

Since $\sin \theta = x/3$, we label the opposite side and the hypotenuse as having lengths x and 3. Then the Pythagorean
Theorem gives the length of the adjacent side as $\sqrt{9 - x^2}$, so we can simply read the value of $\cot \theta$ from the figure:

$\cot \theta = \dfrac{\sqrt{9 - x^2}}{x}$. (Although $\theta > 0$ in the diagram, this expression for $\cot \theta$ is valid even when $\theta < 0$.) Since

$\sin \theta = x/3$, we have $\theta = \sin^{-1}(x/3)$ and so $\displaystyle\int \frac{\sqrt{9 - x^2}}{x^2} \, dx = -\frac{\sqrt{9 - x^2}}{x} - \sin^{-1}\left(\frac{x}{3}\right) + C$.

11. $x = 2 \tan \theta$, where $-\pi/2 < \theta < \pi/2$. Then $dx = 2 \sec^2 \theta \, d\theta$ and

$\sqrt{x^2 + 4} = \sqrt{(2 \tan \theta)^2 + 4} = \sqrt{4 \tan^2 \theta + 4}$

$= \sqrt{4(\tan^2 \theta + 1)} = 2 \sqrt{\sec^2 \theta} = 2 \, |\sec \theta|$

$= 2 \sec \theta$ (since $\sec \theta \ge 0$ for $-\pi/2 < \theta < \pi/2$).

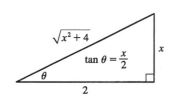

Thus, substitution gives

$$\int \frac{1}{x^2\sqrt{x^2+4}}\,dx = \int \frac{1}{4\tan^2\theta\,(2\sec\theta)}(2\sec^2\theta\,d\theta) = \frac{1}{4}\int \frac{\sec\theta}{\tan^2\theta}\,d\theta = \frac{1}{4}\int \frac{1}{\cos\theta}\cdot\frac{\cos^2\theta}{\sin^2\theta}\,d\theta$$

$$= \frac{1}{4}\int \frac{\cos\theta}{\sin^2\theta}\,d\theta = \frac{1}{4}\int \frac{1}{u^2}\,du \quad [u=\sin\theta,\ du=\cos\theta\,d\theta]$$

$$= \frac{1}{4}\left(-\frac{1}{u}\right)+C = -\frac{1}{4}\frac{1}{\sin\theta}+C = -\frac{1}{4}\cdot\frac{\sqrt{x^2+4}}{x}+C = -\frac{\sqrt{x^2+4}}{4x}+C$$

13. The radicand has the form (variable)^2 − (constant)^2, so we'll use the substitution from Exercise 10; that is,
$t=\sec\theta$, where $0\le\theta<\pi/2$ or $\pi\le\theta<3\pi/2$. Then $dt=\sec\theta\tan\theta\,d\theta$ and
$\sqrt{x^2-1}=\sqrt{\sec^2\theta-1}=\sqrt{\tan^2\theta}=|\tan\theta|=\tan\theta$ (since $\tan\theta\ge 0$ for the specified values of θ).
When $t=\sqrt{2}$, $\theta=\frac{\pi}{4}$; when $t=2$, $\theta=\frac{\pi}{3}$. Thus,

$$\int_{\sqrt{2}}^{2}\frac{1}{t^3\sqrt{t^2-1}}\,dt = \int_{\pi/4}^{\pi/3}\frac{1}{\sec^3\theta\tan\theta}\sec\theta\tan\theta\,d\theta = \int_{\pi/4}^{\pi/3}\frac{1}{\sec^2\theta}\,d\theta = \int_{\pi/4}^{\pi/3}\cos^2\theta\,d\theta$$

$$= \int_{\pi/4}^{\pi/3}\tfrac{1}{2}(1+\cos 2\theta)\,d\theta = \tfrac{1}{2}\left[\theta+\tfrac{1}{2}\sin 2\theta\right]_{\pi/4}^{\pi/3}$$

$$= \tfrac{1}{2}\left[\left(\tfrac{\pi}{3}+\tfrac{1}{2}\cdot\tfrac{\sqrt{3}}{2}\right)-\left(\tfrac{\pi}{4}+\tfrac{1}{2}\cdot 1\right)\right] = \tfrac{1}{2}\left(\tfrac{\pi}{12}+\tfrac{\sqrt{3}}{4}-\tfrac{1}{2}\right) = \tfrac{\pi}{24}+\tfrac{\sqrt{3}}{8}-\tfrac{1}{4}$$

15. (a) $\dfrac{2}{x^2+3x-4} = \dfrac{2}{(x+4)(x-1)} = \dfrac{A}{x+4}+\dfrac{B}{x-1}$

(b) x^2+x+1 is irreducible, so $\dfrac{x^2}{(x-1)(x^2+x+1)} = \dfrac{A}{x-1}+\dfrac{Bx+C}{x^2+x+1}$.

17. $\dfrac{x-9}{(x+5)(x-2)} = \dfrac{A}{x+5}+\dfrac{B}{x-2}$. Multiply both sides by $(x+5)(x-2)$ to get
$x-9 = A(x-2)+B(x+5)$ (∗), or equivalently, $x-9 = (A+B)x-2A+5B$. Equating coefficients of x
on each side of the equation gives us $1=A+B$ (**1**) and equating constants gives us $-9=-2A+5B$ (**2**). Adding
two times (**1**) to (**2**) gives us $-7=7B \Leftrightarrow B=-1$ and hence, $A=2$. [Alternatively, to find the coefficients
A and B, we may use substitution as follows: substitute 2 for x in (∗) to get $-7=7B \Leftrightarrow B=-1$,
then substitute -5 for x in (∗) to get $-14=-7A \Leftrightarrow A=2$.] Thus,

$$\int \frac{x-9}{(x+5)(x-2)}\,dx = \int\left(\frac{2}{x+5}+\frac{-1}{x-2}\right)dx = 2\ln|x+5|-\ln|x-2|+C.$$

To find the constants in problems involving partial fractions, we may use the coefficient comparison method or the substitution method
(as in the solution for Exercise 17) or a combination of both methods.

19. $\dfrac{1}{x^2-1} = \dfrac{1}{(x+1)(x-1)} = \dfrac{A}{x+1}+\dfrac{B}{x-1}$. Multiply both sides by $(x+1)(x-1)$ to get
$1=A(x-1)+B(x+1)$. Substituting 1 for x gives $1=2B \Leftrightarrow B=\frac{1}{2}$.
Substituting -1 for x gives $1=-2A \Leftrightarrow A=-\frac{1}{2}$. Thus,

$$\int_{2}^{3}\frac{1}{x^2-1}\,dx = \int_{2}^{3}\left(\frac{-1/2}{x+1}+\frac{1/2}{x-1}\right)dx = \left[-\tfrac{1}{2}\ln|x+1|+\tfrac{1}{2}\ln|x-1|\right]_{2}^{3}$$

$$= \left(-\tfrac{1}{2}\ln 4+\tfrac{1}{2}\ln 2\right)-\left(-\tfrac{1}{2}\ln 3+\tfrac{1}{2}\ln 1\right) = \tfrac{1}{2}(\ln 2+\ln 3-\ln 4) \quad \left[\text{or } \tfrac{1}{2}\ln\tfrac{3}{2}\right]$$

21. $\dfrac{10}{(x-1)(x^2+9)} = \dfrac{A}{x-1} + \dfrac{Bx+C}{x^2+9}$. Multiply both sides by $(x-1)(x^2+9)$ to get

$10 = A(x^2+9) + (Bx+C)(x-1)$ $(\star)$. Substituting 1 for x gives $10 = 10A$ $\Leftrightarrow$ $A = 1$. Substituting 0 for x

gives $10 = 9A - C$ $\Rightarrow$ $C = 9(1) - 10 = -1$. The coefficients of the x^2-terms in $(\star)$ must be equal, so

$0 = A + B$ $\Rightarrow$ $B = -1$. Thus,

$$\int \frac{10}{(x-1)(x^2+9)}\,dx = \int \left(\frac{1}{x-1} + \frac{-x-1}{x^2+9}\right)dx = \int \left(\frac{1}{x-1} - \frac{x}{x^2+9} - \frac{1}{x^2+9}\right)dx$$

$$= \ln|x-1| - \tfrac{1}{2}\ln(x^2+9) \ \ [\text{let } u = x^2+9] \ -\tfrac{1}{3}\tan^{-1}\left(\tfrac{x}{3}\right) \ [\text{Formula 1}] \ +C$$

23. $\dfrac{x^3+x^2+2x+1}{(x^2+1)(x^2+2)} = \dfrac{Ax+B}{x^2+1} + \dfrac{Cx+D}{x^2+2}$. Multiply both sides by $(x^2+1)(x^2+2)$ to get

$x^3+x^2+2x+1 = (Ax+B)(x^2+2) + (Cx+D)(x^2+1)$ $\Leftrightarrow$

$x^3+x^2+2x+1 = (Ax^3+Bx^2+2Ax+2B) + (Cx^3+Dx^2+Cx+D)$ $\Leftrightarrow$

$x^3+x^2+2x+1 = (A+C)x^3 + (B+D)x^2 + (2A+C)x + (2B+D)$. Comparing coefficients gives us the

following system of equations:

$$A + C = 1 \ \ \textbf{(1)} \qquad B + D = 1 \ \ \textbf{(2)}$$
$$2A + C = 2 \ \ \textbf{(3)} \qquad 2B + D = 1 \ \ \textbf{(4)}$$

Subtracting equation **(1)** from equation **(3)** gives us $A = 1$, so $C = 0$. Subtracting equation **(2)** from equation **(4)**

gives us $B = 0$, so $D = 1$. Thus, $I = \displaystyle\int \frac{x^3+x^2+2x+1}{(x^2+1)(x^2+2)}\,dx = \int \left(\frac{x}{x^2+1} + \frac{1}{x^2+2}\right)dx$. For $\displaystyle\int \frac{x}{x^2+1}\,dx$,

let $u = x^2+1$ so $du = 2x\,dx$ and then $\displaystyle\int \frac{x}{x^2+1}\,dx = \frac{1}{2}\int \frac{1}{u}\,du = \frac{1}{2}\ln|u| + C = \frac{1}{2}\ln(x^2+1) + C$. For

$\displaystyle\int \frac{1}{x^2+2}\,dx$, use Formula 1 with $a = \sqrt{2}$. So $\displaystyle\int \frac{1}{x^2+2}\,dx = \int \frac{1}{x^2+(\sqrt{2})^2}\,dx = \frac{1}{\sqrt{2}}\tan^{-1}\frac{x}{\sqrt{2}} + C$. Thus,

$I = \dfrac{1}{2}\ln(x^2+1) + \dfrac{1}{\sqrt{2}}\tan^{-1}\dfrac{x}{\sqrt{2}} + C$.

25.

$\begin{array}{r} x - 1 \\ x+1\overline{\smash{\big)}\ x^2} \\ \underline{x^2 + x} \\ -x \\ \underline{-x - 1} \\ 1 \end{array}$

By long division, $\dfrac{x^2}{x+1} = x - 1 + \dfrac{1}{x+1}$. Thus,

$$\int \frac{x^2}{x+1}\,dx = \int \left(x - 1 + \frac{1}{x+1}\right)dx$$
$$= \tfrac{1}{2}x^2 - x + \ln|x+1| + C$$

27.

$\begin{array}{r} x \\ x^2+1\overline{\smash{\big)}\ x^3} \\ \underline{x^3 + x} \\ -x \end{array}$

By long division, $\dfrac{x^3}{x^2+1} = x - \dfrac{x}{x^2+1}$. Thus,

$$\int_0^1 \frac{x^3}{x^2+1}\,dx = \int_0^1 x\,dx - \int_0^1 \frac{x\,dx}{x^2+1}$$
$$= \left[\tfrac{1}{2}x^2\right]_0^1 - \tfrac{1}{2}\int_1^2 \frac{1}{u}\,du \ \ [\text{where } u = x^2+1,\ du = 2x\,dx]$$
$$= \tfrac{1}{2} - \left[\tfrac{1}{2}\ln u\right]_1^2 = \tfrac{1}{2} - \tfrac{1}{2}\ln 2 = \tfrac{1}{2}(1 - \ln 2)$$

29. Let $u = \sqrt{x}$, so $u^2 = x$ and $dx = 2u\,du$. Thus,

$$\int_9^{16} \frac{\sqrt{x}}{x-4}\,dx = \int_3^4 \frac{u}{u^2-4}2u\,du = 2\int_3^4 \frac{u^2}{u^2-4}\,du = 2\int_3^4 \left(1 + \frac{4}{u^2-4}\right)du \quad \text{[by long division]}$$

$$= 2 + 8\int_3^4 \frac{du}{(u+2)(u-2)}.$$

Multiply $\dfrac{1}{(u+2)(u-2)} = \dfrac{A}{u+2} + \dfrac{B}{u-2}$ by $(u+2)(u-2)$ to get $1 = A(u-2) + B(u+2)$. Equating

coefficients we get $A + B = 0$ and $-2A + 2B = 1$. Solving gives us $B = \frac{1}{4}$ and $A = -\frac{1}{4}$, so

$\dfrac{1}{(u+2)(u-2)} = \dfrac{-1/4}{u+2} + \dfrac{1/4}{u-2}$ and the last integral is

$$2 + 8\int_3^4 \left(\frac{-1/4}{u+2} + \frac{1/4}{u-2}\right)du = 2 + 8\left[-\tfrac{1}{4}\ln|u+2| + \tfrac{1}{4}\ln|u-2|\right]_3^4$$

$$= 2 + [2\ln|u-2| - 2\ln|u+2|]_3^4 = 2 + 2\left[\ln\left|\frac{u-2}{u+2}\right|\right]_3^4$$

$$= 2 + 2\left(\ln\tfrac{2}{6} - \ln\tfrac{1}{5}\right) = 2 + 2\ln\tfrac{2/6}{1/5}$$

$$= 2 + 2\ln\tfrac{5}{3} \text{ or } 2 + \ln\left(\tfrac{5}{3}\right)^2 = 2 + \ln\tfrac{25}{9}$$

31. $x^2 + x + 1 = x^2 + x + \tfrac{1}{4} + 1 - \tfrac{1}{4}$ [add and subtract the square of one-half the

                                     coefficient of x to complete the square]

$$= x^2 + x + \tfrac{1}{4} + \tfrac{3}{4} = \left(x + \tfrac{1}{2}\right)^2 + \left(\tfrac{\sqrt{3}}{2}\right)^2$$

So $I = \displaystyle\int \frac{dx}{x^2+x+1} = \int \frac{1}{\left(x+\frac{1}{2}\right)^2 + \left(\frac{\sqrt{3}}{2}\right)^2}\,dx$. Now let $u = x + \tfrac{1}{2} \implies du = dx$ and

$$I = \int \frac{1}{u^2 + \left(\frac{\sqrt{3}}{2}\right)^2}\,du = \frac{1}{\frac{\sqrt{3}}{2}}\tan^{-1}\frac{u}{\frac{\sqrt{3}}{2}} + C = \frac{2}{\sqrt{3}}\tan^{-1}\frac{2\left(x+\frac{1}{2}\right)}{\sqrt{3}} + C = \frac{2}{\sqrt{3}}\tan^{-1}\frac{2x+1}{\sqrt{3}} + C.$$

◆5.8◆ Integration Using Tables and Computer Algebra Systems • •

Keep in mind that there are several ways to approach many of these exercises, and different methods can lead to different forms of the answer.

1. Using long division, $\dfrac{x^3 - x^2 + x - 1}{x^2 + 9} = x - 1 - 8\dfrac{x-1}{x^2+9}$.

$$I = \int \left(x - 1 - 8\frac{x-1}{x^2+9}\right)dx = \int (x-1)dx - 8\left[\int \frac{x}{x^2+9}\,dx - \int \frac{1}{x^2+9}\,dx\right].$$

Using Formula 17 with $a = 3$, we have

$$I \overset{17}{=} \tfrac{1}{2}x^2 - x - 8 \cdot \tfrac{1}{2}\ln(x^2+9) + 8 \cdot \tfrac{1}{3}\tan^{-1}(x/3) + C = \tfrac{1}{2}x^2 - x - 4\ln(x^2+9) + \tfrac{8}{3}\tan^{-1}(x/3) + C.$$

3. Let $u = \pi x \implies du = \pi\,dx$, so

$$\int \sec^3(\pi x)dx = \tfrac{1}{\pi}\int \sec^3 u\,du \overset{71}{=} \tfrac{1}{\pi}\left(\tfrac{1}{2}\sec u\tan u + \tfrac{1}{2}\ln|\sec u + \tan u|\right) + C$$

$$= \tfrac{1}{2\pi}\sec\pi x\tan\pi x + \tfrac{1}{2\pi}\ln|\sec\pi x + \tan\pi x| + C$$

5. Let $u = 3x$. Then $du = 3\,dx$, so

$$\int \frac{\sqrt{9x^2 - 1}}{x^2}\,dx = \int \frac{\sqrt{u^2 - 1}}{u^2/9}\frac{du}{3} = 3\int \frac{\sqrt{u^2 - 1}}{u^2}\,du \overset{42}{=} 3\left(-\frac{\sqrt{u^2 - 1}}{u} + \ln\left|u + \sqrt{u^2 - 1}\right| + C\right)$$

$$= -\frac{\sqrt{9x^2 - 1}}{x} + 3\ln\left|3x + \sqrt{9x^2 - 1}\right| + C$$

7. $\int x^3 \sin x\,dx \overset{84}{=} -x^3 \cos x + 3\int x^2 \cos x\,dx$, $\int x^2 \cos x\,dx \overset{85}{=} x^2 \sin x - 2\int x \sin x\,dx$, and

$\int x \sin x\,dx \overset{82}{=} \sin x - x \cos x + C$. Substituting, we get

$\int x^3 \sin x\,dx = -x^3 \cos x + 3\left[x^2 \sin x - 2(\sin x - x \cos x)\right] + C$

$$= -x^3 \cos x + 3x^2 \sin x - 6\sin x + 6x \cos x + C.$$

So $\int_0^\pi x^3 \sin x\,dx = \left[-x^3 \cos x + 3x^2 \sin x - 6\sin x + 6x \cos x\right]_0^\pi$

$$= (-\pi^3 \cdot -1 + 6\pi \cdot -1) - (0) = \pi^3 - 6\pi.$$

9. Let $u = x^2$. Then $du = 2x\,dx$, so

$$\int x \sin^{-1}(x^2)\,dx = \tfrac{1}{2}\int \sin^{-1} u\,du \overset{87}{=} \tfrac{1}{2}\left(u \sin^{-1} u + \sqrt{1 - u^2}\right) + C$$

$$= \tfrac{1}{2}\left[x^2 \sin^{-1}(x^2) + \sqrt{1 - x^4}\right] + C$$

11. $\int_{-2}^1 \sqrt{5 - 4x - x^2}\,dx = \int_{-2}^1 \sqrt{5 - (x^2 + 4x)}\,dx = \int_{-2}^1 \sqrt{5 + 4 - (x^2 + 4x + 4)}\,dx$

$$= \int_{-2}^1 \sqrt{9 - (x + 2)^2}\,dx = \int_0^3 \sqrt{3^2 - u^2}\,du \quad [u = x + 2,\ du = dx]$$

$$\overset{30}{=} \left[\tfrac{u}{2}\sqrt{9 - u^2} + \tfrac{9}{2}\sin^{-1}\left(\tfrac{u}{3}\right)\right]_0^3 = \left[\left(0 + \tfrac{9}{2} \cdot \tfrac{\pi}{2}\right) - (0 + 0)\right] = \tfrac{9\pi}{4}$$

13. Let $u = \sin x$. Then $du = \cos x\,dx$, so

$$\int \sin^2 x \cos x \ln(\sin x)\,dx = \int u^2 \ln u\,du \overset{101}{=} \frac{u^{2+1}}{(2 + 1)^2}\left[(2 + 1)\ln u - 1\right] + C = \tfrac{1}{9}u^3(3\ln u - 1) + C$$

$$= \tfrac{1}{9}\sin^3 x[3\ln(\sin x) - 1] + C.$$

15. $\int_0^{\pi/2} \cos^5 x\,dx \overset{74}{=} \tfrac{1}{5}\left[\cos^4 x \sin x\right]_0^{\pi/2} + \tfrac{4}{5}\int_0^{\pi/2}\cos^3 x\,dx \overset{68}{=} 0 + \tfrac{4}{5}\left[\tfrac{1}{3}\left(2 + \cos^2 x\right)\sin x\right]_0^{\pi/2} = \tfrac{4}{15}(2 - 0) = \tfrac{8}{15}$

17. $\displaystyle \int \frac{x^4\,dx}{\sqrt{x^{10} - 2}} = \int \frac{x^4\,dx}{\sqrt{(x^5)^2 - 2}} = \frac{1}{5}\int \frac{du}{\sqrt{u^2 - 2}} \quad [u = x^5,\ du = 5x^4\,dx]$

$$\overset{43}{=} \tfrac{1}{5}\ln\left|u + \sqrt{u^2 - 2}\right| + C = \tfrac{1}{5}\ln\left|x^5 + \sqrt{x^{10} - 2}\right| + C$$

19. Let $u = 1 + e^x$, so $du = e^x\,dx$. Then

$$\int e^x \ln(1 + e^x)\,dx = \int \ln u\,du \overset{100}{=} u \ln u - u + C$$

$$= (1 + e^x)\ln(1 + e^x) - e^x - 1 + C$$

$$= (1 + e^x)\ln(1 + e^x) - e^x + C_1, \text{ where } C_1 = C - 1$$

21. Let $u = e^x$. Then $x = \ln u$, $dx = du/u$, so

$$\int \sqrt{e^{2x} - 1}\,dx = \int \frac{\sqrt{u^2 - 1}}{u}\,du \overset{41}{=} \sqrt{u^2 - 1} - \cos^{-1}(1/u) + C = \sqrt{e^{2x} - 1} - \cos^{-1}(e^{-x}) + C.$$

23. (a) $\displaystyle \frac{d}{du}\left[\frac{1}{b^3}\left(a + bu - \frac{a^2}{a + bu} - 2a \ln|a + bu|\right) + C\right] = \frac{1}{b^3}\left[b + \frac{ba^2}{(a + bu)^2} - \frac{2ab}{(a + bu)}\right]$

$$= \frac{1}{b^3}\left[\frac{b(a + bu)^2 + ba^2 - (a + bu)2ab}{(a + bu)^2}\right] = \frac{1}{b^3}\left[\frac{b^3 u^2}{(a + bu)^2}\right] = \frac{u^2}{(a + bu)^2}$$

(b) Let $t = a + bu \implies dt = b\,du$. Note that $u = \dfrac{t - a}{b}$ and $du = \dfrac{1}{b}\,dt$.

$$\int \frac{u^2\,du}{(a + bu)^2} = \frac{1}{b^3} \int \frac{(t - a)^2}{t^2}\,dt = \frac{1}{b^3} \int \frac{t^2 - 2at + a^2}{t^2}\,dt$$

$$= \frac{1}{b^3} \int \left(1 - \frac{2a}{t} + \frac{a^2}{t^2}\right) dt = \frac{1}{b^3}\left(t - 2a \ln|t| - \frac{a^2}{t}\right) + C$$

$$= \frac{1}{b^3}\left(a + bu - \frac{a^2}{a + bu} - 2a \ln|a + bu|\right) + C$$

25. Maple, Mathematica and Derive all give $\int x^2\sqrt{5 - x^2}\,dx = -\frac{1}{4}x(5 - x^2)^{3/2} + \frac{5}{8}x\sqrt{5 - x^2} + \frac{25}{8}\sin^{-1}\left(\frac{1}{\sqrt{5}}x\right)$.

Using Formula 31, we get $\int x^2\sqrt{5 - x^2}\,dx = \frac{1}{8}x(2x^2 - 5)\sqrt{5 - x^2} + \frac{1}{8}(5^2)\sin^{-1}\left(\frac{1}{\sqrt{5}}x\right) + C$. But

$-\frac{1}{4}x(5 - x^2)^{3/2} + \frac{5}{8}x\sqrt{5 - x^2} = \frac{1}{8}x\sqrt{5 - x^2}\left[5 - 2(5 - x^2)\right] = \frac{1}{8}x(2x^2 - 5)\sqrt{5 - x^2}$, and the $\sin^{-1}$ terms

are the same in each expression, so the answers are equivalent.

27. Maple and Derive both give $\int \sin^3 x \cos^2 x\,dx = -\frac{1}{5}\sin^2 x \cos^3 x - \frac{2}{15}\cos^3 x$ (although Derive factors the

expression), and Mathematica gives $\int \sin^3 x \cos^2 x\,dx = -\frac{1}{8}\cos x - \frac{1}{48}\cos 3x + \frac{1}{80}\cos 5x$. We can use a CAS to

show that both of these expressions are equal to $-\frac{1}{3}\cos^3 x + \frac{1}{5}\cos^5 x$. Using Formula 86, we write

$$\int \sin^3 x \cos^2 x\,dx = -\frac{1}{5}\sin^2 x \cos^3 x + \frac{2}{5}\int \sin x \cos^2 x\,dx = -\frac{1}{5}\sin^2 x \cos^3 x + \frac{2}{5}\left(-\frac{1}{3}\cos^3 x\right) + C$$

$$= -\frac{1}{5}\sin^2 x \cos^3 x - \frac{2}{15}\cos^3 x + C$$

29. Maple gives $\int x\sqrt{1 + 2x}\,dx = \frac{1}{10}(1 + 2x)^{5/2} - \frac{1}{6}(1 + 2x)^{3/2}$, Mathematica gives $\sqrt{1 + 2x}\left(\frac{2}{5}x^2 + \frac{1}{15}x - \frac{1}{15}\right)$,

and Derive gives $\frac{1}{15}(1 + 2x)^{3/2}(3x - 1)$. The first two expressions can be simplified to Derive's result. If we use

Formula 54, we get

$$\int x\sqrt{1 + 2x}\,dx = \frac{2}{15(2)^2}(3 \cdot 2x - 2 \cdot 1)(1 + 2x)^{3/2} + C = \frac{1}{30}(6x - 2)(1 + 2x)^{3/2} + C$$

$$= \frac{1}{15}(3x - 1)(1 + 2x)^{3/2}$$

31. Maple gives $\int \tan^5 x\,dx = \frac{1}{4}\tan^4 x - \frac{1}{2}\tan^2 x + \frac{1}{2}\ln(1 + \tan^2 x)$, Mathematica

gives $\int \tan^5 x\,dx = \frac{1}{4}[-1 - 2\cos(2x)]\sec^4 x - \ln(\cos x)$, and Derive gives

$\int \tan^5 x\,dx = \frac{1}{4}\tan^4 x - \frac{1}{2}\tan^2 x - \ln(\cos x)$. These expressions are equivalent, and none includes absolute

value bars or a constant of integration. Note that Mathematica's and Derive's expressions suggest that the integral is

undefined where $\cos x < 0$, which is not the case.

Using Formula 75, $\int \tan^5 x\,dx = \frac{1}{5 - 1}\tan^{5-1} x - \int \tan^{5-2} x\,dx = \frac{1}{4}\tan^4 x - \int \tan^3 x\,dx$. Using Formula 69,

$\int \tan^3 x\,dx = \frac{1}{2}\tan^2 x + \ln|\cos x| + C$, so $\int \tan^5 x\,dx = \frac{1}{4}\tan^4 x - \frac{1}{2}\tan^2 x - \ln|\cos x| + C$.

33. Derive gives $I = \int 2^x\sqrt{4^x - 1}\,dx = \dfrac{2^{x-1}\sqrt{2^{2x} - 1}}{\ln 2} - \dfrac{\ln\left(\sqrt{2^{2x} - 1} + 2^x\right)}{2\ln 2}$ immediately. Neither Maple nor

Mathematica is able to evaluate I in its given form. However, if we instead write I as $\int 2^x\sqrt{(2^x)^2 - 1}\,dx$, both

systems give the same answer as Derive (after minor simplification). Our trick works because the CAS now

recognizes 2^x as a promising substitution.

35. Maple gives the antiderivative

$$F(x) = \int \frac{x^2 - 1}{x^4 + x^2 + 1}\, dx = -\tfrac{1}{2}\ln\left(x^2 + x + 1\right) + \tfrac{1}{2}\ln\left(x^2 - x + 1\right).$$

We can see that at 0, this antiderivative is 0. From the graphs, it appears
that F has a maximum at $x = -1$ and a minimum at $x = 1$ [since
$F'(x) = f(x)$ changes sign at these x-values], and that F has inflection
points at $x \approx -1.7$, $x = 0$, and $x \approx 1.7$ [since $f(x)$ has extrema at these
x-values].

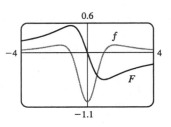

37. Since f is everywhere positive, we know that its antiderivative F is increasing. Maple gives $\int \sin^4 x\, \cos^6 x\, dx = $
$-\tfrac{1}{10}\sin^3 x\, \cos^7 x - \tfrac{3}{80}\sin x\, \cos^7 x + \tfrac{1}{160}\cos^5 x\, \sin x + \tfrac{1}{128}\cos^3 x\, \sin x + \tfrac{3}{256}\cos x\, \sin x + \tfrac{3}{256}x$ and this
expression is 0 at $x = 0$.

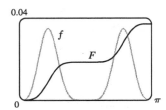

 Approximate Integration · · · · · · · · · · ·

5.9

1. (a) $\Delta x = (b - a)/n = (4 - 0)/2 = 2$

$$L_2 = \sum_{i=1}^{2} f(x_{i-1})\, \Delta x = f(x_0) \cdot 2 + f(x_1) \cdot 2 = 2\left[f(0) + f(2)\right] = 2(0.5 + 2.5) = 6$$

$$R_2 = \sum_{i=1}^{2} f(x_i)\, \Delta x = f(x_1) \cdot 2 + f(x_2) \cdot 2 = 2\left[f(2) + f(4)\right] = 2(2.5 + 3.5) = 12$$

$$M_2 = \sum_{i=1}^{2} f(\overline{x}_i)\Delta x = f(\overline{x}_1) \cdot 2 + f(\overline{x}_2) \cdot 2 = 2\left[f(1) + f(3)\right] \approx 2(1.6 + 3.2) = 9.6$$

(b)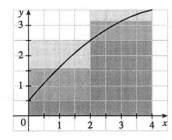

L_2 is an underestimate, since the area under the small rectangles is
less than the area under the curve, and R_2 is an overestimate, since
the area under the large rectangles is greater than the area under the
curve. It appears that M_2 is an overestimate, though it is fairly close
to I. See the solution to Exercise 33 for a proof of the fact that if f is
concave down on $[a, b]$, then the Midpoint Rule is an overestimate of
$\int_a^b f(x)\, dx$.

(c) $T_2 = \left(\tfrac{1}{2}\Delta x\right)[f(x_0) + 2f(x_1) + f(x_2)] = \tfrac{2}{2}[f(0) + 2f(2) + f(4)] = 0.5 + 2(2.5) + 3.5 = 9.$
This approximation is an underestimate, since the graph is concave down. Thus, $T_2 = 9 < I$. See the solution to
Exercise 33 for a general proof of this conclusion.

(d) For any n, we will have $L_n < T_n < I < M_n < R_n$.

3. $f(x) = \cos(x^2)$, $\Delta x = \frac{1-0}{4} = \frac{1}{4}$

(a) $T_4 = \frac{1}{4 \cdot 2}\left[f(0) + 2f\left(\frac{1}{4}\right) + 2f\left(\frac{2}{4}\right) + 2f\left(\frac{3}{4}\right) + f(1)\right] \approx 0.895759$

(b) $M_4 = \frac{1}{4}\left[f\left(\frac{1}{8}\right) + f\left(\frac{3}{8}\right) + f\left(\frac{5}{8}\right) + f\left(\frac{7}{8}\right)\right] \approx 0.908907$

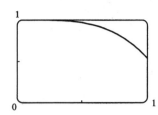

The graph shows that f is concave down on $[0, 1]$. So T_4 is an underestimate and M_4 is an overestimate. We can conclude that $0.895759 < \int_0^1 \cos(x^2)\,dx < 0.908907$.

5. $f(x) = x^2 \sin x$, $\Delta x = \dfrac{b-a}{n} = \dfrac{\pi - 0}{8} = \dfrac{\pi}{8}$

(a) $M_8 = \frac{\pi}{8}\left[f\left(\frac{\pi}{16}\right) + f\left(\frac{3\pi}{16}\right) + f\left(\frac{5\pi}{16}\right) + \cdots + f\left(\frac{15\pi}{16}\right)\right] \approx 5.932957$

(b) $S_8 = \frac{\pi}{8 \cdot 3}\left[f(0) + 4f\left(\frac{\pi}{8}\right) + 2f\left(\frac{2\pi}{8}\right) + 4f\left(\frac{3\pi}{8}\right) + 2f\left(\frac{4\pi}{8}\right) + 4f\left(\frac{5\pi}{8}\right) + 2f\left(\frac{6\pi}{8}\right) + 4f\left(\frac{7\pi}{8}\right) + f(\pi)\right]$
≈ 5.869247

Actual: $\int_0^\pi x^2 \sin x\,dx \overset{84}{=} \left[-x^2 \cos x\right]_0^\pi + 2\int_0^\pi x \cos x\,dx \overset{83}{=} \left[-\pi^2(-1) - 0\right] + 2[\cos x + x \sin x]_0^\pi$
$= \pi^2 + 2[(-1 + 0) - (1 + 0)] = \pi^2 - 4 \approx 5.869604$

Errors: $E_M = \text{actual} - M_8 = \int_0^\pi x^2 \sin x\,dx - M_8 \approx -0.063353$

$E_S = \text{actual} - S_8 = \int_0^\pi x^2 \sin x\,dx - S_8 \approx 0.000357$

7. $f(x) = e^{-x^2}$, $\Delta x = \dfrac{1-0}{10} = \dfrac{1}{10}$

(a) $T_{10} = \frac{1}{10 \cdot 2}\left[f(0) + 2f(0.1) + 2f(0.2) + \cdots + 2f(0.8) + 2f(0.9) + f(1)\right] \approx 0.746211$

(b) $M_{10} = \frac{1}{10}\left[f(0.05) + f(0.15) + f(0.25) + \cdots + f(0.75) + f(0.85) + f(0.95)\right] \approx 0.747131$

(c) $S_{10} = \frac{1}{10 \cdot 3}\left[f(0) + 4f(0.1) + 2f(0.2) + 4f(0.3) + 2f(0.4) + 4f(0.5)\right.$
$\left. + 2f(0.6) + 4f(0.7) + 2f(0.8) + 4f(0.9) + f(1)\right] \approx 0.746825$

9. $f(x) = e^{1/x}$, $\Delta x = \dfrac{2-1}{4} = \dfrac{1}{4}$

(a) $T_4 = \frac{1}{4 \cdot 2}\left[f(1) + 2f(1.25) + 2f(1.5) + 2f(1.75) + f(2)\right] \approx 2.031893$

(b) $M_4 = \frac{1}{4}\left[f(1.125) + f(1.375) + f(1.625) + f(1.875)\right] \approx 2.014207$

(c) $S_4 = \frac{1}{4 \cdot 3}\left[f(1) + 4f(1.25) + 2f(1.5) + 4f(1.75) + f(2)\right] \approx 2.020651$

11. $f(t) = \sin\left(e^{t/2}\right)$, $\Delta t = \dfrac{1/2 - 0}{8} = \dfrac{1}{16}$

(a) $T_8 = \frac{1}{16 \cdot 2}\left[f(0) + 2f\left(\frac{1}{16}\right) + 2f\left(\frac{2}{16}\right) + \cdots + 2f\left(\frac{7}{16}\right) + f\left(\frac{1}{2}\right)\right] \approx 0.451948$

(b) $M_8 = \frac{1}{16}\left[f\left(\frac{1}{32}\right) + f\left(\frac{3}{32}\right) + f\left(\frac{5}{32}\right) + \cdots + f\left(\frac{13}{32}\right) + f\left(\frac{15}{32}\right)\right] \approx 0.451991$

(c) $S_8 = \frac{1}{16 \cdot 3}\left[f(0) + 4f\left(\frac{1}{16}\right) + 2f\left(\frac{2}{16}\right) + \cdots + 4f\left(\frac{7}{16}\right) + f\left(\frac{1}{2}\right)\right] \approx 0.451976$

13. $f(y) = \dfrac{1}{1+y^5}$, $\Delta y = \dfrac{3-0}{6} = \dfrac{1}{2}$

(a) $T_6 = \frac{1}{2 \cdot 2}\left[f(0) + 2f\left(\frac{1}{2}\right) + 2f\left(\frac{2}{2}\right) + 2f\left(\frac{3}{2}\right) + 2f\left(\frac{4}{2}\right) + 2f\left(\frac{5}{2}\right) + f(3)\right] \approx 1.064275$

(b) $M_6 = \frac{1}{2}\left[f\left(\frac{1}{4}\right) + f\left(\frac{3}{4}\right) + f\left(\frac{5}{4}\right) + f\left(\frac{7}{4}\right) + f\left(\frac{9}{4}\right) + f\left(\frac{11}{4}\right)\right] \approx 1.067416$

(c) $S_6 = \frac{1}{2 \cdot 3}\left[f(0) + 4f\left(\frac{1}{2}\right) + 2f\left(\frac{2}{2}\right) + 4f\left(\frac{3}{2}\right) + 2f\left(\frac{4}{2}\right) + 4f\left(\frac{5}{2}\right) + f(3)\right] \approx 1.074915$

15. $f(x) = e^{-x^2}$, $\Delta x = \dfrac{2-0}{10} = \dfrac{1}{5}$

(a) $T_{10} = \frac{1}{5 \cdot 2}\{f(0) + 2[f(0.2) + f(0.4) + \cdots + f(1.8)] + f(2)\} \approx 0.881839$

$M_{10} = \frac{1}{5}[f(0.1) + f(0.3) + f(0.5) + \cdots + f(1.7) + f(1.9)] \approx 0.882202$

(b) $f(x) = e^{-x^2}$, $f'(x) = -2xe^{-x^2}$, $f''(x) = (4x^2 - 2)e^{-x^2}$, $f'''(x) = 4x(3 - 2x^2)e^{-x^2}$. $f'''(x) = 0 \Leftrightarrow$

$x = 0$ or $x = \pm\sqrt{\frac{3}{2}}$. So to find the maximum value of $|f''(x)|$ on $[0, 2]$, we need only consider its values at

$x = 0$, $x = 2$, and $x = \sqrt{\frac{3}{2}}$. $|f''(0)| = 2$, $|f''(2)| \approx 0.2564$ and $\left|f''\left(\sqrt{\frac{3}{2}}\right)\right| \approx 0.8925$. Thus, taking $K = 2$,

$a = 0$, $b = 2$, and $n = 10$ in Theorem 3, we get $|E_T| \leq 2 \cdot 2^3/(12 \cdot 10^2) = \frac{1}{75} = 0.01\overline{3}$, and

$|E_M| \leq |E_T|/2 \leq 0.00\overline{6}$.

(c) Take $K = 2$ [as in part (b)] in Theorem 3. $|E_T| \leq \dfrac{K(b-a)^3}{12n^2} \leq 10^{-5} \Leftrightarrow \dfrac{2(2-0)^3}{12n^2} \leq 10^{-5} \Leftrightarrow$

$\frac{3}{4}n^2 \geq 10^5 \Leftrightarrow n \geq 365.1\ldots \Leftrightarrow n \geq 366$. Take $n = 366$ for T_n. For E_M, again take $K = 2$ in

Theorem 3 to get $|E_M| \leq 10^{-5} \Leftrightarrow \frac{3}{2}n^2 \geq 10^5 \Leftrightarrow n \geq 258.2 \Rightarrow n \geq 259$. Take $n = 259$ for M_n.

17. (a) $T_{10} = \frac{1}{10 \cdot 2}\{f(0) + 2[f(0.1) + f(0.2) + \cdots + f(0.9)] + f(1)\} \approx 1.71971349$

$S_{10} = \frac{1}{10 \cdot 3}[f(0) + 4f(0.1) + 2f(0.2) + 4f(0.3) + \cdots + 4f(0.9) + f(1)] \approx 1.71828278$

Since $I = \int_0^1 e^x \, dx = [e^x]_0^1 = e - 1 \approx 1.71828183$, $E_T = I - T_{10} \approx -0.00143166$ and

$E_S = I - S_{10} \approx -0.00000095$.

(b) $f(x) = e^x \Rightarrow f''(x) = e^x \leq e$ for $0 \leq x \leq 1$. Taking $K = e$, $a = 0$, $b = 1$, and $n = 10$ in Theorem 3, we

get $|E_T| \leq e(1)^3/(12 \cdot 10^2) \approx 0.002265 > 0.00143166$ [actual $|E_T|$ from (a)]. $f^{(4)}(x) = e^x < e$ for

$0 \leq x \leq 1$. Using Theorem 4, we have $|E_S| \leq e(1)^5/(180 \cdot 10^4) \approx 0.0000015 > 0.00000095$ [actual $|E_S|$

from (a)]. We see that the actual errors are about two-thirds the size of the error estimates.

(c) From part (b), we take $K = e$ to get $|E_T| \leq \dfrac{K(b-a)^3}{12n^2} \leq 0.00001 \Rightarrow n^2 \geq \dfrac{e(1^3)}{12(0.00001)} \Rightarrow$

$n \geq 150.5$. Take $n = 151$ for T_n. Now $|E_M| \leq \dfrac{K(b-a)^3}{24n^2} \leq 0.00001 \Rightarrow n \geq 106.4$. Take $n = 107$ for

M_n. Finally, $|E_S| \leq \dfrac{K(b-a)^5}{180n^4} \leq 0.00001 \Rightarrow n^4 \geq \dfrac{e(1^5)}{180(0.00001)} \Rightarrow n \geq 6.23$. Take $n = 8$ for S_n

(since n has to be even for Simpson's Rule).

19. (a) Using a CAS, we differentiate $f(x) = e^{\cos x}$ twice, and find that

$f''(x) = e^{\cos x}(\sin^2 x - \cos x)$. From the graph, we see that the

maximum value of $|f''(x)|$ occurs at the endpoints of the

interval $[0, 2\pi]$. Since $f''(0) = -e$, we can use $K = e$ or $K = 2.8$.

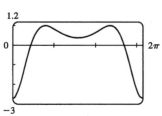

(b) A CAS gives $M_{10} \approx 7.954926518$. (In Maple, use student[middlesum].)

(c) Using Theorem 3 for the Midpoint Rule, with $K = e$, we get $|E_M| \leq \dfrac{e(2\pi - 0)^3}{24 \cdot 10^2} \approx 0.280945995$. With

$K = 2.8$, we get $|E_M| \leq \dfrac{2.8(2\pi - 0)^3}{24 \cdot 10^2} = 0.289391916$.

(d) A CAS gives $I \approx 7.954926521$.

(e) The actual error is only about 3×10^{-9}, much less than the estimate in part (c).

(f) We use the CAS to differentiate twice more, and then graph

$$f^{(4)}(x) = e^{\cos x}\left(\sin^4 x - 6\sin^2 x \cos x + 3 - 7\sin^2 x + \cos x\right).$$

From the graph, we see that the maximum value of $\left|f^{(4)}(x)\right|$ occurs

at the endpoints of the interval $[0, 2\pi]$. Since $f^{(4)}(0) = 4e$, we can use

$K = 4e$ or $K = 10.9$.

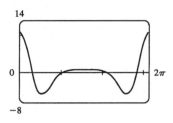

(g) A CAS gives $S_{10} \approx 7.953789422$. (In Maple, use `student[simpson]`.)

(h) Using Theorem 4 with $K = 4e$, we get $|E_S| \leq \dfrac{4e(2\pi - 0)^5}{180 \cdot 10^4} \approx 0.059153618$. With $K = 10.9$, we get

$|E_S| \leq \dfrac{10.9(2\pi - 0)^5}{180 \cdot 10^4} = 0.059299814$.

(i) The actual error is about $7.954926521 - 7.953789422 \approx 0.00114$. This is quite a bit smaller than the estimate in part (h), though the difference is not nearly as great as it was in the case of the Midpoint Rule.

(j) To ensure that $|E_S| \leq 0.0001$, we use Theorem 4: $|E_S| \leq \dfrac{4e(2\pi)^5}{180 \cdot n^4} \leq 0.0001 \;\Rightarrow\; \dfrac{4e(2\pi)^5}{180 \cdot 0.0001} \leq n^4 \;\Rightarrow$

$n^4 \geq 5{,}915{,}362 \;\Leftrightarrow\; n \geq 49.3$. So we must take $n \geq 50$ to ensure that $|I - S_n| \leq 0.0001$. ($K = 10.9$ leads to the same value of n.)

21. $I = \int_0^1 x^3\,dx = \left[\frac{1}{4}x^4\right]_0^1 = 0.25.\; f(x) = x^3$.

$n = 4$: $\quad L_4 = \frac{1}{4}\left[0^3 + \left(\frac{1}{4}\right)^3 + \left(\frac{2}{4}\right)^3 + \left(\frac{3}{4}\right)^3\right] = 0.140625$

$\qquad R_4 = \frac{1}{4}\left[\left(\frac{1}{4}\right)^3 + \left(\frac{2}{4}\right)^3 + \left(\frac{3}{4}\right)^3 + 1^3\right] = 0.390625$

$\qquad T_4 = \frac{1}{4\cdot 2}\left[0^3 + 2\left(\frac{1}{4}\right)^3 + 2\left(\frac{2}{4}\right)^3 + 2\left(\frac{3}{4}\right)^3 + 1^3\right] = 0.265625,$

$\qquad M_4 = \frac{1}{4}\left[\left(\frac{1}{8}\right)^3 + \left(\frac{3}{8}\right)^3 + \left(\frac{5}{8}\right)^3 + \left(\frac{7}{8}\right)^3\right] = 0.2421875,$

$\qquad E_L = I - L_4 = \frac{1}{4} - 0.140625 = 0.109375,\; E_R = \frac{1}{4} - 0.390625 = -0.140625,$

$\qquad E_T = \frac{1}{4} - 0.265625 = -0.015625,\; E_M = \frac{1}{4} - 0.2421875 = 0.0078125$

$n = 8$: $\quad L_8 = \frac{1}{8}\left[f(0) + f\left(\frac{1}{8}\right) + f\left(\frac{2}{8}\right) + \cdots + f\left(\frac{7}{8}\right)\right] \approx 0.191406$

$\qquad R_8 = \frac{1}{8}\left[f\left(\frac{1}{8}\right) + f\left(\frac{2}{8}\right) + \cdots + f\left(\frac{7}{8}\right) + f(1)\right] \approx 0.316406$

$\qquad T_8 = \frac{1}{8\cdot 2}\left\{f(0) + 2\left[f\left(\frac{1}{8}\right) + f\left(\frac{2}{8}\right) + \cdots + f\left(\frac{7}{8}\right)\right] + f(1)\right\} \approx 0.253906$

$\qquad M_8 = \frac{1}{8}\left[f\left(\frac{1}{16}\right) + f\left(\frac{3}{16}\right) + \cdots + f\left(\frac{13}{16}\right) + f\left(\frac{15}{16}\right)\right] = 0.248047$

$\qquad E_L \approx \frac{1}{4} - 0.191406 \approx 0.058594,\; E_R \approx \frac{1}{4} - 0.316406 \approx -0.066406,$

$\qquad E_T \approx \frac{1}{4} - 0.253906 \approx -0.003906,\; E_M \approx \frac{1}{4} - 0.248047 \approx 0.001953.$

$n = 16$: $L_{16} = \frac{1}{16}\left[f(0) + f\left(\frac{1}{16}\right) + f\left(\frac{2}{16}\right) + \cdots + f\left(\frac{15}{16}\right)\right] \approx 0.219727$

$R_{16} = \frac{1}{16}\left[f\left(\frac{1}{16}\right) + f\left(\frac{2}{16}\right) + \cdots + f\left(\frac{15}{16}\right) + f(1)\right] \approx 0.282227$

$T_{16} = \frac{1}{16 \cdot 2}\left\{f(0) + 2\left[f\left(\frac{1}{16}\right) + f\left(\frac{2}{16}\right) + \cdots + f\left(\frac{15}{16}\right)\right] + f(1)\right\} \approx 0.250977$

$M_{16} = \frac{1}{16}\left[f\left(\frac{1}{32}\right) + f\left(\frac{3}{32}\right) + \cdots + f\left(\frac{31}{32}\right)\right] \approx 0.249512$

$E_L \approx \frac{1}{4} - 0.219727 \approx 0.030273$, $E_R \approx \frac{1}{4} - 0.282227 \approx -0.032227$,

$E_T \approx \frac{1}{4} - 0.250977 \approx -0.000977$, $E_M \approx \frac{1}{4} - 0.249512 \approx 0.000488$.

n	L_n	R_n	T_n	M_n
4	0.140625	0.390625	0.265625	0.242188
8	0.191406	0.316406	0.253906	0.248047
16	0.219727	0.282227	0.250977	0.249512

n	E_L	E_R	E_T	E_M
4	0.109375	−0.140625	−0.015625	0.007813
8	0.058594	−0.066406	−0.003906	0.001953
16	0.030273	−0.032227	−0.000977	0.000488

Observations:

1. E_L and E_R are always opposite in sign, as are E_T and E_M.

2. As n is doubled, E_L and E_R are decreased by about a factor of 2, and E_T and E_M are decreased by a factor of about 4.

3. The Midpoint approximation is about twice as accurate as the Trapezoidal approximation.

4. All the approximations become more accurate as the value of n increases.

5. The Midpoint and Trapezoidal approximations are much more accurate than the endpoint approximations.

23. $\Delta x = (4 - 0)/4 = 1$

(a) $T_4 = \frac{1}{2}[f(0) + 2f(1) + 2f(2) + 2f(3) + f(4)] \approx \frac{1}{2}[0 + 2(3) + 2(5) + 2(3) + 1] = 11.5$

(b) $M_4 = 1 \cdot [f(0.5) + f(1.5) + f(2.5) + f(3.5)] \approx 1 + 4.5 + 4.5 + 2 = 12$

(c) $S_4 = \frac{1}{3}[f(0) + 4f(1) + 2f(2) + 4f(3) + f(4)] \approx \frac{1}{3}[0 + 4(3) + 2(5) + 4(3) + 1] = 11.\overline{6}$

25. By the Total Change Theorem, the increase in velocity is equal to $\int_0^6 a(t)\, dt$. We use Simpson's Rule with $n = 6$ and $\Delta t = (6 - 0)/6 = 1$ to estimate this integral:

$\int_0^6 a(t)\, dt \approx S_6 = \frac{1}{3}[a(0) + 4a(1) + 2a(2) + 4a(3) + 2a(4) + 4a(5) + a(6)]$

$\approx \frac{1}{3}[0 + 4(0.5) + 2(4.1) + 4(9.8) + 2(12.9) + 4(9.5) + 0] = \frac{1}{3}(113.2) = 37.7\overline{3}$ ft/s

27. By the Total Change Theorem, the energy used is equal to $\int_0^6 P(t)\, dt$. We use Simpson's Rule with $n = 12$ and $\Delta t = (6 - 0)/12 = \frac{1}{2}$ to estimate this integral:

$\int_0^6 P(t)\, dt \approx S_{12} = \frac{1/2}{3}[P(0) + 4P(0.5) + 2P(1) + 4P(1.5) + 2P(2) + 4P(2.5)$

$+ 2P(3) + 4P(3.5) + 2P(4) + 4P(4.5) + 2P(5) + 4P(5.5) + P(6)]$

$= \frac{1}{6}[1814 + 4(1735) + 2(1686) + 4(1646) + 2(1637) + 4(1609) + 2(1604)$

$+ 4(1611) + 2(1621) + 4(1666) + 2(1745) + 4(1886) + 2052]$

$= \frac{1}{6}(61{,}064) = 10{,}177.\overline{3}$ megawatt-hours.

29. (a) We are given the function values at the endpoints of 8 intervals of length 0.4, so we'll use the Midpoint Rule with $n = 8/2 = 4$ and $\Delta x = (3.2 - 0)/4 = 0.8$.

$$\int_0^{3.2} f(x)\, dx \approx M_4 = 0.8[f(0.4) + f(1.2) + f(2.0) + f(2.8)]$$
$$= 0.8[6.5 + 6.4 + 7.6 + 8.8]$$
$$= 0.8(29.3) = 23.44$$

(b) $-4 \le f''(x) \le 1 \ \Rightarrow \ |f''(x)| \le 4$, so use $K = 4$, $a = 0$, $b = 3.2$, and $n = 4$ in Theorem 3. So

$$|E_M| \le \frac{4(3.2 - 0)^3}{24(4)^2} = \frac{128}{375} = 0.341\overline{3}.$$

31. $I(\theta) = \dfrac{N^2 \sin^2 k}{k^2}$, where $k = \dfrac{\pi N d \sin \theta}{\lambda}$, $N = 10{,}000$, $d = 10^{-4}$, and $\lambda = 632.8 \times 10^{-9}$. So

$I(\theta) = \dfrac{(10^4)^2 \sin^2 k}{k^2}$, where $k = \dfrac{\pi (10^4)(10^{-4}) \sin \theta}{632.8 \times 10^{-9}}$. Now $n = 10$ and $\Delta \theta = \dfrac{10^{-6} - (-10^{-6})}{10} = 2 \times 10^{-7}$,

so $M_{10} = 2 \times 10^{-7}[I(-0.0000009) + I(-0.0000007) + \cdots + I(0.0000009)] \approx 59.4$.

33. Since the Trapezoidal and Midpoint approximations on the interval $[a, b]$ are the sums of the Trapezoidal and Midpoint approximations on the subintervals $[x_{i-1}, x_i]$, $i = 1, 2, \ldots, n$, we can focus our attention on one such interval. The condition $f''(x) < 0$ for $a \le x \le b$ means that the graph of f is concave down as in Figure 5. In that figure, T_n is the area of the trapezoid $AQRD$, $\int_a^b f(x)\, dx$ is the area of the region $AQPRD$, and M_n is the area of the trapezoid $ABCD$, so $T_n < \int_a^b f(x)\, dx < M_n$. In general, the condition $f'' < 0$ implies that the graph of f on $[a, b]$ lies above the chord joining the points $(a, f(a))$ and $(b, f(b))$. Thus, $\int_a^b f(x)\, dx > T_n$. Since M_n is the area under a tangent to the graph, and since $f'' < 0$ implies that the tangent lies above the graph, we also have $M_n > \int_a^b f(x)\, dx$. Thus, $T_n < \int_a^b f(x)\, dx < M_n$.

35. $T_n = \frac{1}{2} \Delta x \left[f(x_0) + 2f(x_1) + \cdots + 2f(x_{n-1}) + f(x_n) \right]$ and

$M_n = \Delta x \left[f(\overline{x}_1) + f(\overline{x}_2) + \cdots + f(\overline{x}_{n-1}) + f(\overline{x}_n) \right]$, where $\overline{x}_i = \frac{1}{2}(x_{i-1} + x_i)$. Now

$$T_{2n} = \frac{1}{2} \left(\tfrac{1}{2} \Delta x \right) [f(x_0) + 2f(\overline{x}_1) + 2f(x_1) + 2f(\overline{x}_2) + 2f(x_2) + \cdots$$
$$+ 2f(\overline{x}_{n-1}) + 2f(x_{n-1}) + 2f(\overline{x}_n) + f(x_n)]$$

so $\quad \frac{1}{2}(T_n + M_n) = \frac{1}{2} T_n + \frac{1}{2} M_n$

$$= \tfrac{1}{4} \Delta x \left[f(x_0) + 2f(x_1) + \cdots + 2f(x_{n-1}) + f(x_n) \right]$$
$$+ \tfrac{1}{4} \Delta x \left[2f(\overline{x}_1) + 2f(\overline{x}_2) + \cdots + 2f(\overline{x}_{n-1}) + 2f(\overline{x}_n) \right]$$

$$= T_{2n}$$

5.10 Improper Integrals · · · · · · · · · · · · · · · · ·

1. (a) Since $\int_1^\infty x^4 e^{-x^4}\, dx$ has an infinite interval of integration, it is an improper integral of Type I.

(b) Since $y = \sec x$ has an infinite discontinuity at $x = \frac{\pi}{2}$, $\int_0^{\pi/2} \sec x\, dx$ is a Type II improper integral.

(c) Since $y = \dfrac{x}{(x-2)(x-3)}$ has an infinite discontinuity at $x = 2$, $\int_0^2 \dfrac{x}{x^2 - 5x + 6}\, dx$ is a Type II improper integral.

(d) Since $\displaystyle\int_{-\infty}^0 \dfrac{1}{x^2 + 5}\, dx$ has an infinite interval of integration, it is an improper integral of Type I.

3. The area under the graph of $y = 1/x^3 = x^{-3}$ between $x = 1$ and $x = t$ is

$A(t) = \int_1^t x^{-3}\, dx = \left[-\frac{1}{2}x^{-2}\right]_1^t = -\frac{1}{2}t^{-2} - \left(-\frac{1}{2}\right) = \frac{1}{2} - 1/(2t^2)$. So the area for $1 \le x \le 10$ is $A(10) = 0.5 - 0.005 = 0.495$, the area for $1 \le x \le 100$ is $A(100) = 0.5 - 0.00005 = 0.49995$, and the area for $1 \le x \le 1000$ is $A(1000) = 0.5 - 0.0000005 = 0.4999995$. The total area under the curve for $x \ge 1$ is $\displaystyle\lim_{t\to\infty} A(t) = \lim_{t\to\infty} \left[\frac{1}{2} - 1/(2t^2)\right] = \frac{1}{2}$.

5. $I = \displaystyle\int_1^\infty \dfrac{1}{(3x+1)^2}\, dx = \lim_{t\to\infty} \int_1^t \dfrac{1}{(3x+1)^2}\, dx$. Now

$\displaystyle\int \dfrac{1}{(3x+1)^2}\, dx = \frac{1}{3}\int \dfrac{1}{u^2}\, du \quad [u = 3x+1,\ du = 3\, dx]$

$= -\dfrac{1}{3u} + C = -\dfrac{1}{3(3x+1)} + C,$

so $I = \displaystyle\lim_{t\to\infty} \left[-\dfrac{1}{3(3x+1)}\right]_1^t = \lim_{t\to\infty}\left[-\dfrac{1}{3(3t+1)} + \dfrac{1}{12}\right] = 0 + \dfrac{1}{12} = \dfrac{1}{12}.$ Convergent

7. $\int_0^\infty e^{-x}\, dx = \displaystyle\lim_{t\to\infty}\int_0^t e^{-x}\, dx = \lim_{t\to\infty}\left[-e^{-x}\right]_0^t = \lim_{t\to\infty}\left(-e^{-t} + 1\right) = 1.$ Convergent

9. $\displaystyle\int_{-\infty}^{-1} \dfrac{1}{\sqrt{2-w}}\, dw = \lim_{t\to-\infty}\int_t^{-1} \dfrac{1}{\sqrt{2-w}}\, dw = \lim_{t\to-\infty}\left[-2\sqrt{2-w}\right]_t^{-1} \quad [u = 2-w,\ du = -dw]$

$= \displaystyle\lim_{t\to-\infty}\left[-2\sqrt{3} + 2\sqrt{2-t}\right] = \infty.$ Divergent

11. $I = \int_{-\infty}^\infty x^3\, dx = I_1 + I_2 = \int_{-\infty}^0 x^3\, dx + \int_0^\infty x^3\, dx$, but $I_1 = \displaystyle\lim_{t\to-\infty}\left[\frac{1}{4}x^4\right]_t^0 = \lim_{t\to-\infty}\left(-\frac{1}{4}t^4\right) = -\infty.$ Since I_1 is divergent, I is divergent, and there is no need to evaluate I_2. Divergent

13. $\int_{-\infty}^\infty x e^{-x^2}\, dx = \int_{-\infty}^0 x e^{-x^2}\, dx + \int_0^\infty x e^{-x^2}\, dx.$

$\int_{-\infty}^0 x e^{-x^2}\, dx = \displaystyle\lim_{t\to-\infty}\left(-\frac{1}{2}\right)\left[e^{-x^2}\right]_t^0 = \lim_{t\to-\infty}\left(-\frac{1}{2}\right)\left(1 - e^{-t^2}\right) = -\frac{1}{2}\cdot 1 = -\frac{1}{2}$, and

$\int_0^\infty x e^{-x^2}\, dx = \displaystyle\lim_{t\to\infty}\left(-\frac{1}{2}\right)\left[e^{-x^2}\right]_0^t = \lim_{t\to\infty}\left(-\frac{1}{2}\right)\left(e^{-t^2} - 1\right) = -\frac{1}{2}\cdot(-1) = \frac{1}{2}.$

Therefore, $\int_{-\infty}^\infty x e^{-x^2}\, dx = -\frac{1}{2} + \frac{1}{2} = 0.$ Convergent

15. $\int_0^\infty \cos x \, dx = \lim\limits_{t\to\infty} [\sin x]_0^t = \lim\limits_{t\to\infty} \sin t$, which does not exist. Divergent

17. $\int_{-\infty}^1 xe^{2x} \, dx = \lim\limits_{t\to-\infty} \int_t^1 xe^{2x} \, dx = \lim\limits_{t\to-\infty} \left[\frac{1}{2}xe^{2x} - \frac{1}{4}e^{2x}\right]_t^1$ (by parts with $u = x$ and $dv = e^{2x}\,dx$)

$$= \lim\limits_{t\to-\infty} \left[\frac{1}{2}e^2 - \frac{1}{4}e^2 - \frac{1}{2}te^{2t} + \frac{1}{4}e^{2t}\right] = \frac{1}{4}e^2 - 0 + 0 = \frac{1}{4}e^2$$

since $\lim\limits_{t\to-\infty} te^{2t} = \lim\limits_{t\to-\infty} \dfrac{t}{e^{-2t}} \overset{\text{H}}{=} \lim\limits_{t\to-\infty} \dfrac{1}{-2e^{-2t}} = \lim\limits_{t\to-\infty} -\frac{1}{2}e^{2t} = 0$. Convergent

19. $\displaystyle\int_1^\infty \frac{\ln x}{x} \, dx = \lim\limits_{t\to\infty} \left[\frac{(\ln x)^2}{2}\right]_1^t$ (by substitution with $u = \ln x$, $du = dx/x$) $= \lim\limits_{t\to\infty} \dfrac{(\ln t)^2}{2} = \infty$. Divergent

21. Integrate by parts with $u = \ln x$, $dv = dx/x^2 \;\Rightarrow\; du = dx/x$, $v = -1/x$.

$$\int_1^\infty \frac{\ln x}{x^2} \, dx = \lim\limits_{t\to\infty} \int_1^t \frac{\ln x}{x^2} \, dx = \lim\limits_{t\to\infty} \left[-\frac{\ln x}{x} - \frac{1}{x}\right]_1^t = \lim\limits_{t\to\infty} \left(-\frac{\ln t}{t} - \frac{1}{t} + 0 + 1\right)$$
$$= -0 - 0 + 0 + 1 = 1$$

since $\lim\limits_{t\to\infty} \dfrac{\ln t}{t} \overset{\text{H}}{=} \lim\limits_{t\to\infty} \dfrac{1/t}{1} = 0$. Convergent

23. There is an infinite discontinuity at the left endpoint of $[0, 3]$.

$$\int_0^3 \frac{dx}{\sqrt{x}} = \lim\limits_{t\to0+} \int_t^3 \frac{dx}{\sqrt{x}} = \lim\limits_{t\to0+} [2\sqrt{x}]_t^3 = \lim\limits_{t\to0+} \left(2\sqrt{3} - 2\sqrt{t}\right) = 2\sqrt{3}.$$ Convergent

25. There is an infinite discontinuity at the right endpoint of $[-1, 0]$.

$$\int_{-1}^0 \frac{dx}{x^2} = \lim\limits_{t\to0-} \int_{-1}^t \frac{dx}{x^2} = \lim\limits_{t\to0-} \left[\frac{-1}{x}\right]_{-1}^t = \lim\limits_{t\to0-} \left[-\frac{1}{t} + \frac{1}{-1}\right] = \infty.$$ Divergent

27. $\int_0^{\pi/4} \csc^2 t \, dt = \lim\limits_{s\to0+} \int_s^{\pi/4} \csc^2 t \, dt = \lim\limits_{s\to0+} [-\cot t]_s^{\pi/4} = \lim\limits_{s\to0+} \left[-\cot\frac{\pi}{4} + \cot s\right] = \infty$. Divergent

29. $\displaystyle\int_{-2}^3 \frac{dx}{x^4} = \int_{-2}^0 \frac{dx}{x^4} + \int_0^3 \frac{dx}{x^4}$, but $\displaystyle\int_{-2}^0 \frac{dx}{x^4} = \lim\limits_{t\to0-} \left[-\frac{x^{-3}}{3}\right]_{-2}^t = \lim\limits_{t\to0-} \left[-\frac{1}{3t^3} - \frac{1}{24}\right] = \infty$. Divergent

31. $I = \int_0^2 z^2 \ln z \, dz = \lim\limits_{t\to0+} \int_t^2 z^2 \ln z \, dz \overset{101}{=} \lim\limits_{t\to0+} \left[\frac{z^3}{3^2}(3\ln z - 1)\right]_t^2$

$$= \lim\limits_{t\to0+} \left[\frac{8}{9}(3\ln 2 - 1) - \frac{1}{9}t^3(3\ln t - 1)\right] = \frac{8}{3}\ln 2 - \frac{8}{9} - \frac{1}{9}\lim\limits_{t\to0+} [t^3(3\ln t - 1)] = \frac{8}{3}\ln 2 - \frac{8}{9} - \frac{1}{9}L.$$

Now $L = \lim\limits_{t\to0+} [t^3(3\ln t - 1)] = \lim\limits_{t\to0+} \dfrac{3\ln t - 1}{t^{-3}} \overset{\text{H}}{=} \lim\limits_{t\to0+} \dfrac{3/t}{-3/t^4} = \lim\limits_{t\to0+} (-t^3) = 0$. Thus, $L = 0$ and

$I = \frac{8}{3}\ln 2 - \frac{8}{9}$. Convergent

33.

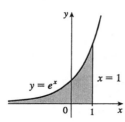

$$\text{Area} = \int_{-\infty}^{1} e^x \, dx = \lim_{t \to -\infty} [e^x]_t^1$$

$$= e - \lim_{t \to -\infty} e^t = e$$

35.

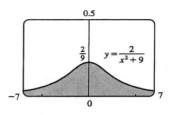

$$\text{Area} = \int_{-\infty}^{\infty} \frac{2}{x^2 + 9} \, dx = 2 \cdot 2 \int_{0}^{\infty} \frac{1}{x^2 + 9} \, dx$$

$$= 4 \lim_{t \to \infty} \int_{0}^{t} \frac{1}{x^2 + 9} \, dx = 4 \lim_{t \to \infty} \left[\frac{1}{3} \tan^{-1} \frac{x}{3} \right]_0^t$$

$$= \frac{4}{3} \lim_{t \to \infty} \left[\tan^{-1} \frac{t}{3} - 0 \right] = \frac{4}{3} \cdot \frac{\pi}{2} = \frac{2\pi}{3}$$

37.

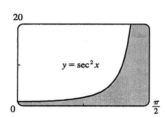

$$\text{Area} = \int_{0}^{\pi/2} \sec^2 x \, dx = \lim_{t \to (\pi/2)^-} \int_{0}^{t} \sec^2 x \, dx$$

$$= \lim_{t \to (\pi/2)^-} [\tan x]_0^t = \lim_{t \to (\pi/2)^-} (\tan t - 0) = \infty$$

Infinite area

39. (a)

t	$\int_1^t g(x) \, dx$
2	0.447453
5	0.577101
10	0.621306
100	0.668479
1000	0.672957
10,000	0.673407

$g(x) = \dfrac{\sin^2 x}{x^2}$. It appears that the integral is convergent.

(c)

Since $\int_1^\infty f(x) \, dx$ is finite and the area under $g(x)$ is less than the area under $f(x)$ on any interval $[1, t]$, $\int_1^\infty g(x) \, dx$ must be finite; that is, the integral is convergent.

(b) $-1 \le \sin x \le 1 \Rightarrow 0 \le \sin^2 x \le 1 \Rightarrow 0 \le \dfrac{\sin^2 x}{x^2} \le \dfrac{1}{x^2}$. Since $\displaystyle\int_1^\infty \frac{1}{x^2} \, dx$ is convergent (Equation 2 with $p = 2 > 1$), $\displaystyle\int_1^\infty \frac{\sin^2 x}{x^2} \, dx$ is convergent by the Comparison Theorem.

41. For $x \ge 1$, $\dfrac{\cos^2 x}{1 + x^2} \le \dfrac{1}{1 + x^2} < \dfrac{1}{x^2}$. $\displaystyle\int_1^\infty \frac{1}{x^2} \, dx$ is convergent by Equation 2 with $p = 2 > 1$, so $\displaystyle\int_1^\infty \frac{\cos^2 x}{1 + x^2} \, dx$ is convergent by the Comparison Theorem.

43. For $x \geq 1$, $x + e^{2x} > e^{2x} > 0 \Rightarrow \dfrac{1}{x + e^{2x}} \leq \dfrac{1}{e^{2x}} = e^{-2x}$ on $[1, \infty)$.

$$\int_1^\infty e^{-2x}\, dx = \lim_{t \to \infty}\left[-\frac{1}{2}e^{-2x}\right]_1^t = \lim_{t \to \infty}\left[-\frac{1}{2}e^{-2t} + \frac{1}{2}e^{-2}\right] = \frac{1}{2}e^{-2}. \text{ Therefore, } \int_1^\infty e^{-2x}\, dx \text{ is convergent,}$$

and by the Comparison Theorem, $\displaystyle\int_1^\infty \frac{dx}{x + e^{2x}}$ is also convergent.

45. $\dfrac{1}{x \sin x} \geq \dfrac{1}{x}$ on $\left(0, \frac{\pi}{2}\right]$ since $0 \leq \sin x \leq 1$. $\displaystyle\int_0^{\pi/2} \frac{dx}{x} = \lim_{t \to 0^+}\int_t^{\pi/2} \frac{dx}{x} = \lim_{t \to 0^+} \left[\ln x\right]_t^{\pi/2}.$

But $\ln t \to -\infty$ as $t \to 0^+$, so $\displaystyle\int_0^{\pi/2} \frac{dx}{x}$ is divergent, and by the Comparison Theorem, $\displaystyle\int_0^{\pi/2} \frac{dx}{x \sin x}$ is also

divergent.

47. $\displaystyle\int_0^\infty \frac{dx}{\sqrt{x}\,(1 + x)} = \int_0^1 \frac{dx}{\sqrt{x}\,(1 + x)} + \int_1^\infty \frac{dx}{\sqrt{x}\,(1 + x)} = \lim_{t \to 0^+}\int_t^1 \frac{dx}{\sqrt{x}\,(1 + x)} + \lim_{t \to \infty}\int_1^t \frac{dx}{\sqrt{x}\,(1 + x)}.$ Now

$$\int \frac{dx}{\sqrt{x}\,(1 + x)} = \int \frac{2u\, du}{u(1 + u^2)} \quad [u = \sqrt{x}, x = u^2, dx = 2u\, du]$$

$$= 2 \int \frac{du}{1 + u^2} = 2 \tan^{-1} u + C = 2 \tan^{-1} \sqrt{x} + C,$$

so $\displaystyle\int_0^\infty \frac{dx}{\sqrt{x}\,(1 + x)} = \lim_{t \to 0^+}\left[2\tan^{-1}\sqrt{x}\right]_t^1 + \lim_{t \to \infty}\left[2\tan^{-1}\sqrt{x}\right]_1^t$

$$= \lim_{t \to 0^+}\left[2\left(\tfrac{\pi}{4}\right) - 2\tan^{-1}\sqrt{t}\right] + \lim_{t \to \infty}\left[2\tan^{-1}\sqrt{t} - 2\left(\tfrac{\pi}{4}\right)\right] = \tfrac{\pi}{2} - 0 + 2\left(\tfrac{\pi}{2}\right) - \tfrac{\pi}{2} = \pi.$$

49. If $p = 1$, then $\displaystyle\int_0^1 \frac{dx}{x^p} = \lim_{t \to 0^+}\int_t^1 \frac{dx}{x} = \lim_{t \to 0^+}\left[\ln x\right]_t^1 = \infty.$ Divergent.

If $p \neq 1$, then $\displaystyle\int_0^1 \frac{dx}{x^p} = \lim_{t \to 0^+}\int_t^1 \frac{dx}{x^p}$ (note that the integral is not improper if $p < 0$)

$$= \lim_{t \to 0^+}\left[\frac{x^{-p+1}}{-p + 1}\right]_t^1 = \lim_{t \to 0^+}\frac{1}{1 - p}\left[1 - \frac{1}{t^{p-1}}\right]$$

If $p > 1$, then $p - 1 > 0$, so $\dfrac{1}{t^{p-1}} \to \infty$ as $t \to 0^+$, and the integral diverges.

If $p < 1$, then $p - 1 < 0$, so $\dfrac{1}{t^{p-1}} \to 0$ as $t \to 0^+$ and $\displaystyle\int_0^1 \frac{dx}{x^p} = \frac{1}{1 - p}\left[\lim_{t \to 0^+}\left(1 - t^{1-p}\right)\right] = \frac{1}{1 - p}.$

Thus, the integral converges if and only if $p < 1$, and in that case its value is $\dfrac{1}{1 - p}$.

51. (a) $I = \int_{-\infty}^\infty x\, dx = \int_{-\infty}^0 x\, dx + \int_0^\infty x\, dx$, and

$\int_0^\infty x\, dx = \lim_{t \to \infty}\int_0^t x\, dx = \lim_{t \to \infty}\left[\frac{1}{2}x^2\right]_0^t = \lim_{t \to \infty}\left[\frac{1}{2}t^2 - 0\right] = \infty$, so I is divergent.

(b) $\int_{-t}^t x\, dx = \left[\frac{1}{2}x^2\right]_{-t}^t = \frac{1}{2}t^2 - \frac{1}{2}t^2 = 0$, so $\lim_{t \to \infty}\int_{-t}^t x\, dx = 0$. Therefore, $\int_{-\infty}^\infty x\, dx \neq \lim_{t \to \infty}\int_{-t}^t x\, dx$.

53. (a) We would expect a small percentage of bulbs to burn out in the first few hundred hours, most of the bulbs to burn out after close to 700 hours, and a few overachievers to burn on and on.

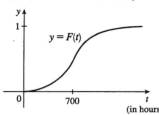

(b) $r(t) = F'(t)$ is the rate at which the fraction $F(t)$ of burnt-out bulbs increases as t increases. This could be interpreted as a fractional burnout rate.

(c) $\int_0^\infty r(t)\,dt = \lim_{x \to \infty} F(x) = 1$, since all of the bulbs will eventually burn out.

55. $I = \displaystyle\int_0^\infty te^{kt}\,dt = \lim_{s \to \infty} \left[\frac{1}{k^2}(kt - 1)e^{kt} \right]_0^s$ (Formula 96, or parts) $= \displaystyle\lim_{s \to \infty} \left[\left(\frac{1}{k}se^{ks} - \frac{1}{k^2}e^{ks} \right) - \left(-\frac{1}{k^2} \right) \right]$.

Since $k < 0$ the first two terms approach 0 (you can verify that the first term does so with l'Hospital's Rule), so the limit is equal to $1/k^2$. Thus, $M = -kI = -k\left(1/k^2\right) = -1/k = -1/(-0.000121) \approx 8264.5$ years.

57. $I = \displaystyle\int_a^\infty \frac{1}{x^2 + 1}\,dx = \lim_{t \to \infty} \int_a^t \frac{1}{x^2 + 1}\,dx = \lim_{t \to \infty} \left[\tan^{-1} x \right]_a^t = \lim_{t \to \infty} \left(\tan^{-1} t - \tan^{-1} a \right) = \frac{\pi}{2} - \tan^{-1} a$.

$I < 0.001 \quad \Rightarrow \quad \frac{\pi}{2} - \tan^{-1} a < 0.001 \quad \Rightarrow \quad \tan^{-1} a > \frac{\pi}{2} - 0.001 \quad \Rightarrow \quad a > \tan\left(\frac{\pi}{2} - 0.001 \right) \approx 1000$.

59. We use integration by parts: let $u = x$, $dv = xe^{-x^2}\,dx \quad \Rightarrow \quad du = dx$, $v = -\frac{1}{2}e^{-x^2}$. So

$$\int_0^\infty x^2 e^{-x^2}\,dx = \lim_{t \to \infty} \left[-\frac{1}{2}xe^{-x^2} \right]_0^t + \frac{1}{2}\int_0^\infty e^{-x^2}\,dx$$

$$= \lim_{t \to \infty} \left[-t \Big/ \left(2e^{t^2} \right) \right] + \frac{1}{2}\int_0^\infty e^{-x^2}\,dx = \frac{1}{2}\int_0^\infty e^{-x^2}\,dx$$

(The limit is 0 by l'Hospital's Rule.)

61. For the first part of the integral, let $x = 2\tan\theta \quad \Rightarrow \quad dx = 2\sec^2\theta\,d\theta$.

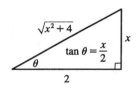

$$\int \frac{1}{\sqrt{x^2 + 4}}\,dx = \int \frac{2\sec^2\theta}{2\sec\theta}\,d\theta = \int \sec\theta\,d\theta = \ln|\sec\theta + \tan\theta|.$$ From the

figure, $\tan\theta = \dfrac{x}{2}$, and $\sec\theta = \dfrac{\sqrt{x^2 + 4}}{2}$. So

$$I = \int_0^\infty \left(\frac{1}{\sqrt{x^2 + 4}} - \frac{C}{x + 2} \right) dx = \lim_{t \to \infty} \left[\ln \left| \frac{\sqrt{x^2 + 4}}{2} + \frac{x}{2} \right| - C \ln|x + 2| \right]_0^t$$

$$= \lim_{t \to \infty} \left[\ln \frac{\sqrt{t^2 + 4} + t}{2} - C \ln(t + 2) - (\ln 1 - C \ln 2) \right]$$

$$= \lim_{t \to \infty} \left[\ln \left(\frac{\sqrt{t^2 + 4} + t}{2(t + 2)^C} \right) + \ln 2^C \right]$$

$$= \ln \left(\lim_{t \to \infty} \frac{t + \sqrt{t^2 + 4}}{(t + 2)^C} \right) + \ln 2^{C-1}$$

Now $L = \displaystyle\lim_{t \to \infty} \frac{t + \sqrt{t^2 + 4}}{(t + 2)^C} \overset{\text{H}}{=} \lim_{t \to \infty} \frac{1 + t/\sqrt{t^2 + 4}}{C(t + 2)^{C-1}} = \frac{2}{C \displaystyle\lim_{t \to \infty} (t + 2)^{C-1}}$.

If $C < 1$, $L = \infty$ and I diverges. If $C = 1$, $L = 2$ and I converges to $\ln 2 + \ln 2^0 = \ln 2$. If $C > 1$, $L = 0$ and I diverges to $-\infty$.

5 Review

— • CONCEPT CHECK • ——————

1. (a) $\sum_{i=1}^{n} f(x_i^*) \Delta x$ is an expression for a Riemann sum of a function f.

   x_i^* is a point in the ith subinterval $[x_{i-1}, x_i]$ and Δx is the length of the subintervals.

   (b) See Figure 1 in Section 5.2.

   (c) In Section 5.2, see Figure 3 and the paragraph above it.

2. (a) See Definition 5.2.2.

   (b) See Figure 2 in Section 5.2.

   (c) In Section 5.2, see Figure 4 and the paragraph above it.

3. (a) See the Evaluation Theorem at the beginning of Section 5.3.

   (b) See the Total Change Theorem after Example 6 in Section 5.3.

4. $\int_{t_1}^{t_2} r(t)\, dt$ represents the change in the amount of water in the reservoir between time t_1 and time t_2.

5. (a) $\int_{60}^{120} v(t)\, dt$ represents the change in position of the particle from $t = 60$ to $t = 120$ seconds.

   (b) $\int_{60}^{120} |v(t)|\, dt$ represents the total distance traveled by the particle from $t = 60$ to 120 seconds.

   (c) $\int_{60}^{120} a(t)\, dt$ represents the change in the velocity of the particle from $t = 60$ to $t = 120$ seconds.

6. (a) $\int f(x)\, dx$ is the family of functions $\{F \mid F' = f\}$. Any two such functions differ by a constant.

   (b) The connection is given by the Evaluation Theorem: $\int_a^b f(x)\, dx = \left[\int f(x)\, dx\right]_a^b$ if f is continuous.

7. See the Fundamental Theorem of Calculus after Example 5 in Section 5.4.

8. (a) See the Substitution Rule (5.5.4). This says that it is permissible to operate with the dx after an integral sign as if it were a differential.

   (b) See Formula 5.6.1 or 5.6.2. We try to choose $u = f(x)$ to be a function that becomes simpler when differentiated (or at least not more complicated) as long as $dv = g'(x)\, dx$ can be readily integrated to give v.

9. See the Midpoint Rule, the Trapezoidal Rule, and Simpson's Rule, as well as their associated error bounds, all in Section 5.9. We would expect the best estimate to be given by Simpson's Rule.

10. See Definitions 1(a), (b), and (c) in Section 5.10.

11. See Definitions 3(b), (a), and (c) in Section 5.10.

12. See the Comparison Theorem after Example 8 in Section 5.10.

13. The precise version of this statement is given by the Fundamental Theorem of Calculus. See the statement of this theorem and the paragraph that follows it in Section 5.4.

———————— ▲ TRUE–FALSE QUIZ ▲ ————————

1. True by Property 2 of the Integral in Section 5.2.

3. True by Property 3 of the Integral in Section 5.2.

5. False.   For example, let $f(x) = x^2$. Then $\int_0^1 \sqrt{x^2}\, dx = \int_0^1 x\, dx = \frac{1}{2}$, but $\sqrt{\int_0^1 x^2\, dx} = \sqrt{\frac{1}{3}} = \frac{1}{\sqrt{3}}$.

7. True by Comparison Property 7 of the Integral in Section 5.2.

9. True.   The integrand is an odd function that is continuous on $[-1, 1]$, so the result follows from Equation 5.5.6(b).

11. False. This is an improper integral, since the denominator vanishes at $x = 1$.

$$\int_0^4 \frac{x}{x^2 - 1}\, dx = \int_0^1 \frac{x}{x^2 - 1}\, dx + \int_1^4 \frac{x}{x^2 - 1}\, dx \text{ and}$$

$$\int_0^1 \frac{x}{x^2 - 1}\, dx = \lim_{t \to 1^-} \int_0^t \frac{x}{x^2 - 1}\, dx = \lim_{t \to 1^-} \left[\tfrac{1}{2}\ln|x^2 - 1|\right]_0^t = \lim_{t \to 1^-} \tfrac{1}{2}\ln|t^2 - 1| = \infty$$

So the integral diverges.

13. False. See the remarks and Figure 4 before Example 1 in Section 5.2, and notice that $y = x - x^3 < 0$ for $1 < x \le 2$.

15. False. For example, the function $y = |x|$ is continuous on $\mathbb{R}$, but has no derivative at $x = 0$.

17. False. See Exercise 51 in Section 5.10.

◆ **EXERCISES** ◆

---

1. (a)

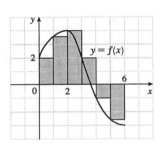

$$L_6 = \sum_{i=1}^6 f(x_{i-1})\,\Delta x \quad [\Delta x = \tfrac{6-0}{6} = 1]$$
$$= f(x_0) \cdot 1 + f(x_1) \cdot 1 + f(x_2) \cdot 1$$
$$\qquad + f(x_3) \cdot 1 + f(x_4) \cdot 1 + f(x_5) \cdot 1$$
$$\approx 2 + 3.5 + 4 + 2 + (-1) + (-2.5) = 8$$

The Riemann sum represents the sum of the areas of the four rectangles above the x-axis minus the sum of the areas of the two rectangles below the x-axis.

(b)

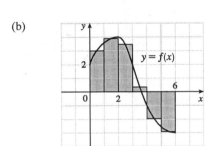

$$M_6 = \sum_{i=1}^6 f(\overline{x}_i)\,\Delta x \quad [\Delta x = \tfrac{6-0}{6} = 1]$$
$$= f(\overline{x}_1) \cdot 1 + f(\overline{x}_2) \cdot 1 + f(\overline{x}_3) \cdot 1$$
$$\qquad + f(\overline{x}_4) \cdot 1 + f(\overline{x}_5) \cdot 1 + f(\overline{x}_6) \cdot 1$$
$$= f(0.5) + f(1.5) + f(2.5) + f(3.5) + f(4.5) + f(5.5)$$
$$\approx 3 + 3.9 + 3.4 + 0.3 + (-2) + (-2.9) = 5.7$$

The Riemann sum represents the sum of the areas of the four rectangles above the x-axis minus the sum of the areas of the two rectangles below the x-axis.

3. $\int_0^1 \left(x + \sqrt{1 - x^2}\right)dx = \int_0^1 x\, dx + \int_0^1 \sqrt{1 - x^2}\, dx = I_1 + I_2.$

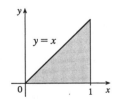

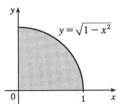

I_1 can be interpreted as the area of the triangle shown in the figure and I_2 can be interpreted as the area of the quarter-circle. Area $= \tfrac{1}{2}(1)(1) + \tfrac{1}{4}(\pi)(1)^2 = \tfrac{1}{2} + \tfrac{\pi}{4}$.

5. $\int_0^6 f(x)\,dx = \int_0^4 f(x)\,dx + \int_4^6 f(x)\,dx \;\Rightarrow\; 10 = 7 + \int_4^6 f(x)\,dx \;\Rightarrow\; \int_4^6 f(x)\,dx = 10 - 7 = 3$

7. First note that either a or b must be the graph of $\int_0^x f(t)\,dt$, since $\int_0^0 f(t)\,dt = 0$, and $c(0) \neq 0$. Now notice that $b > 0$ when c is increasing, and that $c > 0$ when a is increasing. It follows that c is the graph of $f(x)$, b is the graph of $f'(x)$, and a is the graph of $\int_0^x f(t)\,dt$.

9. $\int_1^2 \left(8x^3 + 3x^2\right) dx = \left[\frac{8}{4}x^4 + \frac{3}{3}x^3\right]_1^2 = \left(2 \cdot 2^4 + 2^3\right) - (2 + 1) = 40 - 3 = 37$

11. $\int_0^1 \left(1 - x^9\right) dx = \left[x - \frac{1}{10}x^{10}\right]_0^1 = \left(1 - \frac{1}{10}\right) - 0 = \frac{9}{10}$

13. $\int_1^8 \sqrt[3]{x}\,(x - 1)\,dx = \int_1^8 \left(x^{4/3} - x^{1/3}\right) dx = \left[\frac{3}{7}x^{7/3} - \frac{3}{4}x^{4/3}\right]_1^8 = \left(\frac{3}{7} \cdot 128 - \frac{3}{4} \cdot 16\right) - \left(\frac{3}{7} - \frac{3}{4}\right) = \frac{1209}{28}$

15. $u = x^2 + 1$, $du = 2x\,dx$, so $\displaystyle\int_0^1 \frac{x}{x^2 + 1}\,dx = \int_1^2 \frac{1}{u}\left(\frac{1}{2}\,du\right) = \frac{1}{2}\left[\ln u\right]_1^2 = \frac{1}{2}\ln 2.$

17. Let $u = 1 + 2x^3$. Then $du = 6x^2\,dx$, so
$$\int_0^2 x^2\left(1 + 2x^3\right)^3 dx = \int_1^{17} u^3\left(\tfrac{1}{6}\,du\right) = \left[\tfrac{1}{24}u^4\right]_1^{17} = \tfrac{1}{24}\left(17^4 - 1\right) = 3480.$$

19. $\int_0^1 e^{\pi t}\,dt = \left[\frac{1}{\pi}e^{\pi t}\right]_0^1 = \frac{1}{\pi}(e^{\pi} - 1)$

21. Integrate by parts with $u = x$, $dv = \sec x \tan x\,dx \;\Rightarrow\; du = dx$, $v = \sec x$:
$$\int x \sec x \tan x\,dx = x \sec x - \int \sec x\,dx \overset{14}{=} x \sec x - \ln|\sec x + \tan x| + C.$$

23. Let $u = \dfrac{1}{t}$. Then $du = -\dfrac{1}{t^2}\,dt$, so $\displaystyle\int \frac{\cos(1/t)}{t^2}\,dt = \int \cos u\,(-du) = -\sin u + C = -\sin\left(\frac{1}{t}\right) + C.$

25. Since the degree of the numerator is equal to the degree of the denominator, we first change the form of the integrand by using long division.
$$\int \frac{6x + 1}{3x + 2}\,dx = \int \left(2 - \frac{3}{3x + 2}\right) dx = 2x - 3 \cdot \tfrac{1}{3}\ln|3x + 2| + C = 2x - \ln|3x + 2| + C$$

27. Integrate by parts with $u = x^2$, $dv = e^{-x}dx \;\Rightarrow\; du = 2x\,dx$, $v = -e^{-x}$:
$$I = \int x^2 e^{-x}\,dx = -x^2 e^{-x} + 2\int xe^{-x}dx$$

Now integrate by parts with $u = x$, $dv = e^{-x}dx \;\Rightarrow\; du = dx$, $v = -e^{-x}$:
$$\int xe^{-x}dx = -xe^{-x} + \int e^{-x}dx = -xe^{-x} - e^{-x} + C$$

Thus, $I = -x^2 e^{-x} + 2\left(-xe^{-x} - e^{-x} + C\right) = -x^2 e^{-x} - 2xe^{-x} - 2e^{-x} + C = -e^{-x}\left(x^2 + 2x + 2\right) + C.$

29. $\dfrac{1}{t^2 + 6t + 8} = \dfrac{1}{(t + 2)(t + 4)} = \dfrac{A}{t + 2} + \dfrac{B}{t + 4}$. Multiply both sides by $(t + 2)(t + 4)$
to get $1 = A(t + 4) + B(t + 2)$. Substituting -4 for t gives
$1 = -2B \;\Leftrightarrow\; B = -\frac{1}{2}$. Substituting -2 for t gives $1 = 2A \;\Leftrightarrow\; A = \frac{1}{2}$. Thus,
$$\int \frac{dt}{t^2 + 6t + 8} = \int \left(\frac{1/2}{t + 2} - \frac{1/2}{t + 4}\right) dt = \tfrac{1}{2}\ln|t + 2| - \tfrac{1}{2}\ln|t + 4| + C = \tfrac{1}{2}\ln\left|\frac{t + 2}{t + 4}\right| + C.$$

31. Let $x = 3\sin\theta$, where $-\pi/2 \le \theta \le \pi/2$. Then $dx = 3\cos\theta\, d\theta$ and

$\sqrt{9 - x^2} = \sqrt{9 - 9\sin^2\theta} = \sqrt{9\cos^2\theta} = 3\,|\cos\theta| = 3\cos\theta$ since $\cos\theta \ge 0$ for $-\pi/2 \le \theta \le \pi/2$. When $x = 0,\, 3\sin\theta = 0 \;\Rightarrow\; \theta = 0$ and when $x = 3,\, 3\sin\theta = 3 \;\Rightarrow\; \sin\theta = 1 \;\Rightarrow\; \theta = \frac{\pi}{2}$. Thus,

$$\int_0^3 x^3\sqrt{9 - x^2}\, dx = \int_0^{\pi/2} (3\sin\theta)^3(3\cos\theta)(3\cos\theta\, d\theta) = 3^5\int_0^{\pi/2}\sin^3\theta\cos^2\theta\, d\theta$$

$$= 3^5\int_0^{\pi/2}\sin^2\theta\cos^2\theta(\sin\theta\, d\theta) = 3^5\int_0^{\pi/2}(1 - \cos^2\theta)\cos^2\theta(\sin\theta\, d\theta) = I.$$

Now let $u = \cos\theta$ so that $du = -\sin\theta\, d\theta$. When $\theta = 0,\, u = 1$ and when $\theta = \frac{\pi}{2},\, u = 0$. Substitution gives us

$$I = 3^5\int_1^0 (1 - u^2)u^2(-du) = 3^5\int_0^1 (u^2 - u^4)\, du = 3^5\left[\tfrac{1}{3}u^3 - \tfrac{1}{5}u^5\right]_0^1$$

$$= 3^5\left(\tfrac{1}{3} - \tfrac{1}{5}\right) = 3^5\left(\tfrac{2}{15}\right) = \tfrac{162}{5} = 32.4.$$

Another method: Let $u = 9 - x^2$. Then $du = -2x\, dx$ and $x^2 = 9 - u$, so

$$\int_0^3 x^3\sqrt{9 - x^2}\, dx = \int_0^3 x^2\sqrt{9 - x^2}\,(x\, dx) = \int_9^0 (9 - u)\sqrt{u}\left(-\tfrac{1}{2}\, du\right) = \tfrac{1}{2}\int_0^9 \left(9u^{1/2} - u^{3/2}\right) du$$

$$= \tfrac{1}{2}\left[6u^{3/2} - \tfrac{2}{5}u^{5/2}\right]_0^9 = \tfrac{1}{2}\left[(6\cdot 27 - \tfrac{2}{5}\cdot 243) - 0\right] = \tfrac{1}{2}\left(\tfrac{324}{5}\right) = \tfrac{162}{5} = 32.4$$

33. Let $u = 1 + \sec\theta$. Then $du = \sec\theta\tan\theta\, d\theta$, so

$$\int \frac{\sec\theta\tan\theta}{1 + \sec\theta}\, d\theta = \int \frac{1}{u}\, du = \ln|u| + C = \ln|1 + \sec\theta| + C$$

In Exercise 35, let $f(x)$ denote the integrand and $F(x)$ its antiderivative (with $C = 0$).

35. Let $u = 1 + \sin x$. Then $du = \cos x\, dx$, so

$$\int \frac{\cos x\, dx}{\sqrt{1 + \sin x}} = \int u^{-1/2}\, du = 2u^{1/2} + C = 2\sqrt{1 + \sin x} + C.$$

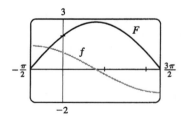

37. From the graph, it appears that the area under the curve $y = x\sqrt{x}$ between $x = 0$ and $x = 4$ is somewhat less than half the area of an 8×4 rectangle, so perhaps about 13 or 14. To find the exact value, we evaluate

$$\int_0^4 x\sqrt{x}\, dx = \int_0^4 x^{3/2}\, dx = \left[\tfrac{2}{5}x^{5/2}\right]_0^4 = \tfrac{2}{5}(4)^{5/2} = \tfrac{64}{5} = 12.8.$$

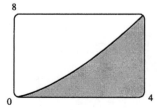

39. By FTC1, $F(x) = \int_1^x \sqrt{1 + t^4}\, dt \;\Rightarrow\; F'(x) = \sqrt{1 + x^4}$.

41. $y = \int_{\sqrt{x}}^{x} \frac{e^t}{t} \, dt = \int_{\sqrt{x}}^{1} \frac{e^t}{t} \, dt + \int_{1}^{x} \frac{e^t}{t} \, dt = -\int_{1}^{\sqrt{x}} \frac{e^t}{t} \, dt + \int_{1}^{x} \frac{e^t}{t} \, dt \quad \Rightarrow$

$\dfrac{dy}{dx} = -\dfrac{d}{dx}\left(\displaystyle\int_{1}^{\sqrt{x}} \frac{e^t}{t} \, dt\right) + \dfrac{d}{dx}\left(\displaystyle\int_{1}^{x} \frac{e^t}{t} \, dt\right)$. Let $u = \sqrt{x}$. Then

$\dfrac{d}{dx}\displaystyle\int_{1}^{\sqrt{x}} \frac{e^t}{t} \, dt = \dfrac{d}{dx}\displaystyle\int_{1}^{u} \frac{e^t}{t} \, dt = \dfrac{d}{du}\left(\displaystyle\int_{1}^{u} \frac{e^t}{t} \, dt\right) \dfrac{du}{dx} = \dfrac{e^u}{u} \cdot \dfrac{1}{2\sqrt{x}} = \dfrac{e^{\sqrt{x}}}{\sqrt{x}} \cdot \dfrac{1}{2\sqrt{x}} = \dfrac{e^{\sqrt{x}}}{2x},$

so $\dfrac{dy}{dx} = -\dfrac{e^{\sqrt{x}}}{2x} + \dfrac{e^x}{x}$.

43. $u = e^x \quad \Rightarrow \quad du = e^x \, dx$, so

$\int e^x \sqrt{1 - e^{2x}} \, dx = \int \sqrt{1 - u^2} \, du \overset{30}{=} \tfrac{1}{2} u \sqrt{1 - u^2} + \tfrac{1}{2}\sin^{-1} u + C = \tfrac{1}{2}\left[e^x \sqrt{1 - e^{2x}} + \sin^{-1}(e^x)\right] + C$

45. $\displaystyle\int \sqrt{x^2 + x + 1} \, dx = \int \sqrt{x^2 + x + \tfrac{1}{4} + \tfrac{3}{4}} \, dx = \int \sqrt{\left(x + \tfrac{1}{2}\right)^2 + \tfrac{3}{4}} \, dx$

$= \displaystyle\int \sqrt{u^2 + \left(\tfrac{\sqrt{3}}{2}\right)^2} \, du \quad [u = x + \tfrac{1}{2}, \, du = dx]$

$\overset{21}{=} \tfrac{1}{2} u \sqrt{u^2 + \tfrac{3}{4}} + \tfrac{3}{8} \ln\left(u + \sqrt{u^2 + \tfrac{3}{4}}\right) + C$

$= \dfrac{2x + 1}{4} \sqrt{x^2 + x + 1} + \tfrac{3}{8} \ln\left(x + \tfrac{1}{2} + \sqrt{x^2 + x + 1}\right) + C$

47. $f(x) = \sqrt{1 + x^4}, \quad \Delta x = \dfrac{b - a}{n} = \dfrac{1 - 0}{10} = \dfrac{1}{10}$.

(a) $T_{10} = \dfrac{1}{10 \cdot 2}\{f(0) + 2\left[f(0.1) + f(0.2) + \cdots + f(0.9)\right] + f(1)\} \approx 1.090608$

(b) $M_{10} = \dfrac{1}{10}\left[f\left(\tfrac{1}{20}\right) + f\left(\tfrac{3}{20}\right) + f\left(\tfrac{5}{20}\right) + \cdots + f\left(\tfrac{19}{20}\right)\right] \approx 1.088840$

(c) $S_{10} = \dfrac{1}{10 \cdot 3}\left[f(0) + 4f(0.1) + 2f(0.2) + \cdots + 4f(0.9) + f(1)\right] \approx 1.089429$

f is concave upward, so the Trapezoidal Rule gives us an overestimate, the Midpoint Rule gives an underestimate, and we cannot tell whether Simpson's Rule gives us an overestimate or an underestimate.

49. $f(x) = \left(1 + x^4\right)^{1/2}, \, f'(x) = \tfrac{1}{2}\left(1 + x^4\right)^{-1/2}\left(4x^3\right) = 2x^3\left(1 + x^4\right)^{-1/2}, \, f''(x) = \left(2x^6 + 6x^2\right)\left(1 + x^4\right)^{-3/2}$.

A graph of f'' on $[0, 1]$ shows that it has its maximum at $x = 1$, so $|f''(x)| \leq f''(1) = \sqrt{8}$ on $[0, 1]$. By taking

$K = \sqrt{8}$, we find that the error in Exercise 47(a) is bounded by $\dfrac{K(b - a)^3}{12n^2} = \dfrac{\sqrt{8}}{1200} \approx 0.0024$, and in (b) by about

$\tfrac{1}{2}(0.0024) = 0.0012$.

Note: Another way to estimate K is to let $x = 1$ in the factor $2x^6 + 6x^2$ (maximizing the numerator) and let $x = 0$

in the factor $\left(1 + x^4\right)^{-3/2}$ (minimizing the denominator). Doing so gives us $K = 8$ and errors of $0.00\overline{6}$ and $0.00\overline{3}$.

Using $K = 8$ for the Trapezoidal Rule, we have $|E_T| \leq \dfrac{K(b - a)^3}{12n^2} \leq 0.00001 \quad \Leftrightarrow \quad \dfrac{8(1 - 0)^3}{12n^2} \leq \dfrac{1}{100,000}$

$\Leftrightarrow \quad n^2 \geq \dfrac{800,000}{12} \quad \Leftrightarrow \quad n \gtrsim 258.2$, so we should take $n = 259$.

For the Midpoint Rule, $|E_M| \leq \dfrac{K(b - a)^3}{24n^2} \leq 0.00001 \quad \Leftrightarrow \quad n^2 \geq \dfrac{800,000}{24} \quad \Leftrightarrow \quad n \gtrsim 182.6$, so we should

take $n = 183$.

51. (a) $f(x) = \sin(\sin x)$. A CAS gives

$$f^{(4)}(x) = \sin(\sin x)\left[\cos^4 x + 7\cos^2 x - 3\right]$$

$$+ \cos(\sin x)\left[6\cos^2 x \sin x + \sin x\right]$$

From the graph, we see that $\left|f^{(4)}(x)\right| < 3.8$ for $x \in [0, \pi]$.

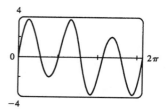

(b) We use Simpson's Rule with $f(x) = \sin(\sin x)$ and $\Delta x = \frac{\pi}{10}$:

$$\int_0^\pi f(x)\,dx \approx \tfrac{\pi}{10 \cdot 3}\left[f(0) + 4f\left(\tfrac{\pi}{10}\right) + 2f\left(\tfrac{2\pi}{10}\right) + \cdots + 4f\left(\tfrac{9\pi}{10}\right) + f(\pi)\right] \approx 1.786721$$

From part (a), we know that $\left|f^{(4)}(x)\right| < 3.8$ on $[0, \pi]$, so we use Theorem 5.9.4 with $K = 3.8$, and estimate the error as $|E_S| \leq \dfrac{3.8(\pi - 0)^5}{180(10)^4} \approx 0.000646$.

(c) If we want the error to be less than 0.00001, we must have $|E_S| \leq \frac{3.8\pi^5}{180n^4} \leq 0.00001$, so

$n^4 \geq \frac{3.8\pi^5}{180(0.00001)} \approx 646{,}041.6 \quad \Rightarrow \quad n \geq 28.35$. Since n must be even for Simpson's Rule, we must have $n \geq 30$ to ensure the desired accuracy.

53. If $1 \leq x \leq 3$, then $\sqrt{1^2 + 3} \leq \sqrt{x^2 + 3} \leq \sqrt{3^2 + 3} \quad \Rightarrow \quad 2 \leq \sqrt{x^2 + 3} \leq 2\sqrt{3}$, so $2(3 - 1) \leq \int_1^3 \sqrt{x^2 + 3}\,dx \leq 2\sqrt{3}(3 - 1)$; that is, $4 \leq \int_1^3 \sqrt{x^2 + 3}\,dx \leq 4\sqrt{3}$.

55. $\displaystyle\int_1^\infty \frac{1}{(2x + 1)^3}\,dx = \lim_{t \to \infty} \int_1^t \frac{1}{(2x + 1)^3}\,dx = \lim_{t \to \infty} \int_1^t \tfrac{1}{2}(2x + 1)^{-3}\,2\,dx$

$$= \lim_{t \to \infty}\left[-\frac{1}{4(2x + 1)^2}\right]_1^t = -\frac{1}{4}\lim_{t \to \infty}\left[\frac{1}{(2t + 1)^2} - \frac{1}{9}\right] = -\frac{1}{4}\left(0 - \frac{1}{9}\right) = \frac{1}{36}$$

57. $\displaystyle\int_{-\infty}^0 e^{-2x}\,dx = \lim_{t \to -\infty} \int_t^0 e^{-2x}\,dx = \lim_{t \to -\infty}\left[-\tfrac{1}{2}e^{-2x}\right]_t^0 = \lim_{t \to -\infty}\left(-\tfrac{1}{2} + \tfrac{1}{2}e^{-2t}\right) = \infty$. Divergent

59. Let $u = \ln x$. Then $du = dx/x$, so $\displaystyle\int \frac{dx}{x\sqrt{\ln x}} = \int \frac{du}{\sqrt{u}} = 2\sqrt{u} + C = 2\sqrt{\ln x} + C$.

Thus, $\displaystyle\int_1^e \frac{dx}{x\sqrt{\ln x}} = \lim_{t \to 1^+} \int_t^e \frac{dx}{x\sqrt{\ln x}} = \lim_{t \to 1^+}\left[2\sqrt{\ln x}\right]_t^e = \lim_{t \to 1^+}\left(2\sqrt{\ln e} - 2\sqrt{\ln t}\right) = 2 \cdot 1 - 2 \cdot 0 = 2$.

61. $\dfrac{x^3}{x^5 + 2} \leq \dfrac{x^3}{x^5} = \dfrac{1}{x^2}$ for x in $[1, \infty)$. $\displaystyle\int_1^\infty \frac{1}{x^2}\,dx$ is convergent by (5.10.2) with $p = 2 > 1$. Therefore,

$\displaystyle\int_1^\infty \frac{x^3}{x^5 + 2}\,dx$ is convergent by the Comparison Theorem.

63. (a) displacement $= \int_0^5 (t^2 - t)\,dt = \left[\tfrac{1}{3}t^3 - \tfrac{1}{2}t^2\right]_0^5 = \tfrac{125}{3} - \tfrac{25}{2} = \tfrac{175}{6} = 29.1\overline{6}$ meters

(b) distance traveled $= \int_0^5 |t^2 - t|\,dt = \int_0^5 |t(t - 1)|\,dt = \int_0^1 (t - t^2)\,dt + \int_1^5 (t^2 - t)\,dt$

$$= \left[\tfrac{1}{2}t^2 - \tfrac{1}{3}t^3\right]_0^1 + \left[\tfrac{1}{3}t^3 - \tfrac{1}{2}t^2\right]_1^5$$

$$= \tfrac{1}{2} - \tfrac{1}{3} - 0 + \left(\tfrac{125}{3} - \tfrac{25}{2}\right) - \left(\tfrac{1}{3} - \tfrac{1}{2}\right) = \tfrac{177}{6} = 29.5 \text{ meters}$$

65. Note that $r(t) = b'(t)$, where $b(t) =$ the number of barrels of oil consumed up to time t. So, by the Total Change Theorem, $\int_0^3 r(t)\,dt = b(3) - b(0)$ represents the number of barrels of oil consumed from Jan. 1, 2000, through Jan. 1, 2003.

67. Both numerator and denominator approach 0 as $a \to 0$, so we use l'Hospital's Rule. (Note that we are differentiating *with respect to a*, since that is the quantity which is changing.) We also use FTC1:

$$\lim_{a \to 0} T(x, t) = \lim_{a \to 0} \frac{C \int_0^a e^{-(x-u)^2/(4kt)}\,du}{a\sqrt{4\pi kt}} \overset{\text{H}}{=} \lim_{a \to 0} \frac{Ce^{-(x-a)^2/(4kt)}}{\sqrt{4\pi kt}} = \frac{Ce^{-x^2/(4kt)}}{\sqrt{4\pi kt}}$$

69. Using FTC1, we differentiate both sides of the given equation, $\int_0^x f(t)\,dt = xe^{2x} + \int_0^x e^{-t} f(t)\,dt$, and get

$$f(x) = e^{2x} + 2xe^{2x} + e^{-x} f(x) \quad \Rightarrow \quad f(x)(1 - e^{-x}) = e^{2x} + 2xe^{2x} \quad \Rightarrow \quad f(x) = \frac{e^{2x}(1 + 2x)}{1 - e^{-x}}.$$

71. Let $u = f(x)$ and $du = f'(x)\,dx$. So $2\int_a^b f(x) f'(x)\,dx = 2\int_{f(a)}^{f(b)} u\,du = \left[u^2\right]_{f(a)}^{f(b)} = [f(b)]^2 - [f(a)]^2$.

73. By the Fundamental Theorem of Calculus,

$$\int_0^\infty f'(x)\,dx = \lim_{t \to \infty} \int_0^t f'(x)\,dx = \lim_{t \to \infty} [f(t) - f(0)] = \lim_{t \to \infty} f(t) - f(0) = 0 - f(0) = -f(0).$$

Focus on Problem Solving

1.

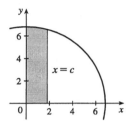 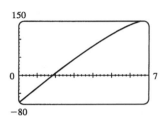

By symmetry, the problem can be reduced to finding the line $x = c$ such that the shaded area is one-third of the area of the quarter-circle. The equation of the circle is $y = \sqrt{49 - x^2}$, so we require that $\int_0^c \sqrt{49 - x^2}\, dx = \frac{1}{3} \cdot \frac{1}{4}\pi(7)^2$
$\Leftrightarrow \left[\frac{1}{2}x\sqrt{49 - x^2} + \frac{49}{2}\sin^{-1}(x/7)\right]_0^c = \frac{49}{12}\pi$ (by Formula 30) $\Leftrightarrow \frac{1}{2}c\sqrt{49 - c^2} + \frac{49}{2}\sin^{-1}(c/7) = \frac{49}{12}\pi$.
This equation would be difficult to solve exactly, so we plot the left-hand side as a function of c, and find that the equation holds for $c \approx 1.85$. So the cuts should be made at distances of about 1.85 inches from the center of the pizza.

3. Differentiating both sides of the equation $x \sin \pi x = \int_0^{x^2} f(t)\, dt$ (using FTC1 and the Chain Rule for the right side) gives $\sin \pi x + \pi x \cos \pi x = 2x f(x^2)$. Letting $x = 2$ so that $f(x^2) = f(4)$, we obtain $\sin 2\pi + 2\pi \cos 2\pi = 4f(4)$, so $f(4) = \frac{1}{4}(0 + 2\pi \cdot 1) = \frac{\pi}{2}$.

5. By FTC2, $\int_0^1 f'(x)\, dx = f(1) - f(0) = 1 - 0 = 1$.

7. By l'Hospital's Rule and the Fundamental Theorem, using the notation $\exp(y) = e^y$,

$$\lim_{x \to 0} \frac{\int_0^x (1 - \tan 2t)^{1/t}\, dt}{x} \overset{\text{H}}{=} \lim_{x \to 0} \frac{(1 - \tan 2x)^{1/x}}{1} = \exp\left(\lim_{x \to 0} \frac{\ln(1 - \tan 2x)}{x}\right)$$

$$\overset{\text{H}}{=} \exp\left(\lim_{x \to 0} \frac{-2\sec^2 2x}{1 - \tan 2x}\right) = \exp\left(\frac{-2 \cdot 1^2}{1 - 0}\right) = e^{-2}.$$

9. Such a function cannot exist. $f'(x) > 3$ for all x means that f is differentiable (and hence continuous) for all x. So by Part 2 of the Fundamental Theorem, $\int_1^4 f'(x)\, dx = f(4) - f(1) = 7 - (-1) = 8$. However, if $f'(x) > 3$ for all x, then $\int_1^4 f'(x)\, dx \geq 3 \cdot (4 - 1) = 9$ by Comparison Property 8 in Section 5.2.
Another solution: By the Mean Value Theorem there exists a number $c \in (1, 4)$ such that
$$f'(c) = \frac{f(4) - f(1)}{4 - 1} = \frac{7 - (-1)}{3} = \frac{8}{3} \quad \Rightarrow \quad 8 = 3f'(c). \text{ But } f'(x) > 3 \quad \Rightarrow \quad 3f'(c) > 9, \text{ so such a}$$
function cannot exist.

11. $f(x) = 2 + x - x^2 = (-x + 2)(x + 1) = 0 \iff x = 2$ or $x = -1$. $f(x) \geq 0$ for $x \in [-1, 2]$ and $f(x) < 0$ everywhere else. The integral $\int_a^b (2 + x - x^2) \, dx$ has a maximum on the interval where the integrand is positive, which is $[-1, 2]$. So $a = -1$, $b = 2$. (Any larger interval gives a smaller integral since $f(x) < 0$ outside $[-1, 2]$. Any smaller interval also gives a smaller integral since $f(x) \geq 0$ in $[-1, 2]$.)

13. By FTC1, $\dfrac{d}{dx} \int_0^x \left(\int_1^{\sin t} \sqrt{1 + u^4} \, du \right) dt = \int_1^{\sin x} \sqrt{1 + u^4} \, du$. Again using FTC1,

$$\frac{d^2}{dx^2} \int_0^x \left(\int_1^{\sin t} \sqrt{1 + u^4} \, du \right) dt = \frac{d}{dx} \int_1^{\sin x} \sqrt{1 + u^4} \, du = \sqrt{1 + \sin^4 x} \cos x.$$

15. The given integral represents the difference of the shaded areas, which appears to be 0. It can be calculated by integrating with respect to either x or y, so we find x in terms of y for each curve:

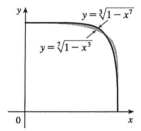

$y = \sqrt[3]{1 - x^7} \implies x = \sqrt[7]{1 - y^3}$ and

$y = \sqrt[7]{1 - x^3} \implies x = \sqrt[3]{1 - y^7}$, so

$\int_0^1 \left(\sqrt[3]{1 - y^7} - \sqrt[7]{1 - y^3} \right) dy = \int_0^1 \left(\sqrt[7]{1 - x^3} - \sqrt[3]{1 - x^7} \right) dx$. But this equation is of the form $z = -z$. So

$\int_0^1 \left(\sqrt[3]{1 - x^7} - \sqrt[7]{1 - x^3} \right) dx = 0.$

17. In accordance with the hint, we let $I_k = \int_0^1 (1 - x^2)^k \, dx$, and we find an expression for I_{k+1} in terms of I_k. We integrate I_{k+1} by parts with $u = (1 - x^2)^{k+1}$, $dv = dx \implies du = (k + 1)(1 - x^2)^k (-2x)$, $v = x$, and then split the remaining integral into identifiable quantities:

$$I_{k+1} = \left[x(1 - x^2)^{k+1} \right]_0^1 + 2(k + 1) \int_0^1 x^2 (1 - x^2)^k \, dx$$

$$= (2k + 2) \int_0^1 (1 - x^2)^k \left[1 - (1 - x^2) \right] dx = (2k + 2)(I_k - I_{k+1})$$

So $I_{k+1} [1 + (2k + 2)] = (2k + 2) I_k \implies I_{k+1} = \dfrac{2k + 2}{2k + 3} I_k.$

Now to complete the proof, we use induction: $I_0 = 1 = \dfrac{2^0 (0!)^2}{1!}$, so the formula holds for $n = 0$. Now suppose it holds for $n = k$. Then

$$I_{k+1} = \frac{2k + 2}{2k + 3} I_k = \frac{2k + 2}{2k + 3} \left[\frac{2^{2k} (k!)^2}{(2k + 1)!} \right] = \frac{2(k + 1) 2^{2k} (k!)^2}{(2k + 3)(2k + 1)!} = \frac{2(k + 1) 2^{2k} (k!)^2}{(2k + 3)(2k + 1)!} \cdot \frac{2(k + 1)}{2k + 2}$$

$$= \frac{[2(k + 1)]^2 2^{2k} (k!)^2}{(2k + 3)(2k + 2)(2k + 1)!} = \frac{2^{2(k+1)} [(k + 1)!]^2}{[2(k + 1) + 1]!}$$

So by induction, the formula holds for all integers $n \geq 0$.

19. (a) The tangent to the curve $y = f(x)$ at $x = x_0$ has the equation $y - f(x_0) = f'(x_0)(x - x_0)$. The y-intercept of this tangent line is $f(x_0) - f'(x_0)x_0$. Thus, L is the distance from the point $(0, f(x_0) - f'(x_0)x_0)$ to

the point $(x_0, f(x_0))$; that is, $L^2 = x_0^2 + [f'(x_0)]^2 x_0^2$, so $[f'(x_0)]^2 = \dfrac{L^2 - x_0^2}{x_0^2}$ and $f'(x_0) = -\dfrac{\sqrt{L^2 - x_0^2}}{x_0}$

for $0 < x_0 < L$.

(b) $\dfrac{dy}{dx} = -\dfrac{\sqrt{L^2 - x^2}}{x} \quad\Rightarrow\quad y = \int \left(-\dfrac{\sqrt{L^2 - x^2}}{x} \right) dx.$

Let $x = L \sin\theta$. Then $dx = L \cos\theta\, d\theta$ and

$$y = \int \frac{-L \cos\theta\, L \cos\theta\, d\theta}{L \sin\theta} = L \int \frac{\sin^2\theta - 1}{\sin\theta}\, d\theta$$

$$= L \int (\sin\theta - \csc\theta)\, d\theta$$

$$= -L \cos\theta - L \ln|\csc\theta - \cot\theta| + C$$

$$= -\sqrt{L^2 - x^2} - L \ln\left(\frac{L}{x} - \frac{\sqrt{L^2 - x^2}}{x} \right) + C.$$

When $x = L$, $y = 0$, and $0 = -0 - L\ln(1 - 0) + C$, so $C = 0$. Therefore,

$$y = -\sqrt{L^2 - x^2} - L \ln\left(\frac{L - \sqrt{L^2 - x^2}}{x} \right)$$

Applications of Integration

6.1 More about Areas • • • • • • • • • • • • • • •

1. $A = \int_{x=0}^{x=4} (y_T - y_B)\, dx = \int_0^4 \left[(5x - x^2) - x \right] dx = \int_0^4 (4x - x^2)\, dx$

$= \left[2x^2 - \tfrac{1}{3}x^3 \right]_0^4 = \left(32 - \tfrac{64}{3} \right) - (0) = \tfrac{32}{3}$

3. $A = \int_{y=-1}^{y=1} (x_R - x_L)\, dy = \int_{-1}^1 \left[e^y - (y^2 - 2) \right] dy$

$= \int_{-1}^1 (e^y - y^2 + 2)\, dy = \left[e^y - \tfrac{1}{3}y^3 + 2y \right]_{-1}^1 = (e^1 - \tfrac{1}{3} + 2) - (e^{-1} + \tfrac{1}{3} - 2) = e - \dfrac{1}{e} + \dfrac{10}{3}$

5. $A = \int_{-1}^2 \left[(9 - x^2) - (x + 1) \right] dx$

$= \int_{-1}^2 (8 - x - x^2)\, dx$

$= \left[8x - \dfrac{x^2}{2} - \dfrac{x^3}{3} \right]_{-1}^2$

$= (16 - 2 - \tfrac{8}{3}) - (-8 - \tfrac{1}{2} + \tfrac{1}{3})$

$= 22 - 3 + \tfrac{1}{2} = \tfrac{39}{2}$

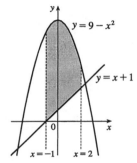
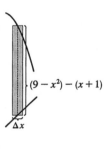

7. The curves intersect when $x = x^2 \;\Rightarrow\; x^2 - x = 0 \;\Leftrightarrow\; x(x - 1) = 0 \;\Leftrightarrow\; x = 0, 1.$

$A = \int_0^1 (x - x^2)\, dx$

$= \left[\tfrac{1}{2}x^2 - \tfrac{1}{3}x^3 \right]_0^1$

$= \tfrac{1}{2} - \tfrac{1}{3}$

$= \tfrac{1}{6}$

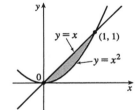
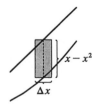

9. The curves intersect when $4x^2 = x^2 + 3 \;\Leftrightarrow\; 3x^2 = 3 \;\Leftrightarrow\; x^2 = 1 \;\Leftrightarrow\; x = \pm 1.$

$A = \int_{-1}^1 \left[(x^2 + 3) - 4x^2 \right] dx$

$= 2 \int_0^1 (3 - 3x^2)\, dx$

$= 2 \left[3x - x^3 \right]_0^1 = 2(3 - 1) = 4$

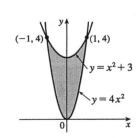
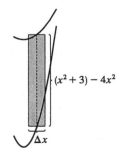

11. The curves intersect when $y^2 = 2y + 3 \iff y^2 - 2y - 3 = 0 \iff (y-3)(y+1) = 0 \iff y = -1, 3$.

$$A = \int_{-1}^{3} [(2y+3) - y^2] \, dy$$
$$= [y^2 + 3y - \tfrac{1}{3}y^3]_{-1}^{3}$$
$$= (9 + 9 - 9) - (1 - 3 + \tfrac{1}{3})$$
$$= \tfrac{32}{3}$$

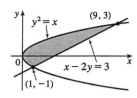

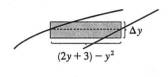

13. The curves intersect when $1 - y^2 = y^2 - 1 \iff 2 = 2y^2 \iff y^2 = 1 \iff y = \pm 1$.

$$A = \int_{-1}^{1} [(1 - y^2) - (y^2 - 1)] \, dy$$
$$= \int_{-1}^{1} 2(1 - y^2) \, dy$$
$$= 2 \cdot 2 \int_{0}^{1} (1 - y^2) \, dy$$
$$= 4[y - \tfrac{1}{3}y^3]_0^1 = 4(1 - \tfrac{1}{3}) = \tfrac{8}{3}$$

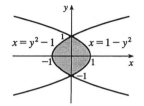

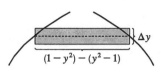

15. The curves intersect when $x^2 = \dfrac{2}{x^2 + 1} \iff x^4 + x^2 = 2 \iff x^4 + x^2 - 2 = 0 \iff$

$(x^2 + 2)(x^2 - 1) = 0 \iff x^2 = 1 \iff x = \pm 1$.

$$A = \int_{-1}^{1} \left(\frac{2}{x^2 + 1} - x^2 \right) dx$$
$$= 2 \int_{0}^{1} \left(\frac{2}{x^2 + 1} - x^2 \right) dx$$
$$= 2[2\tan^{-1} x - \tfrac{1}{3}x^3]_0^1 = 2(2 \cdot \tfrac{\pi}{4} - \tfrac{1}{3})$$
$$= \pi - \tfrac{2}{3} \approx 2.47$$

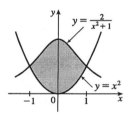

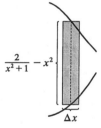

17.

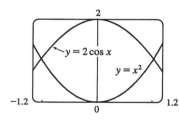

From the graph, we see that the curves intersect at $x = \pm a \approx \pm 1.02$, with $2\cos x > x^2$ on $(-a, a)$. So the area of the region bounded by the curves is

$$A = \int_{-a}^{a} (2\cos x - x^2) \, dx = 2 \int_{0}^{a} (2\cos x - x^2) \, dx$$
$$= 2[2\sin x - \tfrac{1}{3}x^3]_0^a \approx 2.70$$

19.

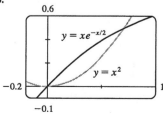

From the graph, we see that the curves intersect at $x = 0$ and $x = a \approx 0.70$, with $xe^{-x/2} > x^2$ on $(0, a)$. So the area of the region bounded by the curves is

$$A = \int_{0}^{a} \left(xe^{-x/2} - x^2 \right) dx$$
$$= \left[4(-\tfrac{1}{2}x - 1)e^{-x/2} - \tfrac{1}{3}x^3 \right]_0^a \quad \text{(Formula 96 with } a = -\tfrac{1}{2})$$
$$\approx 0.08$$

21. As in Example 4, we approximate the distance between the two cars after ten seconds using Simpson's Rule with
$\Delta t = 1\,\text{s} = \frac{1}{3600}\,\text{h}$.

$\text{distance}_{\text{Kelly}} - \text{distance}_{\text{Chris}} = \int_0^{10} v_K\,dt - \int_0^{10} v_C\,dt = \int_0^{10}(v_K - v_C)\,dt \approx S_{10}$

$\qquad = \frac{1}{3\cdot 3600}[(0-0) + 4(22-20) + 2(37-32) + 4(52-46) + 2(61-54) + 4(71-62)$

$\qquad\qquad\qquad +2(80-69) + 4(86-75) + 2(93-81) + 4(98-86) + (102-90)]$

$\qquad = \frac{1}{10{,}800}(242) = \frac{121}{5400}\,\text{mi}$

So after 10 seconds, Kelly's car is about $\dfrac{121}{5400}\,\text{mi}\left(5280\dfrac{\text{ft}}{\text{mi}}\right) \approx 118$ ft ahead of Chris's.

23. If $x =$ distance from left end of pool and $w = w(x) =$ width at x, then Simpson's Rule with $n = 8$ and $\Delta x = 2$
gives

$\qquad \text{Area} = \int_0^{16} w\,dx \approx \frac{2}{3}[0 + 4(6.2) + 2(7.2) + 4(6.8) + 2(5.6) + 4(5.0) + 2(4.8) + 4(4.8) + 0]$

$\qquad\qquad = \frac{2}{3}(126.4) \approx 84\,\text{m}^2$

25. $\cos x = \sin 2x = 2\sin x \cos x \ \Leftrightarrow\ 2\sin x \cos x - \cos x = 0 \ \Leftrightarrow\ \cos x\,(2\sin x - 1) = 0 \ \Leftrightarrow$
$2\sin x = 1$ or $\cos x = 0 \ \Leftrightarrow\ x = \frac{\pi}{6}$ or $\frac{\pi}{2}$.

$\quad A = \int_0^{\pi/6}(\cos x - \sin 2x)\,dx + \int_{\pi/6}^{\pi/2}(\sin 2x - \cos x)\,dx$

$\qquad = \left[\sin x + \tfrac{1}{2}\cos 2x\right]_0^{\pi/6} + \left[-\tfrac{1}{2}\cos 2x - \sin x\right]_{\pi/6}^{\pi/2}$

$\qquad = \left(\tfrac{1}{2} + \tfrac{1}{2}\cdot\tfrac{1}{2}\right) - \left(0 + \tfrac{1}{2}\cdot 1\right)$

$\qquad\qquad + \left[-\tfrac{1}{2}\cdot(-1) - 1\right] - \left(-\tfrac{1}{2}\cdot\tfrac{1}{2} - \tfrac{1}{2}\right)$

$\qquad = \tfrac{3}{4} - \tfrac{1}{2} - \tfrac{1}{2} + \tfrac{3}{4} = \tfrac{1}{2}$

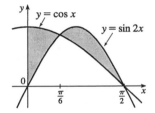

27. Let the equation of the large circle be $x^2 + y^2 = R^2$. Then the equation
of the small circle is $x^2 + (y - b)^2 = r^2$, where $b = \sqrt{R^2 - r^2}$ is the
distance between the centers of the circles. The desired area is

$\quad A = \int_{-r}^{r}\left[(b + \sqrt{r^2 - x^2}) - \sqrt{R^2 - x^2}\right]dx$

$\qquad = 2\int_0^r\left(b + \sqrt{r^2 - x^2} - \sqrt{R^2 - x^2}\right)dx$

$\qquad = 2\int_0^r b\,dx + 2\int_0^r\sqrt{r^2 - x^2}\,dx - 2\int_0^r\sqrt{R^2 - x^2}\,dx$

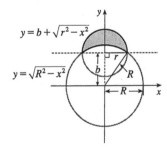

The first integral is just $2br = 2r\sqrt{R^2 - r^2}$. The second integral represents the area of a quarter-circle of radius r,
so its value is $\frac{1}{4}\pi r^2$. To evaluate the other integral, note that

$\quad \int\sqrt{a^2 - x^2}\,dx = \int a^2\cos^2\theta\,d\theta \ \ (x = a\sin\theta,\ dx = a\cos\theta\,d\theta) \ \ = \left(\tfrac{1}{2}a^2\right)\int(1 + \cos 2\theta)\,d\theta$

$\qquad\qquad = \tfrac{1}{2}a^2\left(\theta + \tfrac{1}{2}\sin 2\theta\right) + C = \tfrac{1}{2}a^2(\theta + \sin\theta\,\cos\theta) + C$

$\qquad\qquad = \dfrac{a^2}{2}\arcsin\left(\dfrac{x}{a}\right) + \dfrac{a^2}{2}\left(\dfrac{x}{a}\right)\dfrac{\sqrt{a^2 - x^2}}{a} + C = \dfrac{a^2}{2}\arcsin\left(\dfrac{x}{a}\right) + \dfrac{x}{2}\sqrt{a^2 - x^2} + C$

Thus, the desired area is

$$A = 2r\sqrt{R^2 - r^2} + 2\left(\tfrac{1}{4}\pi r^2\right) - \left[R^2\arcsin(x/R) + x\sqrt{R^2 - x^2}\right]_0^r$$

$$= 2r\sqrt{R^2 - r^2} + \tfrac{1}{2}\pi r^2 - \left[R^2\arcsin(r/R) + r\sqrt{R^2 - r^2}\right] = r\sqrt{R^2 - r^2} + \tfrac{\pi}{2}r^2 - R^2\arcsin(r/R)$$

29. By symmetry of the ellipse about the x- and y-axes,

$$A = 4\int_0^a y\,dx = 4\int_{\pi/2}^0 b\sin\theta\,(-a\sin\theta)\,d\theta \qquad \begin{bmatrix} x = a\cos\theta = 0 & \Rightarrow & \theta = \tfrac{\pi}{2} \text{ and} \\ x = a\cos\theta = a & \Rightarrow & \theta = 0 \end{bmatrix}$$

$$= 4ab\int_0^{\pi/2} \sin^2\theta\,d\theta = 4ab\int_0^{\pi/2} \tfrac{1}{2}(1 - \cos 2\theta)\,d\theta$$

$$= 2ab\left[\theta - \tfrac{1}{2}\sin 2\theta\right]_0^{\pi/2} = 2ab\left(\tfrac{\pi}{2}\right) = \pi ab$$

Note that the formula for the area of a circle, $A = \pi r^2$, is just a special case of this formula with $a = b = r$.

31.

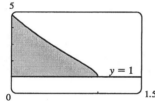

$x = \cos t,\ y = e^t,\ 0 \le t \le \tfrac{\pi}{2}$.

When $x = 0$, $t = \pi/2$; when $x = 1$, $t = 0$.

$A = \int_0^1 (y - 1)\,dx = \int_{\pi/2}^0 (e^t - 1)(-\sin t)\,dt$

$\quad = \int_0^{\pi/2} (e^t\sin t - \sin t)\,dt$

By Example 4 in Section 5.6 or Integration Formula 98, $\int e^x\sin x\,dx = \tfrac{1}{2}e^x(\sin x - \cos x) + C$, so

$$\int_0^{\pi/2}(e^t\sin t - \sin t)\,dt = \left[\tfrac{1}{2}e^t(\sin t - \cos t) + \cos t\right]_0^{\pi/2}$$

$$= \left[\tfrac{1}{2}e^{\pi/2}(1 - 0) + 0\right] - \left[\tfrac{1}{2}e^0(0 - 1) + 1\right]$$

$$= \tfrac{1}{2}e^{\pi/2} - \tfrac{1}{2} = \tfrac{1}{2}\left(e^{\pi/2} - 1\right)$$

33. By symmetry, the area of the region enclosed by the loop is twice the area above the x-axis inside the loop. $y = 0 \iff t^3 - 3t = 0 \iff$ $t(t^2 - 3) = 0 \iff t = 0, \pm\sqrt{3}$. The top half of the loop is described by $x = t^2$, $y = t^3 - 3t$, $-\sqrt{3} \le t \le 0$, so, using the Substitution Rule with $y = t^3 - 3t$ and $dx = 2t\,dt$, we find that

area $= 2\int_0^3 y\,dx = 2\int_0^{-\sqrt{3}}(t^3 - 3t)2t\,dt = 4\int_0^{-\sqrt{3}}(t^4 - 3t^2)\,dt$

$= 4\left[\tfrac{1}{5}t^5 - t^3\right]_0^{-\sqrt{3}} = 4\left[\tfrac{1}{5}\left(-3^{1/2}\right)^5 - \left(-3^{1/2}\right)^3\right]$

$= 4\left[\tfrac{1}{5}\left(-9\sqrt{3}\right) - \left(-3\sqrt{3}\right)\right] = \tfrac{24}{5}\sqrt{3} \approx 8.31$.

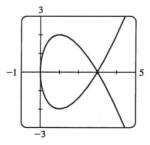

35. We first assume that $c > 0$, since c can be replaced by $-c$ in both equations without changing the graphs, and if $c = 0$ the curves do not enclose a region. We see from the graph that the enclosed area A lies between $x = -c$ and $x = c$, and by symmetry, it is equal to four times the area in the first quadrant.

The enclosed area is

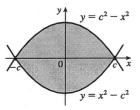

$$A = 4 \int_0^c (c^2 - x^2)\, dx = 4 \left[c^2 x - \tfrac{1}{3} x^3 \right]_0^c$$

$$= 4\left(c^3 - \tfrac{1}{3}c^3\right) = 4\left(\tfrac{2}{3}c^3\right) = \tfrac{8}{3}c^3$$

So $A = 576$ $\Leftrightarrow$ $\tfrac{8}{3}c^3 = 576$ $\Leftrightarrow$ $c^3 = 216$ $\Leftrightarrow$ $c = \sqrt[3]{216} = 6$.

Note that $c = -6$ is another solution, since the graphs are the same.

37.

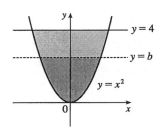

By the symmetry of the problem, we consider only the first quadrant, where $y = x^2$ $\Rightarrow$ $x = \sqrt{y}$. We are looking for a number b such that $\int_0^b \sqrt{y}\, dy = \int_b^4 \sqrt{y}\, dy$ $\Rightarrow$ $\tfrac{2}{3}\left[y^{3/2}\right]_0^b = \tfrac{2}{3}\left[y^{3/2}\right]_b^4$ $\Rightarrow$

$b^{3/2} = 4^{3/2} - b^{3/2}$ $\Rightarrow$ $2b^{3/2} = 8$ $\Rightarrow$ $b^{3/2} = 4$ $\Rightarrow$ $b = 4^{2/3} \approx 2.52$.

39. The area under the graph of f from 0 to t is equal to $\int_0^t f(x)\, dx$, so the requirement is that $\int_0^t f(x)\, dx = t^3$ for all t. We differentiate both sides of this equation with respect to t (with the help of FTC1) to get $f(t) = 3t^2$. This function is positive and continuous, as required.

41. The curve and the line will determine a region when they intersect at two or more points. So we solve the equation

$x/(x^2 + 1) = mx$ $\Rightarrow$ $x = x(mx^2 + m)$ $\Rightarrow$

$x(mx^2 + m) - x = 0$ $\Rightarrow$ $x(mx^2 + m - 1) = 0$ $\Rightarrow$

$x = 0$ or $mx^2 + m - 1 = 0$ $\Rightarrow$ $x = 0$ or $x^2 = \dfrac{1 - m}{m}$ $\Rightarrow$

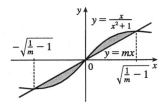

$x = 0$ or $x = \pm\sqrt{\dfrac{1}{m} - 1}$. Note that if $m = 1$, this has only

the solution $x = 0$, and no region is determined. But if $1/m - 1 > 0$ $\Leftrightarrow$ $1/m > 1$ $\Leftrightarrow$ $0 < m < 1$, then there are two solutions. [Another way of seeing this is to observe that the slope of the tangent to $y = x/(x^2 + 1)$ at the origin is $y' = 1$ and therefore we must have $0 < m < 1$.] Note that we cannot just integrate between the positive and negative roots, since the curve and the line cross at the origin. Since mx and $x/(x^2 + 1)$ are both odd functions, the total area is twice the area between the curves on the interval $\left[0, \sqrt{1/m - 1}\right]$. So the total area enclosed is

$$2 \int_0^{\sqrt{1/m-1}} \left[\frac{x}{x^2 + 1} - mx \right] dx = 2\left[\tfrac{1}{2}\ln(x^2 + 1) - \tfrac{1}{2}mx^2 \right]_0^{\sqrt{1/m-1}}$$

$$= \left[\ln(1/m - 1 + 1) - m(1/m - 1)\right] - (\ln 1 - 0)$$

$$= \ln(1/m) - 1 + m = m - \ln m - 1$$

6.2 Volumes • • • • • • • • • • • • • • • • • • •

1. A cross-section is a disk with radius $1/x$, so its area is $A(x) = \pi(1/x)^2$.

$$V = \int_1^2 A(x)\,dx = \int_1^2 \pi\left(\frac{1}{x}\right)^2 dx = \pi \int_1^2 \frac{1}{x^2}\,dx = \pi\left[-\frac{1}{x}\right]_1^2 = \pi\left[-\frac{1}{2} - (-1)\right] = \frac{\pi}{2}$$

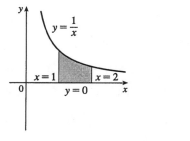

3. A cross-section is a disk with radius $\sqrt{y}$, so its area is $A(y) = \pi\left(\sqrt{y}\right)^2$.

$$V = \int_0^4 A(y)\,dy = \int_0^4 \pi\left(\sqrt{y}\right)^2 dy = \pi \int_0^4 y\,dy = \pi\left[\tfrac{1}{2}y^2\right]_0^4 = 8\pi$$

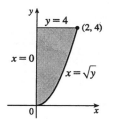

5. A cross-section is a washer (annulus) with inner radius x^2 and outer radius $\sqrt{x}$, so its area is
$A(x) = \pi(\sqrt{x})^2 - \pi(x^2)^2 = \pi(x - x^4)$.

$$V = \int_0^1 A(x)\,dx = \pi \int_0^1 (x - x^4)\,dx = \pi\left[\tfrac{1}{2}x^2 - \tfrac{1}{5}x^5\right]_0^1 = \pi\left(\tfrac{1}{2} - \tfrac{1}{5}\right) = \tfrac{3\pi}{10}$$

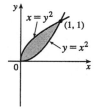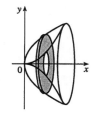

7. A cross-section is a washer with inner radius y^2 and outer radius $2y$, so its area is
$A(y) = \pi(2y)^2 - \pi(y^2)^2 = \pi(4y^2 - y^4)$.

$$V = \int_0^2 A(y)\,dy = \pi \int_0^2 (4y^2 - y^4)\,dy = \pi\left[\tfrac{4}{3}y^3 - \tfrac{1}{5}y^5\right]_0^2 = \pi\left(\tfrac{32}{3} - \tfrac{32}{5}\right) = \tfrac{64\pi}{15}$$

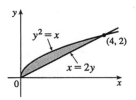
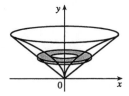

9. A cross-section is a washer with inner radius $1 - \sqrt{x}$ and outer radius $1 - x$, so its area is
$A(x) = \pi(1-x)^2 - \pi(1-\sqrt{x})^2 = \pi\left[(1-2x+x^2) - (1-2\sqrt{x}+x)\right] = \pi(-3x + x^2 + 2\sqrt{x})$.

$$V = \int_0^1 A(x)\,dx = \pi \int_0^1 (-3x + x^2 + 2\sqrt{x})\,dx$$
$$= \pi\left[-\tfrac{3}{2}x^2 + \tfrac{1}{3}x^3 + \tfrac{4}{3}x^{3/2}\right]_0^1 = \pi\left(-\tfrac{3}{2} + \tfrac{5}{3}\right) = \tfrac{\pi}{6}$$

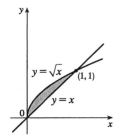
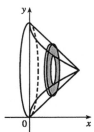

11. $y = x^2 \;\Rightarrow\; x = \sqrt{y}$ for $x \geq 0$. The outer radius is the distance from $x = -1$ to $x = \sqrt{y}$ and the inner radius is the distance from $x = -1$ to $x = y^2$.

$$V = \int_0^1 \pi\left\{\left[\sqrt{y} - (-1)\right]^2 - \left[y^2 - (-1)\right]^2\right\} dy = \pi \int_0^1 \left[(\sqrt{y}+1)^2 - (y^2+1)^2\right] dy$$
$$= \pi \int_0^1 (y + 2\sqrt{y} + 1 - y^4 - 2y^2 - 1)\,dy = \pi \int_0^1 (y + 2\sqrt{y} - y^4 - 2y^2)\,dy$$
$$= \pi\left[\tfrac{1}{2}y^2 + \tfrac{4}{3}y^{3/2} - \tfrac{1}{5}y^5 - \tfrac{2}{3}y^3\right]_0^1 = \pi\left(\tfrac{1}{2} + \tfrac{4}{3} - \tfrac{1}{5} - \tfrac{2}{3}\right) = \tfrac{29}{30}\pi$$

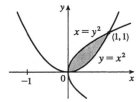
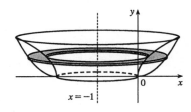

13. $y = \sqrt[3]{x} \iff x = y^3$

A cross-section is a washer with inner radius $8 - 4y$ and outer

radius $8 - y^3$, so its area is $A(y) = \pi (8 - y^3)^2 - \pi (8 - 4y)^2$.

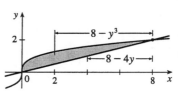

$V = \int_0^2 A(y)\,dy = \pi \int_0^2 \left[(8 - y^3)^2 - (8 - 4y)^2 \right] dy = \pi \int_0^2 \left[(64 - 16y^3 + y^6) - (64 - 64y + 16y^2) \right] dy$

$\quad = \pi \int_0^2 (-16y^3 + y^6 + 64y - 16y^2)\,dy = \pi \left[-4y^4 + \frac{1}{7}y^7 + 32y^2 - \frac{16}{3}y^3 \right]_0^2$

$\quad = \pi \left(-64 + \frac{128}{7} + 128 - \frac{128}{3} \right) = \frac{832}{21}\pi$

15.

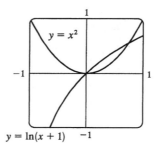

$y = x^2$ and $y = \ln(x + 1)$ intersect at $x = 0$ and at $x = a \approx 0.747$.

$$V = \pi \int_0^a \left\{ [\ln(x + 1)]^2 - (x^2)^2 \right\} dx \approx 0.132$$

17. (a) $\pi \int_0^{\pi/2} \cos^2 x\,dx$ describes the volume of the solid obtained by rotating the region

$\mathcal{R} = \left\{ (x, y) \mid 0 \le x \le \frac{\pi}{2}, 0 \le y \le \cos x \right\}$ of the xy-plane about the x-axis.

(b) $\pi \int_0^1 (y^4 - y^8)\,dy = \pi \int_0^1 \left[(y^2)^2 - (y^4)^2 \right] dy$ describes the volume of the solid obtained by rotating the region

$\mathcal{R} = \left\{ (x, y) \mid 0 \le y \le 1, y^4 \le x \le y^2 \right\}$ of the xy-plane about the y-axis.

19. There are 10 subintervals over the 15-cm length, so we'll use $n = 10/2 = 5$ for the Midpoint Rule.

$V = \int_0^{15} A(x)\,dx \approx M_5 = \frac{15-0}{5}[A(1.5) + A(4.5) + A(7.5) + A(10.5) + A(13.5)]$

$\qquad = 3(18 + 79 + 106 + 128 + 39) = 3 \cdot 370 = 1110\ \text{cm}^3$

21. We'll form a right circular cone with height h and base radius r by revolving the line $y = \frac{r}{h}x$ about the x-axis.

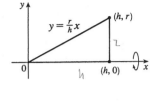

$$V = \pi \int_0^h \left(\frac{r}{h}x\right)^2 dx = \pi \int_0^h \frac{r^2}{h^2}x^2 dx = \pi \frac{r^2}{h^2}\left[\frac{1}{3}x^3\right]_0^h$$

$$= \pi \frac{r^2}{h^2}\left(\frac{1}{3}h^3\right) = \frac{1}{3}\pi r^2 h$$

Another solution: Revolve $x = -\frac{r}{h}y + r$ about the y-axis.

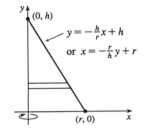

$$V = \pi \int_0^h \left(-\frac{r}{h}y + r\right)^2 dy \overset{*}{=} \pi \int_0^h \left[\frac{r^2}{h^2}y^2 - \frac{2r^2}{h}y + r^2\right] dy$$

$$= \pi \left[\frac{r^2}{3h^2}y^3 - \frac{r^2}{h}y^2 + r^2 y\right]_0^h = \pi\left(\tfrac{1}{3}r^2 h - r^2 h + r^2 h\right) = \tfrac{1}{3}\pi r^2 h$$

* Or use substitution with $u = r - \frac{r}{h}y$ and $du = -\frac{r}{h}dy$ to get

$$\pi \int_r^0 u^2\left(-\frac{h}{r}du\right) = -\pi\frac{h}{r}\left[\frac{1}{3}u^3\right]_r^0 = -\pi\frac{h}{r}\left(-\frac{1}{3}r^3\right) = \frac{1}{3}\pi r^2 h.$$

23. $x^2 + y^2 = r^2 \Leftrightarrow x^2 = r^2 - y^2$

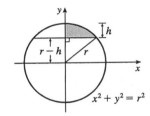

$$V = \pi \int_{r-h}^r (r^2 - y^2)\, dy = \pi\left[r^2 y - \frac{y^3}{3}\right]_{r-h}^r$$

$$= \pi\left\{\left[r^3 - \frac{r^3}{3}\right] - \left[r^2(r-h) - \frac{(r-h)^3}{3}\right]\right\}$$

$$= \pi\left\{\tfrac{2}{3}r^3 - \tfrac{1}{3}(r-h)\left[3r^2 - (r-h)^2\right]\right\}$$

$$= \tfrac{1}{3}\pi\left\{2r^3 - (r-h)\left[3r^2 - (r^2 - 2rh + h^2)\right]\right\}$$

$$= \tfrac{1}{3}\pi\left\{2r^3 - (r-h)\left[2r^2 + 2rh - h^2\right]\right\}$$

$$= \tfrac{1}{3}\pi\left(2r^3 - 2r^3 - 2r^2 h + rh^2 + 2r^2 h + 2rh^2 - h^3\right)$$

$$= \tfrac{1}{3}\pi\left(3rh^2 - h^3\right) = \tfrac{1}{3}\pi h^2(3r - h), \text{ or, equivalently, } \pi h^2\left(r - \frac{h}{3}\right)$$

25. For a cross-section at height y, we see from similar triangles that $\dfrac{\alpha/2}{b/2} = \dfrac{h-y}{h}$, so $\alpha = b\left(1 - \dfrac{y}{h}\right)$.

Similarly, for cross-sections having $2b$ as their base and β replacing α, $\beta = 2b\left(1 - \dfrac{y}{h}\right)$. So

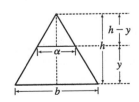

$$V = \int_0^h A(y)\, dy = \int_0^h \left[b\left(1 - \frac{y}{h}\right)\right]\left[2b\left(1 - \frac{y}{h}\right)\right] dy = \int_0^h 2b^2\left(1 - \frac{y}{h}\right)^2 dy$$

$$= 2b^2 \int_0^h \left(1 - \frac{2y}{h} + \frac{y^2}{h^2}\right) dy = 2b^2\left[y - \frac{y^2}{h} + \frac{y^3}{3h^2}\right]_0^h = 2b^2\left[h - h + \tfrac{1}{3}h\right]$$

$$= \tfrac{2}{3}b^2 h \ [= \tfrac{1}{3}Bh \text{ where } B \text{ is the area of the base, as with any pyramid.}]$$

27. A cross-section at height z is a triangle similar to the base, so we'll multiply the legs of the base triangle, 3 and 4, by a proportionality factor of $(5 - z)/5$. Thus, the triangle at height z has area

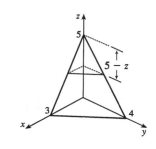

$$A(z) = \frac{1}{2} \cdot 3\left(\frac{5-z}{5}\right) \cdot 4\left(\frac{5-z}{5}\right) = 6\left(1 - \frac{z}{5}\right)^2, \text{ so}$$

$$V = \int_0^5 A(z)\,dz = 6\int_0^5 (1 - z/5)^2\,dz$$

$$= 6\int_1^0 u^2(-5\,du) \quad [u = 1 - z/5,\ du = -\tfrac{1}{5}dz]$$

$$= -30\left[\tfrac{1}{3}u^3\right]_1^0 = -30\left(-\tfrac{1}{3}\right) = 10 \text{ cm}^3$$

29. If l is a leg of the isosceles right triangle and $2y$ is the hypotenuse, then

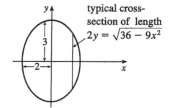

typical cross-section of length $2y = \sqrt{36 - 9x^2}$

$$l^2 + l^2 = (2y)^2 \quad \Rightarrow \quad 2l^2 = 4y^2 \quad \Rightarrow \quad l^2 = 2y^2.$$

$$V = \int_{-2}^2 A(x)\,dx = 2\int_0^2 A(x)\,dx = 2\int_0^2 \tfrac{1}{2}(l)(l)\,dx = 2\int_0^2 y^2\,dx$$

$$= 2\int_0^2 \tfrac{1}{4}(36 - 9x^2)\,dx = \tfrac{9}{2}\int_0^2 (4 - x^2)\,dx = \tfrac{9}{2}\left[4x - \tfrac{1}{3}x^3\right]_0^2$$

$$= \tfrac{9}{2}\left(8 - \tfrac{8}{3}\right) = 24$$

31. The cross-section of the base corresponding to the coordinate y has length $2x = 2\sqrt{y}$. The square has area $A(y) = \left(2\sqrt{y}\right)^2 = 4y$, so

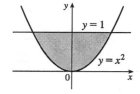

$$V = \int_0^1 A(y)\,dy = \int_0^1 4y\,dy = \left[2y^2\right]_0^1 = 2.$$

33. A typical cross-section perpendicular to the y-axis in the base has length $\ell(y) = 3 - \tfrac{3}{2}y$. This length is the leg of an isosceles right triangle, so

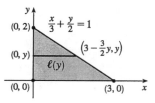

$$A(y) = \tfrac{1}{2}\left[\ell(y)\right]^2 \quad \left[\tfrac{1}{2}bh \text{ with base} = \text{height}\right]$$

$$= \tfrac{1}{2}\left[3\left(1 - \tfrac{1}{2}y\right)\right]^2 = \tfrac{9}{2}\left(1 - \tfrac{1}{2}y\right)^2$$

Thus,

$$V = \int_0^2 A(y)\,dy = \tfrac{9}{2}\int_1^0 u^2(-2\,du) \quad \left[u = 1 - \tfrac{1}{2}y,\ du = -\tfrac{1}{2}\,dy\right]$$

$$= -9\left[\tfrac{1}{3}u^3\right]_1^0 = -9\left(-\tfrac{1}{3}\right) = 3$$

35. (a) The torus is obtained by rotating the circle $(x - R)^2 + y^2 = r^2$ about the
y-axis. Solving for x, we see that the right half of the circle is given by

$x = R + \sqrt{r^2 - y^2} = f(y)$ and the left half by

$x = R - \sqrt{r^2 - y^2} = g(y)$. So

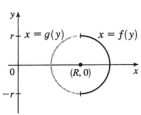

$V = \pi \int_{-r}^{r} \left\{ [f(y)]^2 - [g(y)]^2 \right\} dy$

$= 2\pi \int_{0}^{r} \left[\left(R^2 + 2R\sqrt{r^2 - y^2} + r^2 - y^2 \right) - \left(R^2 - 2R\sqrt{r^2 - y^2} + r^2 - y^2 \right) \right] dy$

$= 2\pi \int_{0}^{r} 4R\sqrt{r^2 - y^2}\, dy = 8\pi R \int_{0}^{r} \sqrt{r^2 - y^2}\, dy$

(b) Observe that the integral represents a quarter of the area of a circle with radius r, so
$8\pi R \int_{0}^{r} \sqrt{r^2 - y^2}\, dy = 8\pi R \cdot \tfrac{1}{4}\pi r^2 = 2\pi^2 r^2 R$.

37. (a) Volume$(S_1) = \int_{0}^{h} A(z)\, dz = $ Volume(S_2) since the cross-sectional area $A(z)$ at height z is the same for both
solids.

(b) By Cavalieri's Principle, the volume of the cylinder in the figure is the same as that of a right circular cylinder
with radius r and height h, that is, $\pi r^2 h$.

39. The volume is obtained by rotating the area common to two circles of
radius r, as shown. The volume of the right half is

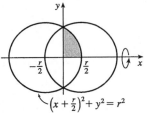

$V_{\text{right}} = \pi \int_{0}^{r/2} y^2\, dx = \pi \int_{0}^{r/2} \left[r^2 - \left(\tfrac{1}{2}r + x \right)^2 \right] dx$

$= \pi \left[r^2 x - \tfrac{1}{3}\left(\tfrac{1}{2}r + x \right)^3 \right]_{0}^{r/2} = \pi \left[\left(\tfrac{1}{2}r^3 - \tfrac{1}{3}r^3 \right) - \left(0 - \tfrac{1}{24}r^3 \right) \right] = \tfrac{5}{24}\pi r^3$

So by symmetry, the total volume is twice this, or $\tfrac{5}{12}\pi r^3$.

Another solution: We observe that the volume is the twice the volume of a cap of a sphere, so we can use the
formula from Exercise 23 with $h = \tfrac{1}{2}r$: $V = 2 \cdot \tfrac{1}{3}\pi h^2 (3r - h) = \tfrac{2}{3}\pi \left(\tfrac{1}{2}r \right)^2 \left(3r - \tfrac{1}{2}r \right) = \tfrac{5}{12}\pi r^3$.

41. Take the x-axis to be the axis of the cylindrical hole of radius r.
A quarter of the cross-section through y, perpendicular to the
y-axis, is the rectangle shown. Using the Pythagorean Theorem
twice, we see that the dimensions of this rectangle are

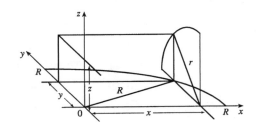

$x = \sqrt{R^2 - y^2}$ and $z = \sqrt{r^2 - y^2}$, so

$\tfrac{1}{4}A(y) = xz = \sqrt{r^2 - y^2}\sqrt{R^2 - y^2}$, and

$V = \int_{-r}^{r} A(y)\, dy = \int_{-r}^{r} 4\sqrt{r^2 - y^2}\sqrt{R^2 - y^2}\, dy$

$= 8\int_{0}^{r} \sqrt{r^2 - y^2}\sqrt{R^2 - y^2}\, dy$

43.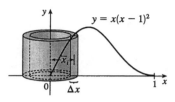

If we were to use the "washer method," we would first have to locate the local maximum point (a, b) of $y = x(x - 1)^2$ using the methods of Chapter 4. Then we would have to solve the equation $y = x(x - 1)^2$ for x in terms of y to obtain the functions $x = g_1(y)$ and $x = g_2(y)$ shown in the figure above. This step would be difficult because it involves the cubic formula. Finally we would find the volume using

$$V = \pi \int_0^b \left\{ [g_1(y)]^2 - [g_2(y)]^2 \right\} dy.$$

Instead, we use cylindrical shells. As in Example 9, we rotate an approximating rectangle with width Δx about the y-axis, to get a cylindrical shell whose average radius is $\overline{x}_i$ and whose volume is

$$2\pi \overline{x}_i \left[\overline{x}_i(\overline{x}_i - 1)^2 \right] \Delta x.$$

So the total volume is

$$V = \lim_{n \to \infty} \sum_{i=1}^n 2\pi \, \overline{x}_i \left[\overline{x}_i(\overline{x}_i - 1)^2 \right] \Delta x = \int_0^1 2\pi x \left[x(x - 1)^2 \right] dx = 2\pi \int_0^1 (x^4 - 2x^3 + x^2) \, dx$$

$$= 2\pi \left[\frac{x^5}{5} - 2\frac{x^4}{4} + \frac{x^3}{3} \right]_0^1 = 2\pi \left(\frac{1}{5} - \frac{1}{2} + \frac{1}{3} \right) = 2\pi \left(\frac{1}{30} \right) = \frac{\pi}{15}.$$

45. Let $y = f(x)$ denote the curve. Using cylindrical shells, $V = \int_2^{10} 2\pi x f(x) \, dx = 2\pi \int_2^{10} x f(x) \, dx = 2\pi I$. Now use Simpson's Rule to approximate I:

$$I \approx S_8 = \frac{10 - 2}{3(8)} \left[2f(2) + 4 \cdot 3f(3) + 2 \cdot 4f(4) + 4 \cdot 5f(5) + 2 \cdot 6f(6) \right.$$
$$\left. + 4 \cdot 7f(7) + 2 \cdot 8f(8) + 4 \cdot 9f(9) + 10f(10) \right]$$
$$\approx \tfrac{1}{3} [2(0) + 12(1.5) + 8(1.9) + 20(2.2) + 12(3.0) + 28(3.8) + 16(4.0) + 36(3.1) + 10(0)]$$
$$= \tfrac{1}{3}(395.2)$$

Thus, $V \approx 2\pi \cdot \tfrac{1}{3}(395.2) \approx 827.7$ or 828 cubic units.

47.

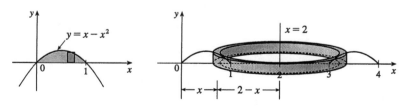

$$V = \int_0^1 (\text{circumference}) \, (\text{height}) \, (\text{thickness}) = \int_0^1 [2\pi(2 - x)] \left(x - x^2 \right) dx$$

$$= 2\pi \int_0^1 \left(x^3 - 3x^2 + 2x \right) dx = 2\pi \left[\tfrac{1}{4}x^4 - x^3 + x^2 \right]_0^1 = 2\pi \left(\tfrac{1}{4} \right) = \tfrac{\pi}{2}$$

See the solution for Exercise 43 as to why the method of cylindrical shells is preferable to slicing.

6.3 Arc Length · · · · · · · · · · · · · · ·

1. $y = 2 - 3x \implies L = \int_{-2}^{1} \sqrt{1 + (dy/dx)^2}\, dx = \int_{-2}^{1} \sqrt{1 + (-3)^2}\, dx = \sqrt{10}\,[1 - (-2)] = 3\sqrt{10}.$

The arc length can be calculated using the distance formula, since the curve is a line segment, so

$$L = [\text{distance from } (-2, 8) \text{ to } (1, -1)] = \sqrt{[1 - (-2)]^2 + [(-1) - 8]^2} = \sqrt{90} = 3\sqrt{10}$$

3. $x = t - t^2$, $y = \frac{4}{3}t^{3/2}$, $1 \le t \le 2$. $dx/dt = 1 - 2t$ and $dy/dt = 2t^{1/2}$, so

$$(dx/dt)^2 + (dy/dt)^2 = (1 - 2t)^2 + \left(2t^{1/2}\right)^2 = 1 - 4t + 4t^2 + 4t = 1 + 4t^2 \text{ and}$$

$$L = \int_a^b \sqrt{(dx/dt)^2 + (dy/dt)^2}\, dt = \int_1^2 \sqrt{1 + 4t^2}\, dt.$$

5. $x = e^t \cos t$, $y = e^t \sin t$, $0 \le t \le \pi$.

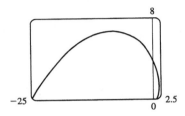

$$\left(\frac{dx}{dt}\right)^2 + \left(\frac{dy}{dt}\right)^2 = \left[e^t(\cos t - \sin t)\right]^2 + \left[e^t(\sin t + \cos t)\right]^2$$

$$= \left(e^t\right)^2 (\cos^2 t - 2\cos t \sin t + \sin^2 t)$$

$$+ \left(e^t\right)^2 (\sin^2 t + 2\sin t \cos t + \cos^2 t)$$

$$= e^{2t}(2\cos^2 t + 2\sin^2 t) = 2e^{2t}$$

Thus, $L = \int_0^\pi \sqrt{2e^{2t}}\, dt = \int_0^\pi \sqrt{2}\, e^t\, dt = \sqrt{2}\,\left[e^t\right]_0^\pi = \sqrt{2}\,(e^\pi - 1).$

7.

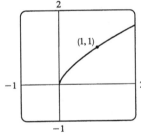

$x = y^{3/2} \implies 1 + (dx/dy)^2 = 1 + \left(\frac{3}{2}y^{1/2}\right)^2 = 1 + \frac{9}{4}y.$

$$L = \int_0^1 \sqrt{1 + \frac{9}{4}y}\, dy = \int_1^{13/4} \sqrt{u}\,\left(\frac{4}{9}\, du\right) \quad \left[u = 1 + \frac{9}{4}y,\ du = \frac{9}{4}\, dy\right]$$

$$= \frac{4}{9} \cdot \frac{2}{3}\left[u^{3/2}\right]_1^{13/4} = \frac{8}{27}\left(\frac{13\sqrt{13}}{8} - 1\right) = \frac{13\sqrt{13} - 8}{27}.$$

9. $x = e^t - t$, $y = 4e^{t/2}$, $-8 \le t \le 3$.

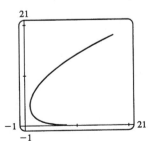

$$(dx/dt)^2 + (dy/dt)^2 = (e^t - 1)^2 + (2e^{t/2})^2 = e^{2t} - 2e^t + 1 + 4e^t$$

$$= e^{2t} + 2e^t + 1 = (e^t + 1)^2$$

$$L = \int_{-8}^{3} \sqrt{(e^t + 1)^2}\, dt = \int_{-8}^{3} (e^t + 1)\, dt = \left[e^t + t\right]_{-8}^{3}$$

$$= (e^3 + 3) - (e^{-8} - 8) = e^3 - e^{-8} + 11$$

11. $x = \ln t$ and $y = e^{-t} \implies \dfrac{dx}{dt} = \dfrac{1}{t}$ and $\dfrac{dy}{dt} = -e^{-t} \implies L = \int_1^2 \sqrt{t^{-2} + e^{-2t}}\, dt.$

Using Simpson's Rule with $n = 10$, $\Delta x = (2 - 1)/10 = 0.1$ and $f(t) = \sqrt{t^{-2} + e^{-2t}}$ we get

$L \approx \frac{0.1}{3}\,[f(1.0) + 4f(1.1) + 2f(1.2) + \cdots + 2f(1.8) + 4f(1.9) + f(2.0)] \approx 0.7314.$

13. $y = \sin x$, $1 + (dy/dx)^2 = 1 + \cos^2 x$, $L = \int_0^\pi \sqrt{1 + \cos^2 x}\, dx$. Let $g(x) = \sqrt{1 + \cos^2 x}$. Then

$$L \approx \tfrac{\pi/10}{3}\left[g(0) + 4g\!\left(\tfrac{\pi}{10}\right) + 2g\!\left(\tfrac{2\pi}{10}\right) + 4g\!\left(\tfrac{3\pi}{10}\right) + 2g\!\left(\tfrac{4\pi}{10}\right) + 4g\!\left(\tfrac{5\pi}{10}\right)\right.$$
$$\left. + 2g\!\left(\tfrac{6\pi}{10}\right) + 4g\!\left(\tfrac{7\pi}{10}\right) + 2g\!\left(\tfrac{8\pi}{10}\right) + 4g\!\left(\tfrac{9\pi}{10}\right) + g(\pi)\right] \approx 3.820$$

15. (a) Let $f(x) = y = x\sqrt[3]{4 - x}$ with $0 \le x \le 4$.

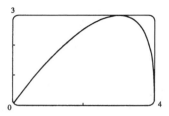

(b) The polygon with one side is just the line segment

joining the points $(0, f(0)) = (0, 0)$ and

$(4, f(4)) = (4, 0)$, and its length is 4. The polygon

with two sides joins the points $(0, 0)$,

$(2, f(2)) = (2, 2\sqrt[3]{2})$ and $(4, 0)$. Its length is

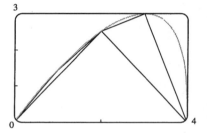

$$\sqrt{(2 - 0)^2 + \left(2\sqrt[3]{2} - 0\right)^2} + \sqrt{(4 - 2)^2 + \left(0 - 2\sqrt[3]{2}\right)^2} = 2\sqrt{4 + 2^{8/3}} \approx 6.43$$

Similarly, the inscribed polygon with four sides joins the points $(0, 0)$, $(1, \sqrt[3]{3})$, $(2, 2\sqrt[3]{2})$, $(3, 3)$, and $(4, 0)$, so its length is

$$\sqrt{1 + \left(\sqrt[3]{3}\right)^2} + \sqrt{1 + \left(2\sqrt[3]{2} - \sqrt[3]{3}\right)^2} + \sqrt{1 + \left(3 - 2\sqrt[3]{2}\right)^2} + \sqrt{1 + 9} \approx 7.50$$

(c) Using the arc length formula with

$$\frac{dy}{dx} = x\left[\tfrac{1}{3}(4 - x)^{-2/3}(-1)\right] + \sqrt[3]{4 - x} = \frac{-x + 3(4 - x)}{3(4 - x)^{2/3}} = \frac{12 - 4x}{3(4 - x)^{2/3}}, \text{ the length of the curve is}$$

$$L = \int_0^4 \sqrt{1 + \left(\frac{dy}{dx}\right)^2}\, dx = \int_0^4 \sqrt{1 + \left[\frac{12 - 4x}{3(4 - x)^{2/3}}\right]^2}\, dx.$$

(d) According to a CAS, the length of the curve is $L \approx 7.7988$. The actual value is larger than any of the approximations in part (b). This is always true, since any approximating straight line between two points on the curve is shorter than the length of the curve between the two points.

17. $x = t^3 \Rightarrow dx/dt = 3t^2$ and $y = t^4 \Rightarrow dy/dt = 4t^3$. So
$L = \int_0^1 \sqrt{9t^4 + 16t^6}\, dt = \int_0^1 \sqrt{t^4(9 + 16t^2)}\, dt = \int_0^1 t^2\sqrt{9 + 16t^2}\, dt$.
Now use Formula 22 from the table of integrals to evaluate L.

$$L = \int_0^4 \left(\tfrac{1}{4}u\right)^2 \sqrt{a^2 + u^2}\left(\tfrac{1}{4}u\right) \quad [a = 3,\, u = 4t,\, du = 4\,dt]$$

$$dt = \tfrac{1}{4}\,du$$

$$= \tfrac{1}{64}\int_0^4 u^2\sqrt{a^2 + u^2}\, du$$

$$= \tfrac{1}{64}\left[\tfrac{u}{8}\left(9 + 2u^2\right)\sqrt{9 + u^2} - \tfrac{81}{8}\ln\left(u + \sqrt{9 + u^2}\right)\right]_0^4$$

$$= \tfrac{1}{64}\left\{\left[\tfrac{1}{2}\cdot 41\cdot 5 - \tfrac{81}{8}\ln(4 + 5)\right] - \left[0 - \tfrac{81}{8}\ln 3\right]\right\}$$

$$= \tfrac{1}{64}\left[\tfrac{205}{2} - \tfrac{81}{8}(2\ln 3) + \tfrac{81}{8}\ln 3\right] \quad \left[\ln 9 = \ln 3^2 = 2\ln 3\right]$$

$$= \tfrac{1}{64}\left(\tfrac{205}{2} - \tfrac{81}{8}\ln 3\right) = \tfrac{205}{128} - \tfrac{81\ln 3}{512} \approx 1.428.$$

19. $y = \ln(\cos x) \Rightarrow y' = \dfrac{1}{\cos x}(-\sin x) = -\tan x \Rightarrow 1 + (y')^2 = 1 + \tan^2 x = \sec^2 x$.

So $L = \int_0^{\pi/4} \sec x\, dx \overset{14}{=} \left[\ln|\sec x + \tan x|\right]_0^{\pi/4} = \ln(\sqrt{2} + 1) - \ln(1 + 0) = \ln(\sqrt{2} + 1) \approx 0.881$.

21. The prey hits the ground when $y = 0 \Leftrightarrow 180 - \tfrac{1}{45}x^2 = 0 \Leftrightarrow x^2 = 45\cdot 180 \Rightarrow x = \sqrt{8100} = 90$, since
x must be positive. $y' = -\tfrac{2}{45}x \Rightarrow 1 + (y')^2 = 1 + \tfrac{4}{45^2}x^2$, so the distance traveled by the prey is

$$L = \int_0^{90}\sqrt{1 + \tfrac{4}{45^2}x^2}\, dx = \int_0^4 \sqrt{1 + u^2}\left(\tfrac{45}{2}\, du\right) \quad [u = \tfrac{2}{45}x,\, du = \tfrac{2}{45}\, dx]$$

$$\overset{21}{=} \tfrac{45}{2}\left[\tfrac{1}{2}u\sqrt{1 + u^2} + \tfrac{1}{2}\ln\left(u + \sqrt{1 + u^2}\right)\right]_0^4$$

$$= \tfrac{45}{2}\left[2\sqrt{17} + \tfrac{1}{2}\ln\left(4 + \sqrt{17}\right)\right] = 45\sqrt{17} + \tfrac{45}{4}\ln\left(4 + \sqrt{17}\right) \approx 209.1 \text{ m}$$

23. The sine wave has amplitude 1 and period 14, since it goes through two periods in a distance of 28 in., so its
equation is $y = 1\sin\left(\tfrac{2\pi}{14}x\right) = \sin\left(\tfrac{\pi}{7}x\right)$. The width w of the flat metal sheet needed to make the panel is the arc
length of the sine curve from $x = 0$ to $x = 28$. We set up the integral to evaluate w using the arc length formula
with $\tfrac{dy}{dx} = \tfrac{\pi}{7}\cos\left(\tfrac{\pi}{7}x\right)$: $L = \int_0^{28}\sqrt{1 + \left[\tfrac{\pi}{7}\cos\left(\tfrac{\pi}{7}x\right)\right]^2}\, dx = 2\int_0^{14}\sqrt{1 + \left[\tfrac{\pi}{7}\cos\left(\tfrac{\pi}{7}x\right)\right]^2}\, dx$. This integral would
be very difficult to evaluate exactly, so we use a CAS, and find that $L \approx 29.36$ inches.

25. $x = a\sin\theta,\, y = b\cos\theta,\, 0 \le \theta \le 2\pi$.

$$\left(\tfrac{dx}{d\theta}\right)^2 + \left(\tfrac{dy}{d\theta}\right)^2 = (a\cos\theta)^2 + (-b\sin\theta)^2 = a^2\cos^2\theta + b^2\sin^2\theta = a^2(1 - \sin^2\theta) + b^2\sin^2\theta$$

$$= a^2 - (a^2 - b^2)\sin^2\theta = a^2 - c^2\sin^2\theta = a^2\left(1 - \tfrac{c^2}{a^2}\sin^2\theta\right) = a^2(1 - e^2\sin^2\theta)$$

So $L = 4\int_0^{\pi/2}\sqrt{a^2\left(1 - e^2\sin^2\theta\right)}\, d\theta$ [by symmetry] $= 4a\int_0^{\pi/2}\sqrt{1 - e^2\sin^2\theta}\, d\theta$.

27. (a) Notice that $0 \le t \le 2\pi$ does not give the complete curve
because $x(0) \ne x(2\pi)$. In fact, we must take $t \in [0, 4\pi]$ in
order to obtain the complete curve, since the first term in each
of the parametric equations has period 2π and the second has
period $\tfrac{2\pi}{11/2} = \tfrac{4\pi}{11}$, and the least common integer multiple of
these two numbers is 4π.

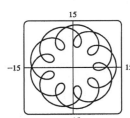

(b) We use the CAS to find the derivatives dx/dt and dy/dt, and then use Formula 1 to find the arc length. Recent versions of Maple express the integral $\int_0^{4\pi} \sqrt{(dx/dt)^2 + (dy/dt)^2}\, dt$ as $88E(2\sqrt{2}\,i)$, where $E(x)$ is the elliptic integral $\displaystyle\int_0^1 \frac{\sqrt{1 - x^2t^2}}{\sqrt{1 - t^2}}\, dt$ and i is the imaginary number $\sqrt{-1}$. Some earlier versions of Maple (as well as Mathematica) cannot do the integral exactly, so we use the command

`evalf(Int(sqrt(diff(x,t)^2+diff(y,t)^2),t=0..4*Pi));` to estimate the length, and find that the arc length is approximately 294.03. Derive's `Para_arc_length` function in the utility file `Int_apps` simplifies the integral to $11\int_0^{4\pi} \sqrt{-4\cos t\, \cos\left(\frac{11t}{2}\right) - 4\sin t\, \sin\left(\frac{11t}{2}\right) + 5}\, dt$.

6.4 Average Value of a Function · · · · · · · · · · ·

1. $g_{\text{ave}} = \frac{1}{\pi/2 - 0} \int_0^{\pi/2} \cos x\, dx = \frac{2}{\pi}[\sin x]_0^{\pi/2} = \frac{2}{\pi}(1 - 0) = \frac{2}{\pi}$

3. $f_{\text{ave}} = \frac{1}{5 - 0} \int_0^5 te^{-t^2}\, dt = \frac{1}{5}\int_0^{-25} e^u\left(-\frac{1}{2}\, du\right)$ $[u = -t^2,\, du = -2t\, dt,\, t\, dt = -\frac{1}{2}\, du]$

$= -\frac{1}{10}[e^u]_0^{-25} = -\frac{1}{10}(e^{-25} - 1) = \frac{1}{10}(1 - e^{-25})$

5. (a) $f_{\text{ave}} = \frac{1}{2 - 0}\int_0^2 (4 - x^2)\, dx$

$= \frac{1}{2}\left[4x - \frac{1}{3}x^3\right]_0^2$

$= \frac{1}{2}\left(8 - \frac{8}{3}\right) = \frac{8}{3}$

(b) $f_{\text{ave}} = f(c) \iff \frac{8}{3} = 4 - c^2 \iff c^2 = \frac{4}{3}$

$\iff c = \frac{2}{\sqrt{3}} \approx 1.15$

(c)

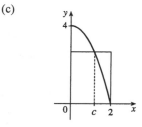

7. (a) $f_{\text{ave}} = \frac{1}{2 - 0}\int_0^2 (x^3 - x + 1)\, dx$

$= \frac{1}{2}\left[\frac{1}{4}x^4 - \frac{1}{2}x^2 + x\right]_0^2$

$= \frac{1}{2}(4 - 2 + 2) = 2$

(b) From the graph, $f(x) = 2$ at $x \approx 1.32$.

(c)

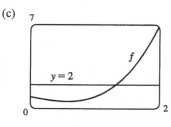

9. f is continuous on $[1, 3]$, so by the Mean Value Theorem for Integrals there exists a number c in $[1, 3]$ such that $\int_1^3 f(x)\, dx = f(c)(3 - 1) \Rightarrow 8 = 2f(c)$; that is, there is a number c such that $f(c) = \frac{8}{2} = 4$.

11. Let $t = 0$ and $t = 12$ correspond to 9 A.M. and 9 P.M., respectively.

$T_{ave} = \frac{1}{12-0} \int_0^{12} \left[50 + 14\sin\frac{1}{12}\pi t\right] dt = \frac{1}{12}\left[50t - 14 \cdot \frac{12}{\pi}\cos\frac{1}{12}\pi t\right]_0^{12}$

$= \frac{1}{12}\left[50 \cdot 12 + 14 \cdot \frac{12}{\pi} + 14 \cdot \frac{12}{\pi}\right] = \left(50 + \frac{28}{\pi}\right)°F \approx 59\ °F$

13. (a) We want to calculate the square root of the average value of $[E(t)]^2 = [155\sin(120\pi t)]^2 = 155^2\sin^2(120\pi t)$.

First, we calculate the average value itself, by integrating $[E(t)]^2$ over one cycle (between $t = 0$ and $t = \frac{1}{60}$, since there are 60 cycles per second) and dividing by $\left(\frac{1}{60} - 0\right)$:

$[E(t)]^2_{ave} = \frac{1}{1/60} \int_0^{1/60} \left[155^2\sin^2(120\pi t)\right] dt = 60 \cdot 155^2 \int_0^{1/60} \frac{1}{2}\left[1 - \cos(240\pi t)\right] dt$

$= 60 \cdot 155^2\left(\frac{1}{2}\right)\left[t - \frac{1}{240\pi}\sin(240\pi t)\right]_0^{1/60} = 60 \cdot 155^2\left(\frac{1}{2}\right)\left[\left(\frac{1}{60} - 0\right) - (0 - 0)\right] = \frac{155^2}{2}$

The RMS value is just the square root of this quantity, which is $\frac{155}{\sqrt{2}} \approx 110$ V.

(b) $220 = \sqrt{[E(t)]^2_{ave}} \quad\Rightarrow$

$220^2 = [E(t)]^2_{ave} = \frac{1}{1/60} \int_0^{1/60} A^2\sin^2(120\pi t)\, dt = 60A^2 \int_0^{1/60} \frac{1}{2}[1 - \cos(240\pi t)]\, dt$

$= 30A^2\left[t - \frac{1}{240\pi}\sin(240\pi t)\right]_0^{1/60} = 30A^2\left[\left(\frac{1}{60} - 0\right) - (0 - 0)\right] = \frac{1}{2}A^2$

Thus, $220^2 = \frac{1}{2}A^2 \quad\Rightarrow\quad A = 220\sqrt{2} \approx 311$ V.

15. $V_{ave} = \frac{1}{5}\int_0^5 V(t)\, dt = \frac{1}{5}\int_0^5 \frac{5}{4\pi}\left[1 - \cos\left(\frac{2}{5}\pi t\right)\right] dt = \frac{1}{4\pi}\int_0^5 \left[1 - \cos\left(\frac{2}{5}\pi t\right)\right] dt$

$= \frac{1}{4\pi}\left[t - \frac{5}{2\pi}\sin\left(\frac{2}{5}\pi t\right)\right]_0^5 = \frac{1}{4\pi}[(5 - 0) - 0] = \frac{5}{4\pi} \approx 0.4$ L

17. Let $F(x) = \int_a^x f(t)\, dt$ for x in $[a, b]$. Then F is continuous on $[a, b]$ and differentiable on (a, b), so by the Mean Value Theorem there is a number c in (a, b) such that $F(b) - F(a) = F'(c)(b - a)$. But $F'(x) = f(x)$ by the Fundamental Theorem of Calculus. Therefore, $\int_a^b f(t)\, dt - 0 = f(c)(b - a)$.

◆6.5◆ Applications to Physics and Engineering · · · · · · ·

1. $W = \int_a^b f(x)\, dx = \int_0^9 \frac{10}{(1+x)^2}\, dx = 10\int_1^{10} \frac{1}{u^2}\, du \quad [u = 1 + x,\ du = dx]$

$= 10\left[-\frac{1}{u}\right]_1^{10} = 10\left(-\frac{1}{10} + 1\right) = 9$ ft-lb

3. $10 = f(x) = kx = \frac{1}{3}k$ [4 inches $= \frac{1}{3}$ foot], so $k = 30$ lb/ft and $f(x) = 30x$. Now 6 inches $= \frac{1}{2}$ foot, so

$W = \int_0^{1/2} 30x\, dx = \left[15x^2\right]_0^{1/2} = \frac{15}{4}$ ft-lb.

5. (a) If $\int_0^{0.12} kx\, dx = 2$ J, then $2 = \left[\frac{1}{2}kx^2\right]_0^{0.12} = \frac{1}{2}k(0.0144) = 0.0072k$ and

$k = \frac{2}{0.0072} = \frac{2500}{9} \approx 277.78$. Thus, the work needed to stretch the spring from 35 cm to 40 cm is

$\int_{0.05}^{0.10} \frac{2500}{9}x\, dx = \left[\frac{1250}{9}x^2\right]_{1/20}^{1/10} = \frac{1250}{9}\left(\frac{1}{100} - \frac{1}{400}\right) = \frac{25}{24} \approx 1.04$ J.

(b) $f(x) = kx$, so $30 = \frac{2500}{9}x$ and $x = \frac{270}{2500}$ m $= 10.8$ cm

In Exercises 7–12, n is the number of subintervals of length Δx, and x_i^* is a sample point in the ith subinterval $[x_{i-1}, x_i]$.

7. The portion of the rope from x ft to $(x + \Delta x)$ ft below the top of the building weighs $\frac{1}{2} \Delta x$ lb and must be lifted x_i^* ft, so its contribution to the total work is $\frac{1}{2} x_i^* \Delta x$ ft-lb. The total work is

$$W = \lim_{n \to \infty} \sum_{i=1}^{n} \tfrac{1}{2} x_i^* \Delta x = \int_0^{50} \tfrac{1}{2} x \, dx = \left[\tfrac{1}{4} x^2\right]_0^{50} = \tfrac{2500}{4} = 625 \text{ ft-lb}$$

Notice that the exact height of the building does not matter (as long as it is more than 50 ft).

9. The work needed to lift the cable is $\lim_{n \to \infty} \sum_{i=1}^{n} 2x_i^* \Delta x = \int_0^{500} 2x \, dx = \left[x^2\right]_0^{500} = 250{,}000$ ft-lb. The work needed to lift the coal is $800 \text{ lb} \cdot 500 \text{ ft} = 400{,}000$ ft-lb. Thus, the total work required is $250{,}000 + 400{,}000 = 650{,}000$ ft-lb.

11. A "slice" of water Δx m thick and lying at a depth of x_i^* m (where $0 \le x_i^* \le \frac{1}{2}$) has volume $(2 \times 1 \times \Delta x)$ m³, a mass of $2000 \Delta x$ kg, weighs about $(9.8)(2000 \Delta x) = 19{,}600 \Delta x$ N, and thus requires about $19{,}600 x_i^* \Delta x$ J of work for its removal. So

$$W = \lim_{n \to \infty} \sum_{i=1}^{n} 19{,}600 x_i^* \Delta x = \int_0^{1/2} 19{,}600 x \, dx = \left[9800 x^2\right]_0^{1/2} = 2450 \text{ J}$$

13. (a) A rectangular "slice" of water Δx m thick and lying x ft above the bottom has width x ft and volume $8x\Delta x$ m³. It weighs about $(9.8 \times 1000)(8x \Delta x)$ N, and must be lifted $(5 - x)$ m by the pump, so the work needed is about $(9.8 \times 10^3)(5 - x)(8x \Delta x)$J. The total work required is

$$W \approx \int_0^3 (9.8 \times 10^3)(5 - x)8x \, dx = (9.8 \times 10^3) \int_0^3 (40x - 8x^2) \, dx = (9.8 \times 10^3) \left[20x^2 - \tfrac{8}{3} x^3\right]_0^3$$

$$= (9.8 \times 10^3)(180 - 72) = (9.8 \times 10^3)(108) = 1058.4 \times 10^3 \approx 1.06 \times 10^6 \text{ J}$$

(b) If only 4.7×10^5 J of work is done, then only the water above a certain level (call it h) will be pumped out. So we use the same formula as in part (a), except that the work is fixed, and we are trying to find the lower limit of integration: $4.7 \times 10^5 \approx \int_h^3 (9.8 \times 10^3)(5 - x)8x \, dx = (9.8 \times 10^3) \left[20x^2 - \tfrac{8}{3} x^3\right]_h^3 \Leftrightarrow$
$\frac{4.7}{9.8} \times 10^2 \approx 48 = (20 \cdot 3^2 - \tfrac{8}{3} \cdot 3^3) - (20h^2 - \tfrac{8}{3} h^3) \Leftrightarrow$
$2h^3 - 15h^2 + 45 = 0$. To find the solution of this equation, we plot
$2h^3 - 15h^2 + 45$ between $h = 0$ and $h = 3$. We see that the equation
is satisfied for $h \approx 2.0$. So the depth of water remaining in the tank is
about 2.0 m.

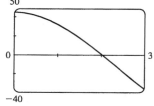

15. $V = \pi r^2 x$, so V is a function of x and P can also be regarded as a function of x. If $V_1 = \pi r^2 x_1$ and $V_2 = \pi r^2 x_2$, then

$$W = \int_{x_1}^{x_2} F(x) \, dx = \int_{x_1}^{x_2} \pi r^2 P(V(x)) \, dx = \int_{x_1}^{x_2} P(V(x)) \, dV(x) \quad [\text{Let } V(x) = \pi r^2 x, \text{ so } dV(x) = \pi r^2 \, dx.]$$

$$= \int_{V_1}^{V_2} P(V) \, dV \text{ by the Substitution Rule.}$$

17. (a) $W = \int_a^b F(r)\, dr = \int_a^b G\frac{m_1 m_2}{r^2}\, dr = Gm_1 m_2 \left[\frac{-1}{r}\right]_a^b = Gm_1 m_2 \left(\frac{1}{a} - \frac{1}{b}\right)$

(b) By part (a), $W = GMm\left(\frac{1}{R} - \frac{1}{R + 1{,}000{,}000}\right)$ where M = mass of earth in kg, R = radius of earth in m, and m = mass of satellite in kg. (Note that 1000 km = 1,000,000 m.) Thus,

$$W = (6.67 \times 10^{-11})(5.98 \times 10^{24})(1000) \times \left(\frac{1}{6.37 \times 10^6} - \frac{1}{7.37 \times 10^6}\right) \approx 8.50 \times 10^9 \text{ J.}$$

In Exercises 19–21, n is the number of subintervals of length Δx and x_i^* is a sample point in the ith subinterval $[x_{i-1}, x_i]$.

19. In the middle of the figure in the text, draw a vertical x-axis that increases in the downward direction. Since $x^2 + y^2 = 10^2$, $y = \sqrt{100 - x^2}$, and the width of the ith rectangular strip is $2y = 2\sqrt{100 - (x_i^*)^2}$. The area of the ith rectangular strip is $2\sqrt{100 - (x_i^*)^2}\,\Delta x$ and the pressure on the strip is $\rho g x_i^*$ [$\rho = 1000$ kg/m^3 and $g = 9.8$ m/s^2]. Thus, the hydrostatic force on the ith strip is the product $\rho g x_i^* 2\sqrt{100 - (x_i^*)^2}\,\Delta x$.

$F = \lim\limits_{n\to\infty} \sum\limits_{i=1}^{n} \rho g x_i^* 2\sqrt{100 - (x_i^*)^2}\,\Delta x = \int_0^{10} \rho g x \cdot 2\sqrt{100 - x^2}\, dx = 9.8 \times 10^3 \int_0^{10} \sqrt{100 - x^2}\, 2x\, dx$

$= 9.8 \times 10^3 \int_{100}^0 u^{1/2}(-du)\quad [u = 100 - x^2,\ du = -2x\, dx]$

$= 9.8 \times 10^3 \int_0^{100} u^{1/2}\, du = 9.8 \times 10^3 \left[\frac{2}{3} u^{3/2}\right]_0^{100} = \frac{2}{3} \cdot 9.8 \times 10^6 \approx 6.5 \times 10^6$ N

21. Using similar triangles, $\dfrac{4 \text{ ft wide}}{8 \text{ ft high}} = \dfrac{a \text{ ft wide}}{x_i^* \text{ ft high}}$, so $a = \frac{1}{2} x_i^*$ and the width of the ith rectangular strip is $12 + 2a = 12 + x_i^*$. The area of the strip is $(12 + x_i^*)\,\Delta x$. The pressure on the strip is δx_i^*.

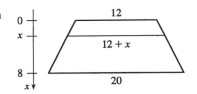

$F = \lim\limits_{n\to\infty} \sum\limits_{i=1}^{n} \delta x_i^*(12 + x_i^*)\,\Delta x = \int_0^8 \delta x \cdot (12 + x)\, dx$

$= \delta \int_0^8 (12x + x^2)\, dx = \delta \left[6x^2 + \frac{x^3}{3}\right]_0^8 = \delta\left(384 + \frac{512}{3}\right)$

$= (62.5)\frac{1664}{3} \approx 3.47 \times 10^4$ lb

23. (a) The area of a strip is $20\,\Delta x$ and the pressure on it is δx_i.

$F = \int_0^3 \delta x 20\, dx = 20\delta \left[\frac{1}{2} x^2\right]_0^3 = 20\delta \cdot \frac{9}{2} = 90\delta = 90(62.5) = 5625$ lb $\approx 5.63 \times 10^3$ lb.

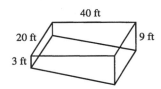

40 ft

20 ft 9 ft

3 ft

(b) $F = \int_0^9 \delta x 20\, dx = 20\delta\left[\frac{1}{2}x^2\right]_0^9 = 20\delta \cdot \frac{81}{2} = 810\delta = 810(62.5) = 50{,}625 \text{ lb} \approx 5.06 \times 10^4 \text{ lb.}$

(c) For the first 3 ft, the length of the side is constant at 40 ft. For $3 < x \le 9$, we can use similar triangles to find the

length a: $\dfrac{a}{40} = \dfrac{9-x}{6} \quad \Rightarrow \quad a = 40 \cdot \dfrac{9-x}{6}.$

$$F = \int_0^3 \delta x 40\, dx + \int_3^9 \delta x (40)\frac{9-x}{6}\, dx = 40\delta\left[\frac{1}{2}x^2\right]_0^3 + \frac{20}{3}\delta\int_3^9\left(9x - x^2\right)dx$$

$$= 180\delta + \frac{20}{3}\delta\left[\frac{9}{2}x^2 - \frac{1}{3}x^3\right]_3^9 = 180\delta + \frac{20}{3}\delta\left[\left(\frac{729}{2} - 243\right) - \left(\frac{81}{2} - 9\right)\right]$$

$$= 180\delta + 600\delta = 780\delta = 780(62.5) = 48{,}750 \text{ lb} = 780(62.5) \approx 4.88 \times 10^4 \text{ lb}$$

(d) For any right triangle with hypotenuse on the bottom,

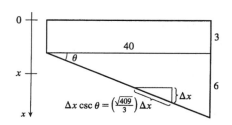

$\csc\theta = \dfrac{\Delta x}{\text{hypotenuse}} \quad \Rightarrow$

$\text{hypotenuse} = \Delta x \csc\theta = \Delta x \dfrac{\sqrt{40^2 + 6^2}}{6} = \dfrac{\sqrt{409}}{3}\Delta x.$

$F = \int_3^9 \delta x 20 \frac{\sqrt{409}}{3}\, dx = \frac{1}{3}\left(20\sqrt{409}\right)\delta\left[\frac{1}{2}x^2\right]_3^9$

$= \frac{1}{3}\cdot 10\sqrt{409}\,\delta(81 - 9)$

$\approx 303{,}356 \text{ lb} \approx 3.03 \times 10^5 \text{ lb}$

25. $m_1 = 4$, $m_2 = 8$; $P_1(-1, 2)$, $P_2(2, 4)$. The total mass of the system is $m = \sum\limits_{i=1}^{2} m_i = m_1 + m_2 = 12$. The

moment of the system about the x-axis is $M_x = \sum\limits_{i=1}^{2} m_i y_i = 4\cdot 2 + 8\cdot 4 = 40$. The moment of the system about

the y-axis is $M_y = \sum\limits_{i=1}^{2} m_i x_i = 4\cdot(-1) + 8\cdot 2 = 12$. $\overline{x} = \dfrac{M_y}{m} = \dfrac{12}{12} = 1$ and $\overline{y} = \dfrac{M_x}{m} = \dfrac{40}{12} = \dfrac{10}{3}$, so the

center of mass of the system is $(\overline{x}, \overline{y}) = \left(1, \frac{10}{3}\right)$.

27. $A = \int_0^2 x^2\, dx = \left[\frac{1}{3}x^3\right]_0^2 = \frac{8}{3}$,

$\overline{x} = \frac{1}{A}\int_0^2 xf(x)\, dx = \frac{1}{A}\int_0^2 x \cdot x^2\, dx = \frac{3}{8}\left[\frac{1}{4}x^4\right]_0^2 = \frac{3}{8}\cdot 4 = \frac{3}{2}$,

$\overline{y} = \frac{1}{A}\int_0^2 \frac{1}{2}[f(x)]^2\, dx = \frac{1}{A}\int_0^2 \frac{1}{2}\left(x^2\right)^2\, dx = \frac{3}{8}\cdot\frac{1}{2}\left[\frac{1}{5}x^5\right]_0^2 = \frac{3}{16}\cdot\frac{32}{5} = \frac{6}{5}$.

Centroid $(\overline{x}, \overline{y}) = \left(\frac{3}{2}, \frac{6}{5}\right) = (1.5, 1.2)$.

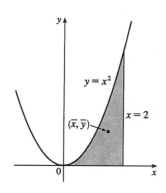

29. $A = \int_0^1 e^x \, dx = [e^x]_0^1 = e - 1,$

$\bar{x} = \frac{1}{A} \int_0^1 x e^x \, dx = \frac{1}{e-1} [x e^x - e^x]_0^1$ [by parts]

$= \frac{1}{e-1} [0 - (-1)] = \frac{1}{e-1},$

$\bar{y} = \frac{1}{A} \int_0^1 \frac{1}{2} (e^x)^2 \, dx = \frac{1}{e-1} \cdot \frac{1}{4} [e^{2x}]_0^1 = \frac{1}{4(e-1)} (e^2 - 1) = \frac{e+1}{4}.$

Centroid $(\bar{x}, \bar{y}) = \left(\frac{1}{e-1}, \frac{e+1}{4} \right) \approx (0.58, 0.93).$

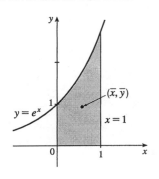

31. By symmetry, $M_y = 0$ and $\bar{x} = 0.$ $A = \frac{1}{2} bh = \frac{1}{2} \cdot 2 \cdot 2 = 2.$

$M_x = \rho \int_{-1}^1 \frac{1}{2} (2 - 2x)^2 \, dx = 2\rho \int_0^1 \frac{1}{2} (2 - 2x)^2 \, dx$

$= \left(2 \cdot 1 \cdot \frac{1}{2} \cdot 2^2 \right) \int_0^1 (1 - x)^2 \, dx$

$= 4 \int_1^0 u^2 (-du)$ $[u = 1 - x, \, du = -dx]$

$= -4 \left[\frac{1}{3} u^3 \right]_1^0 = -4 \left(-\frac{1}{3} \right) = \frac{4}{3}$

$\bar{y} = \frac{1}{m} M_x = \frac{1}{\rho A} M_x = \frac{1}{1 \cdot 2} \cdot \frac{4}{3} = \frac{2}{3}.$ $(\bar{x}, \bar{y}) = \left(0, \frac{2}{3} \right).$

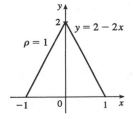

33. (a)

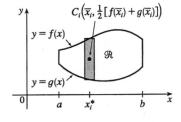

Suppose the region lies between two curves $y = f(x)$ and $y = g(x)$ where $f(x) \geq g(x)$, as illustrated in the figure. Use n subintervals determined by points x_i with $a = x_0 < x_1 < \cdots < x_n = b$ and choose $x_i^* = \bar{x}_i$ to be the midpoint of the ith subinterval; that is, $\bar{x}_i = \frac{1}{2} (x_{i-1} + x_i)$. Then the centroid of the ith approximating rectangle R_i is its center $C_i = \left(\bar{x}_i, \frac{1}{2} [f(\bar{x}_i) + g(\bar{x}_i)] \right)$. Its area is $[f(\bar{x}_i) - g(\bar{x}_i)] \Delta x$, so its mass is $\rho [f(\bar{x}_i) - g(\bar{x}_i)] \Delta x.$

Thus, $M_y(R_i) = \rho [f(\bar{x}_i) - g(\bar{x}_i)] \Delta x \cdot \bar{x}_i = \rho \bar{x}_i [f(\bar{x}_i) - g(\bar{x}_i)] \Delta x$ and

$M_x(R_i) = \rho [f(\bar{x}_i) - g(\bar{x}_i)] \Delta x \cdot \frac{1}{2} [f(\bar{x}_i) + g(\bar{x}_i)] = \rho \cdot \frac{1}{2} \left\{ [f(\bar{x}_i)]^2 - [g(\bar{x}_i)]^2 \right\} \Delta x.$ Summing over i and

taking the limit as $n \to \infty$, we get $M_y = \lim\limits_{n \to \infty} \sum\limits_{i=1}^n \rho \bar{x}_i [f(\bar{x}_i) - g(\bar{x}_i)] \Delta x = \rho \int_a^b x [f(x) - g(x)] \, dx$ and

$M_x = \lim\limits_{n \to \infty} \sum\limits_{i=1}^n \rho \cdot \frac{1}{2} \left[f(\bar{x}_i)^2 - g(\bar{x}_i)^2 \right] \Delta x = \rho \int_a^b \frac{1}{2} \left\{ [f(x)]^2 - [g(x)]^2 \right\} \, dx.$ Thus,

$\bar{x} = \frac{M_y}{m} = \frac{M_y}{\rho A} = \frac{1}{A} \int_a^b x [f(x) - g(x)] \, dx$ and $\bar{y} = \frac{M_x}{m} = \frac{M_x}{\rho A} = \frac{1}{A} \int_a^b \frac{1}{2} \left\{ [f(x)]^2 - [g(x)]^2 \right\} \, dx.$

(b)

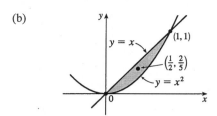

The region is sketched in the figure. We take $f(x) = x$, $g(x) = x^2$, $a = 0$, and $b = 1$ in the formulas in part (a). First we note that the area of the region is

$$A = \int_0^1 (x - x^2)\,dx = \left[\tfrac{1}{2}x^2 - \tfrac{1}{3}x^3\right]_0^1 = \tfrac{1}{6}$$

Therefore,

$$\bar{x} = \frac{1}{A}\int_0^1 x[f(x) - g(x)]\,dx = \frac{1}{1/6}\int_0^1 x(x - x^2)\,dx = 6\int_0^1 (x^2 - x^3)\,dx = 6\left[\tfrac{1}{3}x^3 - \tfrac{1}{4}x^4\right]_0^1 = \frac{1}{2} \text{ and}$$

$$\bar{y} = \frac{1}{A}\int_0^1 \frac{1}{2}\left\{[f(x)]^2 - [g(x)]^2\right\}\,dx = \frac{1}{1/6}\int_0^1 \tfrac{1}{2}(x^2 - x^4)\,dx = 3\left[\tfrac{1}{3}x^3 - \tfrac{1}{5}x^5\right]_0^1 = \frac{2}{5}.$$

The centroid is $\left(\tfrac{1}{2}, \tfrac{2}{5}\right)$.

 Applications to Economics and Biology · · · · · ·

1. $C(2000) = C(0) + \int_0^{2000} C'(x)\,dx = 1{,}500{,}000 + \int_0^{2000} (0.006x^2 - 1.5x + 8)\,dx$

$$= 1{,}500{,}000 + \left[0.002x^3 - 0.75x^2 + 8x\right]_0^{2000} = \$14{,}516{,}000$$

3. If the production level is raised from 1200 units to 1600 units, then the increase in cost is

$$C(1600) - C(1200) = \int_{1200}^{1600} C'(x)\,dx = \int_{1200}^{1600} (74 + 1.1x - 0.002x^2 + 0.00004x^3)\,dx$$

$$= \left[74x + 0.55x^2 - \tfrac{0.002}{3}x^3 + 0.00001x^4\right]_{1200}^{1600}$$

$$= 64{,}331{,}733.33 - 20{,}464{,}800 = \$43{,}866{,}933.33$$

5. $p(x) = 10 \quad \Rightarrow \quad \dfrac{450}{x + 8} = 10 \quad \Rightarrow \quad x + 8 = 45 \quad \Rightarrow \quad x = 37.$

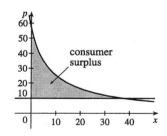

$$\text{Consumer surplus} = \int_0^{37} [p(x) - 10]\,dx = \int_0^{37} \left(\frac{450}{x + 8} - 10\right)dx$$

$$= \left[450\ln(x + 8) - 10x\right]_0^{37}$$

$$= (450\ln 45 - 370) - 450\ln 8$$

$$= 450\ln\left(\tfrac{45}{8}\right) - 370 \approx \$407.25$$

7. $P = p_S(x) = 10 = 5 + \tfrac{1}{10}\sqrt{x} \quad \Rightarrow \quad 50 = \sqrt{x} \quad \Rightarrow \quad x = 2500.$

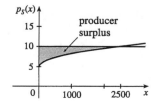

$$\text{Producer surplus} = \int_0^{2500} [P - p_S(x)]\,dx$$

$$= \int_0^{2500} \left(10 - 5 - \tfrac{1}{10}\sqrt{x}\right)dx$$

$$= \left[5x - \tfrac{1}{15}x^{3/2}\right]_0^{2500} \approx \$4166.67$$

9. $p(x) = \dfrac{800{,}000e^{-x/5000}}{x + 20{,}000} = 16 \quad \Rightarrow \quad x = x_1 \approx 3727.04.$

Consumer surplus $= \int_0^{x_1} [p(x) - 16] \, dx \approx \$37{,}753$

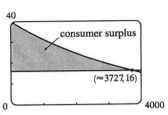

11. $f(8) - f(4) = \int_4^8 f'(t) \, dt = \int_4^8 \sqrt{t} \, dt = \left[\frac{2}{3} t^{3/2} \right]_4^8 = \frac{2}{3} \left(16\sqrt{2} - 8 \right) \approx \$9.75$ million

13. $F = \dfrac{\pi P R^4}{8\eta l} = \dfrac{\pi (4000)(0.008)^4}{8(0.027)(2)} \approx 1.19 \times 10^{-4} \text{ cm}^3/\text{s}$

15. $\int_0^{12} c(t) \, dt = \int_0^{12} \frac{1}{4} t(12 - t) \, dt = \int_0^{12} \left(3t - \frac{1}{4} t^2 \right) dt = \left[\frac{3}{2} t^2 - \frac{1}{12} t^3 \right]_0^{12} = (216 - 144) = 72 \text{ mg} \cdot \text{s/L}.$

Thus, the cardiac output is $F = \dfrac{A}{\int_0^{12} c(t) \, dt} = \dfrac{8 \text{ mg}}{72 \text{ mg} \cdot \text{s/L}} = \dfrac{1}{9} \text{ L/s} = \dfrac{60}{9} \text{ L/min}.$

 Probability · · · · · · · · · · · · · · · · · ·

1. (a) $\displaystyle\int_{30{,}000}^{40{,}000} f(x) \, dx$ is the probability that a randomly chosen tire will have a lifetime between 30,000 and 40,000 miles.

(b) $\displaystyle\int_{25{,}000}^{\infty} f(x) \, dx$ is the probability that a randomly chosen tire will have a lifetime of at least 25,000 miles.

3. (a) In general, we must satisfy the two conditions that are mentioned before Example 1—namely, (1) $f(x) \geq 0$ for all x, and (2) $\int_{-\infty}^{\infty} f(x) \, dx = 1$. Since $f(x) = 0$ or $f(x) = 0.1$, condition (1) is satisfied. For condition (2), we see that $\int_{-\infty}^{\infty} f(x) \, dx = \int_0^{10} 0.1 \, dx = \left[\frac{1}{10} x \right]_0^{10} = 1$. Thus, $f(x)$ is a probability density function for the spinner's values.

(b) Since all the numbers between 0 and 10 are equally likely to be selected, we expect the mean to be halfway between the endpoints of the interval; that is, $x = 5$.

$\mu = \int_{-\infty}^{\infty} x f(x) \, dx = \int_0^{10} x(0.1) \, dx = \left[\frac{1}{20} x^2 \right]_0^{10} = \frac{100}{20} = 5$, as expected.

5. We need to find m so that $\int_m^{\infty} f(t) \, dt = \frac{1}{2} \quad \Rightarrow \quad \lim\limits_{x \to \infty} \int_m^x \frac{1}{5} e^{-t/5} \, dt = \frac{1}{2} \quad \Rightarrow \quad \lim\limits_{x \to \infty} \left[\frac{1}{5}(-5) e^{-t/5} \right]_m^x = \frac{1}{2} \quad \Rightarrow$

$(-1)\left(0 - e^{-m/5} \right) = \frac{1}{2} \quad \Rightarrow \quad e^{-m/5} = \frac{1}{2} \quad \Rightarrow \quad -m/5 = \ln \frac{1}{2} \quad \Rightarrow \quad m = -5 \ln \frac{1}{2} = 5 \ln 2 \approx 3.47$ min.

7. We use an exponential density function with $\mu = 2.5$ min.

(a) $P(X > 4) = \int_4^{\infty} f(t) \, dt = \lim\limits_{x \to \infty} \int_4^x \frac{1}{2.5} e^{-t/2.5} \, dt = \lim\limits_{x \to \infty} \left[-e^{-t/2.5} \right]_4^x = 0 + e^{-4/2.5} \approx 0.202$

(b) $P(0 \leq X \leq 2) = \int_0^2 f(t) \, dt = \left[-e^{-t/2.5} \right]_0^2 = -e^{-2/2.5} + 1 \approx 0.551$

(c) We need to find a value a so that $P(X \geq a) = 0.02$, or, equivalently, $P(0 \leq X \leq a) = 0.98 \quad \Leftrightarrow$

$\int_0^a f(t) \, dt = 0.98 \quad \Leftrightarrow \quad \left[-e^{-t/2.5} \right]_0^a = 0.98 \quad \Leftrightarrow \quad -e^{-a/2.5} + 1 = 0.98 \quad \Leftrightarrow \quad e^{-a/2.5} = 0.02 \quad \Leftrightarrow$

$-a/2.5 = \ln 0.02 \quad \Leftrightarrow \quad a = -2.5 \ln \frac{1}{50} = 2.5 \ln 50 \approx 9.78 \text{ min} \approx 10$ min. The ad should say that if you aren't served within 10 minutes, you get a free hamburger.

9. $P(X \geq 10) = \displaystyle\int_{10}^{\infty} \frac{1}{4.2\sqrt{2\pi}} \exp\left(-\frac{(x - 9.4)^2}{2 \cdot 4.2^2}\right) dx$. To avoid the improper integral we approximate it by the

integral from 10 to 100. Thus, $P(X \geq 10) \approx \displaystyle\int_{10}^{100} \frac{1}{4.2\sqrt{2\pi}} \exp\left(-\frac{(x - 9.4)^2}{2 \cdot 4.2^2}\right) dx \approx 0.443$ (using a calculator

or computer to estimate the integral), so about 44 percent of the households throw out at least 10 lb of paper a week.
Note: We can't evaluate $1 - P(0 \leq X \leq 10)$ for this problem since a significant amount of area lies to the left of
$X = 0$.

11. $P(\mu - 2\sigma \leq X \leq \mu + 2\sigma) = \displaystyle\int_{\mu-2\sigma}^{\mu+2\sigma} \frac{1}{\sigma\sqrt{2\pi}} \exp\left(-\frac{(x-\mu)^2}{2\sigma^2}\right) dx$. Substituting $t = \dfrac{x - \mu}{\sigma}$ and $dt = \dfrac{1}{\sigma} dx$

gives us

$$\int_{-2}^{2} \frac{1}{\sigma\sqrt{2\pi}} e^{-t^2/2}(\sigma\, dt) = \frac{1}{\sqrt{2\pi}} \int_{-2}^{2} e^{-t^2/2}\, dt \approx 0.9545$$

13. (a) First $p(r) = \dfrac{4}{a_0^3} r^2 e^{-2r/a_0} \geq 0$ for $r \geq 0$. Next,

$$\int_{-\infty}^{\infty} p(r)\, dr = \int_{0}^{\infty} \frac{4}{a_0^3} r^2 e^{-2r/a_0}\, dr = \frac{4}{a_0^3} \lim_{t\to\infty} \int_{0}^{t} r^2 e^{-2r/a_0}\, dr$$

As in Exercise 12, we use (2) from that solution (with $b = -2/a_0$) and l'Hospital's Rule to get

$\dfrac{4}{a_0^3}\left[\dfrac{a_0^3}{-8}(-2)\right] = 1$. This satisfies the second condition for a function to be a probability density function.

(b) Using l'Hospital's Rule, $\dfrac{4}{a_0^3} \displaystyle\lim_{r\to\infty} \frac{r^2}{e^{2r/a_0}} = \frac{4}{a_0^3} \lim_{r\to\infty} \frac{2r}{(2/a_0)e^{2r/a_0}} = \frac{2}{a_0^2} \lim_{r\to\infty} \frac{2}{(2/a_0)e^{2r/a_0}} = 0$.

To find the maximum of p, we differentiate:

$$p'(r) = \frac{4}{a_0^3}\left[r^2 e^{-2r/a_0}\left(-\frac{2}{a_0}\right) + e^{-2r/a_0}(2r)\right] = \frac{4}{a_0^3} e^{-2r/a_0}(2r)\left(-\frac{r}{a_0} + 1\right)$$

$p'(r) = 0 \Leftrightarrow r = 0$ or $1 = \dfrac{r}{a_0} \Leftrightarrow r = a_0$ [$a_0 \approx 5.59 \times 10^{-11}$ m]. $p'(r)$ changes from positive to

negative at $r = a_0$, so $p(r)$ has its maximum value at $r = a_0$.

(c) It is fairly difficult to find a viewing rectangle, but knowing the
maximum value from part (b) helps.

$$p(a_0) = \frac{4}{a_0^3} a_0^2 e^{-2a_0/a_0} = \frac{4}{a_0} e^{-2} \approx 9{,}684{,}098{,}979$$

With a maximum of nearly 10 billion and a total area under the curve
of 1, we know that the "hump" in the graph must be extremely
narrow.

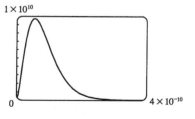

(d) $P(r) = \displaystyle\int_{0}^{r} \frac{4}{a_0^3} s^2 e^{-2s/a_0}\, ds \Rightarrow P(4a_0) = \int_{0}^{4a_0} \frac{4}{a_0^3} s^2 e^{-2s/a_0}\, ds$. Using (2) from the solution to
Exercise 12 (with $b = -2/a_0$),

$$P(4a_0) = \frac{4}{a_0^3}\left[\frac{e^{-2s/a_0}}{-8/a_0^3}\left(\frac{4}{a_0^2}s^2 + \frac{4}{a_0}s + 2\right)\right]_{0}^{4a_0} = \frac{4}{a_0^3}\left(\frac{a_0^3}{-8}\right)\left[e^{-8}(64 + 16 + 2) - 1(2)\right]$$

$$= -\tfrac{1}{2}\left(82e^{-8} - 2\right) = 1 - 41e^{-8} \approx 0.986$$

(e) $\mu = \int_{-\infty}^{\infty} rp(r)\, dr = \dfrac{4}{a_0^3} \lim\limits_{t \to \infty} \int_0^t r^3 e^{-2r/a_0}\, dr$. Integrating by parts three times or using a CAS, we find that

$\int x^3 e^{bx}\, dx = \dfrac{e^{bx}}{b^4}\left(b^3 x^3 - 3b^2 x^2 + 6bx - 6\right)$. So with $b = -\dfrac{2}{a_0}$, we use l'Hospital's Rule, and get

$\mu = \dfrac{4}{a_0^3}\left[\dfrac{a_0^4}{16}(-6)\right] = \tfrac{3}{2}a_0$.

6 Review

─────────── • CONCEPT CHECK • ───────────

1. (a) See Section 6.1, Figure 2 and Equations 6.1.1 and 6.1.2.

   (b) Instead of using "top minus bottom" and integrating from left to right, we use "right minus left" and integrate from bottom to top. See Figures 9 and 10 in Section 6.1.

2. The numerical value of the area represents the number of meters by which Sue is ahead of Kathy after 1 minute.

3. (a) See the discussion in Section 6.2, near Figures 2 and 3, ending in the Definition of Volume.

   (b) See the discussion between Examples 5 and 6 in Section 6.2. If the cross-section is a disk, find the radius in terms of x or y and use $A = \pi(\text{radius})^2$. If the cross-section is a washer, find the inner radius r_{in} and outer radius r_{out} and use $A = \pi\left(r_{out}^2\right) - \pi\left(r_{in}^2\right)$.

4. (a) The length of a curve is defined to be the limit of the lengths of the inscribed polygons, as described near Figure 3 in Section 6.3.

   (b) See Equation 6.3.1.

   (c) See Equations 6.3.2 and 6.3.3.

5. (a) See the boxed equation preceding Example 1 in Section 6.4.

   (b) The Mean Value Theorem for Integrals says that there is a number c at which the value of f is exactly equal to the average value of the function, that is, $f(c) = f_{ave}$. For a geometric interpretation of the Mean Value Theorem for Integrals, see Figure 2 in Section 6.4 and the discussion that accompanies it.

6. $\int_0^6 f(x)\, dx$ represents the amount of work done. Its units are newton-meters, or joules.

7. Let $c(x)$ be the cross-sectional length of the wall (measured parallel to the surface of the fluid) at depth x. Then the hydrostatic force against the wall is given by $F = \int_a^b \delta x c(x)\, dx$, where a and b are the lower and upper limits for x at points of the wall and δ is the weight density of the fluid.

8. (a) The center of mass is the point at which the plate balances horizontally.

   (b) See Equations 6.5.11.

9. See Figure 3 in Section 6.6, and the discussion which precedes it.

10. (a) See the definition before Figure 6 in Section 6.6.

    (b) See the discussion after Figure 6 in Section 6.6.

11. A probability density function f is a function on the domain of a continuous random variable X such that $\int_a^b f(x)\, dx$ measures the probability that X lies between a and b. Such a function f has nonnegative values and satisfies the relation $\int_D f(x)\, dx = 1$, where D is the domain of the corresponding random variable X. If $D = \mathbb{R}$, or if we define $f(x) = 0$ for real numbers $x \notin D$, then $\int_{-\infty}^{\infty} f(x)\, dx = 1$. (Of course, to work with f in this way, we must assume that the integrals of f exist.)

12. (a) $\int_0^{100} f(x)\, dx$ represents the probability that the weight of a randomly chosen female college student is less than 100 pounds.

(b) $\mu = \int_{-\infty}^{\infty} x f(x)\, dx = \int_0^{\infty} x f(x)\, dx$

(c) The median of f is the number m such that $\int_m^{\infty} f(x)\, dx = \frac{1}{2}$.

◆ **EXERCISES** ◆

---

1. $A = \int_0^1 \left[(e^x - 1) - (x^2 - x)\right] dx$

$= \int_0^1 \left(e^x - 1 - x^2 + x\right) dx = \left[e^x - x - \frac{1}{3}x^3 + \frac{1}{2}x^2\right]_0^1$

$= \left(e - 1 - \frac{1}{3} + \frac{1}{2}\right) - (1 - 0 - 0 + 0) = e - \frac{11}{6}$

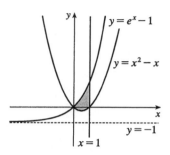

3. $x = 2\theta - \sin\theta \quad \Rightarrow \quad dx = (2 - \cos\theta)\, d\theta$

$A = \int_0^{2\pi} y\, dx = \int_0^{2\pi} \left[(2 - \cos\theta)(2 - \cos\theta)\right] d\theta$

$= \int_0^{2\pi} \left(4 - 4\cos\theta + \cos^2\theta\right) d\theta$

$= \int_0^{2\pi} \left(4 - 4\cos\theta + \frac{1}{2} + \frac{1}{2}\cos 2\theta\right) d\theta$

$= \left[4\theta - 4\sin\theta + \frac{1}{2}\theta + \frac{1}{4}\sin 2\theta\right]_0^{2\pi}$

$= (8\pi - 0 + \pi + 0) - (0) = 9\pi$

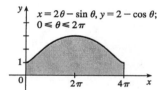

5. (a) Using the Midpoint Rule on $[0, 1]$ with $f(x) = \tan(x^2)$ and $n = 4$, we estimate

$A = \int_0^1 \tan(x^2)\, dx \approx \frac{1}{4}\left[\tan\left(\left(\frac{1}{8}\right)^2\right) + \tan\left(\left(\frac{3}{8}\right)^2\right) + \tan\left(\left(\frac{5}{8}\right)^2\right) + \tan\left(\left(\frac{7}{8}\right)^2\right)\right] \approx \frac{1}{4}(1.53) \approx 0.38$

(b) Using the Midpoint Rule on $[0, 1]$ with $f(x) = \pi \tan^2(x^2)$ (for disks) and $n = 4$, we estimate

$V = \int_0^1 f(x)\, dx \approx \frac{1}{4}\pi\left[\tan^2\left(\left(\frac{1}{8}\right)^2\right) + \tan^2\left(\left(\frac{3}{8}\right)^2\right) + \tan^2\left(\left(\frac{5}{8}\right)^2\right) + \tan^2\left(\left(\frac{7}{8}\right)^2\right)\right] \approx \frac{\pi}{4}(1.114) \approx 0.87$

7. (a) A cross-section is a washer with inner radius x^2 and outer radius x.

$V = \int_0^1 \pi\left[(x)^2 - (x^2)^2\right] dx = \int_0^1 \pi(x^2 - x^4)\, dx = \pi\left[\frac{1}{3}x^3 - \frac{1}{5}x^5\right]_0^1 = \pi\left[\frac{1}{3} - \frac{1}{5}\right] = \frac{2\pi}{15}$

(b) A cross-section is a washer with inner radius y and outer radius $\sqrt{y}$.

$V = \int_0^1 \pi\left[(\sqrt{y})^2 - y^2\right] dy = \int_0^1 \pi(y - y^2)\, dy = \pi\left[\frac{1}{2}y^2 - \frac{1}{3}y^3\right]_0^1 = \pi\left[\frac{1}{2} - \frac{1}{3}\right] = \frac{\pi}{6}$

(c) A cross-section is a washer with inner radius $2 - x$ and outer radius $2 - x^2$.

$V = \int_0^1 \pi\left[(2 - x^2)^2 - (2 - x)^2\right] dx = \int_0^1 \pi(x^4 - 5x^2 + 4x)\, dx = \pi\left[\frac{1}{5}x^5 - \frac{5}{3}x^3 + 2x^2\right]_0^1$

$= \pi\left[\frac{1}{5} - \frac{5}{3} + 2\right] = \frac{8\pi}{15}$

9. (a) The solid is obtained by rotating the region $\mathcal{R} = \{(x, y) \mid 0 \le x \le \frac{\pi}{2}, 0 \le y \le \sqrt{2}\cos x\}$ about the x-axis.

(b) The solid is obtained by rotating the region $\mathcal{R} = \{(x, y) \mid 0 \le x \le 1, 2 - \sqrt{x} \le y \le 2 - x^2\}$ about the x-axis.

Or: The solid is obtained by rotating the region $\mathcal{R} = \{(x, y) \mid 0 \le x \le 1, x^2 \le y \le \sqrt{x}\}$ about the line $y = 2$.

11. Take the base to be the disk $x^2 + y^2 \le 9$. Then $V = \int_{-3}^{3} A(x)\, dx$, where $A(x_0)$ is the area of the isosceles right triangle whose hypotenuse lies along the line $x = x_0$ in the xy-plane. The length of the hypotenuse is $2\sqrt{9 - x^2}$ and the length of each leg is $\sqrt{2}\sqrt{9 - x^2}$. $A(x) = \frac{1}{2}\left(\sqrt{2}\sqrt{9 - x^2}\right)^2 = 9 - x^2$, so

$V = 2\int_{0}^{3} A(x)\, dx = 2\int_{0}^{3}(9 - x^2)\, dx = 2\left[9x - \frac{1}{3}x^3\right]_{0}^{3} = 2(27 - 9) = 36.$

13. Equilateral triangles with sides measuring $\frac{1}{4}x$ meters have height $\frac{1}{4}x \sin 60° = \frac{\sqrt{3}}{8}x$. Therefore,

$A(x) = \frac{1}{2} \cdot \frac{1}{4}x \cdot \frac{\sqrt{3}}{8}x = \frac{\sqrt{3}}{64}x^2$. $V = \int_{0}^{20} A(x)\, dx = \frac{\sqrt{3}}{64}\int_{0}^{20} x^2\, dx = \frac{\sqrt{3}}{64}\left[\frac{1}{3}x^3\right]_{0}^{20} = \frac{8000\sqrt{3}}{64 \cdot 3} = \frac{125\sqrt{3}}{3}$ m³.

15. $x = 3t^2$, $y = 2t^3$, $0 \le t \le 2$.

$L = \int_{0}^{2}\sqrt{(dx/dt)^2 + (dy/dt)^2}\, dt = \int_{0}^{2}\sqrt{(6t)^2 + (6t^2)^2}\, dt = \int_{0}^{2}\sqrt{36t^2 + 36t^4}\, dt = 6\int_{0}^{2} t\sqrt{1 + t^2}\, dt$

$= 6\int_{1}^{5}\sqrt{u}\left(\frac{1}{2}du\right)$ $\left[u = 1 + t^2, du = 2t\, dt\right]$ $= 3\left[\frac{2}{3}u^{3/2}\right]_{1}^{5} = 2(5\sqrt{5} - 1)$

17. $f(x) = kx$ $\Rightarrow$ $30\,\text{N} = k(15 - 12)\,\text{cm}$ $\Rightarrow$ $k = 10\,\text{N/cm} = 1000\,\text{N/m}$. $20\,\text{cm} - 12\,\text{cm} = 0.08\,\text{m}$ $\Rightarrow$

$W = \int_{0}^{0.08} kx\, dx = 1000\int_{0}^{0.08} x\, dx = 500\left[x^2\right]_{0}^{0.08} = 500(0.08)^2 = 3.2\,\text{N-m} = 3.2\,\text{J}.$

19. (a) The parabola has equation $y = ax^2$ with vertex at the origin and passing

through $(4, 4)$. $4 = a \cdot 4^2$ $\Rightarrow$ $a = \frac{1}{4}$ $\Rightarrow$ $y = \frac{1}{4}x^2$ $\Rightarrow$ $x^2 = 4y$

$\Rightarrow$ $x = 2\sqrt{y}$. Each circular disk has radius $2\sqrt{y}$ and is moved $4 - y$ ft.

$W = \int_{0}^{4} \pi\left(2\sqrt{y}\right)^2 62.5\,(4 - y)\, dy = 250\pi\int_{0}^{4} y(4 - y)\, dy$

$= 250\pi\left[2y^2 - \frac{1}{3}y^3\right]_{0}^{4} = 250\pi\left(32 - \frac{64}{3}\right) = \frac{8000\pi}{3} \approx 8378\,\text{ft-lb}$

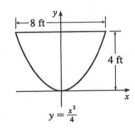

(b) In part (a) we knew the final water level (0) but not the amount of work done. Here we use the same equation, except with the work fixed, and the lower limit of integration (that is, the final water level — call it h) unknown: $W = 4000$ $\Leftrightarrow$ $250\pi\left[2y^2 - \frac{1}{3}y^3\right]_{h}^{4} = 4000$ $\Leftrightarrow$

$\frac{16}{\pi} = \left[(32 - \frac{64}{3}) - (2h^2 - \frac{1}{3}h^3)\right]$ $\Leftrightarrow$ $h^3 - 6h^2 + 32 - \frac{48}{\pi} = 0.$

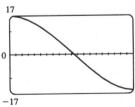

We graph the function $f(h) = h^3 - 6h^2 + 32 - \frac{48}{\pi}$ on the interval $[0, 4]$ to see where it is 0. From the graph, $f(h) = 0$ for $h \approx 2.1$. So the depth of water remaining is about 2.1 ft.

21. As in Example 4 of Section 6.5, $\dfrac{a}{2 - x} = \dfrac{1}{2}$ $\Rightarrow$ $2a = 2 - x$ and

$w = 2(1.5 + a) = 3 + 2a = 3 + 2 - x = 5 - x$. Thus,

$F = \int_{0}^{2} \rho gx(5 - x)\, dx = \rho g\left[\frac{5}{2}x^2 - \frac{1}{3}x^3\right]_{0}^{2} = \rho g\left(10 - \frac{8}{3}\right) = \frac{22}{3}\delta$ $[\rho g = \delta]$ $\approx \frac{22}{3} \cdot 62.5 \approx 458\,\text{lb}.$

23. $x = 100 \implies P = 2000 - 0.1(100) - 0.01(100)^2 = 1890$

$$\text{Consumer surplus} = \int_0^{100} [p(x) - P]\, dx = \int_0^{100} (2000 - 0.1x - 0.01x^2 - 1890)\, dx$$

$$= \left[110x - 0.05x^2 - \tfrac{0.01}{3}x^3\right]_0^{100} = 11{,}000 - 500 - \tfrac{10{,}000}{3} \approx \$7166.67$$

25. $\displaystyle\lim_{h \to 0} f_{\text{ave}} = \lim_{h \to 0} \frac{1}{(x + h) - x} \int_x^{x+h} f(t)\, dt = \lim_{h \to 0} \frac{F(x + h) - F(x)}{h}$, where $F(x) = \int_a^x f(t)\, dt$. But we

recognize this limit as being $F'(x)$ by the definition of a derivative. Therefore, $\displaystyle\lim_{h \to 0} f_{\text{ave}} = F'(x) = f(x)$

by FTC1.

27. $f(x) = \begin{cases} \frac{\pi}{20}\sin\left(\frac{\pi}{10}x\right) & \text{if } 0 \le x \le 10 \\ 0 & \text{if } x < 0 \text{ or } x > 10 \end{cases}$

(a) $f(x) \ge 0$ for all real numbers x and

$$\int_{-\infty}^{\infty} f(x)\, dx = \int_0^{10} \tfrac{\pi}{20}\sin\left(\tfrac{\pi}{10}x\right) dx = \tfrac{\pi}{20} \cdot \tfrac{10}{\pi}\left[-\cos\left(\tfrac{\pi}{10}x\right)\right]_0^{10} = \tfrac{1}{2}(-\cos\pi + \cos 0) = \tfrac{1}{2}(1 + 1) = 1.$$

Therefore, f is a probability density function.

(b) $P(X < 4) = \int_{-\infty}^{4} f(x)\, dx = \int_0^4 \tfrac{\pi}{20}\sin\left(\tfrac{\pi}{10}x\right) dx = \tfrac{1}{2}\left[-\cos\left(\tfrac{\pi}{10}x\right)\right]_0^4 = \tfrac{1}{2}\left(-\cos\tfrac{2\pi}{5} + \cos 0\right)$

$\approx \tfrac{1}{2}(-0.309017 + 1) \approx 0.3455$

(c) $\mu = \int_{-\infty}^{\infty} x f(x)\, dx = \int_0^{10} \tfrac{\pi}{20} x \sin\left(\tfrac{\pi}{10}x\right) dx$

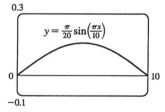

$= \int_0^{\pi} \tfrac{\pi}{20} \cdot \tfrac{10}{\pi} u(\sin u)\left(\tfrac{10}{\pi}\right) du \quad [u = \tfrac{\pi}{10}x,\ du = \tfrac{\pi}{10}\, dx]$

$= \tfrac{5}{\pi}\int_0^{\pi} u \sin u\, du \overset{82}{=} \tfrac{5}{\pi}[\sin u - u\cos u]_0^{\pi} = \tfrac{5}{\pi}[0 - \pi(-1)] = 5$

This answer is expected because the graph of f is symmetric about

the line $x = 5$.

29. (a) The probability density function is $f(t) = \begin{cases} 0 & \text{if } t < 0 \\ \tfrac{1}{8}e^{-t/8} & \text{if } t \ge 0 \end{cases}$

$$P(0 \le X \le 3) = \int_0^3 \tfrac{1}{8}e^{-t/8}\, dt = \left[-e^{-t/8}\right]_0^3 = -e^{-3/8} + 1 \approx 0.3127$$

(b) $P(X > 10) = \int_{10}^{\infty} \tfrac{1}{8}e^{-t/8}\, dt = \lim_{x \to \infty}\left[-e^{-t/8}\right]_{10}^{x} = \lim_{x \to \infty}\left(-e^{-x/8} + e^{-10/8}\right) = 0 + e^{-5/4} \approx 0.2865$

(c) We need to find m such that $P(X \ge m) = \tfrac{1}{2} \implies \int_m^{\infty} \tfrac{1}{8}e^{-t/8}\, dt = \tfrac{1}{2} \implies \lim_{x \to \infty}\left[-e^{-t/8}\right]_m^{x} = \tfrac{1}{2} \implies$

$\displaystyle\lim_{x \to \infty}\left(-e^{-x/8} + e^{-m/8}\right) = \tfrac{1}{2} \implies e^{-m/8} = \tfrac{1}{2} \implies -m/8 = \ln\tfrac{1}{2} \implies$

$m = -8\ln\tfrac{1}{2} = 8\ln 2 \approx 5.55$ minutes.

Focus on Problem Solving

1. $x^2 + y^2 \le 4y \iff x^2 + (y-2)^2 \le 4$, so S is part of a circle, as shown in the diagram. The area of S is

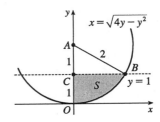

$$\int_0^1 \sqrt{4y - y^2}\, dy \overset{113}{=} \left[\tfrac{y-2}{2}\sqrt{4y - y^2} + 2\cos^{-1}\left(\tfrac{2-y}{2}\right)\right]_0^1 \quad [a = 2]$$

$$= -\tfrac{1}{2}\sqrt{3} + 2\cos^{-1}\left(\tfrac{1}{2}\right) - 2\cos^{-1} 1$$

$$= -\tfrac{\sqrt{3}}{2} + 2\left(\tfrac{\pi}{3}\right) - 2(0) = \tfrac{2\pi}{3} - \tfrac{\sqrt{3}}{2}$$

Another method (without calculus): Note that $\theta = \angle CAB = \tfrac{\pi}{3}$, so the area is

$$(\text{area of sector } OAB) - (\text{area of } \triangle ABC) = \tfrac{1}{2}(2^2)\tfrac{\pi}{3} - \tfrac{1}{2}(1)\sqrt{3} = \tfrac{2\pi}{3} - \tfrac{\sqrt{3}}{2}$$

3. (a) Stacking disks along the y-axis gives us $V = \int_0^h \pi\,[f(y)]^2\, dy$.

(b) Using the Chain Rule, $\dfrac{dV}{dt} = \dfrac{dV}{dh} \cdot \dfrac{dh}{dt} = \pi\,[f(h)]^2 \dfrac{dh}{dt}$.

(c) $kA\sqrt{h} = \pi[f(h)]^2 \dfrac{dh}{dt}$. Set $\dfrac{dh}{dt} = C$: $\pi[f(h)]^2\, C = kA\sqrt{h} \;\Rightarrow\; [f(h)]^2 = \dfrac{kA}{\pi C}\sqrt{h} \;\Rightarrow$

$f(h) = \sqrt{\dfrac{kA}{\pi C}}\, h^{1/4}$; that is, $f(y) = \sqrt{\dfrac{kA}{\pi C}}\, y^{1/4}$. The advantage of having $\dfrac{dh}{dt} = C$ is that the markings on the container are equally spaced.

5. The volume generated from $x = 0$ to $x = b$ is $\int_0^b \pi[f(x)]^2\, dx$. Hence, we are given that $b^2 = \int_0^b \pi[f(x)]^2\, dx$ for all $b > 0$. Differentiating both sides of this equation with respect to b using the Fundamental Theorem of Calculus gives $2b = \pi[f(b)]^2 \;\Rightarrow\; f(b) = \sqrt{2b/\pi}$, since f is positive. Therefore, $f(x) = \sqrt{2x/\pi}$.

7. (a) $V = \pi h^2(r - h/3) = \tfrac{1}{3}\pi h^2(3r - h)$. See the solution to Exercise 6.2.23.

(b) The smaller segment has height $h = 1 - x$ and so by part (a) its volume is $V = \tfrac{1}{3}\pi(1-x)^2[3(1) - (1-x)] = \tfrac{1}{3}\pi(x-1)^2(x+2)$. This volume must be $\tfrac{1}{3}$ of the total volume of the sphere, which is $\tfrac{4}{3}\pi(1)^3$. So $\tfrac{1}{3}\pi(x-1)^2(x+2) = \tfrac{1}{3}\left(\tfrac{4}{3}\pi\right) \;\Rightarrow\; (x^2 - 2x + 1)(x+2) = \tfrac{4}{3} \;\Rightarrow$
$x^3 - 3x + 2 = \tfrac{4}{3} \;\Rightarrow\; 3x^3 - 9x + 2 = 0$. Using Newton's method with $f(x) = 3x^3 - 9x + 2$,
$f'(x) = 9x^2 - 9$, we get $x_{n+1} = x_n - \dfrac{3x_n^3 - 9x_n + 2}{9x_n^2 - 9}$. Taking $x_1 = 0$, we get $x_2 \approx 0.2222$, and $x_3 \approx 0.2261 \approx x_4$, so, correct to four decimal places, $x \approx 0.2261$.

(c) With $r = 0.5$ and $s = 0.75$, the equation $x^3 - 3rx^2 + 4r^3 s = 0$ becomes $x^3 - 3(0.5)x^2 + 4(0.5)^3(0.75) = 0$
$\Rightarrow\; x^3 - \tfrac{3}{2}x^2 + 4\left(\tfrac{1}{8}\right)\tfrac{3}{4} = 0 \;\Rightarrow\; 8x^3 - 12x^2 + 3 = 0$. We use Newton's method with
$f(x) = 8x^3 - 12x^2 + 3$, $f'(x) = 24x^2 - 24x$, so $x_{n+1} = x_n - \dfrac{8x_n^3 - 12x_n^2 + 3}{24x_n^2 - 24x_n}$. Take $x_1 = 0.5$. Then $x_2 \approx 0.6667$, and $x_3 \approx 0.6736 \approx x_4$. So to four decimal places the depth is 0.6736 m.

(d) (i) From part (a) with $r = 5$ in., the volume of water in the bowl is

$$V = \tfrac{1}{3}\pi h^2(3r - h) = \tfrac{1}{3}\pi h^2(15 - h) = 5\pi h^2 - \tfrac{1}{3}\pi h^3. \text{ We are given that } \frac{dV}{dt} = 0.2 \text{ m}^3/\text{s and we want to}$$

find $\dfrac{dh}{dt}$ when $h = 3$. Now $\dfrac{dV}{dt} = 10\pi h \dfrac{dh}{dt} - \pi h^2 \dfrac{dh}{dt}$, so $\dfrac{dh}{dt} = \dfrac{0.2}{\pi(10h - h^2)}$. When $h = 3$, we have

$$\frac{dh}{dt} = \frac{0.2}{\pi(10 \cdot 3 - 3^2)} = \frac{1}{105\pi} \approx 0.003 \text{ in/s}.$$

(ii) From part (a), the volume of water required to fill the bowl from the instant that the water is 4 in. deep is

$$V = \tfrac{1}{2} \cdot \tfrac{4}{3}\pi(5)^3 - \tfrac{1}{3}\pi(4)^2(15 - 4) = \tfrac{2}{3} \cdot 125\pi - \tfrac{16}{3} \cdot 11\pi = \tfrac{74}{3}\pi. \text{ To find the time required to fill the}$$

bowl we divide this volume by the rate: Time $= \dfrac{74\pi/3}{0.2} = \dfrac{370\pi}{3} \approx 387 \text{ s} \approx 6.5 \text{ min}$

9. We are given that the rate of change of the volume of water is $\dfrac{dV}{dt} = -kA(x)$, where k is some positive constant

and $A(x)$ is the area of the surface when the water has depth x. Now we are concerned with the rate of change of

the depth of the water with respect to time, that is, $\dfrac{dx}{dt}$. But by the Chain Rule, $\dfrac{dV}{dt} = \dfrac{dV}{dx}\dfrac{dx}{dt}$, so the first equation

can be written $\dfrac{dV}{dx}\dfrac{dx}{dt} = -kA(x)$ ($\star$). Also, we know that the total volume of water up to a depth x is

$V(x) = \int_0^x A(s)\,ds$, where $A(s)$ is the area of a cross-section of the water at a depth s. Differentiating this

equation with respect to x, we get $dV/dx = A(x)$. Substituting this into equation $\star$, we get

$A(x)(dx/dt) = -kA(x) \quad \Rightarrow \quad dx/dt = -k$, a constant.

11. (a) Choose a vertical x-axis pointing downward with its origin at the surface. In order to calculate the pressure at
depth z, consider n subintervals of the interval $[0, z]$ by points x_i and choose a point $x_i^* \in [x_{i-1}, x_i]$ for each i.
The thin layer of water lying between depth x_{i-1} and depth x_i has a density of approximately $\rho(x_i^*)$, so the
weight of a piece of that layer with unit cross-sectional area is $\rho(x_i^*)g\,\Delta x$. The total weight of a column of
water extending from the surface to depth z (with unit cross-sectional area) would be approximately

$\sum\limits_{i=1}^{n} \rho(x_i^*)g\,\Delta x$. The estimate becomes exact if we take the limit as $n \to \infty$; weight (or force) per unit area at

depth z is $W = \lim\limits_{n \to \infty} \sum\limits_{i=1}^{n} \rho(x_i^*)g\,\Delta x$. In other words, $P(z) = \int_0^z \rho(x)g\,dx$. More generally, if we make no

assumptions about the location of the origin, then $P(z) = P_0 + \int_0^z \rho(x)g\,dx$, where P_0 is the pressure at $x = 0$.
Differentiating, we get $dP/dz = \rho(z)g$.

(b)

$$F = \int_{-r}^{r} P(L + x) \cdot 2\sqrt{r^2 - x^2}\,dx$$

$$= \int_{-r}^{r}\left(P_0 + \int_0^{L+x} \rho_0 e^{z/H}g\,dz\right) \cdot 2\sqrt{r^2 - x^2}\,dx$$

$$= P_0 \int_{-r}^{r} 2\sqrt{r^2 - x^2}\,dx + \rho_0 gH \int_{-r}^{r}\left(e^{(L+x)/H} - 1\right) \cdot 2\sqrt{r^2 - x^2}\,dx$$

$$= (P_0 - \rho_0 gH)\int_{-r}^{r} 2\sqrt{r^2 - x^2}\,dx + \rho_0 gH \int_{-r}^{r} e^{(L+x)/H} \cdot 2\sqrt{r^2 - x^2}\,dx$$

$$= (P_0 - \rho_0 gH)(\pi r^2) + \rho_0 gH e^{L/H}\int_{-r}^{r} e^{x/H} \cdot 2\sqrt{r^2 - x^2}\,dx$$

13. To find the height of the pyramid, we use similar triangles. The first figure shows a cross-section of the pyramid passing through the top and through two opposite corners of the square base. Now $|BD| = b$, since it is a radius of the sphere, which has diameter $2b$ since it is tangent to the opposite sides of the square base. Also, $|AD| = b$ since $\triangle ADB$ is isosceles. So the height is $|AB| = \sqrt{b^2 + b^2} = \sqrt{2}\,b$.

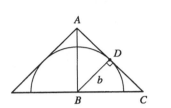 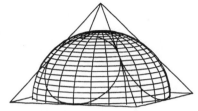 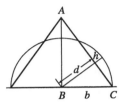

We first observe that the shared volume is equal to half the volume of the sphere, minus the sum of the four equal volumes (caps of the sphere) cut off by the triangular faces of the pyramid. See Exercise 6.2.23 for a derivation of the formula for the volume of a cap of a sphere. To use the formula, we need to find the perpendicular distance h of each triangular face from the surface of the sphere. We first find the distance d from the center of the sphere to one of the triangular faces. The third figure shows a cross-section of the pyramid through the top and through the midpoints of opposite sides of the square base. From similar triangles we find that

$$\frac{d}{b} = \frac{|AB|}{|AC|} = \frac{\sqrt{2}\,b}{\sqrt{b^2 + \left(\sqrt{2}\,b\right)^2}} \quad \Rightarrow \quad d = \frac{\sqrt{2}\,b^2}{\sqrt{3b^2}} = \frac{\sqrt{6}}{3}\,b$$

So $h = b - d = b - \frac{\sqrt{6}}{3}b = \frac{3 - \sqrt{6}}{3}b$. So, using the formula $V = \pi h^2(r - h/3)$ from Exercise 6.2.23 with $r = b$, we find that the volume of each of the caps is

$$\pi\left(\tfrac{3-\sqrt{6}}{3}b\right)^2\left(b - \tfrac{3-\sqrt{6}}{3\cdot 3}b\right) = \tfrac{15-6\sqrt{6}}{9}\cdot\tfrac{6+\sqrt{6}}{9}\pi b^3 = \left(\tfrac{2}{3} - \tfrac{7}{27}\sqrt{6}\right)\pi b^3.$$ So, using our first observation, the

shared volume is $V = \frac{1}{2}\left(\frac{4}{3}\pi b^3\right) - 4\left(\frac{2}{3} - \frac{7}{27}\sqrt{6}\right)\pi b^3 = \left(\frac{28}{27}\sqrt{6} - 2\right)\pi b^3.$

15. $x = \displaystyle\int_1^t \frac{\cos u}{u}\,du$, $y = \displaystyle\int_1^t \frac{\sin u}{u}\,du$, so by FTC1, we have $\dfrac{dx}{dt} = \dfrac{\cos t}{t}$ and $\dfrac{dy}{dt} = \dfrac{\sin t}{t}$. Vertical tangent lines occur when $dx/dt = 0 \;\Leftrightarrow\; \cos t = 0 \;\Leftrightarrow\; t = \frac{\pi}{2} + n\pi$. The parameter value corresponding to the origin, $(x, y) = (0, 0)$, is $t = 1$, so the nearest vertical tangent occurs when $t = \frac{\pi}{2}$. Therefore, the arc length between these

points is $L = \displaystyle\int_1^{\pi/2} \sqrt{\left(\frac{dx}{dt}\right)^2 + \left(\frac{dy}{dt}\right)^2}\,dt = \int_1^{\pi/2} \sqrt{\frac{\cos^2 t}{t^2} + \frac{\sin^2 t}{t^2}}\,dt = \int_1^{\pi/2} \frac{dt}{t} = \Big[\ln t\Big]_1^{\pi/2} = \ln\frac{\pi}{2}.$

17.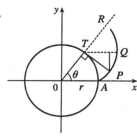

The coordinates of T are $(x_1, y_1) = (r\cos\theta, r\sin\theta)$. Since TP was unwound from arc TA, TP has length $r\theta$. Also $\angle PTQ = \angle PTR - \angle QTR = \frac{1}{2}\pi - \theta$, so P has coordinates

$$x = x_1 + |TP|\cos\angle PTQ = r\cos\theta + r\theta\cos\left(\tfrac{1}{2}\pi - \theta\right) = r(\cos\theta + \theta\sin\theta),$$

$$y = y_1 + |TP|\sin\angle PTQ = r\sin\theta - r\theta\sin\left(\tfrac{1}{2}\pi - \theta\right) = r(\sin\theta - \theta\cos\theta).$$

19. If $h = L$, then

$$P = \frac{\text{area under } y = L\sin\theta}{\text{area of rectangle}} = \frac{\int_0^\pi L\sin\theta\, d\theta}{\pi L} = \frac{[-\cos\theta]_0^\pi}{\pi} = \frac{-(-1)+1}{\pi} = \frac{2}{\pi}$$

If $h = L/2$, then

$$P = \frac{\text{area under } y = \frac{1}{2}L\sin\theta}{\text{area of rectangle}} = \frac{\int_0^\pi \frac{1}{2}L\sin\theta\, d\theta}{\pi L} = \frac{[-\cos\theta]_0^\pi}{2\pi} = \frac{2}{2\pi} = \frac{1}{\pi}$$

Differential Equations

 ## 7.1 Modeling with Differential Equations • • • • • • • •

1. $y = x - x^{-1}$ $\Rightarrow$ $y' = 1 + x^{-2}$. To show that y is a solution of the differential equation, we will substitute the expressions for y and y' in the left-hand side of the equation and show that the left-hand side is equal to the right-hand side.

$$\text{LHS} = xy' + y = x\left(1 + x^{-2}\right) + \left(x - x^{-1}\right) = x + x^{-1} + x - x^{-1} = 2x = \text{RHS}$$

3. (a) $y = \sin kt$ $\Rightarrow$ $y' = k\cos kt$ $\Rightarrow$ $y'' = -k^2\sin kt$. $y'' + 9y = 0$ $\Rightarrow$ $-k^2\sin kt + 9\sin kt = 0$ $\Rightarrow$ $(9 - k^2)\sin kt = 0$ [for all t] $\Rightarrow$ $9 - k^2 = 0$ $\Rightarrow$ $k = \pm 3$

(b) $y = A\sin kt + B\cos kt$ $\Rightarrow$ $y' = Ak\cos kt - Bk\sin kt$ $\Rightarrow$ $y'' = -Ak^2\sin kt - Bk^2\cos kt$.
Thus, $y'' + 9y = 0$ $\Rightarrow$ $-Ak^2\sin kt - Bk^2\cos kt + 9(A\sin kt + B\cos kt) = 0$ $\Rightarrow$
$(9 - k^2)A\sin kt + (9 - k^2)B\cos kt = 0$. The last equation is true for all values of A and B if $k = \pm 3$.

5. (a) $y = e^t$ $\Rightarrow$ $y' = e^t$ $\Rightarrow$ $y'' = e^t$. LHS $= y'' + 2y' + y = e^t + 2e^t + e^t = 4e^t \neq 0$, so $y = e^t$ is not a solution of the differential equation.

(b) $y = e^{-t}$ $\Rightarrow$ $y' = -e^{-t}$ $\Rightarrow$ $y'' = e^{-t}$. LHS $= y'' + 2y' + y = e^{-t} - 2e^{-t} + e^{-t} = 0 = \text{RHS}$, so $y = e^{-t}$ is a solution.

(c) $y = te^{-t}$ $\Rightarrow$ $y' = t(-e^{-t}) + e^{-t}(1) = e^{-t}(1 - t)$ $\Rightarrow$ $y'' = e^{-t}(t - 2)$.

$$\text{LHS} = y'' + 2y' + y = e^{-t}(t - 2) + 2e^{-t}(1 - t) + te^{-t}$$
$$= e^{-t}[(t - 2) + 2(1 - t) + t] = e^{-t}(0) = 0 = \text{RHS},$$

so $y = te^{-t}$ is a solution.

(d) $y = t^2 e^{-t}$ $\Rightarrow$ $y' = te^{-t}(2 - t)$ $\Rightarrow$ $y'' = e^{-t}\left(t^2 - 4t + 2\right)$.

$$\text{LHS} = y'' + 2y' + y = e^{-t}\left(t^2 - 4t + 2\right) + 2te^{-t}(2 - t) + t^2 e^{-t}$$
$$= e^{-t}\left[(t^2 - 4t + 2) + 2t(2 - t) + t^2\right] = e^{-t}(2) \neq 0,$$

so $y = t^2 e^{-t}$ is not a solution.

7. (a) Since the derivative $y' = -y^2$ is always negative (or 0 if $y = 0$), the function y must be decreasing (or equal to 0) on any interval on which it is defined.

(b) $y = \dfrac{1}{x + C}$ $\Rightarrow$ $y' = -\dfrac{1}{(x + C)^2}$. LHS $= y' = -\dfrac{1}{(x + C)^2} = -\left(\dfrac{1}{x + C}\right)^2 = -y^2 = \text{RHS}$

(c) $y = 0$ is a solution of $y' = -y^2$ that is not a member of the family in part (b).

(d) If $y(x) = \dfrac{1}{x + C}$, then $y(0) = \dfrac{1}{0 + C} = \dfrac{1}{C}$. Since $y(0) = 0.5$, $\dfrac{1}{C} = \dfrac{1}{2}$ $\Rightarrow$ $C = 2$, so $y = \dfrac{1}{x + 2}$.

9. (a) $\dfrac{dP}{dt} = 1.2P\left(1 - \dfrac{P}{4200}\right)$. Now $\dfrac{dP}{dt} > 0 \;\Rightarrow\; 1 - \dfrac{P}{4200} > 0$ [assuming that $P > 0$] $\;\Rightarrow\; \dfrac{P}{4200} < 1 \;\Rightarrow\;$

$P < 4200 \;\Rightarrow\;$ the population is increasing for $0 < P < 4200$.

(b) $\dfrac{dP}{dt} < 0 \;\Rightarrow\; P > 4200$

(c) $\dfrac{dP}{dt} = 0 \;\Rightarrow\; P = 4200$ or $P = 0$

11. First graph: This function is increasing *and* also decreasing. But $dy/dt = e^t(y-1)^2 \geq 0$ for all t,

implying that the graph of the solution of the differential equation cannot be decreasing on any interval.

Second graph: When $y = 1$, $dy/dt = 0$, but the graph does not have a horizontal tangent line.

13. (a) P increases most rapidly at the beginning, since there are usually
many simple, easily-learned sub-skills associated with learning a
skill. As t increases, we would expect dP/dt to remain positive, but
decrease. This is because as time progresses, the only points left to
learn are the more difficult ones.

(b) $\dfrac{dP}{dt} = k(M - P)$ is always positive, so the level of performance P is
increasing. As P gets close to M, dP/dt gets close to 0; that is, the
performance levels off, as explained in part (a).

(c)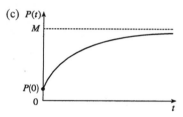

7.2 Direction Fields and Euler's Method • • • • • • • • •

1. (a)

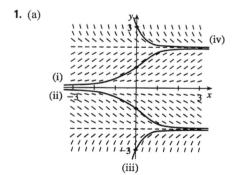

(b) It appears that the constant functions $y = 0$,
$y = -2$, and $y = 2$ are equilibrium solutions.
Note that these three values of y satisfy the given
differential equation $y' = y\left(1 - \tfrac{1}{4}y^2\right)$.

3. $y' = y - 1$. The slopes at each point are independent of x, so the slopes are the same along each line parallel to the
x-axis. Thus, IV is the direction field for this equation. Note that for $y = 1$, $y' = 0$.

5. $y' = y^2 - x^2 = 0 \;\Rightarrow\; y = \pm x$. There are horizontal tangents on these lines only in graph III, so this equation
corresponds to direction field III.

7. (a) $y(0) = 1$ (b) $y(0) = 0$ (c) $y(0) = -1$

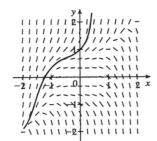   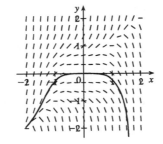   (image for (c))

9. 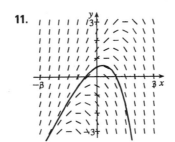   **11.** (image)   **13.**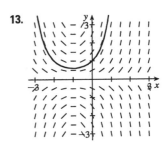

15. In Maple, we can use either `directionfield` (in Maple's share
library) or `plots[fieldplot]` to plot the direction field. To plot the
solution, we can either use the initial-value option in `directionfield`,
or actually solve the equation. In Mathematica, we use
`PlotVectorField` for the direction field, and the
`Plot[Evaluate[...]]` construction to plot the solution, which

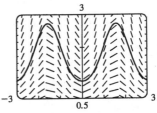

is $y = e^{(1-\cos 2x)/2}$. In Derive, use `Direction_Field` (in utility file ODE_APPR) to plot the direction field.
Then use `DSOLVE1(-y*SIN(2*x),1,x,y,0,1)` (in utility file ODE1) to solve the equation. Simplify each
result.

17.

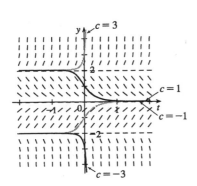

$L = \lim_{t \to \infty} y(t)$ exists for $-2 \le c \le 2$; $L = \pm 2$ for $c = \pm 2$ and
$L = 0$ for $-2 < c < 2$. For other values of c, L does not exist.

19. (a) $y' = F(x, y) = y$ and $y(0) = 1 \Rightarrow x_0 = 0, y_0 = 1$.

(i) $h = 0.4$ and $y_1 = y_0 + hF(x_0, y_0) \Rightarrow y_1 = 1 + 0.4 \cdot 1 = 1.4$. $x_1 = x_0 + h = 0 + 0.4 = 0.4$, so $y_1 = y(0.4) = 1.4$.

(ii) $h = 0.2 \Rightarrow x_1 = 0.2$ and $x_2 = 0.4$, so we need to find y_2.
$y_1 = y_0 + hF(x_0, y_0) = 1 + 0.2y_0 = 1 + 0.2 \cdot 1 = 1.2$,
$y_2 = y_1 + hF(x_1, y_1) = 1.2 + 0.2y_1 = 1.2 + 0.2 \cdot 1.2 = 1.44$.

(iii) $h = 0.1 \Rightarrow x_4 = 0.4$, so we need to find y_4. $y_1 = y_0 + hF(x_0, y_0) = 1 + 0.1y_0 = 1 + 0.1 \cdot 1 = 1.1$,
$y_2 = y_1 + hF(x_1, y_1) = 1.1 + 0.1y_1 = 1.1 + 0.1 \cdot 1.1 = 1.21$,
$y_3 = y_2 + hF(x_2, y_2) = 1.21 + 0.1y_2 = 1.21 + 0.1 \cdot 1.21 = 1.331$,
$y_4 = y_3 + hF(x_3, y_3) = 1.331 + 0.1y_3 = 1.331 + 0.1 \cdot 1.331 = 1.4641$.

(b)

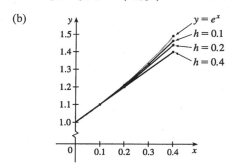

We see that the estimates are underestimates since they are all below the graph of $y = e^x$.

(c) (i) For $h = 0.4$: (exact value) − (approximate value) $= e^{0.4} - 1.4 \approx 0.091$

(ii) For $h = 0.2$: (exact value) − (approximate value) $= e^{0.4} - 1.44 \approx 0.0518$

(iii) For $h = 0.1$: (exact value) − (approximate value) $= e^{0.4} - 1.4641 \approx 0.0277$

Each time the step size is halved, the error estimate also appears to be halved (approximately).

21. $h = 0.5, x_0 = 1, y_0 = 0$, and $F(x, y) = y - 2x$.
Note that $x_1 = x_0 + h = 1 + 0.5 = 1.5, x_2 = 2$, and $x_3 = 2.5$.
$y_1 = y_0 + hF(x_0, y_0) = 0 + 0.5F(1, 0) = 0.5[0 - 2(1)] = -1$.
$y_2 = y_1 + hF(x_1, y_1) = -1 + 0.5F(1.5, -1) = -1 + 0.5[-1 - 2(1.5)] = -3$.
$y_3 = y_2 + hF(x_2, y_2) = -3 + 0.5F(2, -3) = -3 + 0.5[-3 - 2(2)] = -6.5$.
$y_4 = y_3 + hF(x_3, y_3) = -6.5 + 0.5F(2.5, -6.5) = -6.5 + 0.5[-6.5 - 2(2.5)] = -12.25$.

23. $h = 0.1, x_0 = 0, y_0 = 1$, and $F(x, y) = y + xy$.
Note that $x_1 = x_0 + h = 0 + 0.1 = 0.1, x_2 = 0.2, x_3 = 0.3$, and $x_4 = 0.4$.
$y_1 = y_0 + hF(x_0, y_0) = 1 + 0.1F(0, 1) = 1 + 0.1[1 + (0)(1)] = 1.1$.
$y_2 = y_1 + hF(x_1, y_1) = 1.1 + 0.1F(0.1, 1.1) = 1.1 + 0.1[1.1 + (0.1)(1.1)] = 1.221$.
$y_3 = y_2 + hF(x_2, y_2) = 1.221 + 0.1F(0.2, 1.221) = 1.221 + 0.1[1.221 + (0.2)(1.221)] = 1.36752$.
$y_4 = y_3 + hF(x_3, y_3) = 1.36752 + 0.1F(0.3, 1.36752) = 1.36752 + 0.1[1.36752 + (0.3)(1.36752]$
$$= 1.5452976.$$
$y_5 = y_4 + hF(x_4, y_4) = 1.5452976 + 0.1F(0.4, 1.5452976)$
$$= 1.5452976 + 0.1[1.5452976 + (0.4)(1.5452976)] = 1.761639264.$$
Thus, $y(0.5) \approx 1.7616$.

25. (a) $dy/dx + 3x^2 y = 6x^2 \Rightarrow y' = 6x^2 - 3x^2 y$. Store this expression in Y_1 and use the following simple program to evaluate $y(1)$ for each part, using $H = h = 1$ and $N = 1$ for part (i), $H = 0.1$ and $N = 10$ for part (ii), and so forth.

$h \to H: 0 \to X: 3 \to Y$:

For(I, 1, N): $Y + H \times Y_1 \to Y$: $X + H \to X$:

End(loop):

Display Y. [To see all iterations, include this statement in the loop.]

(i) $H = 1, N = 1 \Rightarrow y(1) = 3$ (ii) $H = 0.1, N = 10 \Rightarrow y(1) \approx 2.3928$

(iii) $H = 0.01, N = 100 \Rightarrow y(1) \approx 2.3701$ (iv) $H = 0.001, N = 1000 \Rightarrow y(1) \approx 2.3681$

(b) $y = 2 + e^{-x^3} \Rightarrow y' = -3x^2 e^{-x^3}$

$$\text{LHS} = y' + 3x^2 y = -3x^2 e^{-x^3} + 3x^2 \left(2 + e^{-x^3} \right) = -3x^2 e^{-x^3} + 6x^2 + 3x^2 e^{-x^3} = 6x^2 = \text{RHS}$$

$y(0) = 2 + e^{-0} = 2 + 1 = 3$

(c) The exact value of $y(1)$ is $2 + e^{-1^3} = 2 + e^{-1}$.

(i) For $h = 1$: (exact value) − (approximate value) $= 2 + e^{-1} - 3 \approx -0.6321$

(ii) For $h = 0.1$: (exact value) − (approximate value) $= 2 + e^{-1} - 2.3928 \approx -0.0249$

(iii) For $h = 0.01$: (exact value) − (approximate value) $= 2 + e^{-1} - 2.3701 \approx -0.0022$

(iv) For $h = 0.001$: (exact value) − (approximate value) $= 2 + e^{-1} - 2.3681 \approx -0.0002$

In (ii)–(iv), it seems that when the step size is divided by 10, the error estimate is also divided by 10 (approximately).

27. (a) $R\dfrac{dQ}{dt} + \dfrac{1}{C}Q = E(t)$ becomes

$5Q' + \frac{1}{0.05}Q = 60$ or $Q' + 4Q = 12$.

(b) From the graph, it appears that the limiting value of the charge Q is about 3.

(c) If $Q' = 0$, then $4Q = 12 \Rightarrow Q = 3$ is an equilibrium solution.

(d)

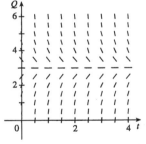

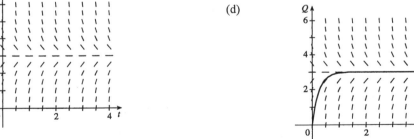

(e) $Q' + 4Q = 12 \Rightarrow Q' = 12 - 4Q$. Now $Q(0) = 0$, so $t_0 = 0$ and $Q_0 = 0$.

$$Q_1 = Q_0 + hF(t_0, Q_0) = 0 + 0.1(12 - 4 \cdot 0) = 1.2$$

$$Q_2 = Q_1 + hF(t_1, Q_1) = 1.2 + 0.1(12 - 4 \cdot 1.2) = 1.92$$

$$Q_3 = Q_2 + hF(t_2, Q_2) = 1.92 + 0.1(12 - 4 \cdot 1.92) = 2.352$$

$$Q_4 = Q_3 + hF(t_3, Q_3) = 2.352 + 0.1(12 - 4 \cdot 2.352) = 2.6112$$

$$Q_5 = Q_4 + hF(t_4, Q_4) = 2.6112 + 0.1(12 - 4 \cdot 2.6112) = 2.76672$$

Thus, $Q_5 = Q(0.5) \approx 2.77$ C.

7.3 Separable Equations • • • • • • • • • • • •

1. $\dfrac{dy}{dx} = y^2 \;\Rightarrow\; \dfrac{dy}{y^2} = dx \; [y \neq 0] \;\Rightarrow\; \displaystyle\int \dfrac{dy}{y^2} = \int dx \;\Rightarrow\; -\dfrac{1}{y} = x + C \;\Rightarrow\; -y = \dfrac{1}{x+C} \;\Rightarrow\;$

$y = \dfrac{-1}{x+C}$, and $y = 0$ is also a solution.

3. $yy' = x \;\Rightarrow\; y\dfrac{dy}{dx} = x \;\Rightarrow\; \int y\, dy = \int x\, dx \;\Rightarrow\; \tfrac{1}{2}y^2 = \tfrac{1}{2}x^2 + C_1 \;\Rightarrow\; y^2 = x^2 + 2C_1 \;\Rightarrow\;$

$x^2 - y^2 = C$ (where $C = -2C_1$). This represents a family of hyperbolas.

5. $\dfrac{dy}{dt} = \dfrac{te^t}{y\sqrt{1+y^2}} \;\Rightarrow\; y\sqrt{1+y^2}\, dy = te^t\, dt \;\Rightarrow\; \int y\sqrt{1+y^2}\, dy = \int te^t\, dt \;\Rightarrow\;$

$\tfrac{1}{3}(1+y^2)^{3/2} = te^t - e^t + C$ [where the first integral is evaluated by substitution and the second by parts] $\;\Rightarrow\;$

$1 + y^2 = [3(te^t - e^t + C)]^{2/3} \;\Rightarrow\; y = \pm\sqrt{[3(te^t - e^t + C)]^{2/3} - 1}$

7. $\dfrac{du}{dt} = 2 + 2u + t + tu \;\Rightarrow\; \dfrac{du}{dt} = (1+u)(2+t) \;\Rightarrow\; \displaystyle\int \dfrac{du}{1+u} = \int (2+t)dt \; [u \neq -1] \;\Rightarrow\;$

$\ln|1 + u| = \tfrac{1}{2}t^2 + 2t + C \;\Rightarrow\; |1 + u| = e^{t^2/2 + 2t + C} = Ke^{t^2/2 + 2t}$, where $K = e^C \;\Rightarrow\;$

$1 + u = \pm Ke^{t^2/2 + 2t} \;\Rightarrow\; u = -1 \pm Ke^{t^2/2 + 2t}$ where $K > 0$. $u = -1$ is also a solution, so

$u = -1 + Ae^{t^2/2 + 2t}$, where A is an arbitrary constant.

9. $\dfrac{dy}{dx} = y^2 + 1,\; y(1) = 0.$ $\displaystyle\int \dfrac{dy}{y^2 + 1} = \int dx \;\Rightarrow\; \tan^{-1} y = x + C.$ $y = 0$ when $x = 1$, so

$1 + C = \tan^{-1} 0 = 0 \;\Rightarrow\; C = -1.$ Thus, $\tan^{-1} y = x - 1$ and $y = \tan(x - 1)$.

11. $xe^{-t}\dfrac{dx}{dt} = t,\; x(0) = 1.$ $\int x\, dx = \int te^t\, dt \;\Rightarrow\; \tfrac{1}{2}x^2 = (t-1)e^t + C$ [by parts]. $x(0) = 1$, so

$\tfrac{1}{2} = (0-1)e^0 + C$ and $C = \tfrac{3}{2}.$ Thus, $\tfrac{1}{2}x^2 = (t-1)e^t + \tfrac{3}{2} \;\Rightarrow\; x^2 = 2(t-1)e^t + 3 \;\Rightarrow\;$

$x = \sqrt{2(t-1)e^t + 3}$ [use the positive square root since $x(0) = +1$].

13. $\dfrac{du}{dt} = \dfrac{2t + \sec^2 t}{2u},\; u(0) = -5.$ $\int 2u\, du = \int (2t + \sec^2 t)\, dt \;\Rightarrow\; u^2 = t^2 + \tan t + C,$ where

$[u(0)]^2 = 0^2 + \tan 0 + C \;\Rightarrow\; C = (-5)^2 = 25.$ Therefore, $u^2 = t^2 + \tan t + 25,$ so $u = \pm\sqrt{t^2 + \tan t + 25}.$

Since $u(0) = -5$, we must have $u = -\sqrt{t^2 + \tan t + 25}.$

15. $\dfrac{dy}{dx} = 4x^3 y,\; y(0) = 7.$ $\dfrac{dy}{y} = 4x^3\, dx$ [if $y \neq 0$] $\;\Rightarrow\; \displaystyle\int \dfrac{dy}{y} = \int 4x^3\, dx \;\Rightarrow\; \ln|y| = x^4 + C \;\Rightarrow\;$

$e^{\ln|y|} = e^{x^4 + C} \;\Rightarrow\; |y| = e^{x^4}e^C \;\Rightarrow\; y = Ae^{x^4};\; y(0) = 7 \;\Rightarrow\; A = 7 \;\Rightarrow\; y = 7e^{x^4}.$

17. (a) $y' = 2x\sqrt{1 - y^2} \;\Rightarrow\; \dfrac{dy}{dx} = 2x\sqrt{1 - y^2} \;\Rightarrow\; \dfrac{dy}{\sqrt{1 - y^2}} = 2x\, dx \;\Rightarrow\; \displaystyle\int \dfrac{dy}{\sqrt{1 - y^2}} = \int 2x\, dx \;\Rightarrow\;$

$\sin^{-1} y = x^2 + C$ for $-\tfrac{\pi}{2} \leq x^2 + C \leq \tfrac{\pi}{2}.$

(b) $y(0) = 0 \;\Rightarrow\; \sin^{-1} 0 = 0^2 + C \;\Rightarrow\; C = 0,$ so $\sin^{-1} y = x^2$

and $y = \sin(x^2)$ for $-\sqrt{\pi/2} \leq x \leq \sqrt{\pi/2}.$

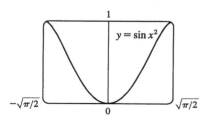

(c) For $\sqrt{1 - y^2}$ to be a real number, we must have $-1 \le y \le 1$; that is, $-1 \le y(0) \le 1$. Thus, the initial-value problem $y' = 2x\sqrt{1 - y^2}$, $y(0) = 2$ does *not* have a solution.

19. $\dfrac{dy}{dx} = \dfrac{\sin x}{\sin y}$, $y(0) = \dfrac{\pi}{2}$. So $\int \sin y\, dy = \int \sin x\, dx$ ⇔

$-\cos y = -\cos x + C$ ⇔ $\cos y = \cos x - C$. From the initial

condition, we need $\cos\frac{\pi}{2} = \cos 0 - C$ ⇒ $0 = 1 - C$ ⇒ $C = 1$,

so the solution is $\cos y = \cos x - 1$. Note that we cannot take $\cos^{-1}$ of

both sides, since that would unnecessarily restrict the solution to the case

where $-1 \le \cos x - 1$ ⇔ $0 \le \cos x$, as $\cos^{-1}$ is defined only on

$[-1, 1]$. Instead we plot the graph using Maple's

`plots[implicitplot]` or Mathematica's

`Plot[Evaluate[···]]`.

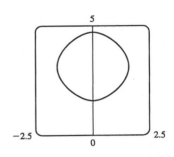

21. (a)

x	y	$y' = 1/y$	x	y	$y' = 1/y$
0	0.5	2	0	-2	-0.5
0	-0.5	-2	0	4	0.25
0	1	1	0	3	$0.\overline{3}$
0	-1	-1	0	0.25	4
0	2	0.5	0	$0.\overline{3}$	3

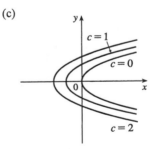

(b) $y' = 1/y$ ⇒ $dy/dx = 1/y$ ⇒

$y\, dy = dx$ ⇒ $\int y\, dy = \int dx$ ⇒

$\frac{1}{2}y^2 = x + c$ ⇒ $y^2 = 2(x + c)$

or $y = \pm\sqrt{2(x + c)}$.

(c)

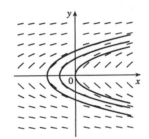

23. The curves $y = kx^2$ form a family of parabolas with axis the y-axis.

Differentiating gives $y' = 2kx$, but $k = y/x^2$, so $y' = 2y/x$. Thus, the

slope of the tangent line at any point (x, y) on one of the parabolas is

$y' = 2y/x$, so the orthogonal trajectories must satisfy $y' = -x/(2y)$

⇔ $2y\, dy = -x\, dx$ ⇔ $y^2 = -x^2/2 + C_1$ ⇔ $x^2 + 2y^2 = C$.

This is a family of ellipses.

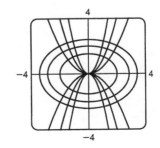

25. Differentiating $y = (x + k)^{-1}$ gives $y' = -\dfrac{1}{(x+k)^2}$, but $k = \dfrac{1}{y} - x$, so

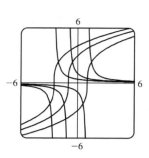

$y' = -\dfrac{1}{(1/y)^2} = -y^2$. Thus, the orthogonal trajectories must satisfy

$y' = -\dfrac{1}{-y^2} = \dfrac{1}{y^2}$ $\Leftrightarrow$ $y^2\,dy = dx$ $\Leftrightarrow$ $\dfrac{y^3}{3} = x + C$ or

$y = [3(x + C)]^{1/3}$

27. From Exercise 7.2.27, $\dfrac{dQ}{dt} = 12 - 4Q$ $\Leftrightarrow$ $\displaystyle\int \dfrac{dQ}{12 - 4Q} = \int dt$ $\Leftrightarrow$ $-\tfrac{1}{4}\ln|12 - 4Q| = t + C$ $\Leftrightarrow$

$\ln|12 - 4Q| = -4t - 4C$ $\Leftrightarrow$ $|12 - 4Q| = e^{-4t-4C}$ $\Leftrightarrow$ $12 - 4Q = Ke^{-4t}$ $[K = \pm e^{-4C}]$ $\Leftrightarrow$

$4Q = 12 - Ke^{-4t}$ $\Leftrightarrow$ $Q = 3 - Ae^{-4t}$ $[A = K/4]$. $Q(0) = 0$ $\Leftrightarrow$ $0 = 3 - A$ $\Leftrightarrow$ $A = 3$ $\Leftrightarrow$

$Q(t) = 3 - 3e^{-4t}$. As $t \to \infty$, $Q(t) \to 3 - 0 = 3$ (the limiting value).

29. $\dfrac{dP}{dt} = k(M - P)$ $\Leftrightarrow$ $\displaystyle\int \dfrac{dP}{P - M} = \int (-k)\,dt$ $\Leftrightarrow$ $\ln|P - M| = -kt + C$ $\Leftrightarrow$ $|P - M| = e^{-kt+C}$

$\Leftrightarrow$ $P - M = Ae^{-kt}$ $[A = \pm e^C]$ $\Leftrightarrow$ $P = M + Ae^{-kt}$. If we assume that performance is at level 0 when

$t = 0$, then $P(0) = 0$ $\Leftrightarrow$ $0 = M + A$ $\Leftrightarrow$ $A = -M$ $\Leftrightarrow$ $P(t) = M - Me^{-kt}$.

$\displaystyle\lim_{t\to\infty} P(t) = M - M \cdot 0 = M$.

31. (a) If $a = b$, then $\dfrac{dx}{dt} = k(a - x)(b - x)^{1/2}$ becomes $\dfrac{dx}{dt} = k(a - x)^{3/2}$. $(a - x)^{-3/2}dx = k\,dt$ $\Rightarrow$

$\int (a - x)^{-3/2}dx = \int k\,dt$ $\Rightarrow$ $2(a - x)^{-1/2} = kt + C$ ([by substitution] $\Rightarrow$ $\dfrac{2}{kt + C} = \sqrt{a - x}$ $\Rightarrow$

$\left(\dfrac{2}{kt + C}\right)^2 = a - x$ $\Rightarrow$ $x(t) = a - \dfrac{4}{(kt + C)^2}$. The initial concentration of HBr is 0, so $x(0) = 0$ $\Rightarrow$

$0 = a - \dfrac{4}{C^2}$ $\Rightarrow$ $\dfrac{4}{C^2} = a$ $\Rightarrow$ $C^2 = \dfrac{4}{a}$ $\Rightarrow$ $C = 2/\sqrt{a}$ (C is positive since

$kt + C = 2(a - x)^{-1/2} > 0$). Thus, $x(t) = a - \dfrac{4}{(kt + 2/\sqrt{a})^2}$.

(b) $\dfrac{dx}{dt} = k(a - x)(b - x)^{1/2}$ $\Rightarrow$ $\dfrac{dx}{(a - x)\sqrt{b - x}} = k\,dt$ $\Rightarrow$ $\displaystyle\int \dfrac{dx}{(a - x)\sqrt{b - x}} = \int k\,dt$ $(\star)$. From the

hint, $u = \sqrt{b - x}$ $\Rightarrow$ $u^2 = b - x$ $\Rightarrow$ $2u\,du = -dx$, so $\displaystyle\int \dfrac{dx}{(a - x)\sqrt{b - x}} = \int \dfrac{-2u\,du}{[a - (b - u^2)]u} =$

$-2\displaystyle\int \dfrac{du}{a - b + u^2} = -2\int \dfrac{du}{\left(\sqrt{a - b}\right)^2 + u^2} \overset{17}{=} -2\left(\dfrac{1}{\sqrt{a - b}}\tan^{-1}\dfrac{u}{\sqrt{a - b}}\right)$. So $(\star)$ becomes

$\dfrac{-2}{\sqrt{a - b}}\tan^{-1}\dfrac{\sqrt{b - x}}{\sqrt{a - b}} = kt + C$. Now $x(0) = 0$ $\Rightarrow$ $C = \dfrac{-2}{\sqrt{a - b}}\tan^{-1}\dfrac{\sqrt{b}}{\sqrt{a - b}}$

and we have $\dfrac{-2}{\sqrt{a - b}}\tan^{-1}\dfrac{\sqrt{b - x}}{\sqrt{a - b}} = kt - \dfrac{2}{\sqrt{a - b}}\tan^{-1}\dfrac{\sqrt{b}}{\sqrt{a - b}}$ $\Rightarrow$

$\dfrac{2}{\sqrt{a - b}}\left(\tan^{-1}\sqrt{\dfrac{b}{a - b}} - \tan^{-1}\sqrt{\dfrac{b - x}{a - b}}\right) = kt$ $\Rightarrow$

$t(x) = \dfrac{2}{k\sqrt{a - b}}\left(\tan^{-1}\sqrt{\dfrac{b}{a - b}} - \tan^{-1}\sqrt{\dfrac{b - x}{a - b}}\right)$.

33. (a) $\dfrac{dC}{dt} = r - kC \ \Rightarrow \ \dfrac{dC}{dt} = -(kC - r) \ \Rightarrow \ \displaystyle\int \dfrac{dC}{kC - r} = \int -dt \ \Rightarrow \ (1/k)\ln|kC - r| = -t + M_1$

$\Rightarrow \ \ln|kC - r| = -kt + M_2 \ \Rightarrow \ |kC - r| = e^{-kt + M_2} \ \Rightarrow \ kC - r = M_3 e^{-kt} \ \Rightarrow$

$kC = M_3 e^{-kt} + r \ \Rightarrow \ C(t) = M_4 e^{-kt} + r/k. \ C(0) = C_0 \ \Rightarrow \ C_0 = M_4 + r/k \ \Rightarrow$

$M_4 = C_0 - r/k \ \Rightarrow \ C(t) = (C_0 - r/k)e^{-kt} + r/k.$

(b) If $C_0 < r/k$, then $C_0 - r/k < 0$ and the formula for $C(t)$ shows that $C(t)$ increases and $\displaystyle\lim_{t \to \infty} C(t) = r/k$.

As t increases, the formula for $C(t)$ shows how the role of C_0 steadily diminishes as that of r/k increases.

35. (a) Let $y(t)$ be the amount of salt (in kg) after t minutes. Then $y(0) = 15$. The amount of liquid in the tank is

1000 L at all times, so the concentration at time t (in minutes) is $y(t)/1000$ kg/L and

$\dfrac{dy}{dt} = -\left[\dfrac{y(t)}{1000}\dfrac{\text{kg}}{\text{L}}\right]\left(10\dfrac{\text{L}}{\text{min}}\right) = -\dfrac{y(t)}{100}\dfrac{\text{kg}}{\text{min}}. \ \displaystyle\int \dfrac{dy}{y} = -\dfrac{1}{100}\int dt \ \Rightarrow \ \ln y = -\dfrac{t}{100} + C,$ and

$y(0) = 15 \ \Rightarrow \ \ln 15 = C,$ so $\ln y = \ln 15 - \dfrac{t}{100}.$ It follows that $\ln\left(\dfrac{y}{15}\right) = -\dfrac{t}{100}$ and $\dfrac{y}{15} = e^{-t/100},$ so

$y = 15e^{-t/100}$ kg.

(b) After 20 minutes, $y = 15e^{-20/100} = 15e^{-0.2} \approx 12.3$ kg.

37. Assume that the raindrop begins at rest, so that $v(0) = 0.$ $dm/dt = km$ and $(mv)' = gm \ \Rightarrow$

$mv' + vm' = gm \ \Rightarrow \ mv' + v(km) = gm \ \Rightarrow \ v' + vk = g \ \Rightarrow \ dv/dt = g - kv \ \Rightarrow$

$\displaystyle\int \dfrac{dv}{g - kv} = \int dt \ \Rightarrow \ -(1/k)\ln|g - kv| = t + C \ \Rightarrow \ \ln|g - kv| = -kt - kC \ \Rightarrow \ g - kv = Ae^{-kt}.$

$v(0) = 0 \ \Rightarrow \ A = g.$ So $kv = g - ge^{-kt} \ \Rightarrow \ v = (g/k)(1 - e^{-kt}).$ Since $k > 0,$ as $t \to \infty,$ $e^{-kt} \to 0$

and therefore, $\displaystyle\lim_{t \to \infty} v(t) = g/k.$

39. (a) The rate of growth of the area is jointly proportional to $\sqrt{A(t)}$ and $M - A(t);$ that is, the rate is proportional to

the product of those two quantities. So for some constant $k,$ $dA/dt = k\sqrt{A}\,(M - A).$ We are interested in the

maximum of the function dA/dt (when the tissue grows the fastest), so we differentiate, using the Chain Rule

and then substituting for dA/dt from the differential equation:

$\dfrac{d}{dt}\left(\dfrac{dA}{dt}\right) = k\left[\sqrt{A}\,(-1)\dfrac{dA}{dt} + (M - A)\cdot\tfrac{1}{2}A^{-1/2}\dfrac{dA}{dt}\right] = \tfrac{1}{2}kA^{-1/2}\dfrac{dA}{dt}[-2A + (M - A)]$

$\qquad = \tfrac{1}{2}kA^{-1/2}\left[k\sqrt{A}(M - A)\right][M - 3A] = \tfrac{1}{2}k^2(M - A)(M - 3A)$

This is 0 when $M - A = 0$ [this situation never actually occurs, since the graph of $A(t)$ is asymptotic to the line

$y = M,$ as in the logistic model] and when $M - 3A = 0 \ \Leftrightarrow \ A(t) = M/3.$ This represents a maximum by

the First Derivative Test, since $\dfrac{d}{dt}\left(\dfrac{dA}{dt}\right)$ goes from positive to negative when $A(t) = M/3.$

(b) From the CAS, we get $A(t) = M\left(\dfrac{Ce^{\sqrt{M}kt} - 1}{Ce^{\sqrt{M}kt} + 1}\right)^2.$ To get C in terms of the initial area A_0 and the maximum

area $M,$ we substitute $t = 0$ and $A = A_0 = A(0)$: $A_0 = M\left(\dfrac{C - 1}{C + 1}\right)^2 \ \Leftrightarrow \ (C + 1)\sqrt{A_0} = (C - 1)\sqrt{M}$

$\Leftrightarrow \ C\sqrt{A_0} + \sqrt{A_0} = C\sqrt{M} - \sqrt{M} \ \Leftrightarrow \ \sqrt{M} + \sqrt{A_0} = C\sqrt{M} - C\sqrt{A_0} \ \Leftrightarrow$

$\sqrt{M} + \sqrt{A_0} = C\left(\sqrt{M} - \sqrt{A_0}\right) \ \Leftrightarrow \ C = \dfrac{\sqrt{M} + \sqrt{A_0}}{\sqrt{M} - \sqrt{A_0}}.$ (Notice that if $A_0 = 0,$ then $C = 1.$)

41. (a) $V = \pi r^2 y \quad \Rightarrow \quad \dfrac{dV}{dt} = \pi r^2 \dfrac{dy}{dt}$ [implicit differentiation] $\Rightarrow$

$$\frac{dy}{dt} = \frac{1}{\pi r^2}\frac{dV}{dt} = \frac{1}{\pi r^2}\left(-a\sqrt{2gy}\right) = \frac{1}{\pi 2^2}\left[-\pi\left(\tfrac{1}{12}\right)^2 \sqrt{2\cdot 32}\sqrt{y}\right] = -\tfrac{1}{72}\sqrt{y}$$

(b) $\dfrac{dy}{dt} = -\tfrac{1}{72}\sqrt{y} \quad \Rightarrow \quad y^{-1/2}\,dy = -\tfrac{1}{72}\,dt \quad \Rightarrow \quad 2\sqrt{y} = -\tfrac{1}{72}t + C.$

$y(0) = 6 \quad \Rightarrow \quad 2\sqrt{6} = 0 + C \quad \Rightarrow \quad C = 2\sqrt{6} \quad \Rightarrow \quad y(t) = \left(-\tfrac{1}{144}t + \sqrt{6}\right)^2.$

(c) We want to find t when $y = 0$, so we set $y = 0 = \left(-\tfrac{1}{144}t + \sqrt{6}\right)^2 \quad \Rightarrow \quad t = 144\sqrt{6} \approx 5$ min 53 s.

7.4 Exponential Growth and Decay · · · · · · · · · · ·

1. The relative growth rate is $\dfrac{1}{P}\dfrac{dP}{dt} = 0.7944$, so $\dfrac{dP}{dt} = 0.7944P$ and, by Theorem 2,

$P(t) = P(0)e^{0.7944t} = 2e^{0.7944t}$. Thus, $P(6) = 2e^{0.7944(6)} \approx 234.99$ or about 235 members.

3. (a) By Theorem 2, $y(t) = y(0)e^{kt} = 500e^{kt}$. Now $y(3) = 500e^{k(3)} = 8000 \quad \Rightarrow \quad e^{3k} = \frac{8000}{500} \quad \Rightarrow$

$3k = \ln 16 \quad \Rightarrow \quad k = (\ln 16)/3$. So $y(t) = 500e^{(\ln 16)t/3} = 500 \cdot 16^{t/3}$

(b) $y(4) = 500 \cdot 16^{4/3} \approx 20{,}159$

(c) $dy/dt = ky \quad \Rightarrow \quad y'(4) = ky(4) = \tfrac{1}{3}\ln 16\left(500 \cdot 16^{4/3}\right)$ [from part (a)] $\approx 18{,}631$ cells/h

(d) $y(t) = 500 \cdot 16^{t/3} = 30{,}000 \quad \Rightarrow \quad 16^{t/3} = 60 \quad \Rightarrow \quad \tfrac{1}{3}t\ln 16 = \ln 60 \quad \Rightarrow \quad t = 3(\ln 60)/(\ln 16) \approx 4.4$ h

5. (a) Let the population (in millions) in the year t be $P(t)$. Since the initial time is the year 1750, we substitute

$t - 1750$ for t in Theorem 2, so the exponential model gives $P(t) = P(1750)e^{k(t-1750)}$. Then

$P(1800) = 980 = 790e^{k(1800-1750)} \quad \Rightarrow \quad \frac{980}{790} = e^{k(50)} \quad \Rightarrow \quad \ln\frac{980}{790} = 50k \quad \Rightarrow$

$k = \tfrac{1}{50}\ln\frac{980}{790} \approx 0.0043104$. So with this model, we have $P(1900) = 790e^{k(1900-1750)} \approx 1508$ million, and

$P(1950) = 790e^{k(1950-1750)} \approx 1871$ million. Both of these estimates are much too low.

(b) In this case, the exponential model gives $P(t) = P(1850)e^{k(t-1850)} \quad \Rightarrow$

$P(1900) = 1650 = 1260e^{k(1900-1850)} \quad \Rightarrow \quad \ln\frac{1650}{1260} = k(50) \quad \Rightarrow \quad k = \tfrac{1}{50}\ln\frac{1650}{1260} \approx 0.005393$. So with

this model, we estimate $P(1950) = 1260e^{k(1950-1850)} \approx 2161$ million. This is still too low, but closer than the

estimate of $P(1950)$ in part (a).

(c) The exponential model gives $P(t) = P(1900)e^{k(t-1900)} \quad \Rightarrow \quad P(1950) = 2560 = 1650e^{k(1950-1900)} \quad \Rightarrow$

$\ln\frac{2560}{1650} = k(50) \quad \Rightarrow \quad k = \tfrac{1}{50}\ln\frac{2560}{1650} \approx 0.008785$. With this model, we estimate

$P(2000) = 1650e^{k(2000-1900)} \approx 3972$ million. This is much too low. The discrepancy is explained by the fact

that the world birth rate (average yearly number of births per person) is about the same as always, whereas the

mortality rate (especially the infant mortality rate) is much lower, owing mostly to advances in medical science

and to the wars in the first part of the twentieth century. The exponential model assumes, among other things,

that the birth and mortality rates will remain constant.

7. (a) If $y = [N_2O_5]$ then by Theorem 2, $\dfrac{dy}{dt} = -0.0005y \Rightarrow y(t) = y(0)e^{-0.0005t} = Ce^{-0.0005t}$.

(b) $y(t) = Ce^{-0.0005t} = 0.9C \Rightarrow e^{-0.0005t} = 0.9 \Rightarrow -0.0005t = \ln 0.9 \Rightarrow$
$t = -2000 \ln 0.9 \approx 211$ s

9. (a) If $y(t)$ is the mass remaining after t days, then $y(t) = y(0)e^{kt} = 100e^{kt}$. $y(30) = 100e^{30k} = \frac{1}{2}(100) \Rightarrow$
$e^{30k} = \frac{1}{2} \Rightarrow k = -(\ln 2)/30 \Rightarrow y(t) = 100e^{-(\ln 2)t/30} = 100 \cdot 2^{-t/30}$

(b) $y(100) = 100 \cdot 2^{-100/30} \approx 9.92$ mg

(c) $100e^{-(\ln 2)t/30} = 1 \Rightarrow -(\ln 2)t/30 = \ln \frac{1}{100} \Rightarrow t = -30 \frac{\ln 0.01}{\ln 2} \approx 199.3$ years

11. Let $y(t)$ be the level of radioactivity. Thus, $y(t) = y(0)e^{-kt}$ and k is determined by using the half-life:
$y(5730) = \frac{1}{2}y(0) \Rightarrow y(0)e^{-k(5730)} = \frac{1}{2}y(0) \Rightarrow e^{-5730k} = \frac{1}{2} \Rightarrow$

$-5730k = \ln \frac{1}{2} \Rightarrow k = -\dfrac{\ln \frac{1}{2}}{5730} = \dfrac{\ln 2}{5730}$. If 74% of the ^{14}C remains, then we know that $y(t) = 0.74y(0)$

$\Rightarrow 0.74 = e^{-t(\ln 2)/5730} \Rightarrow \ln 0.74 = -\dfrac{t\ln 2}{5730} \Rightarrow t = -\dfrac{5730(\ln 0.74)}{\ln 2} \approx 2489 \approx 2500$ years.

13. (a) If $y = u - 75$, $u(0) = 185 \Rightarrow y(0) = 185 - 75 = 110$, and the initial-value problem is $dy/dt = ky$ with
$y(0) = 110$. So the solution is $y(t) = 110e^{kt}$.

(b) $y(30) = 110e^{30k} = 150 - 75 \Rightarrow e^{30k} = \frac{75}{110} = \frac{15}{22} \Rightarrow k = \frac{1}{30}\ln \frac{15}{22}$, so $y(t) = 110e^{\frac{1}{30}t\ln\left(\frac{15}{22}\right)}$ and
$y(45) = 110e^{\frac{45}{30}\ln\left(\frac{15}{22}\right)} \approx 62\,°\mathrm{F}$. Thus, $u(45) \approx 62 + 75 = 137\,°\mathrm{F}$.

(c) $u(t) = 100 \Rightarrow y(t) = 25$. $y(t) = 110e^{\frac{1}{30}t\ln\left(\frac{15}{22}\right)} = 25 \Rightarrow e^{\frac{1}{30}t\ln\left(\frac{15}{22}\right)} = \frac{25}{110} \Rightarrow$

$\frac{1}{30}t\ln \frac{15}{22} = \ln \frac{25}{110} \Rightarrow t = \dfrac{30\ln \frac{25}{110}}{\ln \frac{15}{22}} \approx 116$ min.

15. (a) Let $P(h)$ be the pressure at altitude h. Then $dP/dh = kP \Rightarrow P(h) = P(0)e^{kh} = 101.3e^{kh}$.
$P(1000) = 101.3e^{1000k} = 87.14 \Rightarrow 1000k = \ln\left(\frac{87.14}{101.3}\right) \Rightarrow$
$k = \frac{1}{1000}\ln\left(\frac{87.14}{101.3}\right) \Rightarrow P(h) = 101.3\,e^{\frac{1}{1000}h\ln\left(\frac{87.14}{101.3}\right)}$, so $P(3000) = 101.3e^{3\ln\left(\frac{87.14}{101.3}\right)} \approx 64.5$ kPa.

(b) $P(6187) = 101.3\,e^{\frac{6187}{1000}\ln\left(\frac{87.14}{101.3}\right)} \approx 39.9$ kPa

17. (a) Using $A = A_0\left(1 + \dfrac{r}{n}\right)^{nt}$ with $A_0 = 3000$, $r = 0.05$, and $t = 5$, we have:

(i) Annually: $n = 1$; $\quad A = 3000\left(1 + \frac{0.05}{1}\right)^{1\cdot5} = \$3828.84$

(ii) Semiannually: $n = 2$; $\quad A = 3000\left(1 + \frac{0.05}{2}\right)^{2\cdot5} = \$3840.25$

(iii) Monthly: $n = 12$; $\quad A = 3000\left(1 + \frac{0.05}{12}\right)^{12\cdot5} = \$3850.08$

(iv) Weekly: $n = 52$; $\quad A = 3000\left(1 + \frac{0.05}{52}\right)^{52\cdot5} = \$3851.61$

(v) Daily: $n = 365$; $\quad A = 3000\left(1 + \frac{0.05}{365}\right)^{365\cdot5} = \$3852.01$

(vi) Continuously: $\quad A = 3000e^{(0.05)5} = \$3852.08$

(b) $dA/dt = 0.05A$ and $A(0) = 3000$.

19. (a) $\dfrac{dP}{dt} = kP - m = k\left(P - \dfrac{m}{k}\right)$. Let $y = P - \dfrac{m}{k}$, so $\dfrac{dy}{dt} = \dfrac{dP}{dt}$ and the differential equation becomes

$\dfrac{dy}{dt} = ky$. The solution is $y = y_0 e^{kt}$ $\Rightarrow$ $P - \dfrac{m}{k} = \left(P_0 - \dfrac{m}{k}\right)e^{kt}$ $\Rightarrow$ $P(t) = \dfrac{m}{k} + \left(P_0 - \dfrac{m}{k}\right)e^{kt}$.

(b) There will be an exponential expansion $\Leftrightarrow$ $P_0 - \dfrac{m}{k} > 0$ $\Leftrightarrow$ $m < kP_0$.

(c) The population will be constant if $P_0 - \dfrac{m}{k} = 0$ $\Leftrightarrow$ $m = kP_0$. It will decline if $P_0 - \dfrac{m}{k} < 0$ $\Leftrightarrow$ $m > kP_0$.

(d) $P_0 = 8{,}000{,}000$, $k = \alpha - \beta = 0.016$, $m = 210{,}000$ $\Rightarrow$ $m > kP_0 \ (= 128{,}000)$, so by part (c), the population was declining.

 7.5 **The Logistic Equation** • • • • • • • • • • •

1. (a) $dP/dt = 0.05P - 0.0005P^2 = 0.05P(1 - 0.01P) = 0.05P(1 - P/100)$. Comparing to Equation 1, $dP/dt = kP(1 - P/K)$, we see that the carrying capacity is $K = 100$ and the value of k is 0.05.

(b) The slopes close to 0 occur where P is near 0 or 100. The largest slopes appear to be on the line $P = 50$. The solutions are increasing for $0 < P_0 < 100$ and decreasing for $P_0 > 100$.

(c)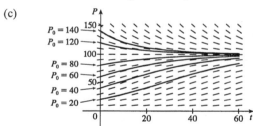

All of the solutions approach $P = 100$ as t increases. As in part (b), the solutions differ since for $0 < P_0 < 100$ they are increasing, and for $P_0 > 100$ they are decreasing. Also, some have an IP and some don't. It appears that the solutions which have $P_0 = 20$ and $P_0 = 40$ have inflection points at $P = 50$.

(d) The equilibrium solutions are $P = 0$ (trivial solution) and $P = 100$. The increasing solutions move away from $P = 0$ and all nonzero solutions approach $P = 100$ as $t \to \infty$.

3. (a) $\dfrac{dy}{dt} = ky\left(1 - \dfrac{y}{K}\right)$ $\Rightarrow$ $y(t) = \dfrac{K}{1 + Ae^{-kt}}$ with $A = \dfrac{K - y(0)}{y(0)}$. With $K = 8 \times 10^7$, $k = 0.71$, and

$y(0) = 2 \times 10^7$, we get the model $y(t) = \dfrac{8 \times 10^7}{1 + 3e^{-0.71t}}$, so $y(1) = \dfrac{8 \times 10^7}{1 + 3e^{-0.71}} \approx 3.23 \times 10^7$ kg.

(b) $y(t) = 4 \times 10^7$ $\Rightarrow$ $\dfrac{8 \times 10^7}{1 + 3e^{-0.71t}} = 4 \times 10^7$ $\Rightarrow$ $2 = 1 + 3e^{-0.71t}$ $\Rightarrow$ $e^{-0.71t} = \frac{1}{3}$ $\Rightarrow$

$-0.71t = \ln\frac{1}{3}$ $\Rightarrow$ $t = \dfrac{\ln 3}{0.71} \approx 1.55$ years

5. (a) We will assume that the difference in the birth and death rates is 20 million/year. Let $t = 0$ correspond to the year 1990 and use a unit of 1 billion for all calculations. $k \approx \dfrac{1}{P}\dfrac{dP}{dt} = \dfrac{1}{5.3}(0.02) = \dfrac{1}{265}$, so

$$\dfrac{dP}{dt} = kP\left(1 - \dfrac{P}{K}\right) = \dfrac{1}{265}P\left(1 - \dfrac{P}{100}\right), \ P \text{ in billions}$$

(b) $A = \dfrac{K - P_0}{P_0} = \dfrac{100 - 5.3}{5.3} = \dfrac{947}{53} \approx 17.8679.$ $P(t) = \dfrac{K}{1 + Ae^{-kt}} = \dfrac{100}{1 + \frac{947}{53}e^{-(1/265)t}},$ so

$P(10) \approx 5.49$ billion.

(c) $P(110) \approx 7.81$, and $P(510) \approx 27.72$. The predictions are 7.81 billion in the year 2100 and 27.72 billion in 2500.

(d) If $K = 50$, then $P(t) = \dfrac{50}{1 + \frac{447}{53}e^{-(1/265)t}}$. So $P(10) \approx 5.48$, $P(110) \approx 7.61$, and $P(510) \approx 22.41$. The predictions become 5.48 billion in the year 2000, 7.61 billion in 2100, and 22.41 billion in the year 2500.

7. (a) Our assumption is that $\dfrac{dy}{dt} = ky(1 - y)$, where y is the fraction of the population that has heard the rumor.

(b) Using the logistic equation (1), $\dfrac{dP}{dt} = kP\left(1 - \dfrac{P}{K}\right)$, we substitute $y = \dfrac{P}{K}$, $P = Ky$, and $\dfrac{dP}{dt} = K\dfrac{dy}{dt}$, to

obtain $K\dfrac{dy}{dt} = k(Ky)(1 - y) \iff \dfrac{dy}{dt} = ky(1 - y)$, our equation in part (a). Now the solution to (1) is

$P(t) = \dfrac{K}{1 + Ae^{-kt}}$, where $A = \dfrac{K - P_0}{P_0}$. We use the same substitution to obtain $Ky = \dfrac{K}{1 + \dfrac{K - Ky_0}{Ky_0}e^{-kt}}$

$\Rightarrow \quad y = \dfrac{y_0}{y_0 + (1 - y_0)e^{-kt}}.$

Alternatively, we could use the same steps as outlined in "The Analytic Solution," following Example 2.

(c) Let t be the number of hours since 8 A.M. Then $y_0 = y(0) = \frac{80}{1000} = 0.08$ and $y(4) = \frac{1}{2}$, so

$\dfrac{1}{2} = y(4) = \dfrac{0.08}{0.08 + 0.92e^{-4k}}$. Thus, $0.08 + 0.92e^{-4k} = 0.16$, $e^{-4k} = \frac{0.08}{0.92} = \frac{2}{23}$, and $e^{-k} = \left(\frac{2}{23}\right)^{1/4}$, so

$y = \dfrac{0.08}{0.08 + 0.92(2/23)^{t/4}} = \dfrac{2}{2 + 23(2/23)^{t/4}}$. Solving this equation for t, we get

$2y + 23y\left(\dfrac{2}{23}\right)^{t/4} = 2 - 2y \Rightarrow \left(\dfrac{2}{23}\right)^{t/4} = \dfrac{2}{23} \cdot \dfrac{1 - y}{y} \Rightarrow \left(\dfrac{2}{23}\right)^{t/4-1} = \dfrac{1 - y}{y}$. It follows that

$\dfrac{t}{4} - 1 = \dfrac{\ln[(1 - y)/y]}{\ln \frac{2}{23}}$, so $t = 4\left[1 + \dfrac{\ln((1 - y)/y)}{\ln \frac{2}{23}}\right]$. When $y = 0.9$, $\dfrac{1 - y}{y} = \frac{1}{9}$, so

$t = 4\left(1 - \dfrac{\ln 9}{\ln \frac{2}{23}}\right) \approx 7.6$ h or 7 h 36 min. Thus, 90% of the population will have heard the rumor by 3:36 P.M.

9. (a) $\dfrac{dP}{dt} = k(P)\left(1 - \dfrac{P}{K}\right) \quad \Rightarrow$

$\dfrac{d^2P}{dt^2} = k\left[P\left(-\dfrac{1}{K}\dfrac{dP}{dt}\right) + \left(1 - \dfrac{P}{K}\right)\dfrac{dP}{dt}\right] = k\dfrac{dP}{dt}\left(-\dfrac{P}{K} + 1 - \dfrac{P}{K}\right)$

$= k\left[kP\left(1 - \dfrac{P}{K}\right)\right]\left(1 - \dfrac{2P}{K}\right) = k^2P\left(1 - \dfrac{P}{K}\right)\left(1 - \dfrac{2P}{K}\right)$

(b) P grows fastest when P' has a maximum, that is, when $P'' = 0$. From part (a), $P'' = 0 \iff P = 0, P = K$, or $P = K/2$. Since $0 < P < K$, we see that $P'' = 0 \iff P = K/2$.

11. (a) The term -15 represents a harvesting of fish at a constant rate—in this case, 15 fish/week. This is the rate at which fish are caught.

(b)

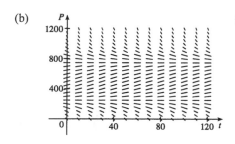

(c) From the graph in part (b), it appears that $P(t) = 250$ and $P(t) = 750$ are the equilibrium solutions. We confirm this analytically by solving the equation $dP/dt = 0$ as follows:

$0.08P(1 - P/1000) - 15 = 0 \Rightarrow$

$0.08P - 0.00008P^2 - 15 = 0 \Rightarrow$

$-0.00008(P^2 - 1000P + 187{,}500) = 0 \Rightarrow$

$(P - 250)(P - 750) = 0 \Rightarrow P = 250$ or 750.

(d)

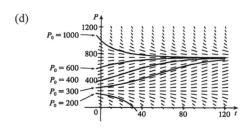

For $0 < P_0 < 250$, $P(t)$ decreases to 0. For $P_0 = 250$, $P(t)$ remains constant. For $250 < P_0 < 750$, $P(t)$ increases and approaches 750. For $P_0 = 750$, $P(t)$ remains constant. For $P_0 > 750$, $P(t)$ decreases and approaches 750.

(e) $\dfrac{dP}{dt} = 0.08P\left(1 - \dfrac{P}{1000}\right) - 15 \Leftrightarrow -\dfrac{100{,}000}{8} \cdot \dfrac{dP}{dt} = (0.08P - 0.00008P^2 - 15) \cdot \left(-\dfrac{100{,}000}{8}\right) \Leftrightarrow$

$-12{,}500\,\dfrac{dP}{dt} = P^2 - 1000P + 187{,}500 \Leftrightarrow \dfrac{dP}{(P - 250)(P - 750)} = -\dfrac{1}{12{,}500}\,dt \Leftrightarrow$

$\displaystyle\int\left(\dfrac{-1/500}{P - 250} + \dfrac{1/500}{P - 750}\right) dP = -\dfrac{1}{12{,}500}\,dt \Leftrightarrow \int\left(\dfrac{1}{P - 250} - \dfrac{1}{P - 750}\right) dP = \dfrac{1}{25}\,dt \Leftrightarrow$

$\ln|P - 250| - \ln|P - 750| = \dfrac{1}{25}t + C \Leftrightarrow \ln\left|\dfrac{P - 250}{P - 750}\right| = \dfrac{1}{25}t + C \Leftrightarrow$

$\left|\dfrac{P - 250}{P - 750}\right| = e^{t/25+C} = ke^{t/25} \Leftrightarrow \dfrac{P - 250}{P - 750} = ke^{t/25} \Leftrightarrow P - 250 = Pke^{t/25} - 750ke^{t/25} \Leftrightarrow$

$P - Pke^{t/25} = 250 - 750ke^{t/25} \Leftrightarrow$

$P(t) = \dfrac{250 - 750ke^{t/25}}{1 - ke^{t/25}}$. If $t = 0$ and $P = 200$, then

$200 = \dfrac{250 - 750k}{1 - k} \Leftrightarrow 200 - 200k = 250 - 750k \Leftrightarrow 550k = 50$

$\Leftrightarrow k = \frac{1}{11}$. Similarly, if $t = 0$ and $P = 300$, then $k = -\frac{1}{9}$. Simplifying

P with these two values of k gives us

$P(t) = \dfrac{250\left(3e^{t/25} - 11\right)}{e^{t/25} - 11}$ and $P(t) = \dfrac{750\left(e^{t/25} + 3\right)}{e^{t/25} + 9}$.

P 1200

0   120
                                                 t

13. (a) $\dfrac{dP}{dt} = (kP)\left(1 - \dfrac{P}{K}\right)\left(1 - \dfrac{m}{P}\right)$. If $m < P < K$, then $dP/dt = (+)(+)(+) = + \;\Rightarrow\; P$ is increasing.

If $0 < P < m$, then $dP/dt = (+)(+)(-) = - \;\Rightarrow\; P$ is decreasing.

(b)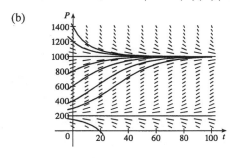

$k = 0.08$, $K = 1000$, and $m = 200 \;\Rightarrow$

$$\frac{dP}{dt} = 0.08P\left(1 - \frac{P}{1000}\right)\left(1 - \frac{200}{P}\right)$$

For $0 < P_0 < 200$, the population dies out. For $P_0 = 200$, the population is steady. For $200 < P_0 < 1000$, the population increases and approaches 1000. For $P_0 > 1000$, the population decreases and approaches 1000.

The equilibrium solutions are $P(t) = 200$ and $P(t) = 1000$.

(c) $\dfrac{dP}{dt} = kP\left(1 - \dfrac{P}{K}\right)\left(1 - \dfrac{m}{P}\right) = kP\left(\dfrac{K-P}{K}\right)\left(\dfrac{P-m}{P}\right) = \dfrac{k}{K}(K-P)(P-m) \;\Leftrightarrow$

$$\int \frac{dP}{(K-P)(P-m)} = \int \frac{k}{K}\,dt.$$

By partial fractions, $\dfrac{1}{(K-P)(P-m)} = \dfrac{A}{K-P} + \dfrac{B}{P-m}$, so $A(P-m) + B(K-P) = 1$.

If $P = m$, $B = \dfrac{1}{K-m}$; if $P = K$, $A = \dfrac{1}{K-m}$, so $\dfrac{1}{K-m}\displaystyle\int\left(\dfrac{1}{K-P} + \dfrac{1}{P-m}\right)dP = \int \dfrac{k}{K}\,dt \;\Rightarrow$

$\dfrac{1}{K-m}\left(-\ln|K-P| + \ln|P-m|\right) = \dfrac{k}{K}t + M$.

But $m < P < K$, so $\dfrac{1}{K-m}\ln\dfrac{P-m}{K-P} = \dfrac{k}{K}t + M \;\Rightarrow\; \ln\dfrac{P-m}{K-P} = (K-m)\dfrac{k}{K}t + M_1 \;\Leftrightarrow$

$\dfrac{P-m}{K-P} = De^{(K-m)(k/K)t} \quad (D = e^{M_1})$. Let $t = 0$: $\dfrac{P_0 - m}{K - P_0} = D$. So $\dfrac{P-m}{K-P} = \dfrac{P_0 - m}{K - P_0}e^{(K-m)(k/K)t}$.

Solving for P, we get $P(t) = \dfrac{m(K - P_0) + K(P_0 - m)e^{(K-m)(k/K)t}}{K - P_0 + (P_0 - m)e^{(K-m)(k/K)t}}$.

(d) If $P_0 < m$, then $P_0 - m < 0$. Let $N(t)$ be the numerator of the expression for $P(t)$ in part (c). Then

$N(0) = P_0(K - m) > 0$, and $P_0 - m < 0 \;\Leftrightarrow\; \displaystyle\lim_{t\to\infty} K(P_0 - m)\,e^{(K-m)(k/K)t} = -\infty \;\Rightarrow$

$\displaystyle\lim_{t\to\infty} N(t) = -\infty$. Since N is continuous, there is a number t such that $N(t) = 0$ and thus $P(t) = 0$. So the

species will become extinct.

15. (a) $dP/dt = kP\cos(rt - \phi) \;\Rightarrow\; (dP)/P = k\cos(rt - \phi)\,dt \;\Rightarrow\; \int (dP)/P = k\int \cos(rt - \phi)\,dt \;\Rightarrow$

$\ln P = (k/r)\sin(rt - \phi) + C$. (Since this is a growth model, $P > 0$ and we can write $\ln P$ instead of $\ln|P|$.)

Since $P(0) = P_0$, we obtain $\ln P_0 = (k/r)\sin(-\phi) + C = -(k/r)\sin\phi + C \;\Rightarrow$

$C = \ln P_0 + (k/r)\sin\phi$. Thus, $\ln P = (k/r)\sin(rt - \phi) + \ln P_0 + (k/r)\sin\phi$, which we can rewrite as

$\ln(P/P_0) = (k/r)[\sin(rt - \phi) + \sin\phi]$ or, after exponentiation, $P(t) = P_0 e^{(k/r)[\sin(rt-\phi)+\sin\phi]}$.

(b) As k increases, the amplitude increases, but the minimum value stays the same.

As r increases, the amplitude and the period decrease.

A change in ϕ produces slight adjustments in the phase shift and amplitude.

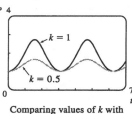

Comparing values of k with $P_0 = 1$, $r = 2$, and $\phi = \pi/2$

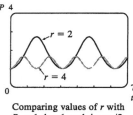

Comparing values of r with $P_0 = 1$, $k = 1$, and $\phi = \pi/2$

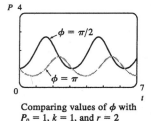

Comparing values of ϕ with $P_0 = 1$, $k = 1$, and $r = 2$

$P(t)$ oscillates between $P_0 e^{(k/r)(1+\sin \phi)}$ and $P_0 e^{(k/r)(-1+\sin \phi)}$ (the extreme values are attained when $rt - \phi$ is an odd multiple of $\frac{\pi}{2}$), so $\lim\limits_{t \to \infty} P(t)$ does not exist.

Predator-Prey Systems • • • • • • • • • • • • •

1. (a) $dx/dt = -0.05x + 0.0001xy$. If $y = 0$, we have $dx/dt = -0.05x$, which indicates that in the absence of y, x declines at a rate proportional to itself. So x represents the predator population and y represents the prey population. The growth of the prey population, $0.1y$ (from $dy/dt = 0.1y - 0.005xy$), is restricted only by encounters with predators (the term $-0.005xy$). The predator population increases only through the term $0.0001xy$; that is, by encounters with the prey and not through additional food sources.

(b) $dy/dt = -0.015y + 0.00008xy$. If $x = 0$, we have $dy/dt = -0.015y$, which indicates that in the absence of x, y would decline at a rate proportional to itself. So y represents the predator population and x represents the prey population. The growth of the prey population, $0.2x$ (from $dx/dt = 0.2x - 0.0002x^2 - 0.006xy = 0.2x(1 - 0.001x) - 0.006xy$), is restricted by a carrying capacity of 1000 [from the term $1 - 0.001x = 1 - x/1000$] and by encounters with predators (the term $-0.006xy$). The predator population increases only through the term $0.00008xy$; that is, by encounters with the prey and not through additional food sources.

3. (a) At $t = 0$, there are about 300 rabbits and 100 foxes. At $t = t_1$, the number of foxes reaches a minimum of about 20 while the number of rabbits is about 1000. At $t = t_2$, the number of rabbits reaches a maximum of about 2400, while the number of foxes rebounds to 100. At $t = t_3$, the number of rabbits decreases to about 1000 and the number of foxes reaches a maximum of about 315. As t increases, the number of foxes decreases greatly to 100, and the number of rabbits decreases to 300 (the initial populations), and the cycle starts again.

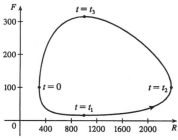

(b)

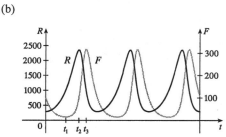

5.

Species 2 graph: closed curve with points labeled $t=2$ (top, around 190), $t=3$ (left around 135), $t=1$ (right around 105), $t=4$ (left around 90), $t=0,5$ (bottom around 40). Axes: Species 2 (vertical, marks 50, 100, 150, 200), Species 1 (horizontal, marks 50, 100, 150, 200, 250).

7. $\dfrac{dW}{dR} = \dfrac{-0.02W + 0.00002RW}{0.08R - 0.001RW}$ $\Leftrightarrow$ $(0.08 - 0.001W)R\,dW = (-0.02 + 0.00002R)W\,dR$ $\Leftrightarrow$

$\dfrac{0.08 - 0.001W}{W}\,dW = \dfrac{-0.02 + 0.00002R}{R}\,dR$ $\Leftrightarrow$ $\displaystyle\int\left(\dfrac{0.08}{W} - 0.001\right)dW = \int\left(-\dfrac{0.02}{R} + 0.00002\right)dR$

$\Leftrightarrow$ $0.08\ln|W| - 0.001W = -0.02\ln|R| + 0.00002R + K$ $\Leftrightarrow$

$0.08\ln W + 0.02\ln R = 0.001W + 0.00002R + K$ $\Leftrightarrow$ $\ln\left(W^{0.08}R^{0.02}\right) = 0.00002R + 0.001W + K$ $\Leftrightarrow$

$W^{0.08}R^{0.02} = e^{0.00002R + 0.001W + K}$ $\Leftrightarrow$ $R^{0.02}W^{0.08} = Ce^{0.00002R}e^{0.001W}$ $\Leftrightarrow$ $\dfrac{R^{0.02}W^{0.08}}{e^{0.00002R}e^{0.001W}} = C.$

In general, if $\dfrac{dy}{dx} = \dfrac{-ry + bxy}{kx - axy}$, then $C = \dfrac{x^r y^k}{e^{bx}e^{ay}}$.

9. (a) Letting $W = 0$ gives us $dR/dt = 0.08R(1 - 0.0002R)$. $dR/dt = 0$ $\Leftrightarrow$ $R = 0$ or 5000. Since $dR/dt > 0$ for $0 < R < 5000$, we would expect the rabbit population to *increase* to 5000 for these values of R. Since $dR/dt < 0$ for $R > 5000$, we would expect the rabbit population to *decrease* to 5000 for these values of R. Hence, in the absence of wolves, we would expect the rabbit population to stabilize at 5000.

(b) R and W are constant $\Rightarrow$ $R' = 0$ and $W' = 0$ $\Rightarrow$

$\begin{cases} 0 = 0.08R(1 - 0.0002R) - 0.001RW \\ 0 = -0.02W + 0.00002RW \end{cases}$ $\Rightarrow$ $\begin{cases} 0 = R[0.08(1 - 0.0002R) - 0.001W] \\ 0 = W(-0.02 + 0.00002R) \end{cases}$

The second equation is true if $W = 0$ or $R = \dfrac{0.02}{0.00002} = 1000$. If $W = 0$ in the first equation, then either $R = 0$ or $R = \dfrac{1}{0.0002} = 5000$ [as in part (a)]. If $R = 1000$, then $0 = 1000[0.08(1 - 0.0002 \cdot 1000) - 0.001W]$ $\Leftrightarrow$ $0 = 80(1 - 0.2) - W$ $\Leftrightarrow$ $W = 64$.

Case (i): $W = 0$, $R = 0$: both populations are zero

Case (ii): $W = 0$, $R = 5000$: see part (a)

Case (iii): $R = 1000$, $W = 64$: the predator/prey interaction balances and the populations are stable.

(c) The populations of wolves and rabbits fluctuate around 64 and 1000, respectively, and eventually stabilize at those values.

(d)

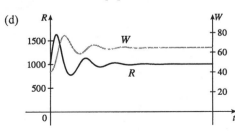

7 Review

— • **CONCEPT CHECK** • —

1. (a) A differential equation is an equation that contains an unknown function and one or more of its derivatives.

   (b) The order of a differential equation is the order of the highest derivative that occurs in the equation.

   (c) An initial condition is a condition of the form $y(t_0) = y_0$.

2. $y' = x^2 + y^2 \geq 0$ for all x and y. $y' = 0$ only at the origin, so there is a horizontal tangent at $(0, 0)$, but nowhere else. The graph of the solution is increasing on every interval.

3. See the paragraph preceding Example 1 in Section 7.2.

4. See the paragraph after Figure 14 in Section 7.2.

5. A separable equation is a first-order differential equation in which the expression for dy/dx can be factored as a function of x times a function of y, that is, $dy/dx = g(x)f(y)$. We can solve the equation by integrating both sides of the equation $dy/f(y) = g(x)dx$ and solving for y.

6. (a) $dy/dt = ky$

   (b) The equation in part (a) is an appropriate model for population growth, assuming that there is enough room and nutrition to support the growth.

   (c) If $y(0) = y_0$, then the solution is $y(t) = y_0 e^{kt}$.

7. (a) $dP/dt = kP(1 - P/K)$, where K is the carrying capacity.

   (b) The equation in part (a) is an appropriate model for population growth, assuming that the population grows at a rate proportional to the size of the population in the beginning, but eventually levels off and approaches its carrying capacity because of limited resources.

8. (a) $dF/dt = kF - aFS$ and $dS/dt = -rS + bFS$.

   (b) In the absence of sharks, an ample food supply would support exponential growth of the fish population, that is, $dF/dt = kF$, where k is a positive constant. In the absence of fish, we assume that the shark population would decline at a rate proportional to itself, that is, $dS/dt = -rS$, where r is a positive constant.

— ▲ **TRUE–FALSE QUIZ** ▲ —

1. **True.** Since $y^4 \geq 0$, $y' = -1 - y^4 < 0$ and the solutions are decreasing functions.

3. **False.** $x + y$ cannot be written in the form $g(x)f(y)$.

5. **True.** By comparing $\dfrac{dy}{dt} = 2y\left(1 - \dfrac{y}{5}\right)$ with the logistic differential equation (7.5.1), we see that the carrying capacity is 5; that is, $\lim\limits_{t \to \infty} y = 5$.

◆ **EXERCISES** ◆

1. (a)

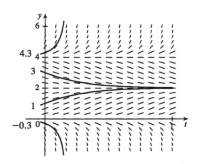

(b) $\lim_{t\to\infty} y(t)$ appears to be finite for $0 \le c \le 4$. In fact $\lim_{t\to\infty} y(t) = 4$ for $c = 4$, $\lim_{t\to\infty} y(t) = 2$ for $0 < c < 4$, and $\lim_{t\to\infty} y(t) = 0$ for $c = 0$. The equilibrium solutions are $y(t) = 0$, $y(t) = 2$, and $y(t) = 4$.

3. (a)

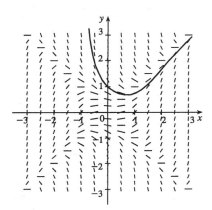

We estimate that when $x = 0.3$, $y = 0.8$, so $y(0.3) \approx 0.8$.

(b) $h = 0.1$, $x_0 = 0$, $y_0 = 1$ and $F(x, y) = x^2 - y^2$. So $y_n = y_{n-1} + 0.1(x_{n-1}^2 - y_{n-1}^2)$.
Thus, $y_1 = 1 + 0.1(0^2 - 1^2) = 0.9$, $y_2 = 0.9 + 0.1(0.1^2 - 0.9^2) = 0.82$,
$y_3 = 0.82 + 0.1(0.2^2 - 0.82^2) = 0.75676$. This is close to our graphical estimate of $y(0.3) \approx 0.8$.

(c) The centers of the horizontal line segments of the direction field are located on the lines $y = x$ and $y = -x$. When a solution curve crosses one of these lines, it has a local maximum or minimum.

5. $(3y^2 + 2y)\, y' = x \cos x \quad \Rightarrow \quad (3y^2 + 2y)\, dy = (x \cos x)\, dx \quad \Rightarrow \quad \int(3y^2 + 2y)\, dy = \int(x \cos x)\, dx \quad \Rightarrow$
$y^3 + y^2 = \cos x + x \sin x + C$. For the last step, use integration by parts or Formula 83 in the Table of Integrals.

7. $xyy' = \ln x \quad \Rightarrow \quad y\, dy = \dfrac{\ln x}{x}\, dx \quad \Rightarrow \quad \int y\, dy = \int \dfrac{\ln x}{x}\, dx$ (Make the substitution $u = \ln x$; then
$du = dx/x$.) So $\int y\, dy = \int u\, du \quad \Rightarrow \quad \tfrac{1}{2}y^2 = \tfrac{1}{2}u^2 + C \quad \Rightarrow \tfrac{1}{2}y^2 = \tfrac{1}{2}(\ln x)^2 + C$. $y(1) = 2 \quad \Rightarrow$
$\tfrac{1}{2}2^2 = \tfrac{1}{2}(\ln 1)^2 + C = C \quad \Leftrightarrow \quad C = 2$. Therefore, $\tfrac{1}{2}y^2 = \tfrac{1}{2}(\ln x)^2 + 2$, or $y = \sqrt{(\ln x)^2 + 4}$. The negative square root is inadmissible, since $y(1) > 0$.

9. The curves $kx^2 + y^2 = 1$ form a family of ellipses for $k > 0$, a family of hyperbolas for $k < 0$, and two parallel lines $y = \pm 1$ for $k = 0$. Solving $kx^2 + y^2 = 1$ for k gives $k = \dfrac{1 - y^2}{x^2}$. Differentiating gives $2kx + 2yy' = 0$ $\Leftrightarrow$

$y' = -\dfrac{kx}{y} = -(1 - y^2)\dfrac{x}{yx^2} = \dfrac{y^2 - 1}{xy}$. Thus, for $k \neq 0$ the orthogonal trajectories must satisfy $y' = -\dfrac{xy}{y^2 - 1}$

$\Rightarrow \dfrac{y^2 - 1}{y}\,dy = -x\,dx \Rightarrow \dfrac{y^2}{2} - \ln|y| = \dfrac{-x^2}{2} + K \Rightarrow y^2 - 2\ln|y| + x^2 = C$. For $k = 0$, the

orthogonal trajectories are given by $x = C_1$ for C_1 an arbitrary constant.

11. (a) $y(t) = y(0)e^{kt} = 1000e^{kt} \Rightarrow y(2) = 1000e^{2k} = 9000 \Rightarrow e^{2k} = 9 \Rightarrow 2k = \ln 9 \Rightarrow$
$k = \frac{1}{2}\ln 9 = \ln 3 \Rightarrow y(t) = 1000e^{(\ln 3)t} = 1000 \cdot 3^t$

(b) $y(3) = 1000 \cdot 3^3 = 27{,}000$

(c) $y'(t) = 1000 \cdot 3^t \cdot \ln 3$, so $y'(3) = 27{,}000 \ln 3 \approx 29{,}663$ bacteria per hour

(d) $1000 \cdot 3^t = 2 \cdot 1000 \Rightarrow 3^t = 2 \Rightarrow t\ln 3 = \ln 2 \Rightarrow t = (\ln 2)/\ln 3 \approx 0.63$ h

13. (a) $C'(t) = -kC(t) \Rightarrow C(t) = C(0)e^{-kt}$ by Theorem 7.4.2. But $C(0) = C_0$, so $C(t) = C_0 e^{-kt}$.

(b) $C(30) = \frac{1}{2}C_0$ since the concentration is reduced by half. Thus, $\frac{1}{2}C_0 = C_0 e^{-30k} \Rightarrow \ln\frac{1}{2} = -30k \Rightarrow$
$k = -\frac{1}{30}\ln\frac{1}{2} = \frac{1}{30}\ln 2$. Since 10% of the original concentration remains if 90% is eliminated, we want the
value of t such that $C(t) = \frac{1}{10}C_0$. Therefore, $\frac{1}{10}C_0 = C_0 e^{-t(\ln 2)/30} \Rightarrow \ln 0.1 = -t(\ln 2)/30 \Rightarrow$
$t = -\frac{30}{\ln 2}\ln 0.1 \approx 100$ h.

15. (a) $\dfrac{dL}{dt} \propto L_\infty - L \Rightarrow \dfrac{dL}{dt} = k(L_\infty - L) \Rightarrow \displaystyle\int \dfrac{dL}{L_\infty - L} = \int k\,dt \Rightarrow -\ln|L_\infty - L| = kt + C \Rightarrow$
$\ln|L_\infty - L| = -kt - C \Rightarrow |L_\infty - L| = e^{-kt-C} \Rightarrow L_\infty - L = Ae^{-kt} \Rightarrow L = L_\infty - Ae^{-kt}$.
At $t = 0$, $L = L(0) = L_\infty - A \Rightarrow A = L_\infty - L(0) \Rightarrow L(t) = L_\infty - [L_\infty - L(0)]e^{-kt}$.

(b) $L_\infty = 53$ cm, $L(0) = 10$ cm, and $k = 0.2 \Rightarrow L(t) = 53 - (53 - 10)e^{-0.2t} = 53 - 43e^{-0.2t}$.

17. Let P be the population and I be the number of infected people. The rate of spread dI/dt is jointly proportional to
I and to $P - I$, so for some constant k, $dI/dt = kI(P - I) \Rightarrow I = \dfrac{I_0 P}{I_0 + (P - I_0)e^{-kPt}}$ (from the
discussion of logistic growth in Section 7.5).
Now, measuring t in days, we substitute $t = 7$, $P = 5000$, $I_0 = 160$ and $I(7) = 1200$ to find k:
$1200 = \dfrac{160 \cdot 5000}{160 + (5000 - 160)e^{-5000 \cdot 7 \cdot k}} \Leftrightarrow k \approx 0.00006448$. So, putting $I = 5000 \times 80\% = 4000$, we solve
for t: $4000 = \dfrac{160 \cdot 5000}{160 + (5000 - 160)e^{-0.00006448 \cdot 5000 \cdot t}} \Leftrightarrow 160 + 4840e^{-0.3224t} = 200 \Leftrightarrow$
$-0.3224t = \ln\frac{40}{4840} \Leftrightarrow t \approx 14.9$. So it takes about 15 days for 80% of the population to be infected.

19. $\dfrac{dh}{dt} = -\dfrac{R}{V}\left(\dfrac{h}{k + h}\right) \Rightarrow \displaystyle\int \dfrac{k + h}{h}\,dh = \int\left(-\dfrac{R}{V}\right)dt \Rightarrow \int\left(1 + \dfrac{k}{h}\right)dh = -\dfrac{R}{V}\int 1\,dt \Rightarrow$
$h + k\ln h = -\dfrac{R}{V}t + C$. This equation gives a relationship between h and t, but it is not possible to isolate h and
express it in terms of t.

21. (a) $dx/dt = 0.4x(1 - 0.000005x) - 0.002xy$, $dy/dt = -0.2y + 0.000008xy$. If $y = 0$, then

$dx/dt = 0.4x(1 - 0.000005x)$, so $dx/dt = 0$ $\Leftrightarrow$ $x = 0$ or $x = 200,000$, which shows that the insect

population increases logistically with a carrying capacity of 200,000. Since $dx/dt > 0$ for $0 < x < 200,000$

and $dx/dt < 0$ for $x > 200,000$, we expect the insect population to stabilize at 200,000.

(b) x and y are constant $\Rightarrow$ $x' = 0$ and $y' = 0$ $\Rightarrow$

$$\begin{cases} 0 = 0.4x(1 - 0.000005x) - 0.002xy \\ 0 = -0.2y + 0.000008xy \end{cases} \Rightarrow \begin{cases} 0 = 0.4x[(1 - 0.000005x) - 0.005y] \\ 0 = y(-0.2 + 0.000008x) \end{cases}$$

The second equation is true if $y = 0$ or $x = \frac{0.2}{0.000008} = 25,000$. If $y = 0$ in the first equation, then either $x = 0$

or $x = \frac{1}{0.000005} = 200,000$. If $x = 25,000$, then $0 = 0.4(25,000)[(1 - 0.000005 \cdot 25,000) - 0.005y]$ $\Rightarrow$

$0 = 10,000[(1 - 0.125) - 0.005y]$ $\Rightarrow$ $0 = 8750 - 50y$ $\Rightarrow$ $y = 175$.

Case (i): $y = 0$, $x = 0$: Zero populations

Case (ii): $y = 0$, $x = 200,000$: In the absence of birds, the insect population is always 200,000.

Case (iii): $x = 25,000$, $y = 175$: The predator/prey interaction balances and the populations are stable.

(c) The populations of the birds and insects fluctuate around 175 and 25,000, respectively, and eventually stabilize

at those values.

(d)

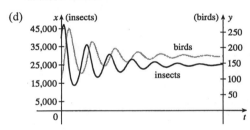

Focus on Problem Solving

1. We use the Fundamental Theorem of Calculus to differentiate the given equation:

$[f(x)]^2 = 100 + \int_0^x \left\{ [f(t)]^2 + [f'(t)]^2 \right\} dt \Rightarrow 2f(x)f'(x) = [f(x)]^2 + [f'(x)]^2 \Rightarrow$

$[f(x)]^2 + [f'(x)]^2 - 2f(x)f'(x) = 0 \Rightarrow [f(x) - f'(x)]^2 = 0 \Leftrightarrow f(x) = f'(x)$. We can solve this as a separable equation, or else use Theorem 7.4.2 with $k = 1$, which says that the solutions are $f(x) = Ce^x$. Now $[f(0)]^2 = 100$, so $f(0) = C = \pm 10$, and hence $f(x) = \pm 10e^x$ are the only functions satisfying the given equation.

3. $f'(x) = \lim\limits_{h \to 0} \dfrac{f(x+h) - f(x)}{h} = \lim\limits_{h \to 0} \dfrac{f(x)\,[f(h) - 1]}{h}$ [since $f(x+h) = f(x)f(h)$]

$= f(x) \lim\limits_{h \to 0} \dfrac{f(h) - 1}{h} = f(x) \lim\limits_{h \to 0} \dfrac{f(h) - f(0)}{h - 0} = f(x)f'(0) = f(x)$

Therefore, $f'(x) = f(x)$ for all x and from Theorem 7.4.2 we get $f(x) = Ae^x$. Now $f(0) = 1 \Rightarrow A = 1 \Rightarrow f(x) = e^x$.

5. Let $y(t)$ denote the temperature of the peach pie t minutes after 5:00 P.M. and R the temperature of the room. In Exercise 7.4.13, Newton's Law of Cooling gives us $dy/dt = k(y - R)$. Solving for y we get $\dfrac{dy}{y - R} = k\,dt \Rightarrow$

$\ln|y - R| = kt + C \Rightarrow |y - R| = e^{kt+C} \Rightarrow y - R = \pm e^{kt} \cdot e^C \Rightarrow y = Me^{kt} + R$, where M is a nonzero constant. We are given temperatures at three times.

$$
\begin{aligned}
y(0) &= 100 &\Rightarrow\quad 100 &= M + R &\Rightarrow\quad R &= 100 - M \\
y(10) &= 80 &\Rightarrow\quad 80 &= Me^{10k} + R &\textbf{(1)}& \\
y(20) &= 65 &\Rightarrow\quad 65 &= Me^{20k} + R &\textbf{(2)}&
\end{aligned}
$$

Substituting $100 - M$ for R in **(1)** and **(2)** gives us

$$-20 = Me^{10k} - M \quad \textbf{(3)} \quad \text{and} \quad -35 = Me^{20k} - M \quad \textbf{(4)}$$

Dividing **(3)** by **(4)** gives us $\dfrac{-20}{-35} = \dfrac{M(e^{10k} - 1)}{M(e^{20k} - 1)} \Rightarrow \dfrac{4}{7} = \dfrac{e^{10k} - 1}{e^{20k} - 1} \Rightarrow 4e^{20k} - 4 = 7e^{10k} - 7 \Rightarrow$

$4e^{20k} - 7e^{10k} + 3 = 0$. This is a quadratic equation in e^{10k}. $\left(4e^{10k} - 3\right)\left(e^{10k} - 1\right) = 0 \Rightarrow e^{10k} = \frac{3}{4}$ or 1

$\Rightarrow 10k = \ln \frac{3}{4}$ or $\ln 1 \Rightarrow k = \frac{1}{10} \ln \frac{3}{4}$ since k is a nonzero constant of proportionality. Substituting $\frac{3}{4}$ for e^{10k} in **(3)** gives us $-20 = M \cdot \frac{3}{4} - M \Rightarrow -20 = -\frac{1}{4}M \Rightarrow M = 80$. Now $R = 100 - M$ so $R = 20\,^\circ$C.

7. (a) While running from $(L, 0)$ to (x, y), the dog travels a distance

$$s = \int_x^L \sqrt{1 + (dy/dx)^2}\, dx = -\int_L^x \sqrt{1 + (dy/dx)^2}\, dx, \text{ so } \frac{ds}{dx} = -\sqrt{1 + (dy/dx)^2}. \text{ The dog and rabbit}$$

run at the same speed, so the rabbit's position when the dog has traveled a distance s is $(0, s)$. Since the dog runs

straight for the rabbit, $\dfrac{dy}{dx} = \dfrac{s - y}{0 - x}$ (see the figure).

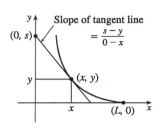

Thus, $s = y - x\dfrac{dy}{dx} \ \Rightarrow \ \dfrac{ds}{dx} = \dfrac{dy}{dx} - \left(x\dfrac{d^2y}{dx^2} + 1\dfrac{dy}{dx} \right) = -x\dfrac{d^2y}{dx^2}$. Equating the two expressions for $\dfrac{ds}{dx}$

gives us $x\dfrac{d^2y}{dx^2} = \sqrt{1 + \left(\dfrac{dy}{dx}\right)^2}$, as claimed.

(b) Letting $z = \dfrac{dy}{dx}$, we obtain the differential equation $x\dfrac{dz}{dx} = \sqrt{1 + z^2}$, or $\dfrac{dz}{\sqrt{1 + z^2}} = \dfrac{dx}{x}$. Integrating:

$$\ln x = \int \frac{dz}{\sqrt{1 + z^2}} \overset{25}{=} \ln\left(z + \sqrt{1 + z^2}\right) + C. \text{ When } x = L, z = dy/dx = 0, \text{ so } \ln L = \ln 1 + C.$$

Therefore, $C = \ln L$, so $\ln x = \ln\left(\sqrt{1 + z^2} + z\right) + \ln L = \ln\left[L\left(\sqrt{1 + z^2} + z\right)\right] \ \Rightarrow$

$$x = L\left(\sqrt{1 + z^2} + z\right) \ \Rightarrow \ \sqrt{1 + z^2} = \frac{x}{L} - z \ \Rightarrow \ 1 + z^2 = \left(\frac{x}{L}\right)^2 - \frac{2xz}{L} + z^2 \ \Rightarrow$$

$$\left(\frac{x}{L}\right)^2 - 2z\left(\frac{x}{L}\right) - 1 = 0 \ \Rightarrow \ z = \frac{(x/L)^2 - 1}{2(x/L)} = \frac{x^2 - L^2}{2Lx} = \frac{x}{2L} - \frac{L}{2}\frac{1}{x} \text{ [for } x > 0\text{]. Since } z = \frac{dy}{dx},$$

$$y = \frac{x^2}{4L} - \frac{L}{2}\ln x + C_1. \text{ Since } y = 0 \text{ when } x = L, \ 0 = \frac{L}{4} - \frac{L}{2}\ln L + C_1 \ \Rightarrow \ C_1 = \frac{L}{2}\ln L - \frac{L}{4}. \text{ Thus,}$$

$$y = \frac{x^2}{4L} - \frac{L}{2}\ln x + \frac{L}{2}\ln L - \frac{L}{4} = \frac{x^2 - L^2}{4L} - \frac{L}{2}\ln\left(\frac{x}{L}\right).$$

(c) As $x \to 0^+$, $y \to \infty$, so the dog never catches the rabbit.

9. (a) We are given that $V = \frac{1}{3}\pi r^2 h$, $dV/dt = 60{,}000\pi$ ft³/h, and $r = 1.5h = \frac{3}{2}h$. So $V = \frac{1}{3}\pi\left(\frac{3}{2}h\right)^2 h = \frac{3}{4}\pi h^3$

$\Rightarrow \dfrac{dV}{dt} = \frac{3}{4}\pi \cdot 3h^2 \dfrac{dh}{dt} = \frac{9}{4}\pi h^2 \dfrac{dh}{dt}$. Therefore, $\dfrac{dh}{dt} = \dfrac{4(dV/dt)}{9\pi h^2} = \dfrac{240{,}000\pi}{9\pi h^2} = \dfrac{80{,}000}{3h^2}$ $(\star)$ $\Rightarrow$

$\int 3h^2\, dh = \int 80{,}000\, dt \ \Rightarrow \ h^3 = 80{,}000t + C.$ When $t = 0$, $h = 60$. Thus, $C = 60^3 = 216{,}000$, so

$h^3 = 80{,}000t + 216{,}000.$ Let $h = 100$. Then $100^3 = 1{,}000{,}000 = 80{,}000t + 216{,}000 \ \Rightarrow$

$80{,}000t = 784{,}000 \ \Rightarrow \ t = 9.8$, so the time required is 9.8 hours.

(b) The floor area of the silo is $F = \pi \cdot 200^2 = 40{,}000\pi$ ft^2, and the area of the base of the pile is

$A = \pi r^2 = \pi\left(\frac{3}{2}h\right)^2 = \frac{9\pi}{4}h^2$. So the area of the floor which is not covered when $h = 60$ is

$F - A = 40{,}000\pi - 8100\pi = 31{,}900\pi \approx 100{,}217$ ft^2. Now $A = \frac{9\pi}{4}h^2 \;\Rightarrow\; dA/dt = \frac{9\pi}{4}\cdot 2h\,(dh/dt)$,

and from $(\star)$ in part (a) we know that when $h = 60$, $dh/dt = \frac{80{,}000}{3(60)^2} = \frac{200}{27}$ ft/h. Therefore,

$dA/dt = \frac{9\pi}{4}(2)(60)\left(\frac{200}{27}\right) = 2000\pi \approx 6283$ ft^2/h.

(c) At $h = 90$ ft, $dV/dt = 60{,}000\pi - 20{,}000\pi = 40{,}000\pi$ ft^3/h. From $(\star)$ in part (a),

$\dfrac{dh}{dt} = \dfrac{4(dV/dt)}{9\pi h^2} = \dfrac{4(40{,}000\pi)}{9\pi h^2} = \dfrac{160{,}000}{9h^2} \;\Rightarrow\; \int 9h^2\,dh = \int 160{,}000\,dt \;\Rightarrow\; 3h^3 = 160{,}000t + C.$

When $t = 0$, $h = 90$; therefore, $C = 3\cdot 729{,}000 = 2{,}187{,}000$. So $3h^3 = 160{,}000t + 2{,}187{,}000$. At the top,

$h = 100 \;\Rightarrow\; 3(100)^3 = 160{,}000t + 2{,}187{,}000 \;\Rightarrow\; t = \frac{813{,}000}{160{,}000} \approx 5.1$. The pile reaches the top after

about 5.1 h.

11. Let $P(a, b)$ be any point on the curve. If m is the slope of the tangent line at P, then $m = y'$ and

an equation of the normal line at P is $y - b = -\dfrac{1}{m}(x - a)$, or equivalently, $y = -\dfrac{1}{m}x + b + \dfrac{a}{m}$.

The y-intercept is always 6, so $b + \dfrac{a}{m} = 6 \;\Rightarrow\; \dfrac{a}{m} = 6 - b \;\Rightarrow\; m = \dfrac{a}{6 - b}$.

We will solve the equivalent differential equation $\dfrac{dy}{dx} = \dfrac{x}{6 - y} \;\Rightarrow\; (6 - y)\,dy = x\,dx \;\Rightarrow$

$\displaystyle\int (6 - y)\,dy = \int x\,dx \;\Rightarrow\; 6y - \tfrac{1}{2}y^2 = \tfrac{1}{2}x^2 + C \;\Rightarrow\; 12y - y^2 = x^2 + K.$ Since $(3, 2)$ is on the curve,

$12(2) - 2^2 = 3^2 + K \;\Rightarrow\; K = 11$. So the curve is given by $12y - y^2 = x^2 + 11 \;\Rightarrow$

$x^2 + y^2 - 12y + 36 = -11 + 36 \;\Rightarrow\; x^2 + (y - 6)^2 = 25$, a circle with center $(0, 6)$ and radius 5.

Infinite Sequences and Series

Sequences · · · · · · · · · · · · · · · ·

1. (a) A sequence is an ordered list of numbers. It can also be defined as a function whose domain is the set of positive integers.

(b) The terms a_n approach 8 as n becomes large. In fact, we can make a_n as close to 8 as we like by taking n sufficiently large.

(c) The terms a_n become large as n becomes large.

3. The first six terms of $a_n = \dfrac{n}{2n+1}$ are: $\dfrac{1}{3}, \dfrac{2}{5}, \dfrac{3}{7}, \dfrac{4}{9}, \dfrac{5}{11}, \dfrac{6}{13}$. It appears that the sequence is approaching $\dfrac{1}{2}$.

$$\lim_{n\to\infty} \frac{n}{2n+1} = \lim_{n\to\infty} \frac{1}{2+1/n} = \frac{1}{2}$$

5. $\left\{1, -\frac{2}{3}, \frac{4}{9}, -\frac{8}{27}, \dots\right\}$. Each term is $-\frac{2}{3}$ times the preceding one, so $a_n = \left(-\frac{2}{3}\right)^{n-1}$.

7. $\{2, 7, 12, 17, \dots\}$. Each term is larger than the preceding one by 5, so
$a_n = a_1 + d(n-1) = 2 + 5(n-1) = 5n - 3$.

9. $a_n = n(n-1)$. $a_n \to \infty$ as $n \to \infty$, so the sequence diverges.

11. $a_n = \dfrac{3+5n^2}{n+n^2} = \dfrac{(3+5n^2)/n^2}{(n+n^2)/n^2} = \dfrac{5+3/n^2}{1+1/n}$, so $a_n \to \dfrac{5+0}{1+0} = 5$ as $n \to \infty$. Converges

13. $a_n = \dfrac{2^n}{3^{n+1}} = \dfrac{1}{3}\left(\dfrac{2}{3}\right)^n$, so $\lim\limits_{n\to\infty} a_n = \dfrac{1}{3}\lim\limits_{n\to\infty} \left(\dfrac{2}{3}\right)^n = \dfrac{1}{3}\cdot 0 = 0$ by (6) with $r = \dfrac{2}{3}$. Converges

15. $a_n = \dfrac{(-1)^{n-1} n}{n^2+1} = \dfrac{(-1)^{n-1}}{n+1/n}$, so $0 \le |a_n| = \dfrac{1}{n+1/n} \le \dfrac{1}{n} \to 0$ as $n \to \infty$, so $a_n \to 0$ by the Squeeze
Theorem and Theorem 4. Converges

17. $a_n = 2 + \cos n\pi$, so
$\{a_n\} = \{2 + \cos\pi, 2 + \cos 2\pi, 2 + \cos 3\pi, 2 + \cos 4\pi, \dots\} = \{2-1, 2+1, 2-1, 2+1, \dots\}$
$= \{1, 3, 1, 3, \dots\}$
This sequence oscillates between 1 and 3, so it diverges.

19. $\lim\limits_{x\to\infty} \dfrac{\ln\left(x^2\right)}{x} = \lim\limits_{x\to\infty} \dfrac{2\ln x}{x} \overset{H}{=} \lim\limits_{x\to\infty} \dfrac{2/x}{1} = 0$, so by Theorem 2, $\left\{\dfrac{\ln\left(n^2\right)}{n}\right\}$ converges to 0.

21. $b_n = \sqrt{n+2} - \sqrt{n} = \left(\sqrt{n+2} - \sqrt{n}\right)\dfrac{\sqrt{n+2}+\sqrt{n}}{\sqrt{n+2}+\sqrt{n}} = \dfrac{2}{\sqrt{n+2}+\sqrt{n}} < \dfrac{2}{\sqrt{n}+\sqrt{n}} = \dfrac{2}{2\sqrt{n}} = \dfrac{1}{\sqrt{n}} \to 0$
as $n \to \infty$. So by the Squeeze Theorem with $a_n = 0$ and $c_n = 1/\sqrt{n}$, $\left\{\sqrt{n+2} - \sqrt{n}\right\}$ converges to 0.

23. $\lim\limits_{x\to\infty} \dfrac{x}{2^x} \overset{H}{=} \lim\limits_{x\to\infty} \dfrac{1}{(\ln 2)2^x} = 0$, so by Theorem 2, $\{n2^{-n}\}$ converges to 0.

25. $0 \le \dfrac{\cos^2 n}{2^n} \le \dfrac{1}{2^n}$ [since $0 \le \cos^2 n \le 1$], so since $\lim\limits_{n\to\infty} \dfrac{1}{2^n} = 0$, $\left\{\dfrac{\cos^2 n}{2^n}\right\}$ converges to 0 by the Squeeze

Theorem.

27.

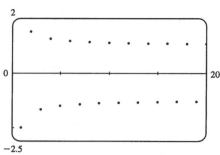

From the graph, we see that the sequence

$\left\{(-1)^n \dfrac{n+1}{n}\right\}$ is divergent, since it oscillates

between 1 and -1 (approximately).

29.

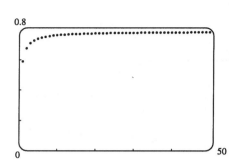

From the graph, it appears that the sequence

converges to about 0.78.

$$\lim_{n\to\infty} \frac{2n}{2n+1} = \lim_{n\to\infty} \frac{2}{2+1/n} = 1, \text{ so}$$

$$\lim_{n\to\infty} \arctan\left(\frac{2n}{2n+1}\right) = \arctan 1 = \frac{\pi}{4}.$$

31.

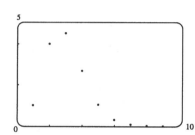

From the graph, it appears that the sequence converges to 0.

$$0 < a_n = \frac{n^3}{n!} = \frac{n}{n} \cdot \frac{n}{(n-1)} \cdot \frac{n}{(n-2)} \cdot \frac{1}{(n-3)} \cdots\cdots \frac{1}{3} \cdot \frac{1}{2} \cdot \frac{1}{1}$$

$$\le \frac{n^2}{((n-1)(n-2)(n-3))} \quad \text{(for } n \ge 4\text{)}$$

$$= \frac{1/n}{(1-1/n)(1-2/n)(1-3/n)} \to 0 \text{ as } n \to \infty$$

So by the Squeeze Theorem, $\{n^3/n!\}$ converges to 0.

33. (a) $a_n = 1000(1.06)^n \;\Rightarrow\; a_1 = 1060, a_2 = 1123.60, a_3 = 1191.02, a_4 = 1262.48,$ and $a_5 = 1338.23.$

(b) $\lim\limits_{n\to\infty} a_n = 1000 \lim\limits_{n\to\infty} (1.06)^n$, so the sequence diverges by (6) with $r = 1.06 > 1.$

35. (a) $a_1 = 1, a_2 = 4 - a_1 = 4 - 1 = 3, a_3 = 4 - a_2 = 4 - 3 = 1, a_4 = 4 - a_3 = 4 - 1 = 3,$

$a_5 = 4 - a_4 = 4 - 3 = 1.$ Since the terms of the sequence alternate between 1 and 3, the sequence is

divergent.

(b) $a_1 = 2, a_2 = 4 - a_1 = 4 - 2 = 2, a_3 = 4 - a_2 = 4 - 2 = 2.$ Since all of the terms are 2, $\lim\limits_{n\to\infty} a_n = 2$ and

hence, the sequence is convergent.

37. (a) Let a_n be the number of rabbit pairs in the nth month. Clearly $a_1 = 1 = a_2$. In the nth month, each pair that is

2 or more months old (that is, a_{n-2} pairs) will produce a new pair to add to the a_{n-1} pairs already present.

Thus, $a_n = a_{n-1} + a_{n-2}$, so that $\{a_n\} = \{f_n\}$, the Fibonacci sequence.

(b) $a_n = \dfrac{f_{n+1}}{f_n} \;\Rightarrow\; a_{n-1} = \dfrac{f_n}{f_{n-1}} = \dfrac{f_{n-1} + f_{n-2}}{f_{n-1}} = 1 + \dfrac{f_{n-2}}{f_{n-1}} = 1 + \dfrac{1}{f_{n-1}/f_{n-2}} = 1 + \dfrac{1}{a_{n-2}}.$ If

$L = \lim\limits_{n\to\infty} a_n$, then $L = \lim\limits_{n\to\infty} a_{n-1}$ and $L = \lim\limits_{n\to\infty} a_{n-2}$, so L must satisfy $L = 1 + \dfrac{1}{L} \;\Rightarrow\;$

$L^2 - L - 1 = 0 \;\Rightarrow\; L = \dfrac{1+\sqrt{5}}{2}$ (since L must be positive).

39. $a_n = \dfrac{1}{2n+3}$ is decreasing since $a_{n+1} = \dfrac{1}{2(n+1)+3} = \dfrac{1}{2n+5} < \dfrac{1}{2n+3} = a_n$ for each $n \geq 1$. The

sequence is bounded since $0 < a_n \leq \frac{1}{5}$ for all $n \geq 1$. Note that $a_1 = \frac{1}{5}$.

41. $a_n = \cos(n\pi/2)$ is not monotonic. The first few terms are $0, -1, 0, 1, 0, -1, 0, 1, \ldots$. In fact, the sequence
consists of the terms $0, -1, 0, 1$ repeated over and over again in that order. The sequence is bounded since $|a_n| \leq 1$
for all $n \geq 1$.

43. Since $\{a_n\}$ is a decreasing sequence, $a_n > a_{n+1}$ for all $n \geq 1$. Because all of its terms lie between 5 and 8, $\{a_n\}$ is
a bounded sequence. By the Monotonic Sequence Theorem, $\{a_n\}$ is convergent; that is, $\{a_n\}$ has a limit L. L must
be less than 8 since $\{a_n\}$ is decreasing, so $5 \leq L < 8$.

45. We show by induction that $\{a_n\}$ is increasing and bounded above by 3.

Let P_n be the proposition that $a_{n+1} > a_n$ and $0 < a_n < 3$. Clearly P_1 is true. Assume that P_n is true. Then

$$a_{n+1} > a_n \quad \Rightarrow \quad \frac{1}{a_{n+1}} < \frac{1}{a_n} \quad \Rightarrow \quad -\frac{1}{a_{n+1}} > -\frac{1}{a_n}.$$

Now $a_{n+2} = 3 - \dfrac{1}{a_{n+1}} > 3 - \dfrac{1}{a_n} = a_{n+1} \quad \Leftrightarrow \quad P_{n+1}$. This proves that $\{a_n\}$ is increasing and bounded above

by 3, so $1 = a_1 < a_n < 3$, that is, $\{a_n\}$ is bounded, and hence convergent by the Monotonic Sequence Theorem. If

$L = \lim\limits_{n\to\infty} a_n$, then $\lim\limits_{n\to\infty} a_{n+1} = L$ also, so L must satisfy $L = 3 - 1/L \quad \Rightarrow \quad L^2 - 3L + 1 = 0 \quad \Rightarrow$

$L = \dfrac{3 \pm \sqrt{5}}{2}$. But $L > 1$, so $L = \dfrac{3 + \sqrt{5}}{2}$.

47. $(0.8)^n < 0.000001 \quad \Rightarrow \quad \ln(0.8)^n < \ln(0.000001) \quad \Rightarrow \quad n\ln(0.8) < \ln(0.000001) \quad \Rightarrow$

$n > \dfrac{\ln(0.000001)}{\ln(0.8)} \quad \Rightarrow \quad n > 61.9$, so n must be at least 62 to satisfy the given inequality.

49. (a) First we show that $a > a_1 > b_1 > b$.

$a_1 - b_1 = \dfrac{a+b}{2} - \sqrt{ab} = \frac{1}{2}\left(a - 2\sqrt{ab} + b\right) = \frac{1}{2}\left(\sqrt{a} - \sqrt{b}\right)^2 > 0$ (since $a > b$) $\Rightarrow \quad a_1 > b_1$. Also

$a - a_1 = a - \frac{1}{2}(a+b) = \frac{1}{2}(a-b) > 0$ and $b - b_1 = b - \sqrt{ab} = \sqrt{b}\left(\sqrt{b} - \sqrt{a}\right) < 0$, so $a > a_1 > b_1 > b$.

In the same way we can show that $a_1 > a_2 > b_2 > b_1$ and so the given assertion is true for $n = 1$. Suppose it is
true for $n = k$, that is, $a_k > a_{k+1} > b_{k+1} > b_k$. Then

$$a_{k+2} - b_{k+2} = \frac{1}{2}(a_{k+1} + b_{k+1}) - \sqrt{a_{k+1}b_{k+1}} = \frac{1}{2}\left(a_{k+1} - 2\sqrt{a_{k+1}b_{k+1}} + b_{k+1}\right)$$

$$= \frac{1}{2}\left(\sqrt{a_{k+1}} - \sqrt{b_{k+1}}\right)^2 > 0$$

$$a_{k+1} - a_{k+2} = a_{k+1} - \frac{1}{2}(a_{k+1} + b_{k+1}) = \frac{1}{2}(a_{k+1} - b_{k+1}) > 0$$

and $b_{k+1} - b_{k+2} = b_{k+1} - \sqrt{a_{k+1}b_{k+1}} = \sqrt{b_{k+1}}\left(\sqrt{b_{k+1}} - \sqrt{a_{k+1}}\right) < 0 \quad \Rightarrow$

$a_{k+1} > a_{k+2} > b_{k+2} > b_{k+1}$, so the assertion is true for $n = k + 1$. Thus, it is true for all n by mathematical
induction.

(b) From part (a) we have $a > a_n > a_{n+1} > b_{n+1} > b_n > b$, which shows that both sequences, $\{a_n\}$ and $\{b_n\}$,
are monotonic and bounded. So they are both convergent by the Monotonic Sequence Theorem.

(c) Let $\lim\limits_{n\to\infty} a_n = \alpha$ and $\lim\limits_{n\to\infty} b_n = \beta$. Then $\lim\limits_{n\to\infty} a_{n+1} = \lim\limits_{n\to\infty} \dfrac{a_n + b_n}{2} \quad \Rightarrow \quad \alpha = \dfrac{\alpha + \beta}{2} \quad \Rightarrow$

$2\alpha = \alpha + \beta \quad \Rightarrow \quad \alpha = \beta$.

8.2 **Series** • • • • • • • • • • • • • • • • •

1. (a) A sequence is an ordered list of numbers whereas a series is the *sum* of a list of numbers.

(b) A series is convergent if the sequence of partial sums is a convergent sequence. A series is divergent if it is not convergent.

3.

n	s_n
1	−2.40000
2	−1.92000
3	−2.01600
4	−1.99680
5	−2.00064
6	−1.99987
7	−2.00003
8	−1.99999
9	−2.00000
10	−2.00000

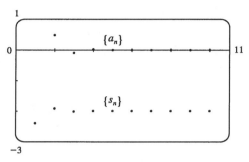

From the graph and the table, it seems that the series converges to −2. In fact, it is a geometric series with $a = -2.4$ and $r = -\frac{1}{5}$, so its sum is

$$\sum_{n=1}^{\infty} \frac{12}{(-5)^n} = \frac{-2.4}{1 - \left(-\frac{1}{5}\right)} = \frac{-2.4}{1.2} = -2.$$ Note that the dot corresponding to

$n = 1$ is part of both $\{a_n\}$ and $\{s_n\}$.

TI-86 Note: To graph $\{a_n\}$ and $\{s_n\}$, set your calculator to Param mode and DrawDot mode. (DrawDot is under GRAPH, MORE, FORMT (F3).) Now under E(t) = make the assignments: xt1=t, yt1=12/(-5)^t, xt2=t, yt2=sum seq(yt1,t,1,t,1). (sum and seq are under LIST, OPS (F5), MORE.) Under WIND use 1,10,1,0,10,1,-3,1,1 to obtain a graph similar to the one above. Then use TRACE (F4) to see the values.

5.

n	s_n
1	1.55741
2	−0.62763
3	−0.77018
4	0.38764
5	−2.99287
6	−3.28388
7	−2.41243
8	−9.21214
9	−9.66446
10	−9.01610

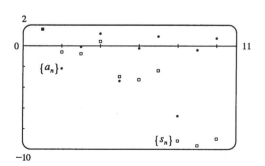

The series $\sum_{n=1}^{\infty} \tan n$ diverges, since its terms do not approach 0.

7.

n	s_n
1	0.64645
2	0.80755
3	0.87500
4	0.91056
5	0.93196
6	0.94601
7	0.95581
8	0.96296
9	0.96838
10	0.97259

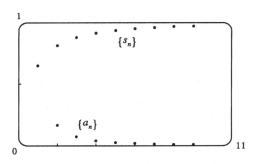

From the graph, it seems that the series converges to 1. To find the sum, we write

$$s_n = \sum_{i=1}^{n} \left(\frac{1}{i^{1.5}} - \frac{1}{(i+1)^{1.5}} \right) = \left(1 - \frac{1}{2^{1.5}} \right) + \left(\frac{1}{2^{1.5}} - \frac{1}{3^{1.5}} \right)$$

$$+ \left(\frac{1}{3^{1.5}} - \frac{1}{4^{1.5}} \right) + \cdots + \left(\frac{1}{n^{1.5}} - \frac{1}{(n+1)^{1.5}} \right) = 1 - \frac{1}{(n+1)^{1.5}}$$

So the sum is $\lim_{n \to \infty} s_n = 1 - 0 = 1$.

9. (a) $\lim_{n \to \infty} a_n = \lim_{n \to \infty} \frac{2n}{3n+1} = \frac{2}{3}$, so the *sequence* $\{a_n\}$ is convergent by (8.1.1).

(b) Since $\lim_{n \to \infty} a_n = \frac{2}{3} \neq 0$, the *series* $\sum_{n=1}^{\infty} a_n$ is divergent by the Test for Divergence (7).

11. $5 - \frac{10}{3} + \frac{20}{9} - \frac{40}{27} + \cdots$ is a geometric series with $a = 5$ and $r = -\frac{2}{3}$. Since $|r| = \frac{2}{3} < 1$, the series converges to $\frac{a}{1-r} = \frac{5}{1-(-2/3)} = \frac{5}{5/3} = 3$.

13. $\sum_{n=1}^{\infty} 5\left(\frac{2}{3}\right)^{n-1}$ is a geometric series with $a = 5$ and $r = \frac{2}{3}$. Since $|r| = \frac{2}{3} < 1$, the series converges to $\frac{a}{1-r} = \frac{5}{1-2/3} = \frac{5}{1/3} = 15$.

15. For $\sum_{n=1}^{\infty} 3^{-n} 8^{n+1} = \sum_{n=1}^{\infty} \left(\frac{1}{3^n} \cdot \frac{8 \cdot 8^n}{1} \right) = \sum_{n=1}^{\infty} 8\left(\frac{8}{3}\right)^n$, $a = \frac{64}{3}$ and $|r| = \frac{8}{3} > 1$, so the series diverges.

17. $\sum_{n=1}^{\infty} \frac{n}{n+5}$ diverges since $\lim_{n \to \infty} a_n = \lim_{n \to \infty} \frac{n}{n+5} = 1 \neq 0$. [Use (7), the Test for Divergence.]

19. Converges. $s_n = \sum_{i=1}^{n} \frac{1}{i(i+2)} = \sum_{i=1}^{n} \left(\frac{1/2}{i} - \frac{1/2}{i+2} \right)$ (using partial fractions) $= \frac{1}{2} \sum_{i=1}^{n} \left(\frac{1}{i} - \frac{1}{i+2} \right)$. The latter sum is a telescoping series:

$$\left(1 - \frac{1}{3} \right) + \left(\frac{1}{2} - \frac{1}{4} \right) + \left(\frac{1}{3} - \frac{1}{5} \right) + \cdots + \left(\frac{1}{n-1} - \frac{1}{n+1} \right) + \left(\frac{1}{n} - \frac{1}{n+2} \right) = 1 + \frac{1}{2} - \frac{1}{n+1} - \frac{1}{n+2}$$

Thus, $\sum_{n=1}^{\infty} \frac{1}{n(n+2)} = \frac{1}{2} \lim_{n \to \infty} \left(1 + \frac{1}{2} - \frac{1}{n+1} - \frac{1}{n+2} \right) = \frac{1}{2} \left(1 + \frac{1}{2} \right) = \frac{3}{4}$.

21. $\sum\limits_{n=1}^{\infty} [2(0.1)^n + (0.2)^n] = 2\sum\limits_{n=1}^{\infty}(0.1)^n + \sum\limits_{n=1}^{\infty}(0.2)^n$. These are convergent geometric series and so by Theorem 8,

their sum is also convergent. $2\left(\frac{0.1}{1-0.1}\right) + \frac{0.2}{1-0.2} = \frac{2}{9} + \frac{1}{4} = \frac{17}{36}$

23. Converges. $s_n = \left(\sin 1 - \sin\frac{1}{2}\right) + \left(\sin\frac{1}{2} - \sin\frac{1}{3}\right) + \cdots + \left(\sin\frac{1}{n} - \sin\frac{1}{n+1}\right) = \sin 1 - \sin\frac{1}{n+1}$, so

$\sum\limits_{n=1}^{\infty}\left(\sin\frac{1}{n} - \sin\frac{1}{n+1}\right) = \lim\limits_{n\to\infty} s_n = \sin 1 - \sin 0 = \sin 1.$

25. Converges. $\sum\limits_{n=1}^{\infty}\frac{3^n + 2^n}{6^n} = \sum\limits_{n=1}^{\infty}\left(\frac{3^n}{6^n} + \frac{2^n}{6^n}\right) = \sum\limits_{n=1}^{\infty}\left[\left(\frac{1}{2}\right)^n + \left(\frac{1}{3}\right)^n\right] = \frac{1/2}{1-1/2} + \frac{1/3}{1-1/3} = 1 + \frac{1}{2} = \frac{3}{2}$

27. $\lim\limits_{n\to\infty} a_n = \lim\limits_{n\to\infty} \arctan n = \frac{\pi}{2} \neq 0$, so the series diverges by the Test for Divergence.

29. $0.\overline{2} = \frac{2}{10} + \frac{2}{10^2} + \cdots$ is a geometric series with $a = \frac{2}{10}$ and $r = \frac{1}{10}$. It converges to $\frac{a}{1-r} = \frac{2/10}{1-1/10} = \frac{2}{9}$.

31. $3.\overline{417} = 3 + \frac{417}{10^3} + \frac{417}{10^6} + \cdots = 3 + \frac{417/10^3}{1-1/10^3} = 3 + \frac{417}{999} = \frac{3414}{999} = \frac{1138}{333}$

33. $\sum\limits_{n=1}^{\infty}\frac{x^n}{3^n} = \sum\limits_{n=1}^{\infty}\left(\frac{x}{3}\right)^n$ is a geometric series with $r = \frac{x}{3}$, so the series converges $\Leftrightarrow |r| < 1 \Leftrightarrow \frac{|x|}{3} < 1 \Leftrightarrow$

$|x| < 3$; that is, $-3 < x < 3$. In that case, the sum of the series is $\frac{a}{1-r} = \frac{x/3}{1-x/3} = \frac{x/3}{1-x/3}\cdot\frac{3}{3} = \frac{x}{3-x}$.

35. $\sum\limits_{n=0}^{\infty}\left(\frac{1}{x}\right)^n = \sum\limits_{n=1}^{\infty}\left(\frac{1}{x}\right)^{n-1}$ is geometric with $r = \frac{1}{x}$, so it converges whenever $\left|\frac{1}{x}\right| < 1 \Leftrightarrow$

$\frac{1}{|x|} < 1 \Leftrightarrow 1 < |x| \Leftrightarrow |x| > 1 \Leftrightarrow x > 1$ or $x < -1$, and the sum is

$\frac{a}{1-r} = \frac{1}{1-1/x} = \frac{1}{1-1/x}\cdot\frac{x}{x} = \frac{x}{x-1}.$

37. After defining f, We use `convert(f,parfrac);` in Maple, `Apart` in Mathematica, or `Expand Rational`

and `Simplify` in Derive to find that the general term is $\frac{1}{(4n+1)(4n-3)} = -\frac{1/4}{4n+1} + \frac{1/4}{4n-3}$. So the nth

partial sum is

$s_n = \sum\limits_{k=1}^{n}\left(-\frac{1/4}{4k+1} + \frac{1/4}{4k-3}\right) = \frac{1}{4}\sum\limits_{k=1}^{n}\left(\frac{1}{4k-3} - \frac{1}{4k+1}\right)$

$= \frac{1}{4}\left[\left(1-\frac{1}{5}\right) + \left(\frac{1}{5}-\frac{1}{9}\right) + \left(\frac{1}{9}-\frac{1}{13}\right) + \cdots + \left(\frac{1}{4n-3}-\frac{1}{4n+1}\right)\right] = \frac{1}{4}\left(1-\frac{1}{4n+1}\right)$

The series converges to $\lim\limits_{n\to\infty} s_n = \frac{1}{4}$. This can be confirmed by directly computing the sum using

`sum(f,1..infinity);` (in Maple), `Sum[f,{n,1,Infinity}]` (in Mathematica), or `Calculus Sum`

(from 1 to ∞) and `Simplify` (in Derive).

39. For $n = 1$, $a_1 = 0$ since $s_1 = 0$. For $n > 1$,

$a_n = s_n - s_{n-1} = \frac{n-1}{n+1} - \frac{(n-1)-1}{(n-1)+1} = \frac{(n-1)n - (n+1)(n-2)}{(n+1)n} = \frac{2}{n(n+1)}$

Also, $\sum\limits_{n=1}^{\infty} a_n = \lim\limits_{n\to\infty} s_n = \lim\limits_{n\to\infty}\frac{1-1/n}{1+1/n} = 1.$

41. (a) The first step in the chain occurs when the local government spends D dollars. The people who receive it spend a fraction c of those D dollars, that is, Dc dollars. Those who receive the Dc dollars spend a fraction c of it, that is, Dc^2 dollars. Continuing in this way, we see that the total spending after n transactions is

$$S_n = D + Dc + Dc^2 + \cdots + Dc^{n-1} = \frac{D(1-c^n)}{1-c} \text{ by (3)}.$$

(b) $\lim\limits_{n\to\infty} S_n = \lim\limits_{n\to\infty} \dfrac{D(1-c^n)}{1-c} = \dfrac{D}{1-c} \lim\limits_{n\to\infty}(1-c^n) = \dfrac{D}{1-c}$ (since $0 < c < 1 \ \Rightarrow \ \lim\limits_{n\to\infty} c^n = 0$)

$= \dfrac{D}{s}$ (since $c + s = 1$) $= kD$ (since $k = 1/s$)

If $c = 0.8$, then $s = 1 - c = 0.2$ and the multiplier is $k = 1/s = 5$.

43. $\sum_{n=2}^{\infty}(1+c)^{-n}$ is a geometric series with $a = (1+c)^{-2}$ and $r = (1+c)^{-1}$, so the series converges when

$|(1+c)^{-1}| < 1 \ \Leftrightarrow \ |1+c| > 1 \ \Leftrightarrow \ 1+c > 1$ or $1+c < -1 \ \Leftrightarrow \ c > 0$ or $c < -2$. We calculate the sum

of the series and set it equal to 2: $\dfrac{(1+c)^{-2}}{1-(1+c)^{-1}} = 2 \ \Leftrightarrow \ \left(\dfrac{1}{1+c}\right)^2 = 2 - 2\left(\dfrac{1}{1+c}\right) \ \Leftrightarrow$

$1 = 2(1+c)^2 - 2(1+c) = 0 \ \Leftrightarrow \ 2c^2 + 2c - 1 = 0 \ \Leftrightarrow \ c = \dfrac{-2 \pm \sqrt{12}}{4} = \dfrac{\pm\sqrt{3}-1}{2}$. However, the negative

root is inadmissible because $-2 < \dfrac{-\sqrt{3}-1}{2} < 0$. So $c = \dfrac{\sqrt{3}-1}{2}$.

45. Let d_n be the diameter of C_n. We draw lines from the centers of the C_i to the center of D (or C), and using the Pythagorean Theorem, we can write $1^2 + \left(1 - \tfrac{1}{2}d_1\right)^2 = \left(1 + \tfrac{1}{2}d_1\right)^2 \ \Leftrightarrow$

$1 = \left(1 + \tfrac{1}{2}d_1\right)^2 - \left(1 - \tfrac{1}{2}d_1\right)^2 = 2d_1$ (difference of squares)

$\Rightarrow \ d_1 = \tfrac{1}{2}$. Similarly,

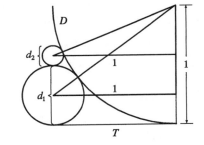

$1 = \left(1 + \tfrac{1}{2}d_2\right)^2 - \left(1 - d_1 - \tfrac{1}{2}d_2\right)^2 = 2d_2 + 2d_1 - d_1^2 - d_1 d_2$

$= (2 - d_1)(d_1 + d_2) \ \Leftrightarrow$

$d_2 = \dfrac{1}{2-d_1} - d_1 = \dfrac{(1-d_1)^2}{2-d_1}, \ 1 = \left(1 + \tfrac{1}{2}d_3\right)^2 - \left(1 - d_1 - d_2 - \tfrac{1}{2}d_3\right)^2 \ \Leftrightarrow \ d_3 = \dfrac{[1-(d_1+d_2)]^2}{2-(d_1+d_2)}$, and in

general, $d_{n+1} = \dfrac{\left(1 - \sum_{i=1}^{n} d_i\right)^2}{2 - \sum_{i=1}^{n} d_i}$. If we actually calculate d_2 and d_3 from the formulas above, we find that they

are $\dfrac{1}{6} = \dfrac{1}{2 \cdot 3}$ and $\dfrac{1}{12} = \dfrac{1}{3 \cdot 4}$ respectively, so we suspect that in general, $d_n = \dfrac{1}{n(n+1)}$. To prove this, we

use induction: assume that for all $k \le n$, $d_k = \dfrac{1}{k(k+1)} = \dfrac{1}{k} - \dfrac{1}{k+1}$. Then

$$\sum_{i=1}^{n} d_i = 1 - \frac{1}{n+1} = \frac{n}{n+1} \quad \text{(telescoping sum). Substituting this into our formula for } d_{n+1}, \text{ we get}$$

$$d_{n+1} = \frac{\left[1 - \dfrac{n}{n+1}\right]^2}{2 - \left(\dfrac{n}{n+1}\right)} = \frac{\dfrac{1}{(n+1)^2}}{\dfrac{n+2}{n+1}} = \frac{1}{(n+1)(n+2)}, \text{ and the induction is complete.}$$

Now, we observe that the partial sums $\sum_{i=1}^{n} d_i$ of the diameters of the circles approach 1 as $n \to \infty$; that is,

$$\sum_{n=1}^{\infty} a_n = \sum_{n=1}^{\infty} \frac{1}{n(n+1)} = 1, \text{ which is what we wanted to prove.}$$

47. The series $1 - 1 + 1 - 1 + 1 - 1 + \cdots$ diverges (geometric series with $r = -1$) so we cannot say that
$0 = 1 - 1 + 1 - 1 + 1 - 1 + \cdots$.

49. Suppose on the contrary that $\sum(a_n + b_n)$ converges. Then $\sum(a_n + b_n)$ and $\sum a_n$ are convergent series. So by Theorem 8, $\sum[(a_n + b_n) - a_n]$ would also be convergent. But $\sum[(a_n + b_n) - a_n] = \sum b_n$, a contradiction, since $\sum b_n$ is given to be divergent.

51. The partial sums $\{s_n\}$ form an increasing sequence, since $s_n - s_{n-1} = a_n > 0$ for all n. Also, the sequence $\{s_n\}$ is bounded since $s_n \leq 1000$ for all n. So by Theorem 8.1.7, the sequence of partial sums converges, that is, the series $\sum a_n$ is convergent.

53. (a) At the first step, only the interval $\left(\frac{1}{3}, \frac{2}{3}\right)$ (length $\frac{1}{3}$) is removed. At the second step, we remove the intervals $\left(\frac{1}{9}, \frac{2}{9}\right)$ and $\left(\frac{7}{9}, \frac{8}{9}\right)$, which have a total length of $2 \cdot \left(\frac{1}{3}\right)^2$. At the third step, we remove 2^2 intervals, each of length $\left(\frac{1}{3}\right)^3$. In general, at the nth step we remove 2^{n-1} intervals, each of length $\left(\frac{1}{3}\right)^n$, for a length of $2^{n-1} \cdot \left(\frac{1}{3}\right)^n = \frac{1}{3}\left(\frac{2}{3}\right)^{n-1}$. Thus, the total length of all removed intervals is $\sum_{n=1}^{\infty} \frac{1}{3}\left(\frac{2}{3}\right)^{n-1} = \frac{1/3}{1 - 2/3} = 1$ (geometric series with $a = \frac{1}{3}$ and $r = \frac{2}{3}$). Notice that at the nth step, the leftmost interval that is removed is $\left(\left(\frac{1}{3}\right)^n, \left(\frac{2}{3}\right)^n\right)$, so we never remove 0, and 0 is in the Cantor set. Also, the rightmost interval removed is $\left(1 - \left(\frac{2}{3}\right)^n, 1 - \left(\frac{1}{3}\right)^n\right)$, so 1 is never removed. Some other numbers in the Cantor set are $\frac{1}{3}, \frac{2}{3}, \frac{1}{9}, \frac{2}{9}, \frac{7}{9}$, and $\frac{8}{9}$.

(b) The area removed at the first step is $\frac{1}{9}$; at the second step, $8 \cdot \left(\frac{1}{9}\right)^2$; at the third step, $(8)^2 \cdot \left(\frac{1}{9}\right)^3$. In general, the area removed at the nth step is $(8)^{n-1}\left(\frac{1}{9}\right)^n = \frac{1}{9}\left(\frac{8}{9}\right)^{n-1}$, so the total area of all removed squares is
$$\sum_{n=1}^{\infty} \frac{1}{9}\left(\frac{8}{9}\right)^{n-1} = \frac{1/9}{1 - 8/9} = 1.$$

55. (a) For $\sum_{n=1}^{\infty} \frac{n}{(n+1)!}$, $s_1 = \frac{1}{1 \cdot 2} = \frac{1}{2}$, $s_2 = \frac{1}{2} + \frac{2}{1 \cdot 2 \cdot 3} = \frac{5}{6}$, $s_3 = \frac{5}{6} + \frac{3}{1 \cdot 2 \cdot 3 \cdot 4} = \frac{23}{24}$,
$s_4 = \frac{23}{24} + \frac{4}{1 \cdot 2 \cdot 3 \cdot 4 \cdot 5} = \frac{119}{120}$. The denominators are $(n+1)!$, so a guess would be $s_n = \frac{(n+1)! - 1}{(n+1)!}$.

(b) For $n = 1$, $s_1 = \frac{1}{2} = \frac{2! - 1}{2!}$, so the formula holds for $n = 1$. Assume $s_k = \frac{(k+1)! - 1}{(k+1)!}$. Then
$$s_{k+1} = \frac{(k+1)! - 1}{(k+1)!} + \frac{k+1}{(k+2)!} = \frac{(k+1)! - 1}{(k+1)!} + \frac{k+1}{(k+1)!(k+2)}$$
$$= \frac{(k+2)! - (k+2) + k+1}{(k+2)!} = \frac{(k+2)! - 1}{(k+2)!}$$

Thus, the formula is true for $n = k + 1$. So by induction, the guess is correct.

(c) $\lim_{n \to \infty} s_n = \lim_{n \to \infty} \frac{(n+1)! - 1}{(n+1)!} = \lim_{n \to \infty} \left[1 - \frac{1}{(n+1)!}\right] = 1$ and so $\sum_{n=1}^{\infty} \frac{n}{(n+1)!} = 1$.

8.3 The Integral and Comparison Tests; Estimating Sums • • •

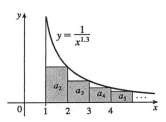

1. The picture shows that $a_2 = \dfrac{1}{2^{1.3}} < \displaystyle\int_1^2 \dfrac{1}{x^{1.3}}\, dx$,

$a_3 = \dfrac{1}{3^{1.3}} < \displaystyle\int_2^3 \dfrac{1}{x^{1.3}}\, dx$, and so on, so $\displaystyle\sum_{n=2}^{\infty} \dfrac{1}{n^{1.3}} < \displaystyle\int_1^{\infty} \dfrac{1}{x^{1.3}}\, dx$. The

integral converges by (5.10.2) with $p = 1.3 > 1$, so the series converges.

3. (a) We cannot say anything about $\sum a_n$. If $a_n > b_n$ for all n and $\sum b_n$ is convergent, then $\sum a_n$ could be
convergent or divergent. (See the note on page 587.)

(b) If $a_n < b_n$ for all n, then $\sum a_n$ is convergent. [This is part (i) of the Comparison Test.]

5. $\displaystyle\sum_{n=1}^{\infty} n^b$ is a p-series with $p = -b$. $\displaystyle\sum_{n=1}^{\infty} b^n$ is a geometric series. By (1), the p-series is convergent if $p > 1$. In this

case, $\displaystyle\sum_{n=1}^{\infty} n^b = \sum_{n=1}^{\infty} (1/n^{-b})$, so $-b > 1$ $\Leftrightarrow$ $b < -1$ are the values for which the series converge. A geometric

series $\displaystyle\sum_{n=1}^{\infty} ar^{n-1}$ converges if $|r| < 1$, so $\displaystyle\sum_{n=1}^{\infty} b^n$ converges if $|b| < 1$ $\Leftrightarrow$ $-1 < b < 1$.

7. The function $f(x) = 1/x^4$ is continuous, positive, and decreasing on $[1, \infty)$, so the Integral Test applies.

$\displaystyle\int_1^{\infty} \dfrac{1}{x^4}\, dx = \lim_{t \to \infty} \int_1^t x^{-4}\, dx = \lim_{t \to \infty} \left[\dfrac{x^{-3}}{-3}\right]_1^t = \lim_{t \to \infty} \left(-\dfrac{1}{3t^3} + \dfrac{1}{3}\right) = \dfrac{1}{3}$. Since this improper integral is

convergent, the series $\displaystyle\sum_{n=1}^{\infty} \dfrac{1}{n^4}$ is also convergent by the Integral Test.

9. $\dfrac{1}{n^2 + n + 1} < \dfrac{1}{n^2}$ for all $n \geq 1$, so $\displaystyle\sum_{n=1}^{\infty} \dfrac{1}{n^2 + n + 1}$ converges by comparison with $\displaystyle\sum_{n=1}^{\infty} \dfrac{1}{n^2}$, which converges

because it is a p-series with $p = 2 > 1$.

11. $1 + \dfrac{1}{8} + \dfrac{1}{27} + \dfrac{1}{64} + \dfrac{1}{125} + \cdots = \displaystyle\sum_{n=1}^{\infty} \dfrac{1}{n^3}$. This is a p-series with $p = 3 > 1$, so it converges by (1).

13. $f(x) = xe^{-x^2}$ is continuous and positive on $[1, \infty)$, and since $f'(x) = e^{-x^2}(1 - 2x^2) < 0$ for
$x > 1$, f is decreasing as well. Thus, we can use the Integral Test.

$\int_1^{\infty} xe^{-x^2}\, dx = \lim_{t \to \infty} \left[-\dfrac{1}{2}e^{-x^2}\right]_1^t = 0 - (-\dfrac{1}{2}e^{-1}) = 1/(2e)$. Since the integral converges, the series converges.

15. $f(x) = \dfrac{1}{x \ln x}$ is continuous and positive on $[2, \infty)$, and also decreasing since $f'(x) = -\dfrac{1 + \ln x}{x^2(\ln x)^2} < 0$ for $x > 2$,

so we can use the Integral Test. $\displaystyle\int_2^{\infty} \dfrac{1}{x \ln x}\, dx = \lim_{t \to \infty} [\ln(\ln x)]_2^t = \lim_{t \to \infty} [\ln(\ln t) - \ln(\ln 2)] = \infty$, so the series

diverges.

17. $\dfrac{5}{2 + 3^n} < \dfrac{5}{3^n}$ for all $n \geq 1$, so $\displaystyle\sum_{n=1}^{\infty} \dfrac{5}{2 + 3^n}$ converges by comparison with $\displaystyle\sum_{n=1}^{\infty} \dfrac{5}{3^n} = 5 \sum_{n=1}^{\infty} \dfrac{1}{3^n}$, which converges

because $\displaystyle\sum_{n=1}^{\infty} \dfrac{1}{3^n}$ is a convergent geometric series with $r = \dfrac{1}{3}$ $(|r| < 1)$.

19. $\dfrac{n+1}{n^2} > \dfrac{n}{n^2} = \dfrac{1}{n}$ for all $n \geq 1$, so $\displaystyle\sum_{n=1}^{\infty} \dfrac{n+1}{n^2}$ diverges by comparison with the harmonic series $\displaystyle\sum_{n=1}^{\infty} \dfrac{1}{n}$.

21. Let $a_n = \dfrac{n^2+1}{n^4+1}$ and $b_n = \dfrac{1}{n^2}$. Then $\sum a_n$ and $\sum b_n$ are series with positive terms and

$$\lim_{n\to\infty}\frac{a_n}{b_n} = \lim_{n\to\infty}\left(\frac{n^2+1}{n^4+1}\cdot\frac{n^2}{1}\right) = \lim_{n\to\infty}\frac{n^4+n^2}{n^4+1} = 1 > 0. \text{ Since } \sum_{n=1}^{\infty}\frac{1}{n^2} \text{ is a convergent } p\text{-series } (p=2>1),$$

so is $\displaystyle\sum_{n=1}^{\infty}\frac{n^2+1}{n^4+1}$ by the Limit Comparison Test.

23. Use the Limit Comparison Test with $a_n = \sin\left(\dfrac{1}{n}\right)$ and $b_n = \dfrac{1}{n}$. Then $\sum a_n$ and $\sum b_n$ are series with positive

terms and $\displaystyle\lim_{n\to\infty}\frac{a_n}{b_n} = \lim_{n\to\infty}\frac{\sin(1/n)}{1/n} = \lim_{\theta\to0}\frac{\sin\theta}{\theta} = 1 > 0.$ Since $\sum_{n=1}^{\infty}b_n$ is the divergent harmonic series,

$\sum_{n=1}^{\infty}\sin(1/n)$ also diverges. (Note that we could also use l'Hospital's Rule to evaluate the limit:

$$\lim_{x\to\infty}\frac{\sin(1/x)}{1/x} \overset{\text{H}}{=} \lim_{x\to\infty}\frac{\cos(1/x)\cdot(-1/x^2)}{-1/x^2} = \lim_{x\to\infty}\cos\frac{1}{x} = \cos0 = 1.)$$

25. We have already shown (in Exercise 15) that when $p=1$ the series $\displaystyle\sum_{n=2}^{\infty}\frac{1}{n(\ln n)^p}$ diverges, so assume that $p\ne1$.

$f(x) = \dfrac{1}{x(\ln x)^p}$ is continuous and positive on $[2,\infty)$, and $f'(x) = -\dfrac{p+\ln x}{x^2(\ln x)^{p+1}} < 0$ if $x > e^{-p}$, so that f is

eventually decreasing and we can use the Integral Test.

$$\int_2^{\infty}\frac{1}{x(\ln x)^p}\,dx = \lim_{t\to\infty}\left[\frac{(\ln x)^{1-p}}{1-p}\right]_2^t \text{ (for } p\ne1) = \lim_{t\to\infty}\left[\frac{(\ln t)^{1-p}}{1-p}\right] - \frac{(\ln2)^{1-p}}{1-p}$$

This limit exists whenever $1-p<0 \iff p>1$, so the series converges for $p>1$.

27. (a) $f(x) = \dfrac{1}{x^2}$ is positive and continuous and $f'(x) = -\dfrac{2}{x^3}$ is negative for $x > 0$, and so the Integral

Test applies. $\displaystyle\sum_{n=1}^{\infty}\frac{1}{n^2} \approx s_{10} = \frac{1}{1^2} + \frac{1}{2^2} + \frac{1}{3^2} + \cdots + \frac{1}{10^2} \approx 1.549768.$

$$R_{10} \le \int_{10}^{\infty}\frac{1}{x^2}\,dx = \lim_{t\to\infty}\left[\frac{-1}{x}\right]_{10}^t = \lim_{t\to\infty}\left(-\frac{1}{t}+\frac{1}{10}\right) = \frac{1}{10}, \text{ so the error is at most } 0.1.$$

(b) $s_{10} + \displaystyle\int_{11}^{\infty}\frac{1}{x^2}\,dx \le s \le s_{10} + \int_{10}^{\infty}\frac{1}{x^2}\,dx \implies s_{10} + \frac{1}{11} \le s \le s_{10} + \frac{1}{10} \implies$

$1.549768 + 0.090909 = 1.640677 \le s \le 1.549768 + 0.1 = 1.649768$, so we get $s \approx 1.64522$ (the average of
1.640677 and 1.649768) with error ≤ 0.005 (the maximum of $1.649768 - 1.64522$ and $1.64522 - 1.640677$,
rounded up).

(c) $R_n \le \displaystyle\int_n^{\infty}\frac{1}{x^2}\,dx = \frac{1}{n}.$ So $R_n < 0.001$ if $\dfrac{1}{n} < \dfrac{1}{1000} \iff n > 1000.$

29. $f(x) = x^{-3/2}$ is positive and continuous and $f'(x) = -\frac{3}{2}x^{-5/2}$ is negative for $x > 0$, so the Integral Test applies.

From the end of Example 7, we see that the error is at most half the length of the interval. From (4), the interval is

$\left(s_n + \int_{n+1}^{\infty}f(x)\,dx, s_n + \int_n^{\infty}f(x)\,dx\right)$, so its length is $\int_n^{\infty}f(x)\,dx - \int_{n+1}^{\infty}f(x)\,dx$. Thus, we need n such that

$$0.01 > \frac{1}{2}\left(\int_n^{\infty}x^{-3/2}\,dx - \int_{n+1}^{\infty}x^{-3/2}\,dx\right) = \frac{1}{2}\left(\lim_{t\to\infty}\left[\frac{-2}{\sqrt{x}}\right]_n^t - \lim_{t\to\infty}\left[\frac{-2}{\sqrt{x}}\right]_{n+1}^t\right) = \frac{1}{\sqrt{n}} - \frac{1}{\sqrt{n+1}}$$

↔ $n > 13.08$ (use a graphing calculator to solve $1/\sqrt{x} - 1/\sqrt{x+1} < 0.01$). Again from the end of Example 7, we approximate s by the midpoint of this interval. In general, the midpoint is

$\frac{1}{2}\left[\left(s_n + \int_{n+1}^{\infty} f(x)\, dx\right) + \left(s_n + \int_{n}^{\infty} f(x)\, dx\right)\right] = s_n + \frac{1}{2}\left(\int_{n+1}^{\infty} f(x)\, dx + \int_{n}^{\infty} f(x)\, dx\right)$. So using $n = 14$,

we have $s \approx s_{14} + \frac{1}{2}\left(\int_{14}^{\infty} x^{-3/2}\, dx + \int_{15}^{\infty} x^{-3/2}\, dx\right) \approx 2.0872 + \frac{1}{\sqrt{14}} + \frac{1}{\sqrt{15}} \approx 2.6127 \approx 2.61$. Any larger

value of n will also work. For instance, $s \approx s_{30} + \frac{1}{\sqrt{30}} + \frac{1}{\sqrt{31}} \approx 2.6124$.

31. $\sum\limits_{n=1}^{10} \dfrac{1}{n^4 + n^2} = \dfrac{1}{2} + \dfrac{1}{20} + \dfrac{1}{90} + \cdots + \dfrac{1}{10{,}100} \approx 0.567975$. Now $\dfrac{1}{n^4 + n^2} < \dfrac{1}{n^4}$, so using the reasoning and

notation of Example 8, the error is $R_{10} \le T_{10} = \sum\limits_{n=11}^{\infty} \dfrac{1}{n^4} \le \int_{10}^{\infty} \dfrac{dx}{x^4} = \lim\limits_{t \to \infty}\left[-\dfrac{x^{-3}}{3}\right]_{10}^{t} = \dfrac{1}{3000} = 0.000\overline{3}$.

33. (a) From the figure, $a_2 + a_3 + \cdots + a_n \le \int_1^n f(x)\, dx$, so with

$f(x) = \dfrac{1}{x}, \dfrac{1}{2} + \dfrac{1}{3} + \dfrac{1}{4} + \cdots + \dfrac{1}{n} \le \int_1^n \dfrac{1}{x}\, dx = \ln n$. Thus,

$s_n = 1 + \dfrac{1}{2} + \dfrac{1}{3} + \dfrac{1}{4} + \cdots + \dfrac{1}{n} \le 1 + \ln n$.

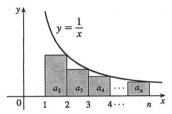

(b) By part (a), $s_{10^6} \le 1 + \ln 10^6 \approx 14.82 < 15$ and $s_{10^9} \le 1 + \ln 10^9 \approx 21.72 < 22$.

35. Since $\dfrac{d_n}{10^n} \le \dfrac{9}{10^n}$ for each n, and since $\sum\limits_{n=1}^{\infty} \dfrac{9}{10^n}$ is a convergent geometric series ($|r| = \frac{1}{10} < 1$),

$0.d_1 d_2 d_3 \ldots = \sum\limits_{n=1}^{\infty} \dfrac{d_n}{10^n}$ will always converge by the Comparison Test.

37. Yes. Since $\sum a_n$ is a convergent series with positive terms, $\lim\limits_{n \to \infty} a_n = 0$ by (8.2.6), and $\sum b_n = \sum \sin(a_n)$ is a

series with positive terms (for large enough n). The Limit Comparison Test gives us

$\lim\limits_{n \to \infty} \dfrac{b_n}{a_n} = \lim\limits_{n \to \infty} \dfrac{\sin(a_n)}{a_n} = 1 > 0$ by Theorem 3.4.2. Thus, $\sum b_n$ is also convergent.

8.4 Other Convergence Tests • • • • • • • • • • • • •

1. (a) An alternating series is a series whose terms are alternately positive and negative.

(b) An alternating series $\sum_{n=1}^{\infty} (-1)^{n-1} b_n$ converges if $0 < b_{n+1} \le b_n$ for all n and $\lim\limits_{n \to \infty} b_n = 0$. (This is the Alternating Series Test.)

(c) The error involved in using the partial sum s_n as an approximation to the total sum s is the remainder $R_n = s - s_n$ and the size of the error is smaller than b_{n+1}; that is, $|R_n| \le b_{n+1}$. (This is the Alternating Series Estimation Theorem.)

3. $\dfrac{4}{7} - \dfrac{4}{8} + \dfrac{4}{9} - \dfrac{4}{10} + \dfrac{4}{11} - \cdots = \sum\limits_{n=1}^{\infty} (-1)^{n-1} \dfrac{4}{n+6}$. Now $b_n = \dfrac{4}{n+6} > 0$, $\{b_n\}$ is decreasing, and

$\lim\limits_{n \to \infty} b_n = 0$, so the series converges by the Alternating Series Test.

5. $b_n = \dfrac{1}{\sqrt{n}} > 0$, $\{b_n\}$ is decreasing, and $\lim\limits_{n \to \infty} b_n = 0$, so the series $\sum\limits_{n=1}^{\infty} \dfrac{(-1)^{n-1}}{\sqrt{n}}$ converges by the Alternating

Series Test.

7. $\displaystyle\sum_{n=1}^{\infty} a_n = \sum_{n=1}^{\infty}(-1)^n \frac{3n-1}{2n+1} = \sum_{n=1}^{\infty}(-1)^n b_n$. Now $\displaystyle\lim_{n\to\infty} b_n = \lim_{n\to\infty}\frac{3-1/n}{2+1/n} = \frac{3}{2} \neq 0$. Since $\displaystyle\lim_{n\to\infty} a_n \neq 0$

(in fact the limit does not exist), the series diverges by the Test for Divergence.

9. $\displaystyle\sum_{n=1}^{\infty}\frac{(-1)^{n-1}}{n} = 1 - \frac{1}{2} + \frac{1}{3} - \frac{1}{4} + \cdots + \frac{1}{49} - \frac{1}{50} + \frac{1}{51} - \frac{1}{52} + \cdots$. The 50th partial sum of this series is an

underestimate, since $\displaystyle\sum_{n=1}^{\infty}\frac{(-1)^{n-1}}{n} = s_{50} + \left(\frac{1}{51} - \frac{1}{52}\right) + \left(\frac{1}{53} - \frac{1}{54}\right) + \cdots$, and the terms in parentheses are

all positive. The result can be seen geometrically in Figure 1.

11. If $p > 0$, $\displaystyle\frac{1}{(n+1)^p} \leq \frac{1}{n^p}$ ($\{1/n^p\}$ is decreasing) and $\displaystyle\lim_{n\to\infty}\frac{1}{n^p} = 0$, so the series converges by the Alternating

Series Test. If $p \leq 0$, $\displaystyle\lim_{n\to\infty}\frac{(-1)^{n-1}}{n^p}$ does not exist, so the series diverges by the Test for Divergence. Thus,

$\displaystyle\sum_{n=1}^{\infty}\frac{(-1)^{n-1}}{n^p}$ converges $\Leftrightarrow$ $p > 0$.

13. Using the Ratio Test with the series $\displaystyle\sum_{n=1}^{\infty}\frac{(-2)^n}{n!}$,

$$\lim_{n\to\infty}\left|\frac{a_{n+1}}{a_n}\right| = \lim_{n\to\infty}\left|a_{n+1}\cdot\frac{1}{a_n}\right| = \lim_{n\to\infty}\left|\frac{(-2)^{n+1}}{(n+1)!}\cdot\frac{n!}{(-2)^n}\right| = \lim_{n\to\infty}\left|\frac{-2}{n+1}\right|$$

$$= 2\lim_{n\to\infty}\frac{1}{n+1} = 2(0) = 0 < 1,$$

so the series is absolutely convergent (and therefore convergent). Now $b_7 = 2^7/7! \approx 0.025 > 0.01$ and

$b_8 = 2^8/8! \approx 0.006 < 0.01$, so by the Alternating Series Estimation Theorem, $n = 7$. (That is, since the 8th term

is less than the desired error, we need to add the first 7 terms to get the sum to the desired accuracy.)

15. The graph gives us an estimate for the sum of the series

$\displaystyle\sum_{n=1}^{\infty}\frac{(-1)^{n-1}}{(2n-1)!}$ of 0.84. $b_5 = \dfrac{1}{(2\cdot 5 - 1)!} = \dfrac{1}{362{,}880} \approx 0.000\,003$,

so $\displaystyle\sum_{n=1}^{\infty}\frac{(-1)^{n-1}}{(2n-1)!} \approx s_4 = \sum_{n=1}^{4}\frac{(-1)^{n-1}}{(2n-1)!} = 1 - \frac{1}{6} + \frac{1}{120} - \frac{1}{5040}$

$\approx 0.841468.$

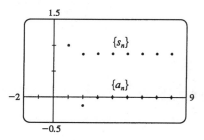

Adding b_5 to s_4 does not change the fourth decimal place of s_4, so

the sum of the series, correct to four decimal places, is 0.8415.

17. $b_6 = \dfrac{1}{2^6 6!} = \dfrac{1}{46{,}080} \approx 0.000\,022$, so

$\displaystyle\sum_{n=0}^{\infty}\frac{(-1)^n}{2^n n!} \approx s_5 = \sum_{n=0}^{5}\frac{(-1)^n}{2^n n!} = 1 - \frac{1}{2} + \frac{1}{8} - \frac{1}{48} + \frac{1}{384} - \frac{1}{3840} \approx 0.606510$. Adding b_6 to s_5 does not change

the fourth decimal place of s_5, so the sum of the series, correct to four decimal places, is 0.6065.

19. Consider the series whose terms are the absolute values of the terms of the given series.

$\displaystyle\sum_{n=1}^{\infty}\left|\frac{(-1)^{n-1}}{\sqrt{n}}\right| = \sum_{n=1}^{\infty}\frac{1}{n^{1/2}}$, which is a divergent p-series ($p = \frac{1}{2} \leq 1$). Thus, $\displaystyle\sum_{n=1}^{\infty}\frac{(-1)^{n-1}}{\sqrt{n}}$ is *not* absolutely

convergent.

21. Using the Ratio Test,

$$\lim_{n\to\infty}\left|\frac{a_{n+1}}{a_n}\right| = \lim_{n\to\infty}\left|\frac{(-3)^{n+1}/(n+1)^3}{(-3)^n/n^3}\right| = \lim_{n\to\infty}\left|\frac{(-3)n^3}{(n+1)^3}\right| = 3\lim_{n\to\infty}\left(\frac{n}{n+1}\right)^3 = 3 > 1, \text{ so the series}$$

diverges.

23. $\left|\dfrac{\sin 2n}{n^2}\right| \le \dfrac{1}{n^2}$ and $\displaystyle\sum_{n=1}^{\infty}\dfrac{1}{n^2}$ converges (p-series, $p = 2 > 1$), so $\displaystyle\sum_{n=1}^{\infty}\dfrac{\sin 2n}{n^2}$ converges absolutely by the

Comparison Test.

25. $\displaystyle\lim_{n\to\infty}\left|\frac{a_{n+1}}{a_n}\right| = \lim_{n\to\infty}\left[\frac{10^{n+1}}{(n+2)4^{2(n+1)+1}}\cdot\frac{(n+1)4^{2n+1}}{10^n}\right] = \lim_{n\to\infty}\left[\frac{10^{n+1}}{(n+2)4^{2n+3}}\cdot\frac{(n+1)4^{2n+1}}{10^n}\right] =$

$\displaystyle\lim_{n\to\infty}\left(\frac{10}{4^2}\cdot\frac{n+1}{n+2}\right) = \frac{5}{8} < 1$, so the series is absolutely convergent by the Ratio Test. Since the terms of this

series are positive, absolute convergence is the same as convergence.

27. $\displaystyle\lim_{n\to\infty}\left|\frac{a_{n+1}}{a_n}\right| = \lim_{n\to\infty}\frac{(n+1)!/[1\cdot3\cdot5\cdot\cdots\cdot(2n-1)(2n+1)]}{n!/[1\cdot3\cdot5\cdot\cdots\cdot(2n-1)]} = \lim_{n\to\infty}\frac{n+1}{2n+1} = \frac{1}{2} < 1$, so the series

converges absolutely by the Ratio Test.

29. By the recursive definition, $\displaystyle\lim_{n\to\infty}\left|\frac{a_{n+1}}{a_n}\right| = \lim_{n\to\infty}\left|\frac{5n+1}{4n+3}\right| = \frac{5}{4} > 1$, so the series diverges by the Ratio Test.

31. (a) $\displaystyle\lim_{n\to\infty}\left|\frac{1/(n+1)^3}{1/n^3}\right| = \lim_{n\to\infty}\frac{n^3}{(n+1)^3} = \lim_{n\to\infty}\frac{1}{(1+1/n)^3} = 1$. Inconclusive.

(b) $\displaystyle\lim_{n\to\infty}\left|\frac{(n+1)}{2^{n+1}}\cdot\frac{2^n}{n}\right| = \lim_{n\to\infty}\frac{n+1}{2n} = \lim_{n\to\infty}\left(\frac{1}{2}+\frac{1}{2n}\right) = \frac{1}{2}$. Conclusive (convergent).

(c) $\displaystyle\lim_{n\to\infty}\left|\frac{(-3)^n}{\sqrt{n+1}}\cdot\frac{\sqrt{n}}{(-3)^{n-1}}\right| = 3\lim_{n\to\infty}\sqrt{\frac{n}{n+1}} = 3\lim_{n\to\infty}\sqrt{\frac{1}{1+1/n}} = 3$. Conclusive (divergent).

(d) $\displaystyle\lim_{n\to\infty}\left|\frac{\sqrt{n+1}}{1+(n+1)^2}\cdot\frac{1+n^2}{\sqrt{n}}\right| = \lim_{n\to\infty}\left[\sqrt{1+\frac{1}{n}}\cdot\frac{1/n^2+1}{1/n^2+(1+1/n)^2}\right] = 1$. Inconclusive.

33. (a) $\displaystyle\lim_{n\to\infty}\left|\frac{a_{n+1}}{a_n}\right| = \lim_{n\to\infty}\left|\frac{x^{n+1}}{(n+1)!}\cdot\frac{n!}{x^n}\right| = \lim_{n\to\infty}\left|\frac{x}{n+1}\right| = |x|\lim_{n\to\infty}\frac{1}{n+1} = |x|\cdot 0 = 0 < 1$, so by the Ratio

Test the series $\displaystyle\sum_{n=0}^{\infty}\frac{x^n}{n!}$ converges for all x.

(b) Since the series of part (a) always converges, we must have $\displaystyle\lim_{n\to\infty}\frac{x^n}{n!} = 0$ by Theorem 8.2.6.

35. (a) $s_5 = \displaystyle\sum_{n=1}^{5}\frac{1}{n2^n} = \frac{1}{2}+\frac{1}{8}+\frac{1}{24}+\frac{1}{64}+\frac{1}{160} = \frac{661}{960} \approx 0.68854$. Now the ratios

$r_n = \dfrac{a_{n+1}}{a_n} = \dfrac{n2^n}{(n+1)2^{n+1}} = \dfrac{n}{2(n+1)}$ form an increasing sequence, since

$r_{n+1} - r_n = \dfrac{n+1}{2(n+2)} - \dfrac{n}{2(n+1)} = \dfrac{(n+1)^2 - n(n+2)}{2(n+1)(n+2)} = \dfrac{1}{2(n+1)(n+2)} > 0$. So by Exercise 34(b),

the error in using s_5 is $R_5 \le \dfrac{a_6}{1 - \lim\limits_{n\to\infty} r_n} = \dfrac{1/(6\cdot2^6)}{1 - 1/2} = \dfrac{1}{192} \approx 0.00521$.

(b) The error in using s_n as an approximation to the sum is $R_n = \dfrac{a_{n+1}}{1 - \frac{1}{2}} = \dfrac{2}{(n+1)2^{n+1}}$. We want

$$R_n < 0.00005 \quad \Leftrightarrow \quad \frac{1}{(n+1)2^n} < 0.00005 \quad \Leftrightarrow \quad (n+1)2^n > 20{,}000.$$ To find such an n we can use trial

and error or a graph. We calculate $(11+1)2^{11} = 24{,}576$, so $s_{11} = \sum\limits_{n=1}^{11} \dfrac{1}{n2^n} \approx 0.693109$ is within 0.00005 of

the actual sum.

8.5 Power Series · · · · · · · · · · · · · · · · ·

1. A power series is a series of the form $\sum_{n=0}^{\infty} c_n x^n = c_0 + c_1 x + c_2 x^2 + c_3 x^3 + \cdots$, where x is a variable and the c_n's are constants called the coefficients of the series.

More generally, a series of the form $\sum_{n=0}^{\infty} c_n (x-a)^n = c_0 + c_1(x-a) + c_2(x-a)^2 + \cdots$ is called a power series in $(x-a)$ or a power series centered at a or a power series about a, where a is a constant.

3. If $a_n = \dfrac{x^n}{\sqrt{n}}$, then $\lim\limits_{n\to\infty} \left| \dfrac{a_{n+1}}{a_n} \right| = \lim\limits_{n\to\infty} \left| \dfrac{x^{n+1}}{\sqrt{n+1}} \cdot \dfrac{\sqrt{n}}{x} \right| = \lim\limits_{n\to\infty} \left| \dfrac{x}{\sqrt{n+1}/\sqrt{n}} \right| = \lim\limits_{n\to\infty} \dfrac{|x|}{\sqrt{1+1/n}} = |x|$.

By the Ratio Test, the series $\sum\limits_{n=1}^{\infty} \dfrac{x^n}{\sqrt{n}}$ converges when $|x| < 1$, so the radius of convergence $R = 1$. When $x = 1$,

the series $\sum\limits_{n=1}^{\infty} \dfrac{1}{\sqrt{n}}$ diverges because it is a p-series with $p = \frac{1}{2} \le 1$. When $x = -1$, the series $\sum\limits_{n=1}^{\infty} \dfrac{(-1)^n}{\sqrt{n}}$

converges by the Alternating Series Test. Thus, the interval of convergence is $I = [-1, 1)$.

5. If $a_n = nx^n$, then $\lim\limits_{n\to\infty} \left| \dfrac{a_{n+1}}{a_n} \right| = \lim\limits_{n\to\infty} \left| \dfrac{(n+1)\,x^{n+1}}{nx^n} \right| = \lim\limits_{n\to\infty} \left| \dfrac{x(n+1)}{n} \right| = |x| \lim\limits_{n\to\infty} \dfrac{n+1}{n} = |x| < 1$ for

convergence (by the Ratio Test), so $R = 1$. When $x = 1$ or -1, $\lim\limits_{n\to\infty} nx^n$ does not exist, so $\sum_{n=0}^{\infty} nx^n$ diverges

for $x = \pm 1$. Thus, $I = (-1, 1)$.

7. If $a_n = \dfrac{x^n}{n!}$, then $\lim\limits_{n\to\infty} \left| \dfrac{a_{n+1}}{a_n} \right| = \lim\limits_{n\to\infty} \left| \dfrac{x^{n+1}}{(n+1)!} \cdot \dfrac{n!}{x^n} \right| = \lim\limits_{n\to\infty} \left| \dfrac{x}{n+1} \right| = |x| \lim\limits_{n\to\infty} \dfrac{1}{n+1} = |x| \cdot 0 = 0 < 1$ for

all x. So, by the Ratio Test, $R = \infty$, and $I = (-\infty, \infty)$.

9. If $a_n = \dfrac{3^n x^n}{(n+1)^2}$, then

$$\lim\limits_{n\to\infty} \left| \dfrac{a_{n+1}}{a_n} \right| = \lim\limits_{n\to\infty} \left| \dfrac{3^{n+1} x^{n+1}}{(n+2)^2} \cdot \dfrac{(n+1)^2}{3^n x^n} \right| = 3\,|x| \lim\limits_{n\to\infty} \left(\dfrac{n+1}{n+2} \right)^2 = 3\,|x| \cdot 1 = 3\,|x|.$$ By the Ratio Test,

the series converges when $3\,|x| < 1 \quad \Leftrightarrow \quad |x| < \frac{1}{3}$, so $R = \frac{1}{3}$. When $x = \frac{1}{3}$,

$$\sum\limits_{n=0}^{\infty} \dfrac{3^n x^n}{(n+1)^2} = \sum\limits_{n=0}^{\infty} \dfrac{1}{(n+1)^2} = \sum\limits_{n=1}^{\infty} \dfrac{1}{n^2},$$ which is a convergent p-series ($p = 2 > 1$). When $x = -\frac{1}{3}$,

$$\sum\limits_{n=0}^{\infty} \dfrac{3^n x^n}{(n+1)^2} = \sum\limits_{n=0}^{\infty} \dfrac{(-1)^n}{(n+1)^2},$$ which converges by the Alternating Series Test. Thus, $I = \left[-\frac{1}{3}, \frac{1}{3}\right]$.

11. If $a_n = (-1)^n \dfrac{x^n}{4^n \ln n}$, then

$$\lim_{n \to \infty} \left| \frac{a_{n+1}}{a_n} \right| = \lim_{n \to \infty} \left| \frac{x^{n+1}}{4^{n+1} \ln(n+1)} \cdot \frac{4^n \ln n}{x^n} \right| = \frac{|x|}{4} \lim_{n \to \infty} \frac{\ln n}{\ln(n+1)} = \frac{|x|}{4} \cdot 1 \ \text{(by l'Hospital's Rule)} = \frac{|x|}{4}.$$

By the Ratio Test, the series converges when $\dfrac{|x|}{4} < 1 \ \Leftrightarrow \ |x| < 4$, so $R = 4$. When $x = -4$,

$$\sum_{n=2}^{\infty} (-1)^n \frac{x^n}{4^n \ln n} = \sum_{n=2}^{\infty} \frac{(-1 \cdot -4)^n}{4^n \ln n} = \sum_{n=2}^{\infty} \frac{1}{\ln n}. \text{ Since } \ln n < n \text{ for } n \ge 2, \ \frac{1}{\ln n} > \frac{1}{n} \text{ and } \sum_{n=2}^{\infty} \frac{1}{n} \text{ is the}$$

divergent harmonic series (without the $n = 1$ term), $\displaystyle\sum_{n=2}^{\infty} \frac{1}{\ln n}$ is divergent by the Comparison Test. When $x = 4$,

$$\sum_{n=2}^{\infty} (-1)^n \frac{x^n}{4^n \ln n} = \sum_{n=2}^{\infty} (-1)^n \frac{1}{\ln n}, \text{ which converges by the Alternating Series Test. Thus, } I = (-4, 4].$$

13. If $a_n = \sqrt{n}\,(x-1)^n$, then $\displaystyle\lim_{n \to \infty} \left| \frac{a_{n+1}}{a_n} \right| = \lim_{n \to \infty} \left| \frac{\sqrt{n+1}\,|x-1|^{n+1}}{\sqrt{n}\,|x-1|^n} \right| = \lim_{n \to \infty} \sqrt{1 + \frac{1}{n}}\,|x-1| = |x-1|.$ By

the Ratio Test, the series converges when $|x-1| < 1$ [so $R = 1$] $\ \Leftrightarrow \ -1 < x - 1 < 1 \ \Leftrightarrow \ 0 < x < 2$.

When $x = 0$, the series becomes $\sum_{n=0}^{\infty} (-1)^n \sqrt{n}$, which diverges by the Test for Divergence. When $x = 2$, the

series becomes $\sum_{n=0}^{\infty} \sqrt{n}$, which also diverges by the Test for Divergence. Thus, $I = (0, 2)$.

15. If $a_n = (-1)^n \dfrac{(x+2)^n}{n 2^n}$, then

$$\lim_{n \to \infty} \left| \frac{a_{n+1}}{a_n} \right| = \lim_{n \to \infty} \left[\frac{|x+2|^{n+1}}{(n+1)\,2^{n+1}} \cdot \frac{n 2^n}{|x+2|^n} \right] = \lim_{n \to \infty} \frac{n}{n+1} \cdot \frac{|x+2|}{2} = \frac{|x+2|}{2}. \text{ By the Ratio Test, the}$$

series converges when $\dfrac{|x+2|}{2} < 1 \ \Leftrightarrow \ |x+2| < 2$ [so $R = 2$] $\ \Leftrightarrow \ -2 < x + 2 < 2 \ \Leftrightarrow \ -4 < x < 0$.

When $x = -4$, the series becomes $\sum_{n=1}^{\infty} (-1)^n \dfrac{(-2)^n}{n 2^n} = \sum_{n=1}^{\infty} \dfrac{2^n}{n 2^n} = \sum_{n=1}^{\infty} \dfrac{1}{n}$, which is the divergent harmonic series.

When $x = 0$, the series is $\sum_{n=1}^{\infty} \dfrac{(-1)^n}{n}$, the alternating harmonic series, which converges by the Alternating Series

Test. Thus, $I = (-4, 0]$.

17. If $a_n = n!(2x-1)^n$, then $\displaystyle\lim_{n \to \infty} \left| \frac{a_{n+1}}{a_n} \right| = \lim_{n \to \infty} \left| \frac{(n+1)!(2x-1)^{n+1}}{n!(2x-1)^n} \right| = \lim_{n \to \infty} (n+1)\,|2x-1| \to \infty \text{ as}$

$n \to \infty$ for all $x \ne \frac{1}{2}$. Since the series diverges for all $x \ne \frac{1}{2}$, $R = 0$ and $I = \left\{ \frac{1}{2} \right\}$.

19. (a) We are given that the power series $\sum_{n=0}^{\infty} c_n x^n$ is convergent for $x = 4$. So by Theorem 3, it must converge for

  at least $-4 < x \le 4$. In particular, it converges when $x = -2$; that is, $\sum_{n=0}^{\infty} c_n (-2)^n$ is convergent.

  (b) It does not follow that $\sum_{n=0}^{\infty} c_n (-4)^n$ is necessarily convergent. [See the comments after Theorem 3 about

  convergence at the endpoint of an interval. An example is $c_n = (-1)^n / (n 4^n)$.]

21. If $a_n = \dfrac{(n!)^k}{(kn)!}x^n$, then

$$\lim_{n\to\infty}\left|\frac{a_{n+1}}{a_n}\right| = \lim_{n\to\infty}\frac{[(n+1)!]^k\,(kn)!}{(n!)^k\,[k(n+1)]!}\,|x| = \lim_{n\to\infty}\frac{(n+1)^k}{(kn+k)(kn+k-1)\cdots(kn+2)(kn+1)}\,|x|$$

$$= \lim_{n\to\infty}\left[\frac{(n+1)}{(kn+1)}\frac{(n+1)}{(kn+2)}\cdots\frac{(n+1)}{(kn+k)}\right]|x|$$

$$= \lim_{n\to\infty}\left[\frac{n+1}{kn+1}\right]\lim_{n\to\infty}\left[\frac{n+1}{kn+2}\right]\cdots\lim_{n\to\infty}\left[\frac{n+1}{kn+k}\right]|x| = \left(\frac{1}{k}\right)^k|x| < 1 \quad\Leftrightarrow$$

$|x| < k^k$ for convergence, and the radius of convergence is $R = k^k$.

23. (a) If $a_n = \dfrac{(-1)^n\,x^{2n+1}}{n!(n+1)!\,2^{2n+1}}$, then

$$\lim_{n\to\infty}\left|\frac{a_{n+1}}{a_n}\right| = \lim_{n\to\infty}\left|\frac{x^{2n+3}}{(n+1)!(n+2)!\,2^{2n+3}}\cdot\frac{n!(n+1)!\,2^{2n+1}}{x^{2n+1}}\right| = \left(\frac{x}{2}\right)^2\lim_{n\to\infty}\frac{1}{(n+1)(n+2)} = 0 \text{ for}$$

all x. So $J_1(x)$ converges for all x and its domain is $(-\infty, \infty)$.

(b), (c) The initial terms of $J_1(x)$ up to $n = 5$ are $a_0 = \dfrac{x}{2}$,

$a_1 = -\dfrac{x^3}{16}$, $a_2 = \dfrac{x^5}{384}$, $a_3 = -\dfrac{x^7}{18{,}432}$,

$a_4 = \dfrac{x^9}{1{,}474{,}560}$, and $a_5 = -\dfrac{x^{11}}{176{,}947{,}200}$. The

partial sums seem to approximate $J_1(x)$ well near the

origin, but as $|x|$ increases, we need to take a large

number of terms to get a good approximation.

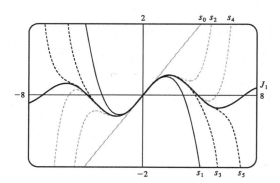

25. $s_{2n-1} = 1 + 2x + x^2 + 2x^3 + x^4 + 2x^5 + \cdots + x^{2n-2} + 2x^{2n-1}$

$$= 1(1+2x) + x^2(1+2x) + x^4(1+2x) + \cdots + x^{2n-2}(1+2x)$$

$$= (1+2x)\left(1 + x^2 + x^4 + \cdots + x^{2n-2}\right)$$

$$= (1+2x)\frac{1-x^{2n}}{1-x^2} \text{ [by (8.2.3) with } r = x^2] \to \frac{1+2x}{1-x^2} \text{ as } n \to \infty \text{ [by (8.2.4)],}$$

when $|x| < 1$. Also $s_{2n} = s_{2n-1} + x^{2n} \to \dfrac{1+2x}{1-x^2}$ since $x^{2n} \to 0$ for $|x| < 1$. Therefore, $s_n \to \dfrac{1+2x}{1-x^2}$ since

s_{2n} and s_{2n-1} both approach $\dfrac{1+2x}{1-x^2}$ as $n \to \infty$. Thus, the interval of convergence is $(-1, 1)$ and

$f(x) = \dfrac{1+2x}{1-x^2}$.

27. For $2 < x < 3$, $\sum c_n x^n$ diverges and $\sum d_n x^n$ converges. By Exercise 8.2.49, $\sum(c_n + d_n)x^n$ diverges. Since

both series converge for $|x| < 2$, the radius of convergence of $\sum(c_n + d_n)x^n$ is 2.

8.6 **Representations of Functions as Power Series** · · · · ·

1. If $f(x) = \sum\limits_{n=0}^{\infty} c_n x^n$ has radius of convergence 10, then $f'(x) = \sum\limits_{n=1}^{\infty} n c_n x^{n-1}$ also has radius of convergence 10 by Theorem 2.

3. Our goal is to write the function in the form $\dfrac{1}{1-r}$, and then use Equation (1) to represent the function as a sum of a power series. $f(x) = \dfrac{1}{1+x} = \dfrac{1}{1-(-x)} = \sum\limits_{n=0}^{\infty} (-x)^n = \sum\limits_{n=0}^{\infty} (-1)^n x^n$ with $|-x| < 1 \iff |x| < 1$, so $R = 1$ and $I = (-1, 1)$.

5. Replacing x with x^3 in (1) gives $f(x) = \dfrac{1}{1-x^3} = \sum\limits_{n=0}^{\infty} (x^3)^n = \sum\limits_{n=0}^{\infty} x^{3n}$. The series converges when $|x^3| < 1$ $\iff |x|^3 < 1 \iff |x| < \sqrt[3]{1} \iff |x| < 1$. Thus, $R = 1$ and $I = (-1, 1)$.

7. If the constant term in the denominator is something other than 1, factor it out of the binomial to obtain a 1.
$f(x) = \dfrac{1}{4+x^2} = \dfrac{1}{4}\left(\dfrac{1}{1+x^2/4}\right) = \dfrac{1}{4}\left(\dfrac{1}{1-(-x^2/4)}\right) = \dfrac{1}{4} \sum\limits_{n=0}^{\infty} \left(-\dfrac{x^2}{4}\right)^n = \sum\limits_{n=0}^{\infty} \dfrac{(-1)^n x^{2n}}{4^{n+1}}$. The series converges when $\left|-\dfrac{x^2}{4}\right| < 1 \iff x^2 < 4 \iff |x| < 2$, so $R = 2$ and $I = (-2, 2)$.

9. $f(x) = \dfrac{1}{x-5} = -\dfrac{1}{5}\left(\dfrac{1}{1-x/5}\right) = -\dfrac{1}{5} \sum\limits_{n=0}^{\infty} \left(\dfrac{x}{5}\right)^n$ or equivalently, $-\sum\limits_{n=0}^{\infty} \dfrac{1}{5^{n+1}} x^n$. The series converges when $\left|\dfrac{x}{5}\right| < 1$; that is, when $|x| < 5$, so $I = (-5, 5)$.

11. (a) $f(x) = \dfrac{1}{(1+x)^2} = \dfrac{d}{dx}\left(\dfrac{-1}{1+x}\right) = -\dfrac{d}{dx}\left[\sum\limits_{n=0}^{\infty} (-1)^n x^n\right]$ [from Exercise 3]

$= \sum\limits_{n=1}^{\infty} (-1)^{n+1} n x^{n-1}$ [from Theorem 2(a)] $= \sum\limits_{n=0}^{\infty} (-1)^n (n+1) x^n$ with $R = 1$.

In the last step, note that we *decreased* the initial value of the summation variable n by 1, and then *increased* each occurrence of n in the term by 1 [also note that $(-1)^{n+2} = (-1)^n$].

(b) $f(x) = \dfrac{1}{(1+x)^3} = -\dfrac{1}{2} \dfrac{d}{dx}\left[\dfrac{1}{(1+x)^2}\right] = -\dfrac{1}{2} \dfrac{d}{dx}\left[\sum\limits_{n=0}^{\infty} (-1)^n (n+1) x^n\right]$ [from part (a)]

$= -\dfrac{1}{2} \sum\limits_{n=1}^{\infty} (-1)^n (n+1) n x^{n-1} = \dfrac{1}{2} \sum\limits_{n=0}^{\infty} (-1)^n (n+2)(n+1) x^n$ with $R = 1$.

(c) $f(x) = \dfrac{x^2}{(1+x)^3} = x^2 \cdot \dfrac{1}{(1+x)^3} = x^2 \cdot \dfrac{1}{2} \sum\limits_{n=0}^{\infty} (-1)^n (n+2)(n+1) x^n$ [from part (b)]

$= \dfrac{1}{2} \sum\limits_{n=0}^{\infty} (-1)^n (n+2)(n+1) x^{n+2}$. To write the power series with x^n rather than x^{n+2}, we will *decrease* each occurrence of n in the term by 2 and *increase* the initial value of the summation variable by 2. This gives us $\dfrac{1}{2} \sum\limits_{n=2}^{\infty} (-1)^n (n)(n-1) x^n$.

13. $f(x) = \ln(5 - x) = -\displaystyle\int \frac{dx}{5 - x} = -\frac{1}{5} \int \frac{dx}{1 - x/5}$

$$= -\frac{1}{5} \int \left[\sum_{n=0}^{\infty} \left(\frac{x}{5} \right)^n \right] dx = C - \frac{1}{5} \sum_{n=0}^{\infty} \frac{x^{n+1}}{5^n(n+1)} = C - \sum_{n=1}^{\infty} \frac{x^n}{n5^n}$$

Putting $x = 0$, we get $C = \ln 5$. The series converges for $|x/5| < 1 \iff |x| < 5$, so $R = 5$.

15. $\dfrac{1}{2 - x} = \dfrac{1}{2(1 - x/2)} = \dfrac{1}{2} \displaystyle\sum_{n=0}^{\infty} \left(\frac{x}{2} \right)^n = \sum_{n=0}^{\infty} \frac{1}{2^{n+1}} x^n$ for $\left| \dfrac{x}{2} \right| < 1 \iff |x| < 2$. Now

$$\frac{1}{(x - 2)^2} = \frac{d}{dx} \left(\frac{1}{2 - x} \right) = \frac{d}{dx} \left(\sum_{n=0}^{\infty} \frac{1}{2^{n+1}} x^n \right) = \sum_{n=1}^{\infty} \frac{n}{2^{n+1}} x^{n-1} = \sum_{n=0}^{\infty} \frac{n+1}{2^{n+2}} x^n. \text{ So}$$

$$f(x) = \frac{x^3}{(x - 2)^2} = x^3 \sum_{n=0}^{\infty} \frac{n+1}{2^{n+2}} x^n = \sum_{n=0}^{\infty} \frac{n+1}{2^{n+2}} x^{n+3} \text{ or } \sum_{n=3}^{\infty} \frac{n-2}{2^{n-1}} x^n \text{ for } |x| < 2. \text{ Thus, } R = 2 \text{ and}$$

$I = (-2, 2)$.

17. $f(x) = \ln(3 + x) = \displaystyle\int \frac{dx}{3 + x} = \frac{1}{3} \int \frac{dx}{1 + x/3} = \frac{1}{3} \int \frac{dx}{1 - (-x/3)} = \frac{1}{3} \int \sum_{n=0}^{\infty} \left(-\frac{x}{3} \right)^n dx$

$$= C + \frac{1}{3} \sum_{n=0}^{\infty} \frac{(-1)^n}{(n+1)3^n} x^{n+1} = \ln 3 + \frac{1}{3} \sum_{n=1}^{\infty} \frac{(-1)^{n-1}}{n3^{n-1}} x^n \quad [C = f(0) = \ln 3]$$

$$= \ln 3 + \sum_{n=1}^{\infty} \frac{(-1)^{n-1}}{n3^n} x^n. \text{ The series converges when } |-x/3| < 1 \iff |x| < 3, \text{ so } R = 3.$$

The terms of the series are $a_0 = \ln 3$, $a_1 = \dfrac{x}{3}$, $a_2 = -\dfrac{x^2}{18}$, $a_3 = \dfrac{x^3}{81}$, $a_4 = -\dfrac{x^4}{324}$, $a_5 = \dfrac{x^5}{1215}, \ldots$.

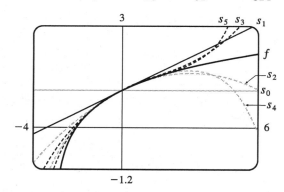

As n increases, $s_n(x)$ approximates f better on the interval of convergence, which is $(-3, 3)$.

19. $f(x) = \ln \left(\dfrac{1 + x}{1 - x} \right) = \ln(1 + x) - \ln(1 - x) = \displaystyle\int \frac{dx}{1 + x} + \int \frac{dx}{1 - x}$

$$= \int \frac{dx}{1 - (-x)} + \int \frac{dx}{1 - x} = \int \left[\sum_{n=0}^{\infty} (-1)^n x^n + \sum_{n=0}^{\infty} x^n \right] dx$$

$$= \int \left[(1 - x + x^2 - x^3 + x^4 - \cdots) + (1 + x + x^2 + x^3 + x^4 + \cdots) \right] dx$$

$$= \int (2 + 2x^2 + 2x^4 + \cdots) dx = \int \sum_{n=0}^{\infty} 2x^{2n} dx = C + \sum_{n=0}^{\infty} \frac{2x^{2n+1}}{2n + 1}$$

But $f(0) = \ln \frac{1}{1} = 0$, so $C = 0$ and we have $f(x) = \displaystyle\sum_{n=0}^{\infty} \frac{2x^{2n+1}}{2n + 1}$ with $R = 1$. If $x = \pm 1$, then

$f(x) = \pm 2 \sum\limits_{n=0}^{\infty} \dfrac{1}{2n+1}$, which both diverge by the Limit Comparison Test with $b_n = \dfrac{1}{n}$. As n increases, $s_n(x)$

approximates f better on the interval of convergence, which is $(-1, 1)$.

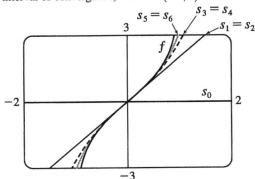

21. $\displaystyle\int \dfrac{dx}{1+x^4} = \int \dfrac{dx}{1-(-x^4)} = \int \sum\limits_{n=0}^{\infty} (-x^4)^n \, dx = \int \sum\limits_{n=0}^{\infty} (-1)^n x^{4n} \, dx = C + \sum\limits_{n=0}^{\infty} \dfrac{(-1)^n x^{4n+1}}{4n+1}$ with $R = 1$.

23. By Example 7, $\arctan x = \sum\limits_{n=0}^{\infty} (-1)^n \dfrac{x^{2n+1}}{2n+1}$, so

$\displaystyle\int \dfrac{\arctan x}{x} \, dx = \int \sum\limits_{n=0}^{\infty} (-1)^n \dfrac{x^{2n}}{2n+1} \, dx = C + \sum\limits_{n=0}^{\infty} (-1)^n \dfrac{x^{2n+1}}{(2n+1)^2}$ with $R = 1$.

25. $\dfrac{1}{1+x^5} = \dfrac{1}{1-(-x^5)} = \sum\limits_{n=0}^{\infty} (-x^5)^n = \sum\limits_{n=0}^{\infty} (-1)^n x^{5n} \ \Rightarrow$

$\displaystyle\int \dfrac{1}{1+x^5} \, dx = \int \sum\limits_{n=0}^{\infty} (-1)^n x^{5n} \, dx = C + \sum\limits_{n=0}^{\infty} (-1)^n \dfrac{x^{5n+1}}{5n+1}$. Thus,

$I = \displaystyle\int_0^{0.2} \dfrac{1}{1+x^5} \, dx = \left[x - \dfrac{x^6}{6} + \dfrac{x^{11}}{11} - \cdots \right]_0^{0.2} = 0.2 - \dfrac{(0.2)^6}{6} + \dfrac{(0.2)^{11}}{11} - \cdots$. The series is alternating, so

if we use the first two terms, the error is at most $(0.2)^{11}/11 \approx 1.9 \times 10^{-9}$. So $I \approx 0.2 - (0.2)^6/6 \approx 0.199989$ to

six decimal places.

27. We substitute x^4 for x in Example 7, and find that

$$\int x^2 \tan^{-1}(x^4) \, dx = \int x^2 \sum\limits_{n=0}^{\infty} (-1)^n \dfrac{(x^4)^{2n+1}}{2n+1} \, dx$$

$$= \int \sum\limits_{n=0}^{\infty} (-1)^n \dfrac{x^{8n+6}}{2n+1} \, dx = C + \sum\limits_{n=0}^{\infty} (-1)^n \dfrac{x^{8n+7}}{(2n+1)(8n+7)}$$

So $\displaystyle\int_0^{1/3} x^2 \tan^{-1}(x^4) \, dx = \left[\dfrac{x^7}{7} - \dfrac{x^{15}}{45} + \cdots \right]_0^{1/3} = \dfrac{1}{7 \cdot 3^7} - \dfrac{1}{45 \cdot 3^{15}} + \cdots$. The series is alternating,

so if we use only one term, the error is at most $1/(45 \cdot 3^{15}) \approx 1.5 \times 10^{-9}$. So

$\int_0^{1/3} x^2 \tan^{-1}(x^4) \, dx \approx 1/(7 \cdot 3^7) \approx 0.000065$ to six decimal places.

29. Using the result of Example 6, $\ln(1-x) = -\sum\limits_{n=1}^{\infty} \dfrac{x^n}{n}$, with $x = -0.1$, we have

$\ln 1.1 = \ln[1-(-0.1)] = 0.1 - \dfrac{0.01}{2} + \dfrac{0.001}{3} - \dfrac{0.0001}{4} + \dfrac{0.00001}{5} - \cdots$. The series is alternating, so if we use only the first four terms, the error is at most $\dfrac{0.00001}{5} = 0.000002$. So

$$\ln 1.1 \approx 0.1 - \frac{0.01}{2} + \frac{0.001}{3} - \frac{0.0001}{4} \approx 0.09531.$$

31. (a) $J_0(x) = \displaystyle\sum_{n=0}^{\infty} \frac{(-1)^n x^{2n}}{2^{2n}(n!)^2}$, $J_0'(x) = \displaystyle\sum_{n=1}^{\infty} \frac{(-1)^n 2n x^{2n-1}}{2^{2n}(n!)^2}$, and $J_0''(x) = \displaystyle\sum_{n=1}^{\infty} \frac{(-1)^n 2n(2n-1)x^{2n-2}}{2^{2n}(n!)^2}$, so

$$x^2 J_0''(x) + x J_0'(x) + x^2 J_0(x) = \sum_{n=1}^{\infty} \frac{(-1)^n 2n(2n-1)x^{2n}}{2^{2n}(n!)^2} + \sum_{n=1}^{\infty} \frac{(-1)^n 2n x^{2n}}{2^{2n}(n!)^2} + \sum_{n=0}^{\infty} \frac{(-1)^n x^{2n+2}}{2^{2n}(n!)^2}$$

$$= \sum_{n=1}^{\infty} \frac{(-1)^n 2n(2n-1)x^{2n}}{2^{2n}(n!)^2} + \sum_{n=1}^{\infty} \frac{(-1)^n 2n x^{2n}}{2^{2n}(n!)^2} + \sum_{n=1}^{\infty} \frac{(-1)^{n-1} x^{2n}}{2^{2n-2}[(n-1)!]^2}$$

$$= \sum_{n=1}^{\infty} \frac{(-1)^n 2n(2n-1)x^{2n}}{2^{2n}(n!)^2} + \sum_{n=1}^{\infty} \frac{(-1)^n 2n x^{2n}}{2^{2n}(n!)^2} + \sum_{n=1}^{\infty} \frac{(-1)^n (-1)^{-1} 2^2 n^2 x^{2n}}{2^{2n}(n!)^2}$$

$$= \sum_{n=1}^{\infty} (-1)^n \left[\frac{2n(2n-1) + 2n - 2^2 n^2}{2^{2n}(n!)^2}\right] x^{2n} = \sum_{n=1}^{\infty} (-1)^n \left[\frac{4n^2 - 2n + 2n - 4n^2}{2^{2n}(n!)^2}\right] x^{2n} = 0$$

(b) $\displaystyle\int_0^1 J_0(x)\,dx = \int_0^1 \left[\sum_{n=0}^{\infty} \frac{(-1)^n x^{2n}}{2^{2n}(n!)^2}\right] dx = \int_0^1 \left(1 - \frac{x^2}{4} + \frac{x^4}{64} - \frac{x^6}{2304} + \cdots\right) dx$

$$= \left[x - \frac{x^3}{3 \cdot 4} + \frac{x^5}{5 \cdot 64} - \frac{x^7}{7 \cdot 2304} + \cdots\right]_0^1 = 1 - \frac{1}{12} + \frac{1}{320} - \frac{1}{16{,}128} + \cdots$$

Since $\frac{1}{16{,}128} \approx 0.000062$, it follows from The Alternating Series Estimation Theorem that, correct to three decimal places, $\int_0^1 J_0(x)\,dx \approx 1 - \frac{1}{12} + \frac{1}{320} \approx 0.920$.

33. (a) $f(x) = \displaystyle\sum_{n=0}^{\infty} \frac{x^n}{n!} \;\Rightarrow\; f'(x) = \sum_{n=1}^{\infty} \frac{n x^{n-1}}{n!} = \sum_{n=1}^{\infty} \frac{x^{n-1}}{(n-1)!} = \sum_{n=0}^{\infty} \frac{x^n}{n!} = f(x)$

(b) By Theorem 7.4.2, the only solution to the differential equation $df(x)/dx = f(x)$ is $f(x) = Ke^x$, but $f(0) = 1$, so $K = 1$ and $f(x) = e^x$.

Or: We could solve the equation $df(x)/dx = f(x)$ as a separable differential equation.

35. If $a_n = \dfrac{x^n}{n^2}$, then by the Ratio Test, $\displaystyle\lim_{n\to\infty} \left|\frac{a_{n+1}}{a_n}\right| = \lim_{n\to\infty} \left|\frac{x^{n+1}}{(n+1)^2} \cdot \frac{n^2}{x^n}\right| = |x| \lim_{n\to\infty} \left(\frac{n}{n+1}\right)^2 = |x| < 1$ for

convergence, so $R = 1$. When $x = \pm 1$, $\displaystyle\sum_{n=1}^{\infty} \left|\frac{x^n}{n^2}\right| = \sum_{n=1}^{\infty} \frac{1}{n^2}$ which is a convergent p-series ($p = 2 > 1$), so the

interval of convergence for f is $[-1, 1]$. By Theorem 2, the radii of convergence of f' and f'' are both 1, so we need

only check the endpoints. $f(x) = \displaystyle\sum_{n=1}^{\infty} \frac{x^n}{n^2} \;\Rightarrow\; f'(x) = \sum_{n=1}^{\infty} \frac{n x^{n-1}}{n^2} = \sum_{n=0}^{\infty} \frac{x^n}{n+1}$, and this series diverges for

$x = 1$ (harmonic series) and converges for $x = -1$ (Alternating Series Test), so the interval of convergence

is $[-1, 1)$. $f''(x) = \displaystyle\sum_{n=1}^{\infty} \frac{n x^{n-1}}{n+1}$ diverges at both 1 and -1 (Test for Divergence) since $\displaystyle\lim_{n\to\infty} \frac{n}{n+1} = 1 \neq 0$, so its

interval of convergence is $(-1, 1)$.

Taylor and Maclaurin Series · · · · · · · · · · · ·

1. Using Theorem 5 with $\sum\limits_{n=0}^{\infty} b_n(x-5)^n$, $b_n = \dfrac{f^{(n)}(a)}{n!}$, so $b_8 = \dfrac{f^{(8)}(5)}{8!}$.

3.

n	$f^{(n)}(x)$	$f^{(n)}(0)$
0	$\cos x$	1
1	$-\sin x$	0
2	$-\cos x$	-1
3	$\sin x$	0
4	$\cos x$	1
⋮	⋮	⋮

We use Equation 7 with $f(x) = \cos x$.

$$\cos x = f(0) + f'(0)x + \frac{f''(0)}{2!}x^2 + \frac{f^{(3)}(0)}{3!}x^3 + \frac{f^{(4)}(0)}{4!}x^4 + \cdots$$

$$= 1 - \frac{x^2}{2!} + \frac{x^4}{4!} - \cdots = \sum_{n=0}^{\infty} \frac{(-1)^n x^{2n}}{(2n)!}$$

If $a_n = \dfrac{(-1)^n x^{2n}}{(2n)!}$, then

$$\lim_{n\to\infty}\left|\frac{a_{n+1}}{a_n}\right| = \lim_{n\to\infty}\left|\frac{x^{2n+2}}{(2n+2)!} \cdot \frac{(2n)!}{x^{2n}}\right| = x^2 \lim_{n\to\infty} \frac{1}{(2n+2)(2n+1)} = 0 < 1 \text{ for all } x.$$

So $R = \infty$ (Ratio Test).

5.

n	$f^{(n)}(x)$	$f^{(n)}(0)$
0	$(1+x)^{-3}$	1
1	$-3(1+x)^{-4}$	-3
2	$12(1+x)^{-5}$	12
3	$-60(1+x)^{-6}$	-60
4	$360(1+x)^{-7}$	360
⋮	⋮	⋮

$$(1+x)^{-3} = f(0) + f'(0)x + \frac{f''(0)}{2!}x^2 + \frac{f'''(0)}{3!}x^3 + \frac{f^{(4)}(0)}{4!}x^4 + \cdots$$

$$= 1 - 3x + \frac{4 \cdot 3}{2!}x^2 - \frac{5 \cdot 4 \cdot 3}{3!}x^3 + \frac{6 \cdot 5 \cdot 4 \cdot 3}{4!}x^4 - \cdots$$

$$= 1 - 3x + \frac{4 \cdot 3 \cdot 2}{2 \cdot 2!}x^2 - \frac{5 \cdot 4 \cdot 3 \cdot 2}{2 \cdot 3!}x^3 + \frac{6 \cdot 5 \cdot 4 \cdot 3 \cdot 2}{2 \cdot 4!}x^4 - \cdots$$

$$= \sum_{n=0}^{\infty} \frac{(-1)^n (n+2)! \, x^n}{2(n!)} = \sum_{n=0}^{\infty} \frac{(-1)^n (n+2)(n+1)x^n}{2}$$

$$\lim_{n\to\infty}\left|\frac{a_{n+1}}{a_n}\right| = \lim_{n\to\infty}\left|\frac{(n+3)(n+2)x^{n+1}}{2} \cdot \frac{2}{(n+2)(n+1)x^n}\right| = |x| \lim_{n\to\infty} \frac{n+3}{n+1} = |x| < 1 \text{ for convergence,}$$

so $R = 1$ (Ratio Test).

7.

n	$f^{(n)}(x)$	$f^{(n)}(2)$
0	$1 + x + x^2$	7
1	$1 + 2x$	5
2	2	2
3	0	0
4	0	0
⋮	⋮	⋮

$$f(x) = 7 + 5(x-2) + \frac{2}{2!}(x-2)^2 + \sum_{n=3}^{\infty} \frac{0}{n!}(x-2)^n$$
$$= 7 + 5(x-2) + (x-2)^2$$

Since $a_n = 0$ for large n, $R = \infty$.

9. Clearly, $f^{(n)}(x) = e^x$, so $f^{(n)}(3) = e^3$ and $e^x = \sum_{n=0}^{\infty} \frac{e^3}{n!}(x-3)^n$. If $a_n = \frac{e^3}{n!}(x-3)^n$, then

$$\lim_{n\to\infty} \left| \frac{a_{n+1}}{a_n} \right| = \lim_{n\to\infty} \left| \frac{e^3(x-3)^{n+1}}{(n+1)!} \cdot \frac{n!}{e^3(x-3)^n} \right| = \lim_{n\to\infty} \frac{|x-3|}{n+1} = 0 < 1 \text{ for all } x, \text{ so } R = \infty.$$

11.

n	$f^{(n)}(x)$	$f^{(n)}(1)$
0	x^{-1}	1
1	$-x^{-2}$	-1
2	$2x^{-3}$	2
3	$-3 \cdot 2x^{-4}$	$-3 \cdot 2$
4	$4 \cdot 3 \cdot 2x^{-5}$	$4 \cdot 3 \cdot 2$
⋮	⋮	⋮

So $f^{(n)}(1) = (-1)^n n!$, and $\frac{1}{x} = \sum_{n=0}^{\infty} \frac{(-1)^n n!}{n!}(x-1)^n = \sum_{n=0}^{\infty} (-1)^n (x-1)^n$. If $a_n = (-1)^n(x-1)^n$ then

$$\lim_{n\to\infty} \left| \frac{a_{n+1}}{a_n} \right| = |x-1| < 1 \text{ for convergence, so } R = 1.$$

13.

n	$f^{(n)}(x)$	$f^{(n)}\left(\frac{\pi}{4}\right)$
0	$\sin x$	$\sqrt{2}/2$
1	$\cos x$	$\sqrt{2}/2$
2	$-\sin x$	$-\sqrt{2}/2$
3	$-\cos x$	$-\sqrt{2}/2$
4	$\sin x$	$\sqrt{2}/2$
⋮	⋮	⋮

$$\sin x = f\left(\tfrac{\pi}{4}\right) + f'\left(\tfrac{\pi}{4}\right)\left(x - \tfrac{\pi}{4}\right) + \frac{f''\left(\tfrac{\pi}{4}\right)}{2!}\left(x - \tfrac{\pi}{4}\right)^2 + \frac{f^{(3)}\left(\tfrac{\pi}{4}\right)}{3!}\left(x - \tfrac{\pi}{4}\right)^3 + \frac{f^{(4)}\left(\tfrac{\pi}{4}\right)}{4!}\left(x - \tfrac{\pi}{4}\right)^4 + \cdots$$

$$= \frac{\sqrt{2}}{2}\left[1 + \left(x - \tfrac{\pi}{4}\right) - \tfrac{1}{2!}\left(x - \tfrac{\pi}{4}\right)^2 - \tfrac{1}{3!}\left(x - \tfrac{\pi}{4}\right)^3 + \tfrac{1}{4!}\left(x - \tfrac{\pi}{4}\right)^4 + \cdots\right]$$

$$= \frac{\sqrt{2}}{2}\left[1 - \tfrac{1}{2!}\left(x - \tfrac{\pi}{4}\right)^2 + \tfrac{1}{4!}\left(x - \tfrac{\pi}{4}\right)^4 - \cdots\right] + \frac{\sqrt{2}}{2}\left[\left(x - \tfrac{\pi}{4}\right) - \tfrac{1}{3!}\left(x - \tfrac{\pi}{4}\right)^3 + \cdots\right]$$

$$= \frac{\sqrt{2}}{2}\sum_{n=0}^{\infty}(-1)^n\left[\tfrac{1}{(2n)!}\left(x - \tfrac{\pi}{4}\right)^{2n} + \tfrac{1}{(2n+1)!}\left(x - \tfrac{\pi}{4}\right)^{2n+1}\right]$$

The series can also be written in the more elegant form $\sin x = \dfrac{\sqrt{2}}{2}\displaystyle\sum_{n=0}^{\infty}\dfrac{(-1)^{n(n-1)/2}\left(x - \tfrac{\pi}{4}\right)^n}{n!}$. If

$a_n = \dfrac{(-1)^{n(n-1)/2}\left(x - \tfrac{\pi}{4}\right)^n}{n!}$, then $\displaystyle\lim_{n\to\infty}\left|\dfrac{a_{n+1}}{a_n}\right| = \lim_{n\to\infty}\dfrac{\left|x - \tfrac{\pi}{4}\right|}{n+1} = 0 < 1$ for all x, so $R = \infty$.

15. If $f(x) = \cos x$, then by Formula 9 with $a = 0$, $|R_n(x)| \le \dfrac{\left|f^{(n+1)}(x)\right|}{(n+1)!}\,|x|^{n+1}$. But $f^{(n+1)}(x) = \pm\sin x$ or

$\pm\cos x$. In each case, $\left|f^{(n+1)}(x)\right| \le 1$, so $|R_n(x)| \le \dfrac{1}{(n+1)!}\,|x|^{n+1} \to 0$ as $n \to \infty$ by Equation 10. So

$\displaystyle\lim_{n\to\infty}R_n(x) = 0$ and, by Theorem 8, the series in Exercise 3 represents $\cos x$ for all x.

17. $\cos x = \displaystyle\sum_{n=0}^{\infty}(-1)^n\dfrac{x^{2n}}{(2n)!} \quad\Rightarrow\quad f(x) = \cos(\pi x) = \displaystyle\sum_{n=0}^{\infty}\dfrac{(-1)^n(\pi x)^{2n}}{(2n)!} = \displaystyle\sum_{n=0}^{\infty}\dfrac{(-1)^n\pi^{2n}x^{2n}}{(2n)!}, \ R = \infty$

19. $\tan^{-1}x = \displaystyle\sum_{n=0}^{\infty}(-1)^n\dfrac{x^{2n+1}}{2n+1} \quad\Rightarrow\quad f(x) = x\tan^{-1}x = x\displaystyle\sum_{n=0}^{\infty}(-1)^n\dfrac{x^{2n+1}}{2n+1} = \displaystyle\sum_{n=0}^{\infty}(-1)^n\dfrac{x^{2n+2}}{2n+1}, \ R = 1$

21. $e^x = \displaystyle\sum_{n=0}^{\infty}\dfrac{x^n}{n!} \quad\Rightarrow\quad f(x) = x^2e^{-x} = x^2\displaystyle\sum_{n=0}^{\infty}\dfrac{(-x)^n}{n!} = \displaystyle\sum_{n=0}^{\infty}\dfrac{(-1)^n\,x^{n+2}}{n!}, \ R = \infty$

23. $\sin^2 x = \tfrac{1}{2}[1 - \cos 2x] = \dfrac{1}{2}\left[1 - \displaystyle\sum_{n=0}^{\infty}\dfrac{(-1)^n(2x)^{2n}}{(2n)!}\right] = 2^{-1}\left[1 - 1 - \displaystyle\sum_{n=1}^{\infty}\dfrac{(-1)^n(2x)^{2n}}{(2n)!}\right]$

$= \displaystyle\sum_{n=1}^{\infty}\dfrac{(-1)^{n+1}2^{2n-1}x^{2n}}{(2n)!}, \ R = \infty$

25.

n	$f^{(n)}(x)$	$f^{(n)}(0)$
0	$(1+x)^{1/2}$	1
1	$\frac{1}{2}(1+x)^{-1/2}$	$\frac{1}{2}$
2	$-\frac{1}{4}(1+x)^{-3/2}$	$-\frac{1}{4}$
3	$\frac{3}{8}(1+x)^{-5/2}$	$\frac{3}{8}$
4	$-\frac{15}{16}(1+x)^{-7/2}$	$-\frac{15}{16}$
$\vdots$	$\vdots$	$\vdots$

So $f^{(n)}(0) = \dfrac{(-1)^{n-1}\, 1 \cdot 3 \cdot 5 \cdot \cdots \cdot (2n-3)}{2^n}$ for $n \geq 2$, and

$$\sqrt{1+x} = 1 + \frac{x}{2} + \sum_{n=2}^{\infty} \frac{(-1)^{n-1} 1 \cdot 3 \cdot 5 \cdot \cdots \cdot (2n-3)}{2^n n!} x^n. \text{ If } a_n = \frac{(-1)^{n-1} 1 \cdot 3 \cdot 5 \cdot \cdots \cdot (2n-3)}{2^n n!} x^n,$$

$$\text{then } \lim_{n \to \infty} \left| \frac{a_{n+1}}{a_n} \right| = \lim_{n \to \infty} \left| \frac{1 \cdot 3 \cdot 5 \cdot \cdots \cdot (2n-3)(2n-1)x^{n+1}}{2^{n+1}(n+1)!} \cdot \frac{2^n n!}{1 \cdot 3 \cdot 5 \cdot \cdots \cdot (2n-3)x^n} \right|$$

$$= \frac{|x|}{2} \lim_{n \to \infty} \frac{2n-1}{n+1} = \frac{|x|}{2} \cdot 2 = |x| < 1 \text{ for convergence, so } R = 1.$$

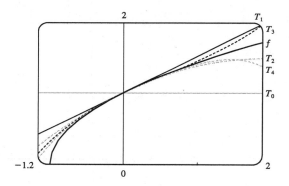

Notice that, as n increases, $T_n(x)$ becomes a better approximation to $f(x)$ for $-1 < x < 1$.

27. $\cos x = \sum_{n=0}^{\infty} (-1)^n \dfrac{x^{2n}}{(2n)!} \quad \Rightarrow \quad f(x) = \cos(x^2) = \sum_{n=0}^{\infty} \dfrac{(-1)^n \left(x^2\right)^{2n}}{(2n)!} = \sum_{n=0}^{\infty} \dfrac{(-1)^n\, x^{4n}}{(2n)!}, \; R = \infty$

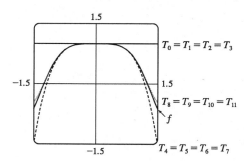

Notice that, as n increases, $T_n(x)$ becomes a better approximation to $f(x)$.

29. $e^x = \sum_{n=0}^{\infty} \dfrac{x^n}{n!}$, so

$$e^{-0.2} = \sum_{n=0}^{\infty} \frac{(-0.2)^n}{n!} = 1 - 0.2 + \frac{1}{2!}(0.2)^2 - \frac{1}{3!}(0.2)^3 + \frac{1}{4!}(0.2)^4 - \frac{1}{5!}(0.2)^5 + \frac{1}{6!}(0.2)^6 - \cdots. \text{ But}$$

$\dfrac{1}{6!}(0.2)^6 = 8.\overline{3} \times 10^{-8}$, so by the Alternating Series Estimation Theorem, $e^{-0.2} \approx \sum_{n=0}^{5} \dfrac{(-0.2)^n}{n!} \approx 0.81873$,

correct to five decimal places.

31. $\displaystyle\int \sin(x^2)\, dx = \int \sum_{n=0}^{\infty} (-1)^n \frac{(x^2)^{2n+1}}{(2n+1)!}\, dx = \int \sum_{n=0}^{\infty} \frac{(-1)^n x^{4n+2}}{(2n+1)!}\, dx = C + \sum_{n=0}^{\infty} \frac{(-1)^n x^{4n+3}}{(4n+3)(2n+1)!}$

33. Using the series from Exercise 25 and substituting x^3 for x, we get

$$\int \sqrt{x^3 + 1}\, dx = \int \left[1 + \frac{x^3}{2} + \sum_{n=2}^{\infty} \frac{(-1)^{n-1} 1 \cdot 3 \cdot 5 \cdots (2n-3)}{2^n n!} x^{3n} \right] dx$$

$$= C + x + \frac{x^4}{8} + \sum_{n=2}^{\infty} \frac{(-1)^{n-1} 1 \cdot 3 \cdot 5 \cdots (2n-3)}{2^n n!(3n+1)} x^{3n+1}$$

35. Using our series from Exercise 31, we get

$$\int_0^1 \sin(x^2)\, dx = \sum_{n=0}^{\infty} \left[\frac{(-1)^n x^{4n+3}}{(4n+3)(2n+1)!} \right]_0^1 = \sum_{n=0}^{\infty} \frac{(-1)^n}{(4n+3)(2n+1)!} \text{ and}$$

$|c_3| = \dfrac{1}{75{,}600} < 0.000014$, so by the Alternating Series Estimation Theorem, we have

$$\int_0^1 \sin(x^2)\, dx \approx \sum_{n=0}^{2} \frac{(-1)^n}{(4n+3)(2n+1)!} = \frac{1}{3} - \frac{1}{42} + \frac{1}{1320} \approx 0.310 \text{ (correct to three decimal places).}$$

37. We first find a series representation for $f(x) = (1+x)^{-1/2}$, and then substitute.

n	$f^{(n)}(x)$	$f^{(n)}(0)$
0	$(1+x)^{-1/2}$	1
1	$-\frac{1}{2}(1+x)^{-3/2}$	$-\frac{1}{2}$
2	$\frac{3}{4}(1+x)^{-5/2}$	$\frac{3}{4}$
3	$-\frac{15}{8}(1+x)^{-7/2}$	$-\frac{15}{8}$
⋮	⋮	⋮

$$\frac{1}{\sqrt{1+x}} = 1 - \frac{x}{2} + \frac{3}{4}\left(\frac{x^2}{2!}\right) - \frac{15}{8}\left(\frac{x^3}{3!}\right) + \cdots \quad\Rightarrow\quad \frac{1}{\sqrt{1+x^3}} = 1 - \frac{1}{2}x^3 + \frac{3}{8}x^6 - \frac{5}{16}x^9 + \cdots \quad\Rightarrow$$

$$\int_0^{0.1} \frac{dx}{\sqrt{1+x^3}} = \left[x - \frac{1}{8}x^4 + \frac{3}{56}x^7 - \frac{1}{32}x^{10} + \cdots \right]_0^{0.1} \approx (0.1) - \frac{1}{8}(0.1)^4, \text{ by the Alternating Series}$$

Estimation Theorem, since $\frac{3}{56}(0.1)^7 \approx 0.0000000054 < 10^{-8}$, which is the maximum desired error. Therefore,

$$\int_0^{0.1} \frac{dx}{\sqrt{1+x^3}} \approx 0.09998750.$$

39. $\lim\limits_{x \to 0} \dfrac{x - \tan^{-1} x}{x^3} = \lim\limits_{x \to 0} \dfrac{x - \left(x - \frac{1}{3}x^3 + \frac{1}{5}x^5 - \frac{1}{7}x^7 + \cdots\right)}{x^3} = \lim\limits_{x \to 0} \dfrac{\frac{1}{3}x^3 - \frac{1}{5}x^5 + \frac{1}{7}x^7 - \cdots}{x^3}$

$\qquad = \lim\limits_{x \to 0} \left(\frac{1}{3} - \frac{1}{5}x^2 + \frac{1}{7}x^4 - \cdots\right) = \frac{1}{3}$

since power series are continuous functions.

41. $\lim\limits_{x \to 0} \dfrac{\sin x - x + \frac{1}{6}x^3}{x^5} = \lim\limits_{x \to 0} \dfrac{\left(x - \frac{1}{3!}x^3 + \frac{1}{5!}x^5 - \frac{1}{7!}x^7 + \cdots\right) - x + \frac{1}{6}x^3}{x^5}$

$\qquad = \lim\limits_{x \to 0} \dfrac{\frac{1}{5!}x^5 - \frac{1}{7!}x^7 + \cdots}{x^5} = \lim\limits_{x \to 0} \left(\dfrac{1}{5!} - \dfrac{x^2}{7!} + \dfrac{x^4}{9!} - \cdots\right) = \dfrac{1}{5!} = \dfrac{1}{120}$

since power series are continuous functions.

43. As in Example 8(a), we have $e^{-x^2} = 1 - \dfrac{x^2}{1!} + \dfrac{x^4}{2!} - \dfrac{x^6}{3!} + \cdots$ and we know that $\cos x = 1 - \dfrac{x^2}{2!} + \dfrac{x^4}{4!} - \cdots$

from Equation 16. Therefore, $e^{-x^2} \cos x = \left(1 - x^2 + \frac{1}{2}x^4 - \cdots\right)\left(1 - \frac{1}{2}x^2 + \frac{1}{24}x^4 - \cdots\right)$. Writing only the

terms with degree ≤ 4, we get $e^{-x^2} \cos x = 1 - \frac{1}{2}x^2 + \frac{1}{24}x^4 - x^2 + \frac{1}{2}x^4 + \frac{1}{2}x^4 + \cdots = 1 - \frac{3}{2}x^2 + \frac{25}{24}x^4 + \cdots$.

45.

$$
\begin{array}{r}
-x + \frac{1}{2}x^2 - \frac{1}{3}x^3 + \cdots \\
1 + x + \frac{1}{2}x^2 + \frac{1}{6}x^3 + \cdots \; \overline{\Big)\; -x - \frac{1}{2}x^2 - \frac{1}{3}x^3 - \cdots} \\
\underline{-x - \;\; x^2 - \frac{1}{2}x^3 - \cdots} \\
\frac{1}{2}x^2 + \frac{1}{6}x^3 - \cdots \\
\underline{\frac{1}{2}x^2 + \frac{1}{2}x^3 + \cdots} \\
-\frac{1}{3}x^3 + \cdots \\
\underline{-\frac{1}{3}x^3 + \cdots} \\
\cdots
\end{array}
$$

From Example 6 in Section 8.6, we have $\ln(1 - x) = -x - \frac{1}{2}x^2 - \frac{1}{3}x^3 - \cdots$, $|x| < 1$. Therefore,

$y = \dfrac{\ln(1 - x)}{e^x} = \dfrac{-x - \frac{1}{2}x^2 - \frac{1}{3}x^3 - \cdots}{1 + x + \frac{1}{2}x^2 + \frac{1}{6}x^3 + \cdots}$. So by the long division above,

$\dfrac{\ln(1 - x)}{e^x} = -x + \dfrac{x^2}{2} - \dfrac{x^3}{3} + \cdots$, $|x| < 1$.

47. $\sum\limits_{n=0}^{\infty} (-1)^n \dfrac{x^{4n}}{n!} = \sum\limits_{n=0}^{\infty} \dfrac{\left(-x^4\right)^n}{n!} = e^{-x^4}$, by (11).

49. $\sum\limits_{n=0}^{\infty} \dfrac{(-1)^n \pi^{2n+1}}{4^{2n+1}(2n+1)!} = \sum\limits_{n=0}^{\infty} \dfrac{(-1)^n \left(\frac{\pi}{4}\right)^{2n+1}}{(2n+1)!} = \sin \frac{\pi}{4} = \frac{1}{\sqrt{2}}$, by (15).

51. $3 + \dfrac{9}{2!} + \dfrac{27}{3!} + \dfrac{81}{4!} + \cdots = \dfrac{3^1}{1!} + \dfrac{3^2}{2!} + \dfrac{3^3}{3!} + \dfrac{3^4}{4!} + \cdots = \displaystyle\sum_{n=1}^{\infty} \dfrac{3^n}{n!} = \sum_{n=0}^{\infty} \dfrac{3^n}{n!} - 1 = e^3 - 1$, by (11).

53. Assume that $|f'''(x)| \leq M$, so $f'''(x) \leq M$ for $a \leq x \leq a + d$. Now $\int_a^x f'''(t)\, dt \leq \int_a^x M\, dt \;\Rightarrow\;$

$f''(x) - f''(a) \leq M(x - a) \;\Rightarrow\; f''(x) \leq f''(a) + M(x - a)$. Thus, $\int_a^x f''(t)\, dt \leq \int_a^x [f''(a) + M(t - a)]\, dt$

$\Rightarrow\; f'(x) - f'(a) \leq f''(a)(x - a) + \frac{1}{2}M(x - a)^2 \;\Rightarrow\; f'(x) \leq f'(a) + f''(a)(x - a) + \frac{1}{2}M(x - a)^2 \;\Rightarrow\;$

$\int_a^x f'(t)\, dt \leq \int_a^x \left[f'(a) + f''(a)(t - a) + \frac{1}{2}M(t - a)^2 \right] dt \;\Rightarrow\;$

$f(x) - f(a) \leq f'(a)(x - a) + \frac{1}{2}f''(a)(x - a)^2 + \frac{1}{6}M(x - a)^3$. So

$f(x) - f(a) - f'(a)(x - a) - \frac{1}{2}f''(a)(x - a)^2 \leq \frac{1}{6}M(x - a)^3$. But

$R_2(x) = f(x) - T_2(x) = f(x) - f(a) - f'(a)(x - a) - \frac{1}{2}f''(a)(x - a)^2$, so $R_2(x) \leq \frac{1}{6}M(x - a)^3$. A similar

argument using $f'''(x) \geq -M$ shows that $R_2(x) \geq -\frac{1}{6}M(x - a)^3$. So $|R_2(x_2)| \leq \frac{1}{6}M\,|x - a|^3$.

Although we have assumed that $x > a$, a similar calculation shows that this inequality is also true if $x < a$.

◆8.8◆ The Binomial Series • • • • • • • • • • • • • •

1. The general binomial series in (2) is

$$(1 + x)^k = \sum_{n=0}^{\infty} \binom{k}{n} x^n = 1 + kx + \frac{k(k - 1)}{2!}x^2 + \frac{k(k - 1)(k - 2)}{3!}x^3 + \cdots .$$

$$(1 + x)^{1/2} = \sum_{n=0}^{\infty} \binom{\frac{1}{2}}{n} x^n = 1 + \left(\tfrac{1}{2}\right)x + \frac{\left(\frac{1}{2}\right)\left(-\frac{1}{2}\right)}{2!}x^2 + \frac{\left(\frac{1}{2}\right)\left(-\frac{1}{2}\right)\left(-\frac{3}{2}\right)}{3!}x^3 + \cdots$$

$$= 1 + \frac{x}{2} - \frac{x^2}{2^2 \cdot 2!} + \frac{1 \cdot 3 \cdot x^3}{2^3 \cdot 3!} - \frac{1 \cdot 3 \cdot 5 \cdot x^4}{2^4 \cdot 4!} + \cdots$$

$$= 1 + \frac{x}{2} + \sum_{n=2}^{\infty} \frac{(-1)^{n-1} 1 \cdot 3 \cdot 5 \cdots (2n - 3) x^n}{2^n \cdot n!} \quad \text{for } |x| < 1, \text{ so } R = 1$$

3. $\dfrac{1}{(2 + x)^3} = \dfrac{1}{[2(1 + x/2)]^3} = \dfrac{1}{8}\left(1 + \dfrac{x}{2}\right)^{-3} = \dfrac{1}{8}\displaystyle\sum_{n=0}^{\infty} \binom{-3}{n}\left(\dfrac{x}{2}\right)^n$. The binomial coefficient is

$$\binom{-3}{n} = \frac{(-3)(-4)(-5) \cdots (-3 - n + 1)}{n!} = \frac{(-3)(-4)(-5) \cdots [-(n + 2)]}{n!}$$

$$= \frac{(-1)^n \cdot 2 \cdot 3 \cdot 4 \cdot 5 \cdots (n + 1)(n + 2)}{2 \cdot n!} = \frac{(-1)^n (n + 1)(n + 2)}{2}$$

Thus, $\dfrac{1}{(2 + x)^3} = \dfrac{1}{8}\displaystyle\sum_{n=0}^{\infty} \dfrac{(-1)^n(n + 1)(n + 2)}{2} \dfrac{x^n}{2^n} = \sum_{n=0}^{\infty} \dfrac{(-1)^n(n + 1)(n + 2)x^n}{2^{n+4}}$ for $\left|\dfrac{x}{2}\right| < 1 \;\Leftrightarrow\;$

$|x| < 2$, so $R = 2$.

5. We must write the binomial in the form $(1+$ expression$)$, so we'll factor out a 4.

$$\frac{x}{\sqrt{4+x^2}} = \frac{x}{\sqrt{4(1+x^2/4)}} = \frac{x}{2\sqrt{1+x^2/4}} = \frac{x}{2}\left(1+\frac{x^2}{4}\right)^{-1/2} = \frac{x}{2}\sum_{n=0}^{\infty}\binom{-\frac{1}{2}}{n}\left(\frac{x^2}{4}\right)^n$$

$$= \frac{x}{2}\left[1+\left(-\tfrac{1}{2}\right)\frac{x^2}{4} + \frac{\left(-\frac{1}{2}\right)\left(-\frac{3}{2}\right)}{2!}\left(\frac{x^2}{4}\right)^2 + \frac{\left(-\frac{1}{2}\right)\left(-\frac{3}{2}\right)\left(-\frac{5}{2}\right)}{3!}\left(\frac{x^2}{4}\right)^3 + \cdots\right]$$

$$= \frac{x}{2} + \frac{x}{2}\sum_{n=1}^{\infty}(-1)^n\frac{1\cdot 3\cdot 5\cdots\cdots(2n-1)}{2^n\cdot 4^n\cdot n!}x^{2n}$$

$$= \frac{x}{2} + \sum_{n=1}^{\infty}(-1)^n\frac{1\cdot 3\cdot 5\cdots\cdots(2n-1)}{n!\,2^{3n+1}}x^{2n+1} \text{ and } \frac{x^2}{4} < 1 \;\Leftrightarrow\; \frac{|x|}{2} < 1 \;\Leftrightarrow\;$$

$|x| < 2$, so $R = 2$.

7. $\dfrac{1}{\sqrt[3]{8+x}} = \dfrac{1}{\sqrt[3]{8(1+x/8)}} = \dfrac{1}{\sqrt[3]{8}\,\sqrt[3]{1+x/8}} = \dfrac{1}{2}\left(1+\dfrac{x}{8}\right)^{-1/3}$

$$= \frac{1}{2}\left[1+\left(-\frac{1}{3}\right)\left(\frac{x}{8}\right) + \frac{\left(-\frac{1}{3}\right)\left(-\frac{4}{3}\right)}{2!}\left(\frac{x}{8}\right)^2 + \frac{\left(-\frac{1}{3}\right)\left(-\frac{4}{3}\right)\left(-\frac{7}{3}\right)}{3!}\left(\frac{x}{8}\right)^3 + \cdots\right]$$

$$= \frac{1}{2}\left[1 + \sum_{n=1}^{\infty}\frac{(-1)^n 1\cdot 4\cdot 7\cdots\cdots(3n-2)}{3^n\cdot n!\,8^n}x^n\right]$$

$$= \frac{1}{2} + \frac{1}{2}\sum_{n=1}^{\infty}\frac{(-1)^n 1\cdot 4\cdot 7\cdots\cdots(3n-2)}{24^n\,n!}x^n \text{ and } \left|\frac{x}{8}\right| < 1 \;\Leftrightarrow\; |x| < 8, \text{ so } R = 8.$$

The three Taylor polynomials are $T_1(x) = \frac{1}{2} - \frac{1}{48}x$, $T_2(x) = \frac{1}{2} - \frac{1}{48}x + \frac{1}{576}x^2$, and
$T_3(x) = \frac{1}{2} - \frac{1}{48}x + \frac{1}{576}x^2 - \frac{7}{41,472}x^3$.

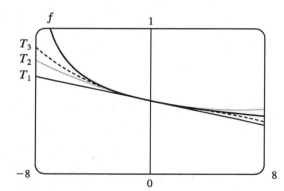

9. (a) $1/\sqrt{1-x^2} = \left[1+\left(-x^2\right)\right]^{-1/2}$

$$= 1 + \left(-\tfrac{1}{2}\right)\left(-x^2\right) + \frac{\left(-\frac{1}{2}\right)\left(-\frac{3}{2}\right)}{2!}\left(-x^2\right)^2 + \frac{\left(-\frac{1}{2}\right)\left(-\frac{3}{2}\right)\left(-\frac{5}{2}\right)}{3!}\left(-x^2\right)^3 + \cdots$$

$$= 1 + \sum_{n=1}^{\infty}\frac{1\cdot 3\cdot 5\cdots\cdots(2n-1)}{2^n\cdot n!}x^{2n}$$

(b) $\sin^{-1}x = \displaystyle\int\frac{1}{\sqrt{1-x^2}}\,dx = C + x + \sum_{n=1}^{\infty}\frac{1\cdot 3\cdot 5\cdots\cdots(2n-1)}{(2n+1)2^n\cdot n!}x^{2n+1}$

$$= x + \sum_{n=1}^{\infty}\frac{1\cdot 3\cdot 5\cdots\cdots(2n-1)}{(2n+1)2^n\cdot n!}x^{2n+1} \text{ since } 0 = \sin^{-1}0 = C.$$

11. (a) $[1 + (-x)]^{-2} = 1 + (-2)(-x) + \dfrac{(-2)(-3)}{2!}(-x)^2 + \dfrac{(-2)(-3)(-4)}{3!}(-x)^3 + \cdots$

$$= 1 + 2x + 3x^2 + 4x^3 + \cdots = \sum_{n=0}^{\infty}(n+1)x^n,$$

so $\dfrac{x}{(1-x)^2} = x\sum_{n=0}^{\infty}(n+1)x^n = \sum_{n=0}^{\infty}(n+1)x^{n+1} = \sum_{n=1}^{\infty}nx^n.$

(b) With $x = \frac{1}{2}$ in part (a), we have $\sum_{n=1}^{\infty}n\left(\frac{1}{2}\right)^n = \sum_{n=1}^{\infty}\dfrac{n}{2^n} = \dfrac{\frac{1}{2}}{\left(1-\frac{1}{2}\right)^2} = \dfrac{\frac{1}{2}}{\frac{1}{4}} = 2.$

13. (a) $(1+x^2)^{1/2} = 1 + \left(\frac{1}{2}\right)x^2 + \dfrac{\left(\frac{1}{2}\right)\left(-\frac{1}{2}\right)}{2!}\left(x^2\right)^2 + \dfrac{\left(\frac{1}{2}\right)\left(-\frac{1}{2}\right)\left(-\frac{3}{2}\right)}{3!}\left(x^2\right)^3 + \cdots$

$$= 1 + \dfrac{x^2}{2} + \sum_{n=2}^{\infty}\dfrac{(-1)^{n-1}\,1\cdot 3\cdot 5\cdot\cdots\cdot(2n-3)}{2^n\cdot n!}x^{2n}$$

(b) The coefficient of x^{10} (corresponding to $n = 5$) in the above Maclaurin series is $\dfrac{f^{(10)}(0)}{10!}$, so

$$\dfrac{f^{(10)}(0)}{10!} = \dfrac{(-1)^4\cdot 1\cdot 3\cdot 5\cdot 7}{2^5\cdot 5!} \quad\Rightarrow\quad f^{(10)}(0) = 10!\left(\dfrac{1\cdot 3\cdot 5\cdot 7}{2^5\cdot 5!}\right) = 99{,}225.$$

15. (a) $g(x) = \sum_{n=0}^{\infty}\binom{k}{n}x^n \quad\Rightarrow\quad g'(x) = \sum_{n=1}^{\infty}\binom{k}{n}nx^{n-1}$, so

$$(1+x)g'(x) = (1+x)\sum_{n=1}^{\infty}\binom{k}{n}nx^{n-1} = \sum_{n=1}^{\infty}\binom{k}{n}nx^{n-1} + \sum_{n=1}^{\infty}\binom{k}{n}nx^n$$

$$= \sum_{n=0}^{\infty}\binom{k}{n+1}(n+1)x^n + \sum_{n=0}^{\infty}\binom{k}{n}nx^n \qquad \begin{bmatrix}\text{Replace } n \text{ with } n+1 \\ \text{in the first series}\end{bmatrix}$$

$$= \sum_{n=0}^{\infty}(n+1)\dfrac{k(k-1)(k-2)\cdots(k-n+1)(k-n)}{(n+1)!}x^n$$

$$\quad + \sum_{n=0}^{\infty}\left[(n)\dfrac{k(k-1)(k-2)\cdots(k-n+1)}{n!}\right]x^n$$

$$= \sum_{n=0}^{\infty}\dfrac{(n+1)k(k-1)(k-2)\cdots(k-n+1)}{(n+1)!}\left[(k-n)+n\right]x^n$$

$$= k\sum_{n=0}^{\infty}\dfrac{k(k-1)(k-2)\cdots(k-n+1)}{n!}x^n = k\sum_{n=0}^{\infty}\binom{k}{n}x^n = kg(x)$$

Thus, $g'(x) = \dfrac{kg(x)}{1+x}.$

(b) $h(x) = (1+x)^{-k}g(x) \quad\Rightarrow$

$$h'(x) = -k(1+x)^{-k-1}g(x) + (1+x)^{-k}g'(x) \quad \text{[Product Rule]}$$

$$= -k(1+x)^{-k-1}g(x) + (1+x)^{-k}\dfrac{kg(x)}{1+x} \quad \text{[from part (a)]}$$

$$= -k(1+x)^{-k-1}g(x) + k(1+x)^{-k-1}g(x) = 0$$

(c) From part (b) we see that $h(x)$ must be constant for $x \in (-1,1)$, so $h(x) = h(0) = 1$ for $x \in (-1,1)$.

Thus, $h(x) = 1 = (1+x)^{-k}g(x) \quad\Leftrightarrow\quad g(x) = (1+x)^k$ for $x \in (-1,1)$.

8.9 Applications of Taylor Polynomials • • • • • • • • •

1. (a)

n	$f^{(n)}(x)$	$f^{(n)}(0)$	$T_n(x)$
0	$\cos x$	1	1
1	$-\sin x$	0	1
2	$-\cos x$	-1	$1 - \frac{1}{2}x^2$
3	$\sin x$	0	$1 - \frac{1}{2}x^2$
4	$\cos x$	1	$1 - \frac{1}{2}x^2 + \frac{1}{24}x^4$
5	$-\sin x$	0	$1 - \frac{1}{2}x^2 + \frac{1}{24}x^4$
6	$-\cos x$	-1	$1 - \frac{1}{2}x^2 + \frac{1}{24}x^4 - \frac{1}{720}x^6$

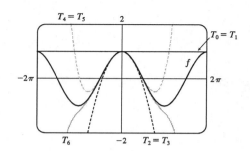

(b)

x	f	$T_0 = T_1$	$T_2 = T_3$	$T_4 = T_5$	T_6
$\frac{\pi}{4}$	0.7071	1	0.6916	0.7074	0.7071
$\frac{\pi}{2}$	0	1	-0.2337	0.0200	-0.0009
π	-1	1	-3.9348	0.1239	-1.2114

(c) As n increases, $T_n(x)$ is a good approximation to $f(x)$ on a larger and larger interval.

3.

n	$f^{(n)}(x)$	$f^{(n)}(1)$
0	$\ln x$	0
1	$1/x$	1
2	$-1/x^2$	-1
3	$2/x^3$	2
4	$-6/x^4$	-6

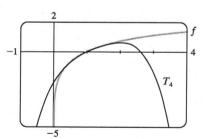

$$T_4(x) = \sum_{n=0}^{4} \frac{f^{(n)}(1)}{n!}(x-1)^n = 0 + (x-1) - \frac{1}{2}(x-1)^2 + \frac{1}{3}(x-1)^3 - \frac{1}{4}(x-1)^4$$

5.

n	$f^{(n)}(x)$	$f^{(n)}\left(\frac{\pi}{6}\right)$
0	$\sin x$	$\frac{1}{2}$
1	$\cos x$	$\frac{\sqrt{3}}{2}$
2	$-\sin x$	$-\frac{1}{2}$
3	$-\cos x$	$-\frac{\sqrt{3}}{2}$

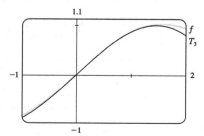

$$T_3(x) = \sum_{n=0}^{3} \frac{f^{(n)}\left(\frac{\pi}{6}\right)}{n!}\left(x-\frac{\pi}{6}\right)^n = \frac{1}{2} + \frac{\sqrt{3}}{2}\left(x-\frac{\pi}{6}\right) - \frac{1}{4}\left(x-\frac{\pi}{6}\right)^2 - \frac{\sqrt{3}}{12}\left(x-\frac{\pi}{6}\right)^3$$

7.

n	$f^{(n)}(x)$	$f^{(n)}(0)$
0	$e^x \sin x$	0
1	$e^x(\sin x + \cos x)$	1
2	$2e^x \cos x$	2
3	$2e^x(\cos x - \sin x)$	2

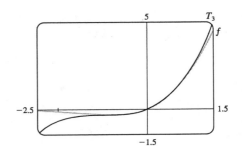

$$T_3(x) = \sum_{n=0}^{3} \frac{f^{(n)}(0)}{n!} x^n = x + x^2 + \tfrac{1}{3}x^3$$

9. In Maple, we can find the Taylor polynomials by the following method: first define f:=sec(x); and then set
T2:=convert(taylor(f,x=0,3),polynom);, T4:=convert(taylor(f,x=0,5),polynom);,
etc. (The third argument in the taylor function is one more than the degree of the desired polynomial). We must
convert to the type polynom because the output of the
taylor function contains an error term which we do not
want. In Mathematica, we use
Tn:=Normal[Series[f,{x,0,n}]], with n=2, 4,
etc. Note that in Mathematica, the "degree" argument is the
same as the degree of the desired polynomial. In Derive,
author sec x, then enter Calculus,Taylor,8,0; and
then simplify the expression. The eighth Taylor polynomial is

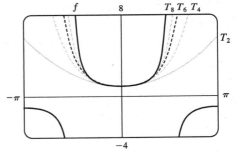

$$T_8(x) = 1 + \tfrac{1}{2}x^2 + \tfrac{5}{24}x^4 + \tfrac{61}{720}x^6 + \tfrac{277}{8064}x^8.$$

11.

$$f(x) = \sqrt{x} \qquad\qquad f(4) = 2$$
$$f'(x) = \tfrac{1}{2}x^{-1/2} \qquad f'(4) = \tfrac{1}{4}$$
$$f''(x) = -\tfrac{1}{4}x^{-3/2} \qquad f''(4) = -\tfrac{1}{32}$$
$$f'''(x) = \tfrac{3}{8}x^{-5/2}$$

(a) $f(x) = \sqrt{x} \approx T_2(x) = 2 + \tfrac{1}{4}(x-4) - \tfrac{1/32}{2!}(x-4)^2 = 2 + \tfrac{1}{4}(x-4) - \tfrac{1}{64}(x-4)^2$

(b) $|R_2(x)| \le \dfrac{M}{3!}|x-4|^3$, where $|f'''(x)| \le M$. Now $4 \le x \le 4.2 \;\Rightarrow\; |x-4| \le 0.2 \;\Rightarrow\;$
$|x-4|^3 \le 0.008$. Since $f'''(x)$ is decreasing on $[4, 4.2]$, we can take $M = |f'''(4)| = \tfrac{3}{8}4^{-5/2} = \tfrac{3}{256}$, so
$|R_2(x)| \le \dfrac{3/256}{6}(0.008) = \dfrac{0.008}{512} = 0.000015625.$

(c) From the graph of $|R_2(x)| = |\sqrt{x} - T_2(x)|$, it seems that the error is
less than 1.52×10^{-5} on $[4, 4.2]$.

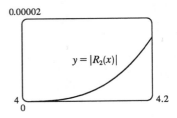

13. $f(x) = e^{x^2}$ $\qquad$ $f(0) = 1$ $\qquad$ $f'''(x) = e^{x^2}(12x + 8x^3)$ $\qquad$ $f'''(0) = 0$

$f'(x) = e^{x^2}(2x)$ $\qquad$ $f'(0) = 0$ $\qquad$ $f^{(4)}(x) = e^{x^2}(12 + 48x^2 + 16x^4)$

$f''(x) = e^{x^2}(2 + 4x^2)$ $\qquad$ $f''(0) = 2$

(a) $f(x) = e^{x^2} \approx T_3(x) = 1 + \frac{2}{2!}x^2 = 1 + x^2$

(b) $|R_3(x)| \le \dfrac{M}{4!}|x|^4$, where $\left|f^{(4)}(x)\right| \le M$.

Now $0 \le x \le 0.1 \;\Rightarrow\; x^4 \le (0.1)^4$, and

letting $x = 0.1$ gives

$|R_3(x)| \le \dfrac{e^{0.01}(12 + 0.48 + 0.0016)}{24}(0.1)^4 \approx$

$0.00006.$

(c)

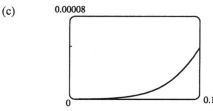

0.00008

0 $\qquad\qquad\qquad\qquad$ 0.1

From the graph of

$|R_3(x)| = \left|e^{x^2} - (1 + x^2)\right|$, it appears that

the error is less than 0.000051 on $[0, 0.1]$.

15. $f(x) = \tan x$ $\qquad$ $f(0) = 0$ $\qquad$ $f'''(x) = 4\sec^2 x \tan^2 x + 2\sec^4 x$ $\qquad$ $f'''(0) = 2$

$f'(x) = \sec^2 x$ $\qquad$ $f'(0) = 1$ $\qquad$ $f^{(4)}(x) = 8\sec^2 x \tan^3 x + 16\sec^4 x \tan x$

$f''(x) = 2\sec^2 x \tan x$ $\qquad$ $f''(0) = 0$

(a) $f(x) = \tan x \approx T_3(x) = x + \frac{1}{3}x^3$

(b) $|R_3(x)| \le \dfrac{M}{4!}|x|^4$, where $\left|f^{(4)}(x)\right| \le M$. Now

$0 \le x \le \frac{\pi}{6} \;\Rightarrow\; x^4 \le \left(\frac{\pi}{6}\right)^4$, and letting $x = \frac{\pi}{6}$

gives

$|R_3(x)| \le \dfrac{8\left(\frac{2}{\sqrt{3}}\right)^2\left(\frac{1}{\sqrt{3}}\right)^3 + 16\left(\frac{2}{\sqrt{3}}\right)^4\left(\frac{1}{\sqrt{3}}\right)}{4!}\left(\frac{\pi}{6}\right)^4$

$= \frac{4\sqrt{3}}{9}\left(\frac{\pi}{6}\right)^4 \approx 0.057859$

(c)

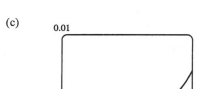

0.01

0 $\qquad\qquad\qquad\qquad$ $\frac{\pi}{6}$

From the graph of

$|R_3(x)| = |\tan x - T_3(3)|$, it seems that

the error is less than 0.006 on $[0, \pi/6]$.

17. From Exercise 5, $\sin x = \frac{1}{2} + \frac{\sqrt{3}}{2}\left(x - \frac{\pi}{6}\right) - \frac{1}{4}\left(x - \frac{\pi}{6}\right)^2 - \frac{\sqrt{3}}{12}\left(x - \frac{\pi}{6}\right)^3 + R_3(x)$, where $|R_3(x)| \le \dfrac{M}{4!}\left|x - \frac{\pi}{6}\right|^4$

with $\left|f^{(4)}(x)\right| = |\sin x| \le M = 1$. Now $x = 35° = (30° + 5°) = \left(\frac{\pi}{6} + \frac{\pi}{36}\right)$ radians, so the error is

$\left|R_3\left(\frac{\pi}{36}\right)\right| \le \dfrac{\left(\frac{\pi}{36}\right)^4}{4!} < 0.000003$. Therefore, to five decimal places,

$\sin 35° \approx \frac{1}{2} + \frac{\sqrt{3}}{2}\left(\frac{\pi}{36}\right) - \frac{1}{4}\left(\frac{\pi}{36}\right)^2 - \frac{\sqrt{3}}{12}\left(\frac{\pi}{36}\right)^3 \approx 0.57358.$

19. All derivatives of e^x are e^x, so $|R_n(x)| \le \dfrac{e^x}{(n+1)!}|x|^{n+1}$, where $0 < x < 0.1$. Letting $x = 0.1$,

$R_n(0.1) \le \dfrac{e^{0.1}}{(n+1)!}(0.1)^{n+1} < 0.00001$, and by trial and error we find that $n = 3$ satisfies this inequality since

$R_3(0.1) < 0.0000046$. Thus, by adding the four terms of the Maclaurin series for e^x corresponding to $n = 0, 1, 2$, and 3, we can estimate $e^{0.1}$ to within 0.00001. (In fact, this sum is $1.105\overline{16}$ and $e^{0.1} \approx 1.10517$.)

21. $\sin x = x - \frac{1}{3!}x^3 + \frac{1}{5!}x^5 - \cdots$. By the Alternating Series Estimation Theorem, the error in the approximation $\sin x = x - \frac{1}{3!}x^3$ is less than

$\left|\frac{1}{5!}x^5\right| < 0.01 \iff |x^5| < 120(0.01) \iff$

$|x| < (1.2)^{1/5} \approx 1.037$. The curves intersect at $x \approx 1.043$, so the graph confirms our estimate. Since both the sine function and the given approximation are odd functions, we need to check the estimate only for $x > 0$. Thus, the desired range of values for x is $-1.037 < x < 1.037$.

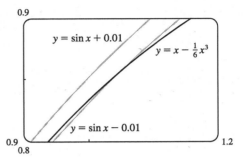

23. Let $s(t)$ be the position function of the car, and for convenience set $s(0) = 0$. The velocity of the car is $v(t) = s'(t)$ and the acceleration is $a(t) = s''(t)$, so the second degree Taylor polynomial is

$T_2(t) = s(0) + v(0)t + \frac{a(0)}{2}t^2 = 20t + t^2$. We estimate the distance travelled during the next second to be

$s(1) \approx T_2(1) = 20 + 1 = 21$ m. The function $T_2(t)$ would not be accurate over a full minute, since the car could not possibly maintain an acceleration of 2 m/s^2 for that long (if it did, its final speed would be

140 m/s ≈ 313 mi/h!)

25. $E = \frac{q}{D^2} - \frac{q}{(D+d)^2} = \frac{q}{D^2} - \frac{q}{D^2(1+d/D)^2} = \frac{q}{D^2}\left[1 - \left(1 + \frac{d}{D}\right)^{-2}\right]$.

We use the Binomial Series to expand $(1 + d/D)^{-2}$:

$$E = \frac{q}{D^2}\left[1 - \left(1 - 2\left(\frac{d}{D}\right) + \frac{2 \cdot 3}{2!}\left(\frac{d}{D}\right)^2 - \frac{2 \cdot 3 \cdot 4}{3!}\left(\frac{d}{D}\right)^3 + \cdots\right)\right]$$

$$= \frac{q}{D^2}\left[2\left(\frac{d}{D}\right) - 3\left(\frac{d}{D}\right)^2 + 4\left(\frac{d}{D}\right)^3 - \cdots\right] \approx \frac{q}{D^2} \cdot 2\left(\frac{d}{D}\right) = 2qd \cdot \frac{1}{D^3}$$

when D is much larger than d; that is, when P is far away from the dipole.

27. Using $f(x) = T_n(x) + R_n(x)$ with $n = 1$ and $x = r$, we have $f(r) = T_1(r) + R_1(r)$, where T_1 is the first-degree Taylor polynomial of f at a. Because $a = x_n$, $f(r) = f(x_n) + f'(x_n)(r - x_n) + R_1(r)$. But r is a root of f, so $f(r) = 0$ and we have $0 = f(x_n) + f'(x_n)(r - x_n) + R_1(r)$. Taking the first two terms to the left side and dividing by $f'(x_n)$, we have $f'(x_n)(x_n - r) - f(x_n) = R_1(r) \Rightarrow x_n - r - \dfrac{f(x_n)}{f'(x_n)} = \dfrac{R_1(r)}{f'(x_n)}$. By the

formula for Newton's method, the left side of the preceding equation is $x_{n+1} - r$, so $|x_{n+1} - r| = \left| \dfrac{R_1(r)}{f'(x_n)} \right|$.

Taylor's Inequality gives us $|R_1(r)| \le \dfrac{|f''(r)|}{2!} |r - x_n|^2$. Combining this inequality with the facts $|f''(x)| \le M$

and $|f'(x)| \ge K$ gives us $|x_{n+1} - r| \le \dfrac{M}{2K} |x_n - r|^2$.

8.10 Using Series to Solve Differential Equations • • • • • •

1. Let $y(x) = \sum\limits_{n=0}^{\infty} c_n x^n$. Then $y'(x) = \sum\limits_{n=1}^{\infty} n c_n x^{n-1}$ and the given equation, $y' - y = 0$, becomes

$\sum\limits_{n=1}^{\infty} n c_n x^{n-1} - \sum\limits_{n=0}^{\infty} c_n x^n = 0$. Replacing n by $n+1$ in the first sum gives $\sum\limits_{n=0}^{\infty} (n+1)c_{n+1} x^n - \sum\limits_{n=0}^{\infty} c_n x^n = 0$,

so $\sum\limits_{n=0}^{\infty} [(n+1)c_{n+1} - c_n] x^n = 0$. Equating coefficients gives $(n+1)c_{n+1} - c_n = 0$, so the recursion relation is

$c_{n+1} = \dfrac{c_n}{n+1}$, $n = 0, 1, 2, \ldots$. Then $c_1 = c_0$, $c_2 = \dfrac{1}{2}c_1 = \dfrac{c_0}{2}$, $c_3 = \dfrac{1}{3}c_2 = \dfrac{1}{3} \cdot \dfrac{1}{2}c_0 = \dfrac{c_0}{3!}$, $c_4 = \dfrac{1}{4}c_3 = \dfrac{c_0}{4!}$, and

in general, $c_n = \dfrac{c_0}{n!}$. Thus, the solution is

$$y(x) = \sum\limits_{n=0}^{\infty} c_n x^n = \sum\limits_{n=0}^{\infty} \dfrac{c_0}{n!} x^n = c_0 \sum\limits_{n=0}^{\infty} \dfrac{x^n}{n!} = c_0 e^x$$

3. Assuming $y(x) = \sum\limits_{n=0}^{\infty} c_n x^n$, we have $y'(x) = \sum\limits_{n=1}^{\infty} n c_n x^{n-1} = \sum\limits_{n=0}^{\infty} (n+1)c_{n+1} x^n$ and

$-x^2 y = -\sum\limits_{n=0}^{\infty} c_n x^{n+2} = -\sum\limits_{n=2}^{\infty} c_{n-2} x^n$. Hence, the equation $y' = x^2 y$ becomes

$\sum\limits_{n=0}^{\infty} (n+1)c_{n+1} x^n - \sum\limits_{n=2}^{\infty} c_{n-2} x^n = 0$ or $c_1 + 2c_2 x + \sum\limits_{n=2}^{\infty} [(n+1)c_{n+1} - c_{n-2}] x^n = 0$. Equating coefficients

gives $c_1 = c_2 = 0$ and $c_{n+1} = \dfrac{c_{n-2}}{n+1}$ for $n = 2, 3, \ldots$. But $c_1 = 0$, so $c_4 = 0$ and $c_7 = 0$ and in general

$c_{3n+1} = 0$. Similarly $c_2 = 0$ so $c_{3n+2} = 0$. Finally $c_3 = \dfrac{c_0}{3}$, $c_6 = \dfrac{c_3}{6} = \dfrac{c_0}{6 \cdot 3} = \dfrac{c_0}{3^2 \cdot 2!}$,

$c_9 = \dfrac{c_6}{9} = \dfrac{c_0}{9 \cdot 6 \cdot 3} = \dfrac{c_0}{3^3 \cdot 3!}$, $\ldots$, and $c_{3n} = \dfrac{c_0}{3^n \cdot n!}$. Thus, the solution is

$$y(x) = \sum\limits_{n=0}^{\infty} c_n x^n = \sum\limits_{n=0}^{\infty} c_{3n} x^{3n} = \sum\limits_{n=0}^{\infty} \dfrac{c_0}{3^n \cdot n!} x^{3n} = c_0 \sum\limits_{n=0}^{\infty} \dfrac{x^{3n}}{3^n n!} = c_0 \sum\limits_{n=0}^{\infty} \dfrac{(x^3/3)^n}{n!} = c_0 e^{x^3/3}$$

5. Let $y(x) = \sum\limits_{n=0}^{\infty} c_n x^n$. Then $3xy'(x) = 3x \sum\limits_{n=1}^{\infty} n c_n x^{n-1} = \sum\limits_{n=0}^{\infty} 3n c_n x^n$,

$y''(x) = \sum\limits_{n=2}^{\infty} n(n-1) c_n x^{n-2} = \sum\limits_{n=0}^{\infty} (n+2)(n+1) c_{n+2} x^n$, and the equation

$y'' + 3xy' + 3y = 0$ becomes $\sum\limits_{n=0}^{\infty} (n+2)(n+1) c_{n+2} x^n + \sum\limits_{n=0}^{\infty} 3n c_n x^n + \sum\limits_{n=0}^{\infty} 3 c_n x^n = 0 \iff$

$\sum\limits_{n=0}^{\infty} [(n+2)(n+1) c_{n+2} + 3n c_n + 3 c_n] x^n = 0$. Thus, the recursion relation is

$c_{n+2} = \dfrac{-3n c_n - 3 c_n}{(n+2)(n+1)} = \dfrac{-3 c_n (n+1)}{(n+2)(n+1)} = -\dfrac{3 c_n}{n+2}$ for $n = 0, 1, 2, \ldots$. Given c_0 and c_1, $c_2 = -\dfrac{3 c_0}{2}$,

$c_4 = -\dfrac{3 c_2}{4} = (-1)^2 \dfrac{3^2 c_0}{2^2 \cdot 2!}$, $c_6 = -\dfrac{3 c_4}{6} = (-1)^3 \dfrac{3^3 c_0}{2^3 \cdot 3!}, \ldots, c_{2n} = (-1)^n \dfrac{3^n c_0}{2^n n!}$ or, equivalently,

$c_0 \left(-\dfrac{3}{2}\right)^n \dfrac{1}{n!}$. Also, $c_3 = -\dfrac{3 c_1}{3}$, $c_5 = -\dfrac{3 c_3}{5} = (-1)^2 \dfrac{3^2 c_1}{5 \cdot 3}$, $c_7 = -\dfrac{3 c_5}{7} = (-1)^3 \dfrac{3^3 c_1}{7 \cdot 5 \cdot 3}, \ldots,$

$c_{2n+1} = (-1)^n \dfrac{3^n c_1}{(2n+1)(2n-1) \cdots \cdots 5 \cdot 3}$. Since $(2n+1)(2n-1) \cdots \cdots 5 \cdot 3$ can be written as

$$\dfrac{(2n+1)\ (2n)\ (2n-1)\ (2n-2) \cdots \cdots 5 \cdot\ 4\ \cdot\ 3\ \cdot\ 2}{(2 \cdot n)\ \ \cdot\ \ [2(n-1)]\ \ \cdot\ \ (2 \cdot 2)\ \cdot\ (2 \cdot 1)} = \dfrac{(2n+1)!}{2^n \cdot n!},$$

c_{2n+1} can be written as $(-1)^n \dfrac{3^n c_1 2^n n!}{(2n+1)!} = c_1 \dfrac{(-6)^n n!}{(2n+1)!}$. Thus, the solution is

$$y(x) = \sum\limits_{n=0}^{\infty} c_{2n} x^{2n} + \sum\limits_{n=0}^{\infty} c_{2n+1} x^{2n+1} = c_0 \sum\limits_{n=0}^{\infty} \left(-\dfrac{3}{2}\right)^n \dfrac{1}{n!} x^{2n} + c_1 \sum\limits_{n=0}^{\infty} \dfrac{(-6)^n n!}{(2n+1)!} x^{2n+1}$$

Note that the c_0-term can be written as $c_0 \sum\limits_{n=0}^{\infty} \left(-\dfrac{3x^2}{2}\right)^n \dfrac{1}{n!} = c_0 e^{-3x^2/2}$.

7. Let $y(x) = \sum\limits_{n=0}^{\infty} c_n x^n$. Then $-xy'(x) = -x \sum\limits_{n=1}^{\infty} n c_n x^{n-1} = -\sum\limits_{n=1}^{\infty} n c_n x^n = -\sum\limits_{n=0}^{\infty} n c_n x^n$,

$y''(x) = \sum\limits_{n=0}^{\infty} (n+2)(n+1) c_{n+2} x^n$, and the equation $y'' - xy' - y = 0$ becomes

$\sum\limits_{n=0}^{\infty} [(n+2)(n+1) c_{n+2} - n c_n - c_n] x^n = 0$. Thus, the recursion relation is

$c_{n+2} = \dfrac{n c_n + c_n}{(n+2)(n+1)} = \dfrac{c_n(n+1)}{(n+2)(n+1)} = \dfrac{c_n}{n+2}$ for $n = 0, 1, 2, \ldots$. One of the given conditions is

$y(0) = 1$. But $y(0) = \sum\limits_{n=0}^{\infty} c_n (0)^n = c_0 + 0 + 0 + \cdots = c_0$, so $c_0 = 1$. Hence, $c_2 = \dfrac{c_0}{2} = \dfrac{1}{2}$, $c_4 = \dfrac{c_2}{4} = \dfrac{1}{2 \cdot 4}$,

$c_6 = \dfrac{c_4}{6} = \dfrac{1}{2 \cdot 4 \cdot 6}, \ldots, c_{2n} = \dfrac{1}{2^n n!}$. The other given condition is $y'(0) = 0$. But

$y'(0) = \sum\limits_{n=1}^{\infty} n c_n (0)^{n-1} = c_1 + 0 + 0 + \cdots = c_1$, so $c_1 = 0$. By the recursion relation, $c_3 = \dfrac{c_1}{3} = 0$, $c_5 = 0, \ldots,$

$c_{2n+1} = 0$ for $n = 0, 1, 2, \ldots$. Thus, the solution to the initial-value problem is

$$y(x) = \sum\limits_{n=0}^{\infty} c_n x^n = \sum\limits_{n=0}^{\infty} c_{2n} x^{2n} = \sum\limits_{n=0}^{\infty} \dfrac{x^{2n}}{2^n n!} = \sum\limits_{n=0}^{\infty} \dfrac{(x^2/2)^n}{n!} = e^{x^2/2}$$

9. Assuming that $y(x) = \sum\limits_{n=0}^{\infty} c_n x^n$, we have $xy = x \sum\limits_{n=0}^{\infty} c_n x^n = \sum\limits_{n=0}^{\infty} c_n x^{n+1}$,

$$x^2 y' = x^2 \sum\limits_{n=1}^{\infty} n c_n x^{n-1} = \sum\limits_{n=0}^{\infty} n c_n x^{n+1},$$

$$y''(x) = \sum\limits_{n=2}^{\infty} n(n-1) c_n x^{n-2} = \sum\limits_{n=-1}^{\infty} (n+3)(n+2) c_{n+3} x^{n+1} \quad \text{[replace } n \text{ with } n+3]$$

$$= 2c_2 + \sum\limits_{n=0}^{\infty} (n+3)(n+2) c_{n+3} x^{n+1},$$

and the equation $y'' + x^2 y' + xy = 0$ becomes $2c_2 + \sum\limits_{n=0}^{\infty} [(n+3)(n+2) c_{n+3} + n c_n + c_n] x^{n+1} = 0$.

So $c_2 = 0$ and the recursion relation is $c_{n+3} = \dfrac{-n c_n - c_n}{(n+3)(n+2)} = -\dfrac{(n+1) c_n}{(n+3)(n+2)}$, $n = 0, 1, 2, \ldots$.

But $c_0 = y(0) = 0 = c_2$ and by the recursion relation, $c_{3n} = c_{3n+2} = 0$ for $n = 0, 1, 2, \ldots$.

Also, $c_1 = y'(0) = 1$, so

$$c_4 = -\frac{2c_1}{4 \cdot 3} = -\frac{2}{4 \cdot 3}, \; c_7 = -\frac{5c_4}{7 \cdot 6} = (-1)^2 \frac{2 \cdot 5}{7 \cdot 6 \cdot 4 \cdot 3} = (-1)^2 \frac{2^2 5^2}{7!}, \; \ldots,$$

$c_{3n+1} = (-1)^n \dfrac{2^2 5^2 \cdot \cdots \cdot (3n-1)^2}{(3n+1)!}$. Thus, the solution is

$$y(x) = \sum\limits_{n=0}^{\infty} c_n x^n = x + \sum\limits_{n=1}^{\infty} \left[(-1)^n \frac{2^2 5^2 \cdot \cdots \cdot (3n-1)^2 x^{3n+1}}{(3n+1)!} \right]$$

 8 Review

━━━━━━━━━━━━━━━━━━━━━━━━━━━━━ • **CONCEPT CHECK** • ━━━━━━━━━━━━━━━━━━━━━━━━━━━━━

1. (a) See Definition 8.1.1.

    (b) See Definition 8.2.2.

    (c) The terms of the sequence $\{a_n\}$ approach 3 as n becomes large.

    (d) By adding sufficiently many terms of the series, we can make the partial sums as close to 3 as we like.

2. (a) See the definition on page 569.

    (b) A sequence is monotonic if it is either increasing or decreasing.

    (c) By Theorem 8.1.7, every bounded, monotonic sequence is convergent.

3. (a) See (4) in Section 8.2.

    (b) See (1) in Section 8.3.

4. If $\sum a_n = 3$, then $\lim\limits_{n \to \infty} a_n = 0$ and $\lim\limits_{n \to \infty} s_n = 3$.

5. (a) See the Test for Divergence on page 578.   (b) See the Integral Test on page 584.

    (c) See the Comparison Test on page 586.   (d) See the Limit Comparison Test on page 588.

    (e) See the Alternating Series Test on page 593.   (f) See the Ratio Test on page 597.

6. (a) See the definition on page 595.

    (b) By (8.4.1), it is convergent.

7. (a) Use (4) in Section 8.3.

(b) See Example 8 in Section 8.3.

(c) By adding terms until you reach the desired accuracy given by the Alternating Series Estimation Theorem on page 594.

8. (a) $\sum_{n=0}^{\infty} c_n (x - a)^n$

(b) Given the power series $\sum_{n=0}^{\infty} c_n (x - a)^n$, the radius of convergence is:

(i) 0 if the series converges only when $x = a$

(ii) ∞ if the series converges for all x, or

(iii) a positive number R such that the series converges if $|x - a| < R$ and diverges if $|x - a| > R$.

(c) The interval of convergence of a power series is the interval that consists of all values of x for which the series converges. Corresponding to the cases in part (b), the interval of convergence is: (i) the single point $\{a\}$, (ii) all real numbers, that is, the real number line $(-\infty, \infty)$, or (iii) an interval with endpoints $a - R$ and $a + R$ which can contain neither, either, or both of the endpoints. In this case, we must test the series for convergence at each endpoint to determine the interval of convergence.

9. (a), (b) See Theorem 8.6.2.

10. (a) $T_n(x) = \sum_{i=0}^{n} \dfrac{f^{(i)}(a)}{i!} (x - a)^i$

(b) $\sum_{n=0}^{\infty} \dfrac{f^{(n)}(a)}{n!} (x - a)^n$

(c) $\sum_{n=0}^{\infty} \dfrac{f^{(n)}(0)}{n!} x^n$ [$a = 0$ in part (b)]

(d) See Theorem 8.7.8.

(e) See Taylor's Inequality (8.7.9).

11. (a) – (e) See the table on page 618.

12. See the Binomial Series (8.8.2) for the expansion. The radius of convergence for the binomial series is 1.

▲ TRUE–FALSE QUIZ ▲

1. False. See Note 2 after Theorem 8.2.6.

3. False. For example, take $c_n = (-1)^n / (n6^n)$.

5. False, since $\lim\limits_{n \to \infty} \left| \dfrac{a_{n+1}}{a_n} \right| = \lim\limits_{n \to \infty} \left| \dfrac{n^3}{(n+1)^3} \right| = \lim\limits_{n \to \infty} \dfrac{1}{(1 + 1/n)^3} = 1.$

7. False. See the note after Example 4 in Section 8.3.

9. True. See (6) in Section 8.1.

11. True. By Theorem 8.7.5 the coefficient of x^3 is $\dfrac{f'''(0)}{3!} = \dfrac{1}{3} \Rightarrow f'''(0) = 2.$

Or: Use Theorem 8.6.2 to differentiate f three times.

13. False. For example, let $a_n = b_n = (-1)^n$. Then $\{a_n\}$ and $\{b_n\}$ are divergent, but $a_n b_n = 1$, so $\{a_n b_n\}$ is convergent.

15. True by Theorem 8.4.1. $\left[\sum (-1)^n a_n \text{ is absolutely convergent and hence convergent.} \right]$

17. False. The Integral Test tells us that the series $\sum_{n=1}^{\infty} a_n$ also converges, but its value is not equal to the value of $\int_1^{\infty} f(x) \, dx$. In fact, a picture like Figure 2 on page 584 shows that the sum of the series is larger than the value of the integral.

◆ **EXERCISES** ◆

1. $\left\{\dfrac{2+n^3}{1+2n^3}\right\}$ converges since $\displaystyle\lim_{n\to\infty}\dfrac{2+n^3}{1+2n^3}=\lim_{n\to\infty}\dfrac{2/n^3+1}{1/n^3+2}=\dfrac{1}{2}$.

3. $\displaystyle\lim_{n\to\infty}a_n=\lim_{n\to\infty}\dfrac{n^3}{1+n^2}=\lim_{n\to\infty}\dfrac{n}{1/n^2+1}=\infty$, so the sequence diverges.

5. $\{\sin n\}$ is divergent since $\displaystyle\lim_{n\to\infty}\sin n$ does not exist.

7. $\left\{\left(1+\dfrac{3}{n}\right)^{4n}\right\}$ is convergent. Let $y=\left(1+\dfrac{3}{x}\right)^{4x}$. Then

$$\lim_{x\to\infty}\ln y=\lim_{x\to\infty}4x\ln(1+3/x)=\lim_{x\to\infty}\dfrac{\ln(1+3/x)}{1/(4x)}\overset{\text{H}}{=}\lim_{x\to\infty}\dfrac{\dfrac{1}{1+3/x}\left(-\dfrac{3}{x^2}\right)}{-1/(4x^2)}=\lim_{x\to\infty}\dfrac{12}{1+3/x}=12$$

so $\displaystyle\lim_{x\to\infty}y=\lim_{n\to\infty}\left(1+\dfrac{3}{n}\right)^{4n}=e^{12}$.

9. $\dfrac{n}{n^3+1}<\dfrac{n}{n^3}=\dfrac{1}{n^2}$, so $\displaystyle\sum_{n=1}^{\infty}\dfrac{n}{n^3+1}$ converges by the Comparison Test with the convergent p-series

$\displaystyle\sum_{n=1}^{\infty}\dfrac{1}{n^2}$ $(p=2>1)$.

11. $\displaystyle\lim_{n\to\infty}\left|\dfrac{a_{n+1}}{a_n}\right|=\lim_{n\to\infty}\left[\dfrac{(n+1)^3}{5^{n+1}}\cdot\dfrac{5^n}{n^3}\right]=\lim_{n\to\infty}\left(1+\dfrac{1}{n}\right)^3\cdot\dfrac{1}{5}=\dfrac{1}{5}<1$, so $\displaystyle\sum_{n=1}^{\infty}\dfrac{n^3}{5^n}$ converges by the Ratio Test.

13. $\left|\dfrac{\sin n}{1+n^2}\right|\le\dfrac{1}{1+n^2}<\dfrac{1}{n^2}$ and since $\displaystyle\sum_{n=1}^{\infty}\dfrac{1}{n^2}$ converges (p-series with $p=2>1$), so does $\displaystyle\sum_{n=1}^{\infty}\left|\dfrac{\sin n}{1+n^2}\right|$ by the

Comparison Test, and so does $\displaystyle\sum_{n=1}^{\infty}\dfrac{\sin n}{1+n^2}$ by Theorem 8.4.1.

15. Let $b_n=\dfrac{\sqrt{n}}{n+1}>0$. Then $0\le\displaystyle\lim_{n\to\infty}b_n=\lim_{n\to\infty}\dfrac{\sqrt{n}}{n+1}\le\lim_{n\to\infty}\dfrac{\sqrt{n}}{n}=\lim_{n\to\infty}\dfrac{1}{\sqrt{n}}=0$, so $\displaystyle\lim_{n\to\infty}b_n=0$. If

$f(x)=\dfrac{\sqrt{x}}{x+1}$ for $x>0$, then $f'(x)=\dfrac{(x+1)\cdot\frac{1}{2\sqrt{x}}-\sqrt{x}\cdot1}{(x+1)^2}=\dfrac{(x+1)-2x}{2\sqrt{x}\,(x+1)^2}=\dfrac{1-x}{2\sqrt{x}\,(x+1)^2}$, so

$f'(x)<0$ for $x>1$. It follows that $f(1)>f(2)>f(3)>\cdots$; that is, $b_n>b_{n+1}$ for all n. Thus,

$\displaystyle\sum_{n=1}^{\infty}(-1)^{n-1}\dfrac{\sqrt{n}}{n+1}$ converges by the Alternating Series Test.

17. $\displaystyle\lim_{n\to\infty}\left|\dfrac{a_{n+1}}{a_n}\right|=\lim_{n\to\infty}\dfrac{1\cdot3\cdot5\cdots\cdots(2n-1)(2n+1)}{5^{n+1}(n+1)!}\cdot\dfrac{5^n n!}{1\cdot3\cdot5\cdots\cdots(2n-1)}=\lim_{n\to\infty}\dfrac{2n+1}{5(n+1)}=\dfrac{2}{5}<1$, so

the series converges by the Ratio Test.

19. This is a convergent geometric series with $r=\frac{4}{5}$.

$$\sum_{n=1}^{\infty}\dfrac{2^{2n+1}}{5^n}=\sum_{n=1}^{\infty}\dfrac{\left(2^2\right)^n\cdot2^1}{5^n}=2\sum_{n=1}^{\infty}\dfrac{4^n}{5^n}=2\sum_{n=1}^{\infty}\left(\dfrac{4}{5}\right)^n=2\left(\dfrac{\frac{4}{5}}{1-\frac{4}{5}}\right)=2(4)=8.$$

21. $\displaystyle\sum_{n=1}^{\infty}\left[\tan^{-1}(n+1)-\tan^{-1}n\right]=\lim_{n\to\infty}\left[(\tan^{-1}2-\tan^{-1}1)+(\tan^{-1}3-\tan^{-1}2)+\cdots\right.$

$$\left.+\,(\tan^{-1}(n+1)-\tan^{-1}n)\right]$$

$$=\lim_{n\to\infty}\left[\tan^{-1}(n+1)-\tan^{-1}1\right]=\tfrac{\pi}{2}-\tfrac{\pi}{4}=\tfrac{\pi}{4}$$

23. $1.2345345345\ldots=1.2+0.0\overline{345}=\dfrac{12}{10}+\dfrac{345/10{,}000}{1-1/1000}=\dfrac{12}{10}+\dfrac{345}{9990}=\dfrac{4111}{3330}$

25. $b_8 = \dfrac{1}{8^5} = \dfrac{1}{32{,}768} \approx 0.000\,031$, so

$$\sum_{n=1}^{\infty} \frac{(-1)^{n+1}}{n^5} \approx s_7 = \sum_{n=1}^{7} \frac{(-1)^{n+1}}{n^5} = 1 - \frac{1}{32} + \frac{1}{243} - \frac{1}{1024} + \frac{1}{3125} - \frac{1}{7776} + \frac{1}{16{,}807} \approx 0.972140.$$

Subtracting b_8 from s_7 does not change the fourth decimal place of s_7, so the sum of the series, correct to four decimal places, is 0.9721.

27. $\displaystyle\sum_{n=1}^{\infty} \frac{1}{2+5^n} \approx \sum_{n=1}^{8} \frac{1}{2+5^n} \approx 0.18976224$. To estimate the error, note that $\dfrac{1}{2+5^n} < \dfrac{1}{5^n}$, so the remainder term is

$$R_8 = \sum_{n=9}^{\infty} \frac{1}{2+5^n} < \sum_{n=9}^{\infty} \frac{1}{5^n} = \frac{1/5^9}{1-1/5} = 6.4 \times 10^{-7} \quad \text{(geometric series with } a = \tfrac{1}{5^9} \text{ and } r = \tfrac{1}{5}\text{)}.$$

29. Use the Limit Comparison Test. $\displaystyle\lim_{n\to\infty} \left| \frac{\left(\frac{n+1}{n}\right)a_n}{a_n} \right| = \lim_{n\to\infty} \frac{n+1}{n} = \lim_{n\to\infty} \left(1 + \frac{1}{n} \right) = 1 > 0$. Since $\sum |a_n|$ is

convergent, so is $\displaystyle\sum \left| \left(\frac{n+1}{n} \right) a_n \right|$, by the Limit Comparison Test.

31. $\displaystyle\lim_{n\to\infty} \left| \frac{a_{n+1}}{a_n} \right| = \lim_{n\to\infty} \left[\frac{|x+2|^{n+1}}{(n+1)\,4^{n+1}} \cdot \frac{n4^n}{|x+2|^n} \right] = \lim_{n\to\infty} \left[\frac{n}{n+1} \cdot \frac{|x+2|}{4} \right] = \frac{|x+2|}{4} < 1 \iff |x+2| < 4$,

so $R = 4$. $|x+2| < 4 \iff -4 < x+2 < 4 \iff -6 < x < 2$. If $x = -6$, then the series becomes

$\displaystyle\sum_{n=1}^{\infty} \frac{(-4)^n}{n4^n} = \sum_{n=1}^{\infty} \frac{(-1)^n}{n}$, the alternating harmonic series, which converges by the Alternating Series Test. When

$x = 2$, the series becomes the harmonic series $\displaystyle\sum_{n=1}^{\infty} \frac{1}{n}$, which diverges. Thus, $I = [-6, 2)$.

33. $\displaystyle\lim_{n\to\infty} \left| \frac{a_{n+1}}{a_n} \right| = \lim_{n\to\infty} \left| \frac{2^{n+1}(x-3)^{n+1}}{\sqrt{n+4}} \cdot \frac{\sqrt{n+3}}{2^n(x-3)^n} \right| = 2\,|x-3| \lim_{n\to\infty} \sqrt{\frac{n+3}{n+4}} = 2\,|x-3| < 1 \iff$

$|x-3| < \tfrac{1}{2}$ $\left[\text{so } R = \tfrac{1}{2} \right] \iff -\tfrac{1}{2} < x - 3 < \tfrac{1}{2} \iff \tfrac{5}{2} < x < \tfrac{7}{2}$. When $x = \tfrac{5}{2}$, the series becomes

$\displaystyle\sum_{n=0}^{\infty} \frac{(-1)^n}{\sqrt{n+3}}$, which is a convergent alternating series. When $x = \tfrac{7}{2}$, the series becomes $\displaystyle\sum_{n=0}^{\infty} \frac{1}{\sqrt{n+3}} = \sum_{n=3}^{\infty} \frac{1}{n^{1/2}}$,

which diverges $\left(p = \tfrac{1}{2} \leq 1 \right)$. Thus, $I = \left[\tfrac{5}{2}, \tfrac{7}{2} \right)$.

35.

$f(x) = \sin x$	$f\left(\frac{\pi}{6}\right) = \frac{1}{2}$	$f'''(x) = -\cos x$	$f'''\left(\frac{\pi}{6}\right) = -\frac{\sqrt{3}}{2}$
$f'(x) = \cos x$	$f'\left(\frac{\pi}{6}\right) = \frac{\sqrt{3}}{2}$	$f^{(4)}(x) = \sin x$	$f^{(4)}\left(\frac{\pi}{6}\right) = \frac{1}{2}$
$f''(x) = -\sin x$	$f''\left(\frac{\pi}{6}\right) = -\frac{1}{2}$	$\vdots$	$\vdots$

Note that $f^{(2n)}\left(\frac{\pi}{6}\right) = (-1)^n \cdot \frac{1}{2}$ and $f^{(2n+1)}\left(\frac{\pi}{6}\right) = (-1)^n \cdot \frac{\sqrt{3}}{2}$.

$$\sin x = \sum_{n=0}^{\infty} \frac{f^{(n)}\left(\frac{\pi}{6}\right)}{n!} \left(x - \frac{\pi}{6} \right)^n = \sum_{n=0}^{\infty} \frac{(-1)^n}{2(2n)!} \left(x - \frac{\pi}{6} \right)^{2n} + \sum_{n=0}^{\infty} \frac{(-1)^n \sqrt{3}}{2(2n+1)!} \left(x - \frac{\pi}{6} \right)^{2n+1}$$

$$= \frac{1}{2} \sum_{n=0}^{\infty} (-1)^n \left[\frac{1}{(2n)!} \left(x - \frac{\pi}{6} \right)^{2n} + \frac{\sqrt{3}}{(2n+1)!} \left(x - \frac{\pi}{6} \right)^{2n+1} \right]$$

37. $\dfrac{1}{1+x} = \dfrac{1}{1-(-x)} = \displaystyle\sum_{n=0}^{\infty} (-x)^n = \sum_{n=0}^{\infty} (-1)^n x^n$ for $|x| < 1 \implies \dfrac{x^2}{1+x} = \displaystyle\sum_{n=0}^{\infty} (-1)^n x^{n+2}$ with $R = 1$.

39. $\dfrac{1}{1-x} = \displaystyle\sum_{n=0}^{\infty} x^n$ for $|x| < 1 \implies \ln(1-x) = -\displaystyle\int \frac{dx}{1-x} = -\int \sum_{n=0}^{\infty} x^n\, dx = C - \sum_{n=0}^{\infty} \frac{x^{n+1}}{n+1}$.

$\ln(1-0) = C - 0 \implies C = 0 \implies \ln(1-x) = -\displaystyle\sum_{n=0}^{\infty} \frac{x^{n+1}}{n+1} = -\sum_{n=1}^{\infty} \frac{x^n}{n}$ with $R = 1$.

41. $\sin x = \sum\limits_{n=0}^{\infty} \dfrac{(-1)^n x^{2n+1}}{(2n+1)!} \quad \Rightarrow \quad \sin(x^4) = \sum\limits_{n=0}^{\infty} \dfrac{(-1)^n (x^4)^{2n+1}}{(2n+1)!} = \sum\limits_{n=0}^{\infty} \dfrac{(-1)^n x^{8n+4}}{(2n+1)!}$ for all x, so the radius of convergence is ∞.

43. $f(x) = \dfrac{1}{\sqrt[4]{16-x}} = \dfrac{1}{\sqrt[4]{16(1-x/16)}} = \dfrac{1}{\sqrt[4]{16}\left(1-\frac{1}{16}x\right)^{1/4}} = \frac{1}{2}\left(1 - \frac{1}{16}x\right)^{-1/4}$

$= \dfrac{1}{2}\left[1 + \left(-\frac{1}{4}\right)\left(-\frac{x}{16}\right) + \dfrac{\left(-\frac{1}{4}\right)\left(-\frac{5}{4}\right)}{2!}\left(-\frac{x}{16}\right)^2 + \dfrac{\left(-\frac{1}{4}\right)\left(-\frac{5}{4}\right)\left(-\frac{9}{4}\right)}{3!}\left(-\frac{x}{16}\right)^3 + \cdots\right]$

$= \dfrac{1}{2} + \sum\limits_{n=1}^{\infty} \dfrac{1\cdot5\cdot9\cdots\cdots(4n-3)}{2\cdot4^n\cdot n!\cdot16^n}x^n = \dfrac{1}{2} + \sum\limits_{n=1}^{\infty} \dfrac{1\cdot5\cdot9\cdots\cdots(4n-3)}{2^1\cdot2^{2n}\cdot n!\cdot2^{4n}}x^n$

$= \dfrac{1}{2} + \sum\limits_{n=1}^{\infty} \dfrac{1\cdot5\cdot9\cdots\cdots(4n-3)}{2^{6n+1}\cdot n!}x^n$

for $\left|-\dfrac{x}{16}\right| < 1 \quad \Rightarrow \quad |x| < 16 \quad \Rightarrow \quad R = 16$.

45. $e^x = \sum\limits_{n=0}^{\infty} \dfrac{x^n}{n!}$, so $\dfrac{e^x}{x} = \dfrac{1}{x}\sum\limits_{n=0}^{\infty} \dfrac{x^n}{n!} = \sum\limits_{n=0}^{\infty} \dfrac{x^{n-1}}{n!} = x^{-1} + \sum\limits_{n=1}^{\infty} \dfrac{x^{n-1}}{n!} = \dfrac{1}{x} + \sum\limits_{n=1}^{\infty} \dfrac{x^{n-1}}{n!}$ and

$\displaystyle\int \dfrac{e^x}{x}\,dx = C + \ln|x| + \sum\limits_{n=1}^{\infty} \dfrac{x^n}{n\cdot n!}$.

47. (a)

$f(x) = x^{1/2} \qquad\qquad f(1) = 1 \qquad\qquad f'''(x) = \frac{3}{8}x^{-5/2} \qquad f'''(1) = \frac{3}{8}$

$f'(x) = \frac{1}{2}x^{-1/2} \qquad\quad f'(1) = \frac{1}{2} \qquad\quad f^{(4)}(x) = -\frac{15}{16}x^{-7/2}$

$f''(x) = -\frac{1}{4}x^{-3/2} \qquad f''(1) = -\frac{1}{4}$

$\sqrt{x} \approx T_3(x) = 1 + \dfrac{1/2}{1!}(x-1) - \dfrac{1/4}{2!}(x-1)^2 + \dfrac{3/8}{3!}(x-1)^3$

$= 1 + \frac{1}{2}(x-1) - \frac{1}{8}(x-1)^2 + \frac{1}{16}(x-1)^3$

(b)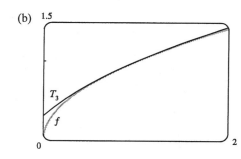

(c) $|R_3(x)| \le \dfrac{M}{4!}|x-1|^4$, where $\left|f^{(4)}(x)\right| \le M$ with

$f^{(4)}(x) = -\frac{15}{16}x^{-7/2}$. Now $0.9 \le x \le 1.1 \quad \Rightarrow$

$-0.1 \le x - 1 \le 0.1 \quad \Rightarrow \quad |x-1| \le 0.1 \quad \Rightarrow$

$(x-1)^4 \le (0.1)^4$, and letting $x = 0.9$ gives

$M = \dfrac{15}{16(0.9)^{7/2}}$, so

$|R_3(x)| \le \dfrac{15}{16(0.9)^{7/2}4!}(0.1)^4 \approx 0.000005648$

$\approx 0.000006 = 6 \times 10^{-6}.$

(d)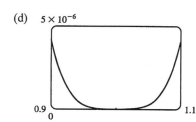

From the graph of $|R_3(x)| = |\sqrt{x} - T_3(x)|$, it appears that the error is less than 4.7×10^{-6} on $[0.9, 1.1]$.

49. $\sin x = \sum\limits_{n=0}^{\infty} (-1)^n \dfrac{x^{2n+1}}{(2n+1)!} = x - \dfrac{x^3}{3!} + \dfrac{x^5}{5!} - \dfrac{x^7}{7!} + \cdots$, so $\sin x - x = -\dfrac{x^3}{3!} + \dfrac{x^5}{5!} - \dfrac{x^7}{7!} + \cdots$ and

$\dfrac{\sin x - x}{x^3} = -\dfrac{1}{3!} + \dfrac{x^2}{5!} - \dfrac{x^4}{7!} + \cdots$ and $\lim\limits_{x \to 0} \dfrac{\sin x - x}{x^3} = \lim\limits_{x \to 0} \left(-\dfrac{1}{6} + \dfrac{x^2}{120} - \dfrac{x^4}{5040} + \cdots \right) = -\dfrac{1}{6}$.

51. Let $y(x) = \sum\limits_{n=0}^{\infty} c_n x^n$. Then $xy' = x \sum\limits_{n=1}^{\infty} n c_n x^{n-1} = \sum\limits_{n=0}^{\infty} n c_n x^n$,

$y''(x) = \sum\limits_{n=2}^{\infty} n(n-1) c_n x^{n-2} = \sum\limits_{n=0}^{\infty} (n+2)(n+1) c_{n+2} x^n$, and the equation $y'' + xy' + y = 0$ becomes

$\sum\limits_{n=0}^{\infty} [(n+2)(n+1)c_{n+2} + nc_n + c_n] x^n = 0$. Thus, the recursion relation is

$c_{n+2} = \dfrac{-nc_n - c_n}{(n+2)(n+1)} = \dfrac{-c_n(n+1)}{(n+2)(n+1)} = -\dfrac{c_n}{n+2}$ for $n = 0, 1, 2, \ldots$. But $c_0 = y(0) = 0$, so $c_{2n} = 0$

for $n = 0, 1, 2, \ldots$. Also, $c_1 = y'(0) = 1$, so $c_3 = -\dfrac{c_1}{3} = -\dfrac{1}{3}$, $c_5 = \dfrac{(-1)^2}{3 \cdot 5} = \dfrac{(-1)^2 2^2 2!}{5!}$,

$c_7 = \dfrac{(-1)^3}{3 \cdot 5 \cdot 7} = \dfrac{(-1)^3 2^3 3!}{7!}, \ldots, c_{2n+1} = \dfrac{(-1)^n 2^n n!}{(2n+1)!} = \dfrac{(-2)^n n!}{(2n+1)!}$ for $n = 0, 1, 2, \ldots$. Note that

$2^n n! = (2 \cdot 1) \cdot (2 \cdot 2) \cdot (2 \cdot 3) \cdot \cdots \cdot (2 \cdot n)$. Thus, the solution to the initial-value problem is

$$y(x) = \sum\limits_{n=0}^{\infty} c_n x^n = \sum\limits_{n=0}^{\infty} \dfrac{(-2)^n n!}{(2n+1)!} x^{2n+1}$$

53. (a) From Formula 14a in Appendix C, with $x = y = \theta$, we get $\tan 2\theta = \dfrac{2\tan\theta}{1 - \tan^2\theta}$, so $\cot 2\theta = \dfrac{1 - \tan^2\theta}{2\tan\theta}$ $\Rightarrow$

$2\cot 2\theta = \dfrac{1 - \tan^2\theta}{\tan\theta} = \cot\theta - \tan\theta$. Replacing θ by $\frac{1}{2}x$, we get $2\cot x = \cot\frac{1}{2}x - \tan\frac{1}{2}x$,

or $\tan\frac{1}{2}x = \cot\frac{1}{2}x - 2\cot x$.

(b) From part (a), $\tan\dfrac{x}{2^n} = \cot\dfrac{x}{2^n} - 2\cot\dfrac{x}{2^{n-1}}$, so the nth partial sum of $\sum\limits_{n=1}^{\infty} \dfrac{1}{2^n}\tan\dfrac{x}{2^n}$ is

$s_n = \dfrac{\tan(x/2)}{2} + \dfrac{\tan(x/4)}{4} + \dfrac{\tan(x/8)}{8} + \cdots + \dfrac{\tan(x/2^n)}{2^n}$

$= \left[\dfrac{\cot(x/2)}{2} - \cot x \right] + \left[\dfrac{\cot(x/4)}{4} - \dfrac{\cot(x/2)}{2} \right] + \left[\dfrac{\cot(x/8)}{8} - \dfrac{\cot(x/4)}{4} \right] + \cdots$

$\qquad + \left[\dfrac{\cot(x/2^n)}{2^n} - \dfrac{\cot(x/2^{n-1})}{2^{n-1}} \right] = -\cot x + \dfrac{\cot(x/2^n)}{2^n}$ [telescoping sum]

Now $\dfrac{\cot(x/2^n)}{2^n} = \dfrac{\cos(x/2^n)}{2^n \sin(x/2^n)} = \dfrac{\cos(x/2^n)}{x} \cdot \dfrac{x/2^n}{\sin(x/2^n)} \to \dfrac{1}{x} \cdot 1 = \dfrac{1}{x}$ as $n \to \infty$ since $x/2^n \to 0$

for $x \neq 0$. Therefore, if $x \neq 0$ and $x \neq n\pi$, then

$\sum\limits_{n=1}^{\infty} \dfrac{1}{2^n}\tan\dfrac{x}{2^n} = \lim\limits_{n \to \infty}\left(-\cot x + \dfrac{1}{2^n}\cot\dfrac{x}{2^n} \right) = -\cot x + \dfrac{1}{x}$.

If $x = 0$, then all terms in the series are 0, so the sum is 0.

Focus on Problem Solving

1. It would be far too much work to compute 15 derivatives of f. The key idea is to remember that $f^{(n)}(0)$ occurs in the coefficient of x^n in the Maclaurin series of f. We start with the Maclaurin series for sin:

$$\sin x = x - \frac{x^3}{3!} + \frac{x^5}{5!} - \cdots. \text{ Then } \sin(x^3) = x^3 - \frac{x^9}{3!} + \frac{x^{15}}{5!} - \cdots \text{ and so the coefficient of } x^{15} \text{ is}$$

$$\frac{f^{(15)}(0)}{15!} = \frac{1}{5!}. \text{ Therefore, } f^{(15)}(0) = \frac{15!}{5!} = 6 \cdot 7 \cdot 8 \cdot 9 \cdot 10 \cdot 11 \cdot 12 \cdot 13 \cdot 14 \cdot 15 = 10{,}897{,}286{,}400.$$

3. (a) Let $a = \arctan x$ and $b = \arctan y$. Then, from Formula 14b in Appendix C,

$$\tan(a - b) = \frac{\tan a - \tan b}{1 + \tan a \tan b} = \frac{\tan(\arctan x) - \tan(\arctan y)}{1 + \tan(\arctan x)\tan(\arctan y)} = \frac{x - y}{1 + xy} \quad \Rightarrow$$

$$\arctan x - \arctan y = a - b = \arctan \frac{x - y}{1 + xy} \quad \text{since} \quad -\frac{\pi}{2} < \arctan x - \arctan y < \frac{\pi}{2}$$

(b) From part (a) we have

$$\arctan \frac{120}{119} - \arctan \frac{1}{239} = \arctan \frac{\frac{120}{119} - \frac{1}{239}}{1 + \frac{120}{119} \cdot \frac{1}{239}} = \arctan \frac{\frac{28{,}561}{28{,}441}}{\frac{28{,}561}{28{,}441}} = \arctan 1 = \frac{\pi}{4}$$

(c) Replacing y by $-y$ in the formula of part (a), we get $\arctan x + \arctan y = \arctan \dfrac{x + y}{1 - xy}$. So

$$4 \arctan \frac{1}{5} = 2\left(\arctan \frac{1}{5} + \arctan \frac{1}{5}\right) = 2 \arctan \frac{\frac{1}{5} + \frac{1}{5}}{1 - \frac{1}{5} \cdot \frac{1}{5}} = 2 \arctan \frac{5}{12} = \arctan \frac{5}{12} + \arctan \frac{5}{12}$$

$$= \arctan \frac{\frac{5}{12} + \frac{5}{12}}{1 - \frac{5}{12} \cdot \frac{5}{12}} = \arctan \frac{120}{119}$$

Thus, from part (b), we have $4 \arctan \frac{1}{5} - \arctan \frac{1}{239} = \arctan \frac{120}{119} - \arctan \frac{1}{239} = \frac{\pi}{4}$.

(d) From Example 7 in Section 8.6 we have $\arctan x = x - \dfrac{x^3}{3} + \dfrac{x^5}{5} - \dfrac{x^7}{7} + \dfrac{x^9}{9} - \dfrac{x^{11}}{11} + \cdots$, so

$$\arctan \frac{1}{5} = \frac{1}{5} - \frac{1}{3 \cdot 5^3} + \frac{1}{5 \cdot 5^5} - \frac{1}{7 \cdot 5^7} + \frac{1}{9 \cdot 5^9} - \frac{1}{11 \cdot 5^{11}} + \cdots$$

This is an alternating series and the size of the terms decreases to 0, so by the Alternating Series Estimation Theorem, the sum lies between s_5 and s_6, that is, $0.197395560 < \arctan \frac{1}{5} < 0.197395562$.

(e) From the series in part (d) we get $\arctan \dfrac{1}{239} = \dfrac{1}{239} - \dfrac{1}{3 \cdot 239^3} + \dfrac{1}{5 \cdot 239^5} - \cdots$. The third term is less than 2.6×10^{-13}, so by the Alternating Series Estimation Theorem, we have, to nine decimal places, $\arctan \frac{1}{239} \approx s_2 \approx 0.004184076$. Thus, $0.004184075 < \arctan \frac{1}{239} < 0.004184077$.

(f) From part (c) we have $\pi = 16 \arctan \frac{1}{5} - 4 \arctan \frac{1}{239}$, so from parts (d) and (e) we have

$16(0.197395560) - 4(0.004184077) < \pi < 16(0.197395562) - 4(0.004184075) \quad \Rightarrow$
$3.141592652 < \pi < 3.141592692$. So, to 7 decimal places, $\pi \approx 3.1415927$.

5. (a) At each stage, each side is replaced by four shorter sides, each of
length $\frac{1}{3}$ of the side length at the preceding stage. Writing s_0 and ℓ_0
for the number of sides and the length of the side of the initial
triangle, we generate the table at right. In general, we have
$s_n = 3 \cdot 4^n$ and $\ell_n = \left(\frac{1}{3}\right)^n$, so the length of the perimeter at the nth
stage of construction is $p_n = s_n \ell_n = 3 \cdot 4^n \cdot \left(\frac{1}{3}\right)^n = 3 \cdot \left(\frac{4}{3}\right)^n$.

$s_0 = 3$	$\ell_0 = 1$
$s_1 = 3 \cdot 4$	$\ell_1 = 1/3$
$s_2 = 3 \cdot 4^2$	$\ell_2 = 1/3^2$
$s_3 = 3 \cdot 4^3$	$\ell_3 = 1/3^3$
$\vdots$	$\vdots$

(b) $p_n = \dfrac{4^n}{3^{n-1}} = 4\left(\dfrac{4}{3}\right)^{n-1}$. Since $\frac{4}{3} > 1$, $p_n \to \infty$ as $n \to \infty$.

(c) The area of each of the small triangles added at a given stage is one-ninth of the area of the triangle added at the
preceding stage. Let a be the area of the original triangle. Then the area a_n of each of the small triangles added
at stage n is $a_n = a \cdot \dfrac{1}{9^n} = \dfrac{a}{9^n}$. Since a small triangle is added to each side at every
stage, it follows that the total area A_n added to the figure at the nth stage is

$A_n = s_{n-1} \cdot a_n = 3 \cdot 4^{n-1} \cdot \dfrac{a}{9^n} = a \cdot \dfrac{4^{n-1}}{3^{2n-1}}$. Then the total area enclosed by the snowflake curve is

$A = a + A_1 + A_2 + A_3 + \cdots = a + a \cdot \dfrac{1}{3} + a \cdot \dfrac{4}{3^3} + a \cdot \dfrac{4^2}{3^5} + a \cdot \dfrac{4^3}{3^7} + \cdots$. After the first term, this is a

geometric series with common ratio $\frac{4}{9}$, so $A = a + \dfrac{a/3}{1 - \frac{4}{9}} = a + \dfrac{a}{3} \cdot \dfrac{9}{5} = \dfrac{8a}{5}$. But the area of the original

equilateral triangle with side 1 is $a = \frac{1}{2} \cdot 1 \cdot \sin\frac{\pi}{3} = \frac{\sqrt{3}}{4}$. So the area enclosed by the snowflake curve is
$\frac{8}{5} \cdot \frac{\sqrt{3}}{4} = \frac{2\sqrt{3}}{5}$.

7. We start with the geometric series $\sum\limits_{n=0}^{\infty} x^n = \dfrac{1}{1-x}$, $|x| < 1$, and differentiate:

$\sum\limits_{n=1}^{\infty} nx^{n-1} = \dfrac{d}{dx}\left(\sum\limits_{n=0}^{\infty} x^n\right) = \dfrac{d}{dx}\left(\dfrac{1}{1-x}\right) = \dfrac{1}{(1-x)^2}$ for $|x| < 1$ $\Rightarrow$

$\sum\limits_{n=1}^{\infty} nx^n = x \sum\limits_{n=1}^{\infty} nx^{n-1} = \dfrac{x}{(1-x)^2}$ for $|x| < 1$. Differentiate again:

$\sum\limits_{n=1}^{\infty} n^2 x^{n-1} = \dfrac{d}{dx}\dfrac{x}{(1-x)^2} = \dfrac{(1-x)^2 - x \cdot 2(1-x)(-1)}{(1-x)^4} = \dfrac{x+1}{(1-x)^3}$ $\Rightarrow$ $\sum\limits_{n=1}^{\infty} n^2 x^n = \dfrac{x^2+x}{(1-x)^3}$ $\Rightarrow$

$\sum\limits_{n=1}^{\infty} n^3 x^{n-1} = \dfrac{d}{dx}\dfrac{x^2+x}{(1-x)^3} = \dfrac{(1-x)^3(2x+1) - (x^2+x)3(1-x)^2(-1)}{(1-x)^6} = \dfrac{x^2+4x+1}{(1-x)^4}$ $\Rightarrow$

$\sum\limits_{n=1}^{\infty} n^3 x^n = \dfrac{x^3+4x^2+x}{(1-x)^4}$, $|x| < 1$. The radius of convergence is 1 because that is the radius of convergence for

the geometric series we started with. If $x = \pm 1$, the series is $\sum n^3(\pm 1)^n$, which diverges by the Test For

Divergence, so the interval of convergence is $(-1, 1)$.

9. $u = 1 + \dfrac{x^3}{3!} + \dfrac{x^6}{6!} + \dfrac{x^9}{9!} + \cdots$, $v = x + \dfrac{x^4}{4!} + \dfrac{x^7}{7!} + \dfrac{x^{10}}{10!} + \cdots$, $w = \dfrac{x^2}{2!} + \dfrac{x^5}{5!} + \dfrac{x^8}{8!} + \cdots$. The key idea is to

differentiate: $\dfrac{du}{dx} = \dfrac{3x^2}{3!} + \dfrac{6x^5}{6!} + \dfrac{9x^8}{9!} + \cdots = \dfrac{x^2}{2!} + \dfrac{x^5}{5!} + \dfrac{x^8}{8!} + \cdots = w$. Similarly,

$\dfrac{dv}{dx} = 1 + \dfrac{x^3}{3!} + \dfrac{x^6}{6!} + \dfrac{x^9}{9!} + \cdots = u$, and $\dfrac{dw}{dx} = x + \dfrac{x^4}{4!} + \dfrac{x^7}{7!} + \dfrac{x^{10}}{10!} + \cdots = v$. So $u' = w$, $v' = u$, and $w' = v$.

Now differentiate the left hand side of the desired equation:

$$\frac{d}{dx}\left(u^3 + v^3 + w^3 - 3uvw\right) = 3u^2 u' + 3v^2 v' + 3w^2 w' - 3\left(u'vw + uv'w + uvw'\right)$$

$$= 3u^2 w + 3v^2 u + 3w^2 v - 3\left(vw^2 + u^2 w + uv^2\right) = 0 \quad \Rightarrow$$

$u^3 + v^3 + w^3 - 3uvw = C$. To find the value of the constant C, we put $x = 0$ in the last equation and get

$1^3 + 0^3 + 0^3 - 3\left(1 \cdot 0 \cdot 0\right) = C \quad \Rightarrow \quad C = 1$, so $u^3 + v^3 + w^3 - 3uvw = 1$.

11. If L is the length of a side of the equilateral triangle, then the area is $A = \frac{1}{2} L \cdot \frac{\sqrt{3}}{2} L = \frac{\sqrt{3}}{4} L^2$ and so $L^2 = \frac{4}{\sqrt{3}} A$.

Let r be the radius of one of the circles. When there are n rows of circles, the figure shows that

$L = \sqrt{3}r + r + (n-2)(2r) + r + \sqrt{3}r = r\left(2n - 2 + 2\sqrt{3}\right)$, so $r = \dfrac{L}{2\left(n + \sqrt{3} - 1\right)}$. The number of circles is

$1 + 2 + \cdots + n = \dfrac{n(n+1)}{2}$ and so the total area of the circles is

$A_n = \dfrac{n(n+1)}{2}\pi r^2 = \dfrac{n(n+1)}{2}\pi \dfrac{L^2}{4\left(n + \sqrt{3} - 1\right)^2} = \dfrac{n(n+1)}{2}\pi \dfrac{4A/\sqrt{3}}{4\left(n + \sqrt{3} - 1\right)^2}$

$= \dfrac{n(n+1)}{\left(n + \sqrt{3} - 1\right)^2}\dfrac{\pi A}{2\sqrt{3}} \quad \Rightarrow$

$\dfrac{A_n}{A} = \dfrac{n(n+1)}{\left(n + \sqrt{3} - 1\right)^2}\dfrac{\pi}{2\sqrt{3}}$

$= \dfrac{1 + 1/n}{\left[1 + \left(\sqrt{3} - 1\right)/n\right]^2}\dfrac{\pi}{2\sqrt{3}} \to \dfrac{\pi}{2\sqrt{3}}$ as $n \to \infty$

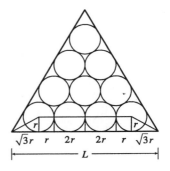

13. Call the series S. We group the terms according to the number of digits in their denominators:

$$S = \underbrace{\left(1 + \tfrac{1}{2} + \cdots + \tfrac{1}{8} + \tfrac{1}{9}\right)}_{g_1} + \underbrace{\left(\tfrac{1}{11} + \cdots + \tfrac{1}{99}\right)}_{g_2} + \underbrace{\left(\tfrac{1}{111} + \cdots + \tfrac{1}{999}\right)}_{g_3} + \cdots$$

Now in the group g_n, there are 9^n terms, since we have 9 choices for each of the n digits in the denominator. Furthermore, each term in g_n is less than $\frac{1}{10^{n-1}}$. So $g_n < 9^n \cdot \frac{1}{10^{n-1}} = 9\left(\frac{9}{10}\right)^{n-1}$. Now $\sum_{n=1}^{\infty} 9\left(\frac{9}{10}\right)^{n-1}$ is a geometric series with $a = 9$ and $r = \frac{9}{10} < 1$. Therefore, by the Comparison Test,

$$S = \sum_{n=1}^{\infty} g_n < \sum_{n=1}^{\infty} 9\left(\frac{9}{10}\right)^{n-1} = \frac{9}{1 - 9/10} = 90.$$

15. Let $f(x) = \sum_{m=0}^{\infty} c_m x^m$ and $g(x) = e^{f(x)} = \sum_{n=0}^{\infty} d_n x^n$. Then $g'(x) = \sum_{n=0}^{\infty} n d_n x^{n-1}$, so $n d_n$ occurs as the coefficient of x^{n-1}. But also

$$g'(x) = e^{f(x)} f'(x) = \left(\sum_{n=0}^{\infty} d_n x^n\right)\left(\sum_{m=1}^{\infty} m c_m x^{m-1}\right)$$

$$= \left(d_0 + d_1 x + d_2 x^2 + \cdots + d_{n-1} x^{n-1} + \cdots\right)\left(c_1 + 2c_2 x + 3c_3 x^2 + \cdots + n c_n x^{n-1} + \cdots\right)$$

so the coefficient of x^{n-1} is $c_1 d_{n-1} + 2c_2 d_{n-2} + 3c_3 d_{n-3} + \cdots + n c_n d_0 = \sum_{i=1}^{n} i c_i d_{n-i}$. Therefore, $n d_n = \sum_{i=1}^{n} i c_i d_{n-i}$.

Appendixes

 A **Intervals, Inequalities, and Absolute Values** · · · · · · ·

1. $|5 - 23| = |-18| = 18$

3. $\left|\sqrt{5} - 5\right| = -\left(\sqrt{5} - 5\right) = 5 - \sqrt{5}$ because $\sqrt{5} - 5 < 0$.

5. For $x < 2$, $x - 2 < 0$, so $|x - 2| = -(x - 2) = 2 - x$.

7. $|x + 1| = \begin{cases} x + 1 & \text{for } x + 1 \geq 0 \iff x \geq -1 \\ -(x + 1) & \text{for } x + 1 < 0 \iff x < -1 \end{cases}$

9. $\left|x^2 + 1\right| = x^2 + 1$ (since $x^2 + 1 \geq 0$ for all x).

11. $2x + 7 > 3 \iff 2x > -4 \iff x > -2$, so $x \in (-2, \infty)$.

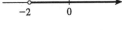

13. $1 - x \leq 2 \iff -x \leq 1 \iff x \geq -1$, so $x \in [-1, \infty)$.

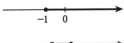

15. $0 \leq 1 - x < 1 \iff -1 \leq -x < 0 \iff 1 \geq x > 0$, so $x \in (0, 1]$.

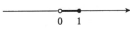

17. $(x - 1)(x - 2) > 0$.   *Case 1:* (both factors are positive, so their product is positive)
$$x - 1 > 0 \iff x > 1, \text{ and } x - 2 > 0 \iff x > 2, \text{ so } x \in (2, \infty).$$
   Case 2: (both factors are negative, so their product is positive)
$$x - 1 < 0 \iff x < 1, \text{ and } x - 2 < 0 \iff x < 2, \text{ so } x \in (-\infty, 1).$$
Thus, the solution set is $(-\infty, 1) \cup (2, \infty)$.

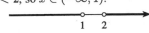

19. $x^2 < 3 \iff x^2 - 3 < 0 \iff \left(x - \sqrt{3}\right)\left(x + \sqrt{3}\right) < 0$. *Case 1:* $x > \sqrt{3}$ and $x < -\sqrt{3}$, which is impossible.
Case 2: $x < \sqrt{3}$ and $x > -\sqrt{3}$. Thus, the solution set is $\left(-\sqrt{3}, \sqrt{3}\right)$.
Another method: $x^2 < 3 \iff |x| < \sqrt{3} \iff -\sqrt{3} < x < \sqrt{3}$.

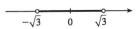

21. $x^3 - x^2 \leq 0 \iff x^2(x - 1) \leq 0$. Since $x^2 \geq 0$ for all x, the inequality is satisfied when $x - 1 \leq 0 \iff x \leq 1$. Thus, the solution set is $(-\infty, 1]$.

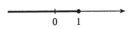

23. $x^3 > x \iff x^3 - x > 0 \iff x(x^2 - 1) > 0 \iff x(x-1)(x+1) > 0$. Constructing a table:

Interval	x	$x-1$	$x+1$	$x(x-1)(x+1)$
$x < -1$	$-$	$-$	$-$	$-$
$-1 < x < 0$	$-$	$-$	$+$	$+$
$0 < x < 1$	$+$	$-$	$+$	$-$
$x > 1$	$+$	$+$	$+$	$+$

Since $x^3 > x$ when the last column is positive, the solution set is $(-1, 0) \cup (1, \infty)$.

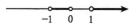

$$-1 \quad 0 \quad 1$$

25. $1/x < 4$. This is clearly true for $x < 0$. So suppose $x > 0$. then $1/x < 4 \iff 1 < 4x \iff \frac{1}{4} < x$. Thus, the solution set is $(-\infty, 0) \cup \left(\frac{1}{4}, \infty\right)$.

$$0 \quad \frac{1}{4}$$

27. $C = \frac{5}{9}(F - 32) \implies F = \frac{9}{5}C + 32$. So $50 \le F \le 95 \implies 50 \le \frac{9}{5}C + 32 \le 95 \implies 18 \le \frac{9}{5}C \le 63 \implies 10 \le C \le 35$. So the interval is $[10, 35]$.

29. (a) Let T represent the temperature in degrees Celsius and h the height in km. $T = 20$ when $h = 0$ and T decreases by $10\,^\circ$C for every km ($1\,^\circ$C for each 100-m rise). Thus, $T = 20 - 10h$ when $0 \le h \le 12$.

(b) From part (a), $T = 20 - 10h \implies 10h = 20 - T \implies h = 2 - T/10$. So $0 \le h \le 5 \implies 0 \le 2 - T/10 \le 5 \implies -2 \le -T/10 \le 3 \implies -20 \le -T \le 30 \implies 20 \ge T \ge -30 \implies -30 \le T \le 20$. Thus, the range of temperatures (in $^\circ$C) to be expected is $[-30, 20]$.

31. $|x + 3| = |2x + 1| \iff$ either $x + 3 = 2x + 1$ or $x + 3 = -(2x + 1)$. In the first case, $x = 2$, and in the second case, $x + 3 = -2x - 1 \iff 3x = -4 \iff x = -\frac{4}{3}$. So the solutions are $-\frac{4}{3}$ and 2.

33. By Property 5 of absolute values, $|x| < 3 \iff -3 < x < 3$, so $x \in (-3, 3)$.

35. $|x - 4| < 1 \iff -1 < x - 4 < 1 \iff 3 < x < 5$, so $x \in (3, 5)$.

37. $|x + 5| \ge 2 \iff x + 5 \ge 2$ or $x + 5 \le -2 \iff x \ge -3$ or $x \le -7$, so $x \in (-\infty, -7] \cup [-3, \infty)$.

39. $|2x - 3| \le 0.4 \iff -0.4 \le 2x - 3 \le 0.4 \iff 2.6 \le 2x \le 3.4 \iff 1.3 \le x \le 1.7$, so $x \in [1.3, 1.7]$.

41. $a(bx - c) \ge bc \iff bx - c \ge \dfrac{bc}{a} \iff bx \ge \dfrac{bc}{a} + c = \dfrac{bc + ac}{a} \iff x \ge \dfrac{bc + ac}{ab}$

43. $|ab| = \sqrt{(ab)^2} = \sqrt{a^2 b^2} = \sqrt{a^2}\sqrt{b^2} = |a|\,|b|$

◆ B ◆ Coordinate Geometry · · · · · · · · · · · · · · ·

1. From the Distance Formula with $x_1 = 1$, $x_2 = 4$, $y_1 = 1$, $y_2 = 5$, we find the distance from $(1, 1)$ to $(4, 5)$ to be $\sqrt{(4-1)^2 + (5-1)^2} = \sqrt{3^2 + 4^2} = \sqrt{25} = 5$.

3. With $P(-3, 3)$ and $Q(-1, -6)$, the slope m of the line through P and Q is $m = \dfrac{-6 - 3}{-1 - (-3)} = -\dfrac{9}{2}$.

5. Using $A(-2, 9)$, $B(4, 6)$, $C(1, 0)$, and $D(-5, 3)$, we have

$|AB| = \sqrt{[4-(-2)]^2 + (6-9)^2} = \sqrt{6^2 + (-3)^2} = \sqrt{45} = \sqrt{9}\sqrt{5} = 3\sqrt{5}$,

$|BC| = \sqrt{(1-4)^2 + (0-6)^2} = \sqrt{(-3)^2 + (-6)^2} = \sqrt{45} = \sqrt{9}\sqrt{5} = 3\sqrt{5}$,

$|CD| = \sqrt{(-5-1)^2 + (3-0)^2} = \sqrt{(-6)^2 + 3^2} = \sqrt{45} = \sqrt{9}\sqrt{5} = 3\sqrt{5}$, and

$|DA| = \sqrt{[-2-(-5)]^2 + (9-3)^2} = \sqrt{3^2 + 6^2} = \sqrt{45} = \sqrt{9}\sqrt{5} = 3\sqrt{5}$. So all sides are of equal length and

we have a rhombus. Moreover, $m_{AB} = \dfrac{6-9}{4-(-2)} = -\dfrac{1}{2}$, $m_{BC} = \dfrac{0-6}{1-4} = 2$, $m_{CD} = \dfrac{3-0}{-5-1} = -\dfrac{1}{2}$, and

$m_{DA} = \dfrac{9-3}{-2-(-5)} = 2$, so the sides are perpendicular. Thus, A, B, C, and D are vertices of a square.

7. $x = 3$

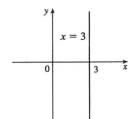

9. $xy = 0 \iff x = 0$ or $y = 0$. The graph consists of the coordinate axes.

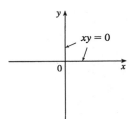

11. By the point-slope form of the equation of a line, an equation of the line through $(2, -3)$ with slope 6 is
$y - (-3) = 6(x-2)$ or $y = 6x - 15$.

13. The slope of the line through $(2, 1)$ and $(1, 6)$ is $m = \dfrac{6-1}{1-2} = -5$, so an equation of the line is
$y - 1 = -5(x-2)$ or $y = -5x + 11$.

15. By the slope-intercept form of the equation of a line, an equation of the line is $y = 3x - 2$.

17. Since the line passes through $(1, 0)$ and $(0, -3)$, its slope is $m = \dfrac{-3-0}{0-1} = 3$, so an equation is $y = 3x - 3$.

Another method: From Exercise 46, $\dfrac{x}{1} + \dfrac{y}{-3} = 1 \implies -3x + y = -3 \implies y = 3x - 3$.

19. Since $m = 0$, $y - 5 = 0(x-4)$ or $y = 5$.

21. Putting the line $x + 2y = 6$ into its slope-intercept form gives us $y = -\frac{1}{2}x + 3$, so we see that this line has
slope $-\frac{1}{2}$. Thus, we want the line of slope $-\frac{1}{2}$ that passes through the point $(1, -6)$: $y - (-6) = -\frac{1}{2}(x-1) \iff$
$y = -\frac{1}{2}x - \frac{11}{2}$.

23. $2x + 5y + 8 = 0 \iff y = -\frac{2}{5}x - \frac{8}{5}$. Since this line has slope $-\frac{2}{5}$, a line perpendicular to it would have slope $\frac{5}{2}$,
so the required line is $y - (-2) = \frac{5}{2}[x - (-1)] \iff y = \frac{5}{2}x + \frac{1}{2}$.

25. $x + 3y = 0 \iff y = -\frac{1}{3}x$, so the slope is $-\frac{1}{3}$ and the y-intercept is 0.

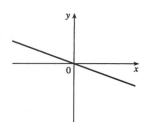

27. $3x - 4y = 12 \iff y = \frac{3}{4}x - 3$, so the slope is $\frac{3}{4}$ and the y-intercept is -3.

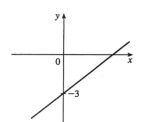

29. $\{(x, y) \mid x < 0\}$

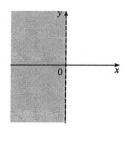

31. $\left\{(x, y) \,\middle|\, |x| \le 2\right\} = \{(x, y) \mid -2 \le x \le 2\}$

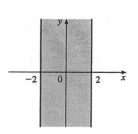

33. $\{(x, y) \mid 0 \le y \le 4, x \le 2\}$

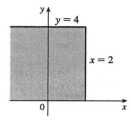

35. $\{(x, y) \mid 1 + x \le y \le 1 - 2x\}$

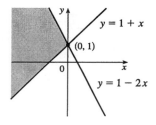

37. An equation of the circle with center $(3, -1)$ and radius 5 is $(x - 3)^2 + (y + 1)^2 = 5^2 = 25$.

39. $x^2 + y^2 - 4x + 10y + 13 = 0 \iff x^2 - 4x + y^2 + 10y = -13 \iff$
$(x^2 - 4x + 4) + (y^2 + 10y + 25) = -13 + 4 + 25 = 16 \iff (x - 2)^2 + (y + 5)^2 = 4^2$. Thus, we have a circle with center $(2, -5)$ and radius 4.

41. $2x - y = 4 \iff y = 2x - 4 \implies m_1 = 2$ and $6x - 2y = 10 \iff 2y = 6x - 10 \iff y = 3x - 5 \implies m_2 = 3$. Since $m_1 \ne m_2$, the two lines are not parallel. To find the point of intersection: $2x - 4 = 3x - 5 \iff x = 1 \implies y = -2$. Thus, the point of intersection is $(1, -2)$.

43. Let M be the point $\left(\dfrac{x_1 + x_2}{2}, \dfrac{y_1 + y_2}{2}\right)$. Then

$$|MP_1|^2 = \left(x_1 - \frac{x_1 + x_2}{2}\right)^2 + \left(y_1 - \frac{y_1 + y_2}{2}\right)^2 = \left(\frac{x_1 - x_2}{2}\right)^2 + \left(\frac{y_1 - y_2}{2}\right)^2 \text{ and}$$

$$|MP_2|^2 = \left(x_2 - \frac{x_1 + x_2}{2}\right)^2 + \left(y_2 - \frac{y_1 + y_2}{2}\right)^2 = \left(\frac{x_2 - x_1}{2}\right)^2 + \left(\frac{y_2 - y_1}{2}\right)^2. \text{ Hence, } |MP_1| = |MP_2|; \text{ that}$$

is, M is equidistant from P_1 and P_2.

45. With $A(1, 4)$ and $B(7, -2)$, the slope of segment AB is $\frac{-2-4}{7-1} = -1$, so its perpendicular bisector has slope 1. The midpoint of AB is $\left(\frac{1+7}{2}, \frac{4+(-2)}{2}\right) = (4, 1)$, so an equation of the perpendicular bisector is $y - 1 = 1(x - 4)$ or $y = x - 3$.

47. If $P(x, y)$ is any point on the parabola, then the distance from P to the focus is $|PF| = \sqrt{x^2 + (y - p)^2}$ and the distance from P to the directrix is $|y + p|$. (Figure 14 in the text illustrates the case where $p > 0$.) The defining property of a parabola is that these distances are equal: $\sqrt{x^2 + (y - p)^2} = |y + p|$. We get an equivalent equation by squaring and simplifying: $x^2 + (y - p)^2 = |y + p|^2 = (y + p)^2 \iff x^2 + y^2 - 2py + p^2 = y^2 + 2py + p^2 \iff x^2 = 4py$. Thus, an equation of a parabola with focus $(0, p)$ and directrix $y = -p$ is $x^2 = 4py$.

49. See Figure 20 in the text. $P(x, y)$ is a point on the ellipse when $|PF_1| + |PF_2| = 2a$; that is, $\sqrt{(x + c)^2 + y^2} + \sqrt{(x - c)^2 + y^2} = 2a$ or $\sqrt{(x + c)^2 + y^2} = 2a - \sqrt{(x + c)^2 + y^2}$. Squaring both sides, we have $x^2 - 2cx + c^2 + y^2 = 4a^2 - 4a\sqrt{(x + c)^2 + y^2} + x^2 + 2cx + c^2 + y^2$, which simplifies to $a\sqrt{(x + c)^2 + y^2} = a^2 + cx$. We square again: $a^2 \left(x^2 + 2cx + c^2 + y^2\right) = a^4 + 2a^2 cx + c^2 x^2$, which becomes $(a^2 - c^2)x^2 + a^2 y^2 = a^2(a^2 - c^2)$. From triangle $F_1 F_2 P$ in Figure 20, we see that $2c < 2a$, so $c < a$ and, therefore, $a^2 - c^2 > 0$. For convenience, let $b^2 = a^2 - c^2$. Then the equation of the ellipse becomes $b^2 x^2 + a^2 y^2 = a^2 b^2$ or, if both sides are divided by $a^2 b^2$, $\dfrac{x^2}{a^2} + \dfrac{y^2}{b^2} = 1$.

51. From Figure 23 in the text, $|PF_1| - |PF_2| = \pm 2a \iff \sqrt{(x + c)^2 + y^2} - \sqrt{(x - c)^2 + y^2} = \pm 2a \iff$
$\sqrt{(x + c)^2 + y^2} = \sqrt{(x - c)^2 + y^2} \pm 2a \iff$
$(x + c)^2 + y^2 = (x - c)^2 + y^2 + 4a^2 \pm 4a\sqrt{(x - c)^2 + y^2} \iff 4cx - 4a^2 = \pm 4a\sqrt{(x - c)^2 + y^2} \iff$
$c^2 x^2 - 2a^2 cx + a^4 = a^2 \left(x^2 - 2cx + c^2 + y^2\right) \iff (c^2 - a^2)x^2 - a^2 y^2 = a^2(c^2 - a^2) \iff$
$b^2 x^2 - a^2 y^2 = a^2 b^2$ [where $b^2 = c^2 - a^2$] $\iff \dfrac{x^2}{a^2} - \dfrac{y^2}{b^2} = 1$.

53.

$x + 4y = 8$, $x = 2y^2 - 8$. Substitute x from the second equation into the first: $(2y^2 - 8) + 4y = 8 \iff 2y^2 + 4y - 16 = 0$
$\iff y^2 + 2y - 8 = 0 \iff (y + 4)(y - 2) = 0 \iff y = -4$ or 2. So the points of intersection are $(24, -4)$ and $(0, 2)$.

55. Differentiating implicitly, $\dfrac{x^2}{a^2} + \dfrac{y^2}{b^2} = 1 \;\Rightarrow\; \dfrac{2x}{a^2} + \dfrac{2yy'}{b^2} = 0 \;\Rightarrow\; y' = -\dfrac{b^2 x}{a^2 y}$ $(y \neq 0)$. Thus, the slope of

the tangent line at P is $-\dfrac{b^2 x_1}{a^2 y_1}$. The slope of $F_1 P$ is $\dfrac{y_1}{x_1 + c}$ and of $F_2 P$ is $\dfrac{y_1}{x_1 - c}$. By the formula from Focus on Problem Solving 3 (Problem 15), we have

$$
\begin{aligned}
\tan \alpha &= \frac{\dfrac{y_1}{x_1 + c} + \dfrac{b^2 x_1}{a^2 y_1}}{1 - \dfrac{b^2 x_1 y_1}{a^2 y_1 (x_1 + c)}} = \frac{a^2 y_1^2 + b^2 x_1 (x_1 + c)}{a^2 y_1 (x_1 + c) - b^2 x_1 y_1} \\[2mm]
&= \frac{a^2 b^2 + b^2 c x_1}{c^2 x_1 y_1 + a^2 c y_1} \quad \left[\begin{matrix}\text{using } b^2 x_1^2 + a^2 y_1^2 = a^2 b^2 \\ \text{and } a^2 - b^2 = c^2\end{matrix}\right] \\[2mm]
&= \frac{b^2 (c x_1 + a^2)}{c y_1 (c x_1 + a^2)} \\[2mm]
&= \frac{b^2}{c y_1}
\end{aligned}
$$

and

$$
\begin{aligned}
\tan \beta &= \frac{-\dfrac{b^2 x_1}{a^2 y_1} - \dfrac{y_1}{x_1 - c}}{1 - \dfrac{b^2 x_1 y_1}{a^2 y_1 (x_1 - c)}} = \frac{-a^2 y_1^2 - b^2 x_1 (x_1 - c)}{a^2 y_1 (x_1 - c) - b^2 x_1 y_1} \\[2mm]
&= \frac{-a^2 b^2 + b^2 c x_1}{c^2 x_1 y_1 - a^2 c y_1} = \frac{b^2 (c x_1 - a^2)}{c y_1 (c x_1 - a^2)} \\[2mm]
&= \frac{b^2}{c y_1}
\end{aligned}
$$

Thus, $\alpha = \beta$.

 Trigonometry • • • • • • • • • • • • • •

1. (a) $210° = 210\left(\frac{\pi}{180}\right) = \frac{7\pi}{6}$ rad $\qquad$ (b) $9° = 9\left(\frac{\pi}{180}\right) = \frac{\pi}{20}$ rad

3. (a) 4π rad $= 4\pi\left(\frac{180}{\pi}\right) = 720°$ $\qquad\qquad$ (b) $-\frac{3\pi}{8}$ rad $= -\frac{3\pi}{8}\left(\frac{180}{\pi}\right) = -67.5°$

5. Using Formula 3, $a = r\theta = 36 \cdot \frac{\pi}{12} = 3\pi$ cm.

7. Using Formula 3, $\theta = a/r = \frac{1}{1.5} = \frac{2}{3}$ rad $= \frac{2}{3}\left(\frac{180}{\pi}\right) = \left(\frac{120}{\pi}\right)° \approx 38.2°$.

9. (a) $\qquad\qquad\qquad\qquad\qquad\qquad\qquad$ (b)

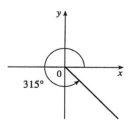

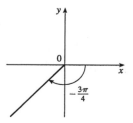

11.

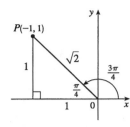

From the diagram we see that a point on the terminal side is $P(-1, 1)$. Therefore, taking $x = -1$, $y = 1$, $r = \sqrt{2}$ in the definitions of the trigonometric ratios, we have $\sin \frac{3\pi}{4} = \frac{1}{\sqrt{2}}$, $\cos \frac{3\pi}{4} = -\frac{1}{\sqrt{2}}$, $\tan \frac{3\pi}{4} = -1$, $\csc \frac{3\pi}{4} = \sqrt{2}$, $\sec \frac{3\pi}{4} = -\sqrt{2}$, and $\cot \frac{3\pi}{4} = -1$.

13. $\sin \theta = y/r = \frac{3}{5}$ $\Rightarrow$ $y = 3$, $r = 5$, and $x = \sqrt{r^2 - y^2} = 4$ (since $0 < \theta < \frac{\pi}{2}$). Therefore taking $x = 4$, $y = 3$, $r = 5$ in the definitions of the trigonometric ratios, we have $\cos \theta = \frac{4}{5}$, $\tan \theta = \frac{3}{4}$, $\csc \theta = \frac{5}{3}$, $\sec \theta = \frac{5}{4}$, and $\cot \theta = \frac{4}{3}$.

15. $\sin 35° = \dfrac{x}{10}$ $\Rightarrow$ $x = 10 \sin 35° \approx 5.73576$ cm

17. $\tan \frac{2\pi}{5} = \dfrac{x}{8}$ $\Rightarrow$ $x = 8 \tan \frac{2\pi}{5} \approx 24.62147$ cm

19.

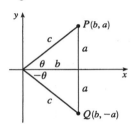

(a) From the diagram we see that $\sin \theta = \dfrac{y}{r} = \dfrac{a}{c}$, and
$$\sin(-\theta) = \dfrac{-a}{c} = -\dfrac{a}{c} = -\sin \theta.$$

(b) Again from the diagram we see that $\cos \theta = \dfrac{x}{r} = \dfrac{b}{c} = \cos(-\theta)$.

21. Using (12a), we have $\sin\left(\frac{\pi}{2} + x\right) = \sin \frac{\pi}{2} \cos x + \cos \frac{\pi}{2} \sin x = 1 \cdot \cos x + 0 \cdot \sin x = \cos x$.

23. Using (6), we have $\sin \theta \cot \theta = \sin \theta \cdot \dfrac{\cos \theta}{\sin \theta} = \cos \theta$.

25. Using (14a), we have $\tan 2\theta = \tan(\theta + \theta) = \dfrac{\tan \theta + \tan \theta}{1 - \tan \theta \tan \theta} = \dfrac{2 \tan \theta}{1 - \tan^2 \theta}$.

27. Since $\sin x = \frac{1}{3}$ we can label the opposite side as having length 1, the hypotenuse as having length 3, and use the Pythagorean Theorem to get that the adjacent side has length $\sqrt{8}$. Then, from the diagram, $\cos x = \frac{\sqrt{8}}{3}$. Similarly we have that $\sin y = \frac{3}{5}$. Now use (12a):

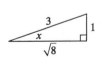

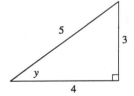

$\sin(x + y) = \sin x \cos y + \cos x \sin y = \frac{1}{3} \cdot \frac{4}{5} + \frac{\sqrt{8}}{3} \cdot \frac{3}{5} = \frac{4}{15} + \frac{3\sqrt{8}}{15} = \frac{4 + 6\sqrt{2}}{15}$.

29. $2 \cos x - 1 = 0$ $\Leftrightarrow$ $\cos x = \frac{1}{2}$ $\Rightarrow$ $x = \frac{\pi}{3}, \frac{5\pi}{3}$ for $x \in [0, 2\pi]$.

31. Using (15a), we have $\sin 2x = \cos x$ $\Leftrightarrow$ $2 \sin x \cos x - \cos x = 0$ $\Leftrightarrow$ $\cos x(2 \sin x - 1) = 0$ $\Leftrightarrow$ $\cos x = 0$ or $2 \sin x - 1 = 0$ $\Rightarrow$ $x = \frac{\pi}{2}, \frac{3\pi}{2}$ or $\sin x = \frac{1}{2}$ $\Rightarrow$ $x = \frac{\pi}{6}$ or $\frac{5\pi}{6}$. Therefore, the solutions are $x = \frac{\pi}{6}, \frac{\pi}{2}, \frac{5\pi}{6}, \frac{3\pi}{2}$.

33. We know that $\sin x = \frac{1}{2}$ when $x = \frac{\pi}{6}$ or $\frac{5\pi}{6}$, and from Figure 13(a), we see that $\sin x \leq \frac{1}{2}$ $\Rightarrow$ $0 \leq x \leq \frac{\pi}{6}$ or $\frac{5\pi}{6} \leq x \leq 2\pi$ for $x \in [0, 2\pi]$.

35. $\tan x = -1$ when $x = \frac{3\pi}{4}, \frac{7\pi}{4}$, and $\tan x = 1$ when $x = \frac{\pi}{4}$ or $\frac{5\pi}{4}$. From Figure 14 we see that $-1 < \tan x < 1$ $\Rightarrow$ $0 \leq x < \frac{\pi}{4}, \frac{3\pi}{4} < x < \frac{5\pi}{4}$, and $\frac{7\pi}{4} < x \leq 2\pi$.

37. $y = \cos\left(x - \frac{\pi}{3}\right)$. We start with the graph of
$y = \cos x$ and shift it $\frac{\pi}{3}$ units to the right.

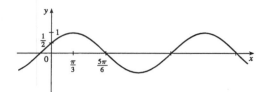

39. $y = \frac{1}{3}\tan\left(x - \frac{\pi}{2}\right)$. We start with the graph of
$y = \tan x$, shift it $\frac{\pi}{2}$ units to the right and
compress it to $\frac{1}{3}$ of its original vertical size.

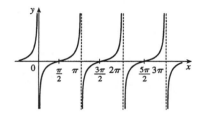

41. (a) $\sin^{-1}(0.5) = \frac{\pi}{6}$ since $\sin\frac{\pi}{6} = 0.5$ and $-\frac{\pi}{2} \le \frac{\pi}{6} \le \frac{\pi}{2}$.

   (b) $\arctan(-1) = -\frac{\pi}{4}$ since $\tan\left(-\frac{\pi}{4}\right) = -1$ and $-\frac{\pi}{4}$ is in $\left(-\frac{\pi}{2}, \frac{\pi}{2}\right)$.

43. (a) $\sin\left(\sin^{-1} 0.7\right) = 0.7$ since 0.7 is in $[-1, 1]$.

   (b) $\arcsin\left(\sin\frac{5\pi}{4}\right) = \arcsin\left(-\frac{1}{\sqrt{2}}\right) = -\frac{\pi}{4}$

45. Let $y = \sin^{-1} x$. Then $-\frac{\pi}{2} \le y \le \frac{\pi}{2}$ $\Rightarrow$ $\cos y \ge 0$, so $\cos\left(\sin^{-1} x\right) = \cos y = \sqrt{1 - \sin^2 y} = \sqrt{1 - x^2}$.

47. $g(x) = \sin^{-1}(3x + 1)$.

   Domain $(g) = \{x \mid -1 \le 3x + 1 \le 1\} = \{x \mid -2 \le 3x \le 0\} = \{x \mid -\frac{2}{3} \le x \le 0\} = \left[-\frac{2}{3}, 0\right]$.
   Range $(g) = \{y \mid -\frac{\pi}{2} \le y \le \frac{\pi}{2}\} = \left[-\frac{\pi}{2}, \frac{\pi}{2}\right]$.

49. From the figure in the text, we see that $x = b\cos\theta$, $y = b\sin\theta$, and from the distance formula we have that the
   distance c from (x, y) to $(a, 0)$ is $c = \sqrt{(x - a)^2 + (y - 0)^2}$ $\Rightarrow$

$$c^2 = (b\cos\theta - a)^2 + (b\sin\theta)^2 = b^2\cos^2\theta - 2ab\cos\theta + a^2 + b^2\sin^2\theta$$

$$= a^2 + b^2\left(\cos^2\theta + \sin^2\theta\right) - 2ab\cos\theta = a^2 + b^2 - 2ab\cos\theta \quad \text{[by (7)]}$$

51. Using the Law of Cosines, we have $c^2 = 1^2 + 1^2 - 2(1)(1)\cos(\alpha - \beta) = 2\left[1 - \cos(\alpha - \beta)\right]$. Now, using the
   distance formula, $c^2 = |AB|^2 = (\cos\alpha - \cos\beta)^2 + (\sin\alpha - \sin\beta)^2$. Equating these two expressions for c^2, we
   get $2\left[1 - \cos(\alpha - \beta)\right] = \cos^2\alpha + \sin^2\alpha + \cos^2\beta + \sin^2\beta - 2\cos\alpha\cos\beta - 2\sin\alpha\sin\beta$ $\Rightarrow$
   $1 - \cos(\alpha - \beta) = 1 - \cos\alpha\cos\beta - \sin\alpha\sin\beta$ $\Rightarrow$ $\cos(\alpha - \beta) = \cos\alpha\cos\beta + \sin\alpha\sin\beta$.

53. In Exercise 52 we used the subtraction formula for cosine to prove the addition formula for cosine. Using that
   formula with $x = \frac{\pi}{2} - \alpha$, $y = \beta$, we get $\cos\left[\left(\frac{\pi}{2} - \alpha\right) + \beta\right] = \cos\left(\frac{\pi}{2} - \alpha\right)\cos\beta - \sin\left(\frac{\pi}{2} - \alpha\right)\sin\beta$ $\Rightarrow$
   $\cos\left[\frac{\pi}{2} - (\alpha - \beta)\right] = \cos\left(\frac{\pi}{2} - \alpha\right)\cos\beta - \sin\left(\frac{\pi}{2} - \alpha\right)\sin\beta$. Now we use the identities given in the problem,
   $\cos\left(\frac{\pi}{2} - \theta\right) = \sin\theta$ and $\sin\left(\frac{\pi}{2} - \theta\right) = \cos\theta$, to get $\sin(\alpha - \beta) = \sin\alpha\cos\beta - \cos\alpha\sin\beta$.

◆D Precise Definitions of Limits • • • • • • • • •

1. On the left side of $x = 2$, we need $|x - 2| < \left|\frac{10}{7} - 2\right| = \frac{4}{7}$. On the right side, we need $|x - 2| < \left|\frac{10}{3} - 2\right| = \frac{4}{3}$.
   For both of these conditions to be satisfied at once, we need the more restrictive of the two to hold, that is,
   $|x - 2| < \frac{4}{7}$. So we can choose $\delta = \frac{4}{7}$, or any smaller positive number.

3. $\left|\sqrt{4x+1}-3\right| < 0.5$ ⟺ $2.5 < \sqrt{4x+1} < 3.5$. We plot the
three parts of this inequality on the same screen and identify the
x-coordinates of the points of intersection using the cursor. It appears

that the inequality holds for $1.3125 \leq x \leq 2.8125$. Since

$|2-1.3125| = 0.6875$ and $|2-2.8125| = 0.8125$, we choose

$0 < \delta < \min\{0.6875, 0.8125\} = 0.6875$.

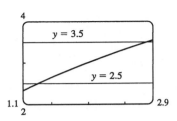

5. For $\varepsilon = 1$, the definition of a limit requires that we find δ such that $\left|(4+x-3x^3)-2\right| < 1$ ⟺
$1 < 4+x-3x^3 < 3$ whenever $0 < |x-1| < \delta$. If we plot the graphs of $y = 1$, $y = 4+x-3x^3$ and $y = 3$ on
the same screen, we see that we need $0.86 \leq x \leq 1.11$. So since $|1-0.86| = 0.14$ and $|1-1.11| = 0.11$, we
choose $\delta = 0.11$ (or any smaller positive number). For $\varepsilon = 0.1$, we must find δ such that
$\left|(4+x-3x^3)-2\right| < 0.1$ ⟺ $1.9 < 4+x-3x^3 < 2.1$ whenever $0 < |x-1| < \delta$. From the graph, we see
that we need $0.988 \leq x \leq 1.012$. So since $|1-0.988| = 0.012$ and $|1-1.012| = 0.012$, we choose $\delta = 0.012$
(or any smaller positive number) for the inequality to hold.

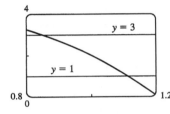 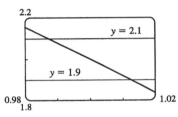

7. Given $\varepsilon > 0$, we need $\delta > 0$ such that if $|x| < \delta$ then $\left|x^3-0\right| < \varepsilon$ ⟺ $|x|^3 < \varepsilon$ ⟺ $|x| < \sqrt[3]{\varepsilon}$. Take
$\delta = \sqrt[3]{\varepsilon}$. Then $|x-0| < \delta$ ⟹ $\left|x^3-0\right| < \delta^3 = \varepsilon$. Thus, $\lim\limits_{x\to 0} x^3 = 0$ by the definition of a limit.

9. Given $\varepsilon > 0$, we need $\delta > 0$ such that if $|x-2| < \delta$, then

$|(3x-2)-4| < \varepsilon$ ⟺ $|3x-6| < \varepsilon \Leftrightarrow 3|x-2| < \varepsilon \Leftrightarrow$

$|x-2| < \varepsilon/3$. So if we choose $\delta = \varepsilon/3$, then

$|x-2| < \delta$ ⟹ $|(3x-2)-4| < \varepsilon$. Thus, $\lim\limits_{x\to 2}(3x-2) = 4$ by

the definition of a limit.

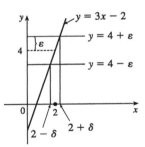

11. (a) $A = \pi r^2$ and $A = 1000$ cm^2 ⟹ $\pi r^2 = 1000$ ⟹ $r^2 = \frac{1000}{\pi}$ ⟹
$r = \sqrt{\frac{1000}{\pi}}$ $(r > 0)$ ≈ 17.8412 cm.

(b) $|A-1000| \leq 5$ ⟹ $-5 \leq \pi r^2 - 1000 \leq 5$ ⟹ $1000-5 \leq \pi r^2 \leq 1000+5$ ⟹
$\sqrt{\frac{995}{\pi}} \leq r \leq \sqrt{\frac{1005}{\pi}}$ ⟹ $17.7966 \leq r \leq 17.8858$. $\sqrt{\frac{1000}{\pi}} - \sqrt{\frac{995}{\pi}} \approx 0.04466$ and
$\sqrt{\frac{1005}{\pi}} - \sqrt{\frac{1000}{\pi}} \approx 0.04455$. So if the machinist gets the radius within 0.0445 cm of 17.8412, the area will be
within 5 cm^2 of 1000.

(c) x is the radius, $f(x)$ is the area, a is the target radius given in part (a), L is the target area (1000), ε is the
tolerance in the area (5), and δ is the tolerance in the radius given in part (b).

13. $\left| \dfrac{6x^2 + 5x - 3}{2x^2 - 1} - 3 \right| < 0.2 \quad \Leftrightarrow \quad 2.8 < \dfrac{6x^2 + 5x - 3}{2x^2 - 1} < 3.2.$ So

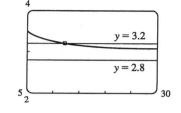

we graph the three parts of this inequality on the same screen, and

find that the curve $y = \dfrac{6x^2 + 5x - 3}{2x^2 - 1}$ seems to lie between the

lines $y = 2.8$ and $y = 3.2$ whenever $x > 12.5$. So we can choose

$N = 13$ (or any larger number), so that the inequality holds

whenever $x \geq N$.

15. (a) $1/x^2 < 0.0001 \quad \Leftrightarrow \quad x^2 > 1/0.0001 = 10{,}000 \quad \Leftrightarrow \quad x > 100 \quad (x > 0)$

(b) If $\varepsilon > 0$ is given, then $1/x^2 < \varepsilon \quad \Leftrightarrow \quad x^2 > 1/\varepsilon \quad \Leftrightarrow \quad x > 1/\sqrt{\varepsilon}.$ Let $N = 1/\sqrt{\varepsilon}$. Then $x > N \quad \Rightarrow$

$x > \dfrac{1}{\sqrt{\varepsilon}} \quad \Rightarrow \quad \left| \dfrac{1}{x^2} - 0 \right| = \dfrac{1}{x^2} < \varepsilon,$ so $\displaystyle\lim_{x \to \infty} \dfrac{1}{x^2} = 0.$

17. (a)

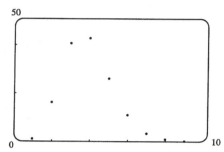

From the graph, it appears that the sequence $\left\{ \dfrac{n^5}{n!} \right\}$ converges to 0, that is, $\displaystyle\lim_{n \to \infty} \dfrac{n^5}{n!} = 0.$

(b)

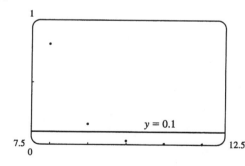

From the first graph, it seems that the smallest possible value of N corresponding to $\varepsilon = 0.1$ is 9, since $n^5/n! < 0.1$ whenever $n \geq 10$, but $9^5/9! > 0.1$. From the second graph, it seems that for $\varepsilon = 0.001$, the smallest possible value for N is 11 since $n^5/n! < 0.001$ whenever $n \geq 12$.

19. If $\displaystyle\lim_{n \to \infty} |a_n| = 0$, then $\displaystyle\lim_{n \to \infty} (-|a_n|) = 0$, and since $-|a_n| \leq a_n \leq |a_n|$, we have that $\displaystyle\lim_{n \to \infty} a_n = 0$ by the

Squeeze Theorem.

Sigma Notation • • • • • • • • • • • • • • •

1. $\displaystyle\sum_{i=1}^{5} \sqrt{i} = \sqrt{1} + \sqrt{2} + \sqrt{3} + \sqrt{4} + \sqrt{5}$

3. $\displaystyle\sum_{i=4}^{6} 3^i = 3^4 + 3^5 + 3^6$

5. $\displaystyle\sum_{k=0}^{4} \frac{2k-1}{2k+1} = -1 + \frac{1}{3} + \frac{3}{5} + \frac{5}{7} + \frac{7}{9}$

7. $\displaystyle\sum_{i=1}^{n} i^{10} = 1^{10} + 2^{10} + 3^{10} + \cdots + n^{10}$

9. $\displaystyle\sum_{j=0}^{n-1} (-1)^j = 1 - 1 + 1 - 1 + \cdots + (-1)^{n-1}$

11. $1 + 2 + 3 + 4 + \cdots + 10 = \displaystyle\sum_{i=1}^{10} i$

13. $\dfrac{1}{2} + \dfrac{2}{3} + \dfrac{3}{4} + \dfrac{4}{5} + \cdots + \dfrac{19}{20} = \displaystyle\sum_{i=1}^{19} \frac{i}{i+1}$

15. $2 + 4 + 6 + 8 + \cdots + 2n = \displaystyle\sum_{i=1}^{n} 2i$

17. $1 + 2 + 4 + 8 + 16 + 32 = \displaystyle\sum_{i=0}^{5} 2^i$

19. $x + x^2 + x^3 + \cdots + x^n = \displaystyle\sum_{i=1}^{n} x^i$

21. $\displaystyle\sum_{i=4}^{8} (3i - 2) = [3(4) - 2] + [3(5) - 2] + [3(6) - 2] + [3(7) - 2] + [3(8) - 2] = 10 + 13 + 16 + 19 + 22 = 80$

23. $\displaystyle\sum_{j=1}^{6} 3^{j+1} = 3^2 + 3^3 + 3^4 + 3^5 + 3^6 + 3^7 = 9 + 27 + 81 + 243 + 729 + 2187 = 3276$

(For a more general method, see Exercise 47.)

25. $\displaystyle\sum_{n=1}^{20} (-1)^n = -1 + 1 - 1 + 1 - 1 + 1 - 1 + 1 - 1 + 1 - 1 + 1 - 1 + 1 - 1 + 1 - 1 + 1 - 1 + 1 = 0$

27. $\displaystyle\sum_{i=0}^{4} \left(2^i + i^2\right) = (1+0) + (2+1) + (4+4) + (8+9) + (16+16) = 61$

29. $\displaystyle\sum_{i=1}^{n} 2i = 2 \sum_{i=1}^{n} i = 2 \cdot \frac{n(n+1)}{2}$ [by Theorem 3(c)] $= n(n+1)$

31. $\displaystyle\sum_{i=1}^{n} \left(i^2 + 3i + 4\right) = \sum_{i=1}^{n} i^2 + 3 \sum_{i=1}^{n} i + \sum_{i=1}^{n} 4 = \frac{n(n+1)(2n+1)}{6} + \frac{3n(n+1)}{2} + 4n$

$\qquad = \tfrac{1}{6}\left[(2n^3 + 3n^2 + n) + (9n^2 + 9n) + 24n\right] = \tfrac{1}{6}\left(2n^3 + 12n^2 + 34n\right)$

$\qquad = \tfrac{1}{3}n\left(n^2 + 6n + 17\right)$

33. $\displaystyle\sum_{i=1}^{n} (i+1)(i+2) = \sum_{i=1}^{n} (i^2 + 3i + 2) = \sum_{i=1}^{n} i^2 + 3 \sum_{i=1}^{n} i + \sum_{i=1}^{n} 2$

$\qquad = \dfrac{n(n+1)(2n+1)}{6} + \dfrac{3n(n+1)}{2} + 2n = \dfrac{n(n+1)}{6}\left[(2n+1) + 9\right] + 2n$

$\qquad = \dfrac{n(n+1)}{3}(n+5) + 2n = \dfrac{n}{3}\left[(n+1)(n+5) + 6\right] = \dfrac{n}{3}\left(n^2 + 6n + 11\right)$

35. $\displaystyle\sum_{i=1}^{n} (i^3 - i - 2) = \sum_{i=1}^{n} i^3 - \sum_{i=1}^{n} i - \sum_{i=1}^{n} 2 = \left[\frac{n(n+1)}{2}\right]^2 - \frac{n(n+1)}{2} - 2n$

$\qquad = \tfrac{1}{4}n(n+1)\left[n(n+1) - 2\right] - 2n = \tfrac{1}{4}n(n+1)(n+2)((n-1)) - 2n$

$\qquad = \tfrac{1}{4}n\left[(n+1)((n-1)(n+2)) - 8\right] = \tfrac{1}{4}n\left[(n^2-1)(n+2) - 8\right] = \tfrac{1}{4}n(n^3 + 2n^2 - n - 10)$

37. By Theorem 2(a) and Example 3, $\displaystyle\sum_{i=1}^{n} c = c \sum_{i=1}^{n} 1 = cn$.

39. $\displaystyle\sum_{i=1}^{n} \left[(i+1)^4 - i^4\right] = (2^4 - 1^4) + (3^4 - 2^4) + (4^4 - 3^4) + \cdots + \left[(n+1)^4 - n^4\right]$

$$= (n+1)^4 - 1^4 = n^4 + 4n^3 + 6n^2 + 4n$$

On the other hand,

$$\sum_{i=1}^{n} \left[(i+1)^4 - i^4\right] = \sum_{i=1}^{n} (4i^3 + 6i^2 + 4i + 1) = 4\sum_{i=1}^{n} i^3 + 6\sum_{i=1}^{n} i^2 + 4\sum_{i=1}^{n} i + \sum_{i=1}^{n} 1$$

$$= 4S + n(n+1)(2n+1) + 2n(n+1) + n \quad \left[\text{where } S = \sum_{i=1}^{n} i^3\right]$$

$$= 4S + 2n^3 + 3n^2 + n + 2n^2 + 2n + n = 4S + 2n^3 + 5n^2 + 4n$$

Thus, $n^4 + 4n^3 + 6n^2 + 4n = 4S + 2n^3 + 5n^2 + 4n$, from which it follows that

$$4S = n^4 + 2n^3 + n^2 = n^2(n^2 + 2n + 1) = n^2(n+1)^2 \text{ and } S = \left[\frac{n(n+1)}{2}\right]^2.$$

41. (a) $\displaystyle\sum_{i=1}^{n} \left[i^4 - (i-1)^4\right] = (1^4 - 0^4) + (2^4 - 1^4) + (3^4 - 2^4) + \cdots + \left[n^4 - (n-1)^4\right] = n^4 - 0 = n^4$

(b) $\displaystyle\sum_{i=1}^{100} \left(5^i - 5^{i-1}\right) = (5^1 - 5^0) + (5^2 - 5^1) + (5^3 - 5^2) + \cdots + (5^{100} - 5^{99}) = 5^{100} - 5^0 = 5^{100} - 1$

(c) $\displaystyle\sum_{i=3}^{99} \left(\frac{1}{i} - \frac{1}{i+1}\right) = \left(\frac{1}{3} - \frac{1}{4}\right) + \left(\frac{1}{4} - \frac{1}{5}\right) + \left(\frac{1}{5} - \frac{1}{6}\right) + \cdots + \left(\frac{1}{99} - \frac{1}{100}\right) = \frac{1}{3} - \frac{1}{100} = \frac{97}{300}$

(d) $\displaystyle\sum_{i=1}^{n} (a_i - a_{i-1}) = (a_1 - a_0) + (a_2 - a_1) + (a_3 - a_2) + \cdots + (a_n - a_{n-1}) = a_n - a_0$

43. $\displaystyle\lim_{n\to\infty} \sum_{i=1}^{n} \frac{1}{n}\left(\frac{i}{n}\right)^2 = \lim_{n\to\infty} \frac{1}{n^3}\sum_{i=1}^{n} i^2 = \lim_{n\to\infty} \frac{1}{n^3}\frac{n(n+1)(2n+1)}{6} = \lim_{n\to\infty} \frac{1}{6}\left(1 + \frac{1}{n}\right)\left(2 + \frac{1}{n}\right)$

$$= \tfrac{1}{6}(1)(2) = \tfrac{1}{3}$$

45. $\displaystyle\lim_{n\to\infty} \sum_{i=1}^{n} \frac{2}{n}\left[\left(\frac{2i}{n}\right)^3 + 5\left(\frac{2i}{n}\right)\right] = \lim_{n\to\infty} \sum_{i=1}^{n} \left[\frac{16}{n^4}i^3 + \frac{20}{n^2}i\right] = \lim_{n\to\infty} \left[\frac{16}{n^4}\sum_{i=1}^{n} i^3 + \frac{20}{n^2}\sum_{i=1}^{n} i\right]$

$$= \lim_{n\to\infty} \left[\frac{16}{n^4}\frac{n^2(n+1)^2}{4} + \frac{20}{n^2}\frac{n(n+1)}{2}\right] = \lim_{n\to\infty} \left[\frac{4(n+1)^2}{n^2} + \frac{10n(n+1)}{n^2}\right]$$

$$= \lim_{n\to\infty} \left[4\left(1 + \frac{1}{n}\right)^2 + 10\left(1 + \frac{1}{n}\right)\right] = 4 \cdot 1 + 10 \cdot 1 = 14$$

47. Let $\displaystyle S = \sum_{i=1}^{n} ar^{i-1} = a + ar + ar^2 + \cdots + ar^{n-1}$. Multiplying both sides by r gives us

$rS = ar + ar^2 + \cdots + ar^{n-1} + ar^n$. Subtracting the first equation from the second, we find

$(r-1)S = ar^n - a = a(r^n - 1)$, so $\displaystyle S = \frac{a(r^n - 1)}{r - 1}$ (since $r \neq 1$).

49. $\displaystyle\sum_{i=1}^{n}\left(2i+2^i\right)=2\sum_{i=1}^{n}i+\sum_{i=1}^{n}2\cdot 2^{i-1}=2\frac{n(n+1)}{2}+\frac{2(2^n-1)}{2-1}=2^{n+1}+n^2+n-2.$

For the first sum we have used Theorem 3(c), and for the second, Exercise 47 with $a=r=2$.

Integration of Rational Functions by Partial Fractions   •   •   •

1. $\dfrac{5}{2x^2-3x-2}=\dfrac{5}{(2x+1)(x-2)}=\dfrac{A}{2x+1}+\dfrac{B}{x-2}$

3. $\dfrac{1}{x^4-x^3}=\dfrac{1}{x^3(x-1)}=\dfrac{A}{x}+\dfrac{B}{x^2}+\dfrac{C}{x^3}+\dfrac{D}{x-1}$

5. Since the degree of the numerator is greater than or equal to the degree of the denominator, we first perform long

division. $\dfrac{x^2+1}{x^2-1}=1+\dfrac{2}{x^2-1}=1+\dfrac{2}{(x-1)(x+1)}=1+\dfrac{A}{x-1}+\dfrac{B}{x+1}$

7. $\dfrac{x^2-2}{x(x^2+2)}=\dfrac{A}{x}+\dfrac{Bx+C}{x^2+2}$

9. $\dfrac{x^3+x^2+1}{x^4+x^3+2x^2}=\dfrac{x^3+x^2+1}{x^2(x^2+x+2)}=\dfrac{A}{x}+\dfrac{B}{x^2}+\dfrac{Cx+D}{x^2+x+2}$

11. $\displaystyle\int\frac{x^2+2}{x+2}\,dx=\int\left(x-2+\frac{6}{x+2}\right)dx$ [division] $=\frac{1}{2}x^2-2x+6\ln|x+2|+C$

13. $\dfrac{4x-1}{(x-1)(x+2)}=\dfrac{A}{x-1}+\dfrac{B}{x+2}\quad\Rightarrow\quad 4x-1=A(x+2)+B(x-1).$ Take $x=1$ to get $3=3A$, then

$x=-2$ to get $-9=-3B\quad\Rightarrow\quad A=1,\ B=3.$ Now

$$\int_2^4\frac{4x-1}{(x-1)(x+2)}\,dx=\int_2^4\left(\frac{1}{x-1}+\frac{3}{x+2}\right)dx=\left[\ln(x-1)+3\ln(x+2)\right]_2^4$$

$$=\ln 3+3\ln 6-\ln 1-3\ln 4=\ln\left(3\cdot 6^3\right)-\ln 4^3=\ln\tfrac{81}{8}.$$

15. $\dfrac{2x+3}{(x+1)^2}=\dfrac{A}{x+1}+\dfrac{B}{(x+1)^2}\quad\Rightarrow\quad 2x+3=A(x+1)+B.$ Take $x=-1$ to get $B=1$, and equate

coefficients of x to get $A=2$. Now

$$\int_0^1\frac{2x+3}{(x+1)^2}\,dx=\int_0^1\left[\frac{2}{x+1}+\frac{1}{(x+1)^2}\right]dx=\left[2\ln(x+1)-\frac{1}{x+1}\right]_0^1$$

$$=2\ln 2-\tfrac{1}{2}-(2\ln 1-1)=2\ln 2+\tfrac{1}{2}$$

17. $\dfrac{4y^2-7y-12}{y(y+2)(y-3)}=\dfrac{A}{y}+\dfrac{B}{y+2}+\dfrac{C}{y-3}\quad\Rightarrow\quad 4y^2-7y-12=A(y+2)(y-3)+By(y-3)+Cy(y+2).$

Setting $y=0$ gives $-12=-6A$, so $A=2$. Setting $y=-2$ gives $18=10B$, so $B=\frac{9}{5}$. Setting $y=3$ gives

$3=15C$, so $C=\frac{1}{5}$. Now

$$\int_1^2\frac{4y^2-7y-12}{y(y+2)(y-3)}\,dy=\int_1^2\left(\frac{2}{y}+\frac{9/5}{y+2}+\frac{1/5}{y-3}\right)dy=\left[2\ln|y|+\tfrac{9}{5}\ln|y+2|+\tfrac{1}{5}\ln|y-3|\right]_1^2$$

$$=2\ln 2+\tfrac{9}{5}\ln 4+\tfrac{1}{5}\ln 1-2\ln 1-\tfrac{9}{5}\ln 3-\tfrac{1}{5}\ln 2$$

$$=2\ln 2+\tfrac{18}{5}\ln 2-\tfrac{1}{5}\ln 2-\tfrac{9}{5}\ln 3=\tfrac{27}{5}\ln 2-\tfrac{9}{5}\ln 3=\tfrac{9}{5}(3\ln 2-\ln 3)=\tfrac{9}{5}\ln\tfrac{8}{3}$$

19. $\dfrac{1}{(x+5)^2\,(x-1)} = \dfrac{A}{x+5} + \dfrac{B}{(x+5)^2} + \dfrac{C}{x-1}$ $\Rightarrow$ $1 = A(x+5)(x-1) + B(x-1) + C(x+5)^2$. Setting

$x = -5$ gives $1 = -6B$, so $B = -\frac{1}{6}$. Setting $x = 1$ gives $1 = 36C$, so $C = \frac{1}{36}$. Setting $x = -2$ gives

$1 = A(3)(-3) + B(-3) + C(3^2) = -9A - 3B + 9C = -9A + \frac{1}{2} + \frac{1}{4} = -9A + \frac{3}{4}$, so $9A = -\frac{1}{4}$ and

$A = -\frac{1}{36}$. Now

$$\int \frac{1}{(x+5)^2\,(x-1)}\,dx = \int \left[\frac{-1/36}{x+5} - \frac{1/6}{(x+5)^2} + \frac{1/36}{x-1} \right] dx$$

$$= -\tfrac{1}{36} \ln|x+5| + \frac{1}{6(x+5)} + \tfrac{1}{36} \ln|x-1| + C$$

21. Complete the square: $x^2 + x + 1 = \left(x + \frac{1}{2}\right)^2 + \frac{3}{4}$ and let $u = x + \frac{1}{2}$. Then

$$\int_0^1 \frac{x}{x^2+x+1}\,dx = \int_{1/2}^{3/2} \frac{u-1/2}{u^2+3/4}\,du = \int_{1/2}^{3/2} \frac{u}{u^2+3/4}\,du - \frac{1}{2}\int_{1/2}^{3/2} \frac{1}{u^2+3/4}\,du$$

$$= \left[\tfrac{1}{2}\ln\!\left(u^2+\tfrac{3}{4}\right) - \tfrac{1}{2}\tfrac{1}{\sqrt{3}/2}\tan^{-1}\!\left(\tfrac{2}{\sqrt{3}}u\right) \right]_{1/2}^{3/2}$$

$$= \tfrac{1}{2}\ln 3 - \tfrac{1}{\sqrt{3}}\left(\tfrac{\pi}{3} - \tfrac{\pi}{6}\right) = \ln\sqrt{3} - \tfrac{\pi}{6\sqrt{3}}.$$

23. $\dfrac{3x^2 - 4x + 5}{(x-1)(x^2+1)} = \dfrac{A}{x-1} + \dfrac{Bx+C}{x^2+1}$ $\Rightarrow$ $3x^2 - 4x + 5 = A(x^2+1) + (Bx+C)(x-1)$. Take $x = 1$ to

get $4 = 2A$ or $A = 2$. Now $(Bx+C)(x-1) = 3x^2 - 4x + 5 - 2(x^2+1) = x^2 - 4x + 3$. Equating coefficients

of x^2 and then comparing the constant terms, we get $B = 1$ and $C = -3$. Hence,

$$\int \frac{3x^2-4x+5}{(x-1)(x^2+1)}\,dx = \int \left[\frac{2}{x-1} + \frac{x-3}{x^2+1} \right] dx = 2\ln|x-1| + \int \frac{x\,dx}{x^2+1} - 3\int \frac{dx}{x^2+1}$$

$$= 2\ln|x-1| + \tfrac{1}{2}\ln\!\left(x^2+1\right) - 3\tan^{-1}x + C$$

$$= \ln(x-1)^2 + \ln\sqrt{x^2+1} - 3\tan^{-1}x + C$$

25. $\dfrac{1}{x^3-1} = \dfrac{1}{(x-1)(x^2+x+1)} = \dfrac{A}{x-1} + \dfrac{Bx+C}{x^2+x+1}$ $\Rightarrow$ $1 = A(x^2+x+1) + (Bx+C)(x-1)$. Take

$x = 1$ to get $A = \frac{1}{3}$. Equating coefficients of x^2 and then comparing the constant terms, we get $0 = \frac{1}{3} + B$,

$1 = \frac{1}{3} - C$, so $B = -\frac{1}{3}$, $C = -\frac{2}{3}$ $\Rightarrow$

$$\int \frac{1}{x^3-1}\,dx = \int \frac{\frac{1}{3}}{x-1}\,dx + \int \frac{-\frac{1}{3}x - \frac{2}{3}}{x^2+x+1}\,dx = \tfrac{1}{3}\ln|x-1| - \frac{1}{3}\int \frac{x+2}{x^2+x+1}\,dx$$

$$= \tfrac{1}{3}\ln|x-1| - \frac{1}{3}\int \frac{x+1/2}{x^2+x+1}\,dx - \frac{1}{3}\int \frac{(3/2)\,dx}{(x+1/2)^2+3/4}$$

$$= \tfrac{1}{3}\ln|x-1| - \tfrac{1}{6}\ln\!\left(x^2+x+1\right) - \tfrac{1}{2}\left(\tfrac{2}{\sqrt{3}}\right)\tan^{-1}\!\left(\frac{x+\frac{1}{2}}{\sqrt{3}/2}\right) + K$$

$$= \tfrac{1}{3}\ln|x-1| - \tfrac{1}{6}\ln\!\left(x^2+x+1\right) - \tfrac{1}{\sqrt{3}}\tan^{-1}\!\left(\tfrac{1}{\sqrt{3}}(2x+1)\right) + K$$

27. $\dfrac{2t^3 - t^2 + 3t - 1}{(t^2 + 1)(t^2 + 2)} = \dfrac{At + B}{t^2 + 1} + \dfrac{Ct + D}{t^2 + 2} \;\Rightarrow$

$$2t^3 - t^2 + 3t - 1 = (At + B)\left(t^2 + 2\right) + (Ct + D)\left(t^2 + 1\right)$$
$$= (A + C)t^3 + (B + D)t^2 + (2A + C)t + (2B + D) \quad \Rightarrow$$

$A + C = 2,\ B + D = -1,\ 2A + C = 3,$ and $2B + D = -1 \;\Rightarrow\; A = 1,\ C = 1,\ B = 0,$ and $D = -1.$ Now

$$\int \frac{2t^3 - t^2 + 3t - 1}{(t^2 + 1)\,(t^2 + 2)}\,dt = \int \left(\frac{t}{t^2 + 1} + \frac{t - 1}{t^2 + 2}\right) dt = \frac{1}{2}\int \frac{2t\,dt}{t^2 + 1} + \frac{1}{2}\int \frac{2t\,dt}{t^2 + 2} - \int \frac{dt}{t^2 + 2}$$

$$= \tfrac{1}{2}\ln\!\left(t^2 + 1\right) + \tfrac{1}{2}\ln\!\left(t^2 + 2\right) - \tfrac{1}{\sqrt{2}}\tan^{-1}\!\left(\tfrac{1}{\sqrt{2}}t\right) + C$$

$$\text{or } \tfrac{1}{2}\ln\!\left(\left(t^2 + 1\right)\left(t^2 + 2\right)\right) - \tfrac{\sqrt{2}}{2}\tan^{-1}\!\left(\tfrac{1}{\sqrt{2}}t\right) + C$$

29.

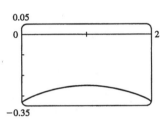

From the graph, we see that the integral will be negative, and we guess that the area is about the same as that of a rectangle with width 2 and height 0.3, so we estimate the integral to be $-(2 \cdot 0.3) = -0.6.$ Now

$$\frac{1}{x^2 - 2x - 3} = \frac{1}{(x - 3)(x + 1)} = \frac{A}{x - 3} + \frac{B}{x + 1} \quad\Leftrightarrow$$

$1 = (A + B)x + A - 3B,$ so $A = -B$ and $A - 3B = 1 \;\Leftrightarrow\; A = \tfrac{1}{4}$

and $B = -\tfrac{1}{4},$ so the integral becomes

$$\int_0^2 \frac{dx}{x^2 - 2x - 3} = \frac{1}{4}\int_0^2 \frac{dx}{x - 3} - \frac{1}{4}\int_0^2 \frac{dx}{x + 1} = \tfrac{1}{4}\left[\ln|x - 3| - \ln|x + 1|\right]_0^2$$

$$= \frac{1}{4}\left[\ln\left|\frac{x - 3}{x + 1}\right|\right]_0^2 = \tfrac{1}{4}\left(\ln\tfrac{1}{3} - \ln 3\right) = -\tfrac{1}{2}\ln 3 \approx -0.55$$

31. $\dfrac{P + S}{P\left[(r - 1)P - S\right]} = \dfrac{A}{P} + \dfrac{B}{(r - 1)P - S} \;\Rightarrow\; P + S = A\left[(r - 1)P - S\right] + BP = \left[(r - 1)A + B\right]P - AS$

$\Rightarrow\;\; (r - 1)A + B = 1,\ -A = 1 \;\Rightarrow\; A = -1,\ B = r.$ Now

$$t = \int \frac{P + S}{P\left[(r - 1)P - S\right]}\,dP = \int \left[\frac{-1}{P} + \frac{r}{(r - 1)P - S}\right] dP = -\int \frac{dP}{P} + \frac{r}{r - 1}\int \frac{r - 1}{(r - 1)P - S}\,dP$$

so $t = -\ln P + \dfrac{r}{r - 1}\ln|(r - 1)P - S| + C.$ Here $r = 0.10$ and $S = 900,$ so

$$t = -\ln P + \tfrac{0.1}{-0.9}\ln|-0.9P - 900| + C = -\ln P - \tfrac{1}{9}\ln(|-1|\,|0.9P + 900|)$$

$$= -\ln P - \tfrac{1}{9}\ln(0.9P + 900) + C$$

When $t = 0,$ $P = 10{,}000,$ so $0 = -\ln 10{,}000 - \tfrac{1}{9}\ln(9900) + C.$ Thus, $C = \ln 10{,}000 + \tfrac{1}{9}\ln 9900\ [\approx 10.2326],$ so our equation becomes

$$t = \ln 10{,}000 - \ln P + \tfrac{1}{9}\ln 9900 - \tfrac{1}{9}\ln(0.9P + 900) = \ln\frac{10{,}000}{P} + \frac{1}{9}\ln\frac{9900}{0.9P + 900}$$

$$= \ln\frac{10{,}000}{P} + \frac{1}{9}\ln\frac{1100}{0.1P + 100} = \ln\frac{10{,}000}{P} + \frac{1}{9}\ln\frac{11{,}000}{P + 1000}$$

33. (a) In Maple, we define $f(x)$, and then use `convert(f,parfrac,x);` to obtain

$$f(x) = \frac{24{,}110/4879}{5x+2} - \frac{668/323}{2x+1} - \frac{9438/80{,}155}{3x-7} + \frac{((22{,}098x + 48{,}935)/260{,}015)}{x^2+x+5}.$$

In Mathematica, we use the command `Apart`, and in Derive, we use `Expand`.

(b) $\displaystyle\int f(x)\,dx = \frac{24{,}110}{4879} \cdot \frac{1}{5}\ln|5x+2| - \frac{668}{323}\cdot\frac{1}{2}\ln|2x+1| - \frac{9438}{80{,}155}\cdot\frac{1}{3}\ln|3x-7|$

$$+ \frac{1}{260{,}015}\int \frac{22{,}098\left(x+\frac{1}{2}\right) + 37{,}886}{\left(x+\frac{1}{2}\right)^2 + \frac{19}{4}}\,dx + C$$

$$= \frac{24{,}110}{4879}\cdot\frac{1}{5}\ln|5x+2| - \frac{668}{323}\cdot\frac{1}{2}\ln|2x+1| - \frac{9438}{80{,}155}\cdot\frac{1}{3}\ln|3x-7|$$

$$+ \frac{1}{260{,}015}\left[22{,}098\cdot\frac{1}{2}\ln(x^2+x+5) + 37{,}886\cdot\sqrt{\frac{4}{19}}\tan^{-1}\left(\frac{1}{\sqrt{19/4}}\left(x+\frac{1}{2}\right)\right)\right] + C$$

$$= \frac{4822}{4879}\ln|5x+2| - \frac{334}{323}\ln|2x+1| - \frac{3146}{80{,}155}\ln|3x-7| + \frac{11{,}049}{260{,}015}\ln(x^2+x+5)$$

$$+ \frac{75{,}772}{260{,}015\sqrt{19}}\tan^{-1}\left[\frac{1}{\sqrt{19}}(2x+1)\right] + C$$

Using a CAS, we get

$$\frac{4822\ln(5x+2)}{4879} - \frac{334\ln(2x+1)}{323} - \frac{3146\ln(3x-7)}{80{,}155}$$

$$+ \frac{11{,}049\ln(x^2+x+5)}{260{,}015} + \frac{3988\sqrt{19}}{260{,}115}\tan^{-1}\left[\frac{\sqrt{19}}{19}(2x+1)\right]$$

The main difference in this answer is that the absolute value signs and the constant of integration have been omitted. Also, the fractions have been reduced and the denominators rationalized.

35. There are only finitely many values of x where $Q(x) = 0$ (assuming that Q is not the zero polynomial). At all other values of x, $F(x)/Q(x) = G(x)/Q(x)$, so $F(x) = G(x)$. In other words, the values of F and G agree at all except perhaps finitely many values of x. By continuity of F and G, the polynomials F and G must agree at those values of x too.

More explicitly: if a is a value of x such that $Q(a) = 0$, then $Q(x) \neq 0$ for all x sufficiently close to a. Thus,

$$F(a) = \lim_{x\to a} F(x) \text{ (by continuity of } F) \ = \lim_{x\to a} G(x) \text{ [whenever } Q(x) \neq 0]$$

$$= G(a) \text{ (by continuity of } G).$$

H **Polar Coordinates** • • • • • • • • • • • •

H.1 **Curves in Polar Coordinates** • • • • • • • • • • •

1. (a) By adding 2π to $\frac{\pi}{2}$, we obtain the point $\left(1, \frac{5\pi}{2}\right)$. The direction opposite $\frac{\pi}{2}$ is $\frac{3\pi}{2}$, so $\left(-1, \frac{3\pi}{2}\right)$ is a point that satisfies the $r < 0$ requirement.

(b) $\left(-2, \frac{\pi}{4}\right)$

(c) $(3, 2)$

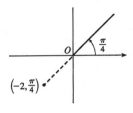

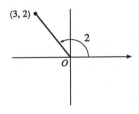

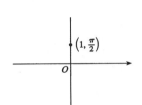

$\left(2, \frac{5\pi}{4}\right), \left(-2, \frac{9\pi}{4}\right)$

$(3, 2+2\pi), (-3, 2+\pi)$

3. (a)

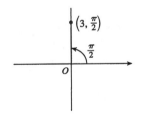

$x = 3\cos\frac{\pi}{2} = 3(0) = 0$ and $y = 3\sin\frac{\pi}{2} = 3(1) = 3$ give us the Cartesian coordinates $(0, 3)$.

(b)

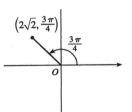

$x = 2\sqrt{2}\cos\frac{3\pi}{4}$
$= 2\sqrt{2}\left(-\frac{1}{\sqrt{2}}\right) = -2$ and
$y = 2\sqrt{2}\sin\frac{3\pi}{4} = 2\sqrt{2}\left(\frac{1}{\sqrt{2}}\right) = 2$
give us $(-2, 2)$.

(c)

$x = -1\cos\frac{\pi}{3} = -\frac{1}{2}$ and
$y = -1\sin\frac{\pi}{3} = -\frac{\sqrt{3}}{2}$ give
us $\left(-\frac{1}{2}, -\frac{\sqrt{3}}{2}\right)$.

5. (a) $x = 1$ and $y = 1 \Rightarrow r = \sqrt{1^2 + 1^2} = \sqrt{2}$ and $\theta = \tan^{-1}\left(\frac{1}{1}\right) = \frac{\pi}{4}$. Since $(1, 1)$ is in the first quadrant, the polar coordinates are (i) $\left(\sqrt{2}, \frac{\pi}{4}\right)$ and (ii) $\left(-\sqrt{2}, \frac{5\pi}{4}\right)$.

(b) $x = 2\sqrt{3}$ and $y = -2 \Rightarrow r = \sqrt{\left(2\sqrt{3}\right)^2 + (-2)^2} = \sqrt{12 + 4} = \sqrt{16} = 4$ and
$\theta = \tan^{-1}\left(-\frac{2}{2\sqrt{3}}\right) = \tan^{-1}\left(-\frac{1}{\sqrt{3}}\right) = -\frac{\pi}{6}$. Since $\left(2\sqrt{3}, -2\right)$ is in the fourth quadrant and $0 \le \theta \le 2\pi$, the polar coordinates are (i) $\left(4, \frac{11\pi}{6}\right)$ and (ii) $\left(-4, \frac{5\pi}{6}\right)$.

7. $r > 1$

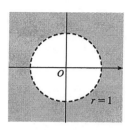

9. $0 \le r \le 2$, $\frac{\pi}{2} \le \theta \le \pi$

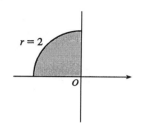

$r = 2$

11. $2 < r < 3$, $\frac{5\pi}{3} \le \theta \le \frac{7\pi}{3}$

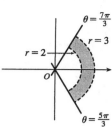

$\theta = \frac{7\pi}{3}$

$r = 3$

$r = 2$

$\theta = \frac{5\pi}{3}$

13. $r = 3\sin\theta \;\Rightarrow\; r^2 = 3r\sin\theta \;\Leftrightarrow\; x^2 + y^2 = 3y \;\Leftrightarrow\; x^2 + y^2 - 3y = 0 \;\Leftrightarrow\; x^2 + y^2 - 3y + \frac{9}{4} = \frac{9}{4}$
$\Leftrightarrow\; x^2 + \left(y - \frac{3}{2}\right)^2 = \left(\frac{3}{2}\right)^2$, a circle of radius $\frac{3}{2}$ centered at $\left(0, \frac{3}{2}\right)$. The first two equations are actually equivalent since $r^2 = 3r\sin\theta \;\Rightarrow\; r(r - 3\sin\theta) = 0 \;\Rightarrow\; r = 0$ or $r = 3\sin\theta$. But $r = 3\sin\theta$ gives the point $r = 0$ (the pole) when $\theta = 0$. Thus, the single equation $r = 3\sin\theta$ is equivalent to the compound condition ($r = 0$ or $r = 3\sin\theta$).

15. $r^2 = \sin 2\theta = 2\sin\theta\cos\theta \;\Leftrightarrow\; r^2 \cdot r^2 = r^2 \cdot 2\sin\theta\cos\theta \;\Leftrightarrow\; r^4 = 2r\sin\theta\, r\cos\theta \;\Leftrightarrow\;$
$\left(r^2\right)^2 = 2(r\sin\theta)(r\cos\theta) \;\Leftrightarrow\; \left(x^2 + y^2\right)^2 = 2yx$

17. $y = 5 \;\Leftrightarrow\; r\sin\theta = 5 \;$ (or $r = 5\csc\theta$)

19. $x^2 + y^2 = 25 \;\Leftrightarrow\; r^2 = 25 \;\Rightarrow\; r = 5$

21. (a) The description leads immediately to the polar equation $\theta = \frac{\pi}{6}$, and the Cartesian equation
$y = \tan\left(\frac{\pi}{6}\right)x = \frac{1}{\sqrt{3}}x$ is slightly more difficult to derive.

(b) The easier description here is the Cartesian equation $x = 3$.

23. As in Example 4, $r = 5$ represents the circle with center O and radius 5.

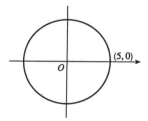

$(5, 0)$

25. $r = \sin\theta \;\Leftrightarrow\; r^2 = r\sin\theta \;\Leftrightarrow\; x^2 + y^2 = y \;\Leftrightarrow\;$
$x^2 + \left(y - \frac{1}{2}\right)^2 = \left(\frac{1}{2}\right)^2$. The reasoning here is the same as in Exercise 13. This is a circle of radius $\frac{1}{2}$ centered at $\left(0, \frac{1}{2}\right)$.

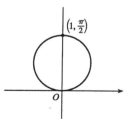

$\left(1, \frac{\pi}{2}\right)$

27. $r = \theta$, $\theta \ge 0$

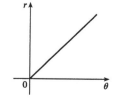

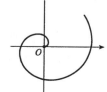

29. $r = 1 - 2\cos\theta$

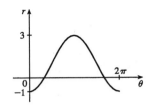

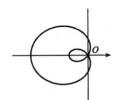

31. $r = 2\cos 4\theta$

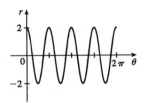

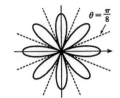

33. $r^2 = 4\cos 2\theta$

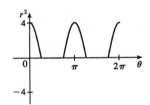

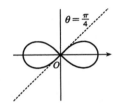

35. For $\theta = 0$, π, and 2π, r has its minimum value of about 0.5.
For $\theta = \frac{\pi}{2}$ and $\frac{3\pi}{2}$, r attains its maximum value of 2. We see that
the graph has a similar shape for $0 \le \theta \le \pi$ and $\pi \le \theta \le 2\pi$.

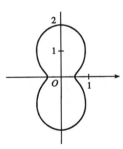

37. $x = (r)\cos\theta = (4 + 2\sec\theta)\cos\theta = 4\cos\theta + 2$. Now, $r \to \infty$ $\Rightarrow$ $(4 + 2\sec\theta) \to \infty$ $\Rightarrow$ $\theta \to \left(\frac{\pi}{2}\right)^-$ or
$\theta \to \left(\frac{3\pi}{2}\right)^+$ (since we need only consider $0 \le \theta < 2\pi$), so $\lim\limits_{r \to \infty} x = \lim\limits_{\theta \to \pi/2^-} (4\cos\theta + 2) = 2$. Also, $r \to -\infty$
$\Rightarrow$ $(4 + 2\sec\theta) \to -\infty$ $\Rightarrow$ $\theta \to \left(\frac{\pi}{2}\right)^+$ or $\theta \to \left(\frac{3\pi}{2}\right)^-$, so $\lim\limits_{r \to -\infty} x = \lim\limits_{\theta \to \pi/2^+} (4\cos\theta + 2) = 2$. Therefore,
$\lim\limits_{r \to \pm\infty} x = 2$ $\Rightarrow$ $x = 2$ is a vertical asymptote.

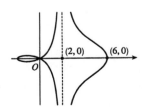

39. (a) We see that the curve crosses itself at the origin, where $r = 0$ (in fact the inner loop corresponds to negative r-values,) so we solve the equation of the limaçon for $r = 0$ $\Leftrightarrow$ $c \sin \theta = -1$ $\Leftrightarrow$ $\sin \theta = -1/c$. Now if $|c| < 1$, then this equation has no solution and hence there is no inner loop. But if $c < -1$, then on the interval $(0, 2\pi)$ the equation has the two solutions $\theta = \sin^{-1}(-1/c)$ and $\theta = \pi - \sin^{-1}(-1/c)$, and if $c > 1$, the solutions are $\theta = \pi + \sin^{-1}(1/c)$ and $\theta = 2\pi - \sin^{-1}(1/c)$. In each case, $r < 0$ for θ between the two solutions, indicating a loop.

(b) For $0 < c < 1$, the dimple (if it exists) is characterized by the fact that y has a local maximum at $\theta = \frac{3\pi}{2}$. So we determine for what c-values $\dfrac{d^2 y}{d\theta^2}$ is negative at $\theta = \frac{3\pi}{2}$, since by the Second Derivative Test this indicates a

maximum: $y = r \sin \theta = \sin \theta + c \sin^2 \theta$ $\Rightarrow$ $\dfrac{dy}{d\theta} = \cos \theta + 2c \sin \theta \cos \theta = \cos \theta + c \sin 2\theta$ $\Rightarrow$

$\dfrac{d^2 y}{d\theta^2} = -\sin \theta + 2c \cos 2\theta$. At $\theta = \frac{3\pi}{2}$, this is equal to $-(-1) + 2c(-1) = 1 - 2c$, which is negative only for $c > \frac{1}{2}$. A similar argument shows that for $-1 < c < 0$, y only has a local minimum at $\theta = \frac{\pi}{2}$ (indicating a dimple) for $c < -\frac{1}{2}$.

41. Using Equation 3 with $r = 3 \cos \theta$ and $dr/d\theta = -3 \sin \theta$, we have

$$\frac{dy}{dx} = \frac{dy/d\theta}{dx/d\theta} = \frac{(dr/d\theta)(\sin \theta) + r \cos \theta}{(dr/d\theta)(\cos \theta) - r \sin \theta} = \frac{-3 \sin \theta \sin \theta + 3 \cos \theta \cos \theta}{-3 \sin \theta \cos \theta - 3 \cos \theta \sin \theta} = \frac{3(\cos^2 \theta - \sin^2 \theta)}{-3(2 \sin \theta \cos \theta)}$$

$$= -\frac{\cos 2\theta}{\sin 2\theta} = -\cot 2\theta = \frac{1}{\sqrt{3}} \text{ when } \theta = \frac{\pi}{3}$$

Another solution: $r = 3 \cos \theta$ $\Rightarrow$ $x = r \cos \theta = 3 \cos^2 \theta$, $y = r \sin \theta = 3 \sin \theta \cos \theta$ $\Rightarrow$

$$\frac{dy}{dx} = \frac{dy/d\theta}{dx/d\theta} = \frac{-3 \sin^2 \theta + 3 \cos^2 \theta}{-6 \cos \theta \sin \theta} = \frac{\cos 2\theta}{-\sin 2\theta} = -\cot 2\theta = \frac{1}{\sqrt{3}} \text{ when } \theta = \frac{\pi}{3}$$

43. $r = 1 + \cos \theta$ $\Rightarrow$ $x = r \cos \theta = \cos \theta + \cos^2 \theta$, $y = r \sin \theta = \sin \theta + \sin \theta \cos \theta$ $\Rightarrow$

$$\frac{dy}{dx} = \frac{dy/d\theta}{dx/d\theta} = \frac{\cos \theta + \cos^2 \theta - \sin^2 \theta}{-\sin \theta - 2 \cos \theta \sin \theta} = \frac{\cos \theta + \cos 2\theta}{-\sin \theta - \sin 2\theta}$$

When $\theta = \dfrac{\pi}{6}$, $\dfrac{dy}{dx} = \dfrac{\frac{\sqrt{3}}{2} + \frac{1}{2}}{-\frac{1}{2} - \frac{\sqrt{3}}{2}} = \dfrac{\frac{\sqrt{3}}{2} + \frac{1}{2}}{-\left(\frac{1}{2} + \frac{\sqrt{3}}{2}\right)} = -1$.

45. $r = 3 \cos \theta$ $\Rightarrow$ $x = r \cos \theta = 3 \cos \theta \cos \theta$, $y = r \sin \theta = 3 \cos \theta \sin \theta$ $\Rightarrow$ $dy/d\theta = -3 \sin^2 \theta + 3 \cos^2 \theta = 3 \cos 2\theta = 0$ $\Rightarrow$ $2\theta = \frac{\pi}{2}$ or $\frac{3\pi}{2}$ $\Leftrightarrow$ $\theta = \frac{\pi}{4}$ or $\frac{3\pi}{4}$. So the tangent is horizontal at $\left(\frac{3}{\sqrt{2}}, \frac{\pi}{4}\right)$ and $\left(-\frac{3}{\sqrt{2}}, \frac{3\pi}{4}\right)$ $\left[\text{same as } \left(\frac{3}{\sqrt{2}}, -\frac{\pi}{4}\right)\right]$. $dx/d\theta = -6 \sin \theta \cos \theta = -3 \sin 2\theta = 0$ $\Rightarrow$ $2\theta = 0$ or π $\Leftrightarrow$ $\theta = 0$ or $\frac{\pi}{2}$. So the tangent is vertical at $(3, 0)$ and $\left(0, \frac{\pi}{2}\right)$.

47. $r = 1 + \cos\theta \Rightarrow x = r\cos\theta = \cos\theta(1 + \cos\theta), \; y = r\sin\theta = \sin\theta(1 + \cos\theta) \Rightarrow$

$dy/d\theta = (1 + \cos\theta)\cos\theta - \sin^2\theta = 2\cos^2\theta + \cos\theta - 1 = (2\cos\theta - 1)(\cos\theta + 1) = 0 \Rightarrow \cos\theta = \frac{1}{2}$ or

$-1 \Rightarrow \theta = \frac{\pi}{3}, \pi,$ or $\frac{5\pi}{3} \Rightarrow$ horizontal tangent at $\left(\frac{3}{2}, \frac{\pi}{3}\right), (0, \pi),$ and $\left(\frac{3}{2}, \frac{5\pi}{3}\right).$

$dx/d\theta = -(1 + \cos\theta)\sin\theta - \cos\theta\sin\theta = -\sin\theta(1 + 2\cos\theta) = 0 \Rightarrow \sin\theta = 0$ or $\cos\theta = -\frac{1}{2} \Rightarrow$

$\theta = 0, \pi, \frac{2\pi}{3},$ or $\frac{4\pi}{3} \Rightarrow$ vertical tangent at $(2, 0), \left(\frac{1}{2}, \frac{2\pi}{3}\right),$ and $\left(\frac{1}{2}, \frac{4\pi}{3}\right).$ Note that the tangent is horizontal, not

vertical when $\theta = \pi$, since $\lim\limits_{\theta \to \pi} \dfrac{dy/d\theta}{dx/d\theta} = 0.$

49. $r = a\sin\theta + b\cos\theta \Rightarrow r^2 = ar\sin\theta + br\cos\theta \Rightarrow x^2 + y^2 = ay + bx \Rightarrow$

$x^2 - bx + \left(\frac{1}{2}b\right)^2 + y^2 - ay + \left(\frac{1}{2}a\right)^2 = \left(\frac{1}{2}b\right)^2 + \left(\frac{1}{2}a\right)^2 \Rightarrow \left(x - \frac{1}{2}b\right)^2 + \left(y - \frac{1}{2}a\right)^2 = \frac{1}{4}\left(a^2 + b^2\right),$ and this

is a circle with center $\left(\frac{1}{2}b, \frac{1}{2}a\right)$ and radius $\frac{1}{2}\sqrt{a^2 + b^2}.$

Note for Exercises 51–54: Maple is able to plot polar curves using the `polarplot` command, or using the `coords=polar` option
in a regular `plot` command. In Mathematica, use `PolarPlot`. In Derive, change to `Polar` under `Options State`. If your
graphing device cannot plot polar equations, you must convert to parametric equations. For example, in Exercise 51,
$x = r\cos\theta = [1 + 2\sin(\theta/2)]\cos\theta, y = r\sin\theta = [1 + 2\sin(\theta/2)]\sin\theta.$

51. $r = 1 + 2\sin(\theta/2).$ The parameter interval
is $[0, 4\pi].$

53. $r = e^{\sin\theta} - 2\cos(4\theta).$ The parameter interval
is $[0, 2\pi].$

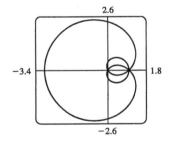

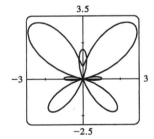

55.

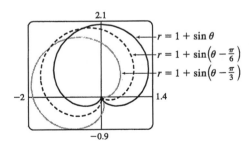

It appears that the graph of $r = 1 + \sin\left(\theta - \frac{\pi}{6}\right)$ is the same shape as the graph of $r = 1 + \sin\theta$, but rotated
counterclockwise about the origin by $\frac{\pi}{6}$. Similarly, the graph of $r = 1 + \sin\left(\theta - \frac{\pi}{3}\right)$ is rotated by $\frac{\pi}{3}$. In general, the
graph of $r = f(\theta - \alpha)$ is the same shape as that of $r = f(\theta)$, but rotated counterclockwise through α about the
origin. That is, for any point (r_0, θ_0) on the curve $r = f(\theta)$, the point $(r_0, \theta_0 + \alpha)$ is on the curve $r = f(\theta - \alpha),$
since $r_0 = f(\theta_0) = f((\theta_0 + \alpha) - \alpha).$

57. (a) $r = \sin n\theta$. From the graphs, it seems that when n is even, the number of loops in the curve (called a rose) is $2n$, and when n is odd, the number of loops is simply n.

This is because in the case of n odd, every point on the graph is traversed twice, due to the fact that

$$r(\theta + \pi) = \sin\left[n(\theta + \pi)\right] = \sin n\theta \cos n\pi + \cos n\theta \sin n\pi = \begin{cases} \sin n\theta & \text{if } n \text{ is even} \\ -\sin n\theta & \text{if } n \text{ is odd} \end{cases}$$

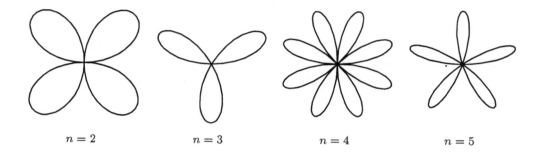

$n = 2$ $n = 3$ $n = 4$ $n = 5$

(b) The graph of $r = |\sin n\theta|$ has $2n$ loops whether n is odd or even, since $r(\theta + \pi) = r(\theta)$.

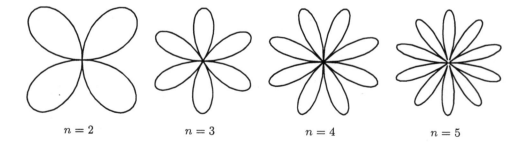

$n = 2$ $n = 3$ $n = 4$ $n = 5$

59. $r = \dfrac{1 - a\cos\theta}{1 + a\cos\theta}$. We start with $a = 0$, since in this case the curve is simply the circle $r = 1$.

As a increases, the graph moves to the left, and its right side becomes flattened. As a increases through about 0.4, the right side seems to grow a dimple, which upon closer investigation (with narrower θ-ranges) seems to appear at $a \approx 0.42$ (the actual value is $\sqrt{2} - 1$). As $a \to 1$, this dimple becomes more pronounced, and the curve begins to stretch out horizontally, until at $a = 1$ the denominator vanishes at $\theta = \pi$, and the dimple becomes an actual cusp. For $a > 1$ we must choose our parameter interval carefully, since $r \to \infty$ as $1 + a\cos\theta \to 0$ $\Leftrightarrow$ $\theta \to \pm\cos^{-1}(-1/a)$. As a increases from 1, the curve splits into two parts. The left part has a loop, which grows larger as a increases, and the right part grows broader vertically, and its left tip develops a dimple when $a \approx 2.42$ (actually, $\sqrt{2} + 1$). As a increases, the dimple grows more and more pronounced. If $a < 0$, we get the same graph as we do for the corresponding positive a-value, but with a rotation through π about the pole, as happened when c was replaced with $-c$ in Exercise 58.

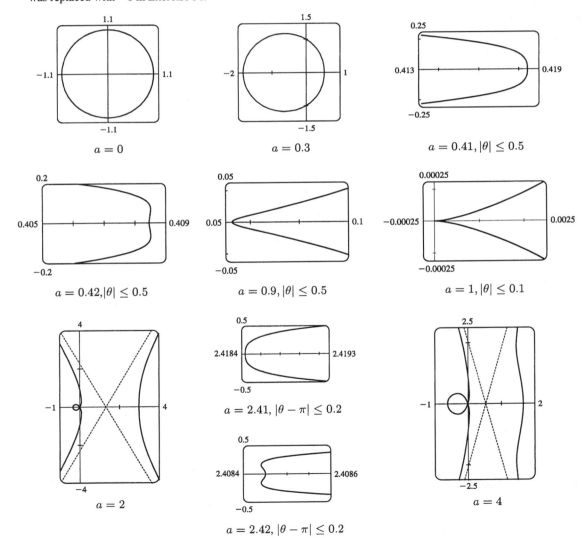

$a = 0$

$a = 0.3$

$a = 0.41, |\theta| \le 0.5$

$a = 0.42, |\theta| \le 0.5$

$a = 0.9, |\theta| \le 0.5$

$a = 1, |\theta| \le 0.1$

$a = 2$

$a = 2.41, |\theta - \pi| \le 0.2$

$a = 2.42, |\theta - \pi| \le 0.2$

$a = 4$

61.

$$\tan \psi = \tan(\phi - \theta) = \frac{\tan \phi - \tan \theta}{1 + \tan \phi \tan \theta} = \frac{\dfrac{dy}{dx} - \tan \theta}{1 + \dfrac{dy}{dx} \tan \theta} = \frac{\dfrac{dy/d\theta}{dx/d\theta} - \tan \theta}{1 + \dfrac{dy/d\theta}{dx/d\theta} \tan \theta}$$

$$= \frac{\dfrac{dy}{d\theta} - \dfrac{dx}{d\theta} \tan \theta}{\dfrac{dx}{d\theta} + \dfrac{dy}{d\theta} \tan \theta} = \frac{\left(\dfrac{dr}{d\theta} \sin \theta + r \cos \theta\right) - \tan \theta \left(\dfrac{dr}{d\theta} \cos \theta - r \sin \theta\right)}{\left(\dfrac{dr}{d\theta} \cos \theta - r \sin \theta\right) + \tan \theta \left(\dfrac{dr}{d\theta} \sin \theta + r \cos \theta\right)}$$

$$= \frac{r \cos \theta + r \cdot \dfrac{\sin^2 \theta}{\cos \theta}}{\dfrac{dr}{d\theta} \cos \theta + \dfrac{dr}{d\theta} \cdot \dfrac{\sin^2 \theta}{\cos \theta}} = \frac{r \cos^2 \theta + r \sin^2 \theta}{\dfrac{dr}{d\theta} \cos^2 \theta + \dfrac{dr}{d\theta} \sin^2 \theta} = \frac{r}{dr/d\theta}$$

Areas and Lengths in Polar Coordinates • • • • • • • • • •

1. $r = \sqrt{\theta}, \ 0 \le \theta \le \frac{\pi}{4}. \ A = \int_0^{\pi/4} \frac{1}{2} r^2 \, d\theta = \int_0^{\pi/4} \frac{1}{2} \left(\sqrt{\theta}\right)^2 d\theta = \int_0^{\pi/4} \frac{1}{2} \theta \, d\theta = \left[\frac{1}{4}\theta^2\right]_0^{\pi/4} = \frac{1}{64}\pi^2$

3. $r = \sin \theta, \ \frac{\pi}{3} \le \theta \le \frac{2\pi}{3}.$

$$A = \int_{\pi/3}^{2\pi/3} \frac{1}{2} \sin^2 \theta \, d\theta = \frac{1}{4} \int_{\pi/3}^{2\pi/3} (1 - \cos 2\theta) \, d\theta = \frac{1}{4}\left[\theta - \frac{1}{2}\sin 2\theta\right]_{\pi/3}^{2\pi/3}$$

$$= \frac{1}{4}\left[\frac{2\pi}{3} - \frac{1}{2}\sin \frac{4\pi}{3} - \frac{\pi}{3} + \frac{1}{2}\sin \frac{2\pi}{3}\right] = \frac{1}{4}\left[\frac{2\pi}{3} - \frac{1}{2}\left(-\frac{\sqrt{3}}{2}\right) - \frac{\pi}{3} + \frac{1}{2}\left(\frac{\sqrt{3}}{2}\right)\right] = \frac{1}{4}\left(\frac{\pi}{3} + \frac{\sqrt{3}}{2}\right) = \frac{\pi}{12} + \frac{\sqrt{3}}{8}$$

5. $r = \theta, \ 0 \le \theta \le \pi. \ A = \int_0^\pi \frac{1}{2}\theta^2 \, d\theta = \left[\frac{1}{6}\theta^3\right]_0^\pi = \frac{1}{6}\pi^3$

7. $r = 4 + 3\sin \theta, \ -\frac{\pi}{2} \le \theta \le \frac{\pi}{2}.$

$$A = \int_{-\pi/2}^{\pi/2} \frac{1}{2}(4 + 3\sin \theta)^2 d\theta = \frac{1}{2} \int_{-\pi/2}^{\pi/2} (16 + 24\sin \theta + 9\sin^2 \theta) \, d\theta$$

$$= \frac{1}{2} \int_{-\pi/2}^{\pi/2} (16 + 9\sin^2 \theta) \, d\theta \quad \text{[by Theorem 5.5.6(b)]}$$

$$= \frac{1}{2} \cdot 2 \int_0^{\pi/2} \left[16 + 9 \cdot \frac{1}{2}(1 - \cos 2\theta)\right] d\theta \quad \text{[by Theorem 5.5.6(a)]}$$

$$= \int_0^{\pi/2} \left(\frac{41}{2} - \frac{9}{2}\cos 2\theta\right) d\theta = \left[\frac{41}{2}\theta - \frac{9}{4}\sin 2\theta\right]_0^{\pi/2} = \left(\frac{41\pi}{4} - 0\right) - (0 - 0) = \frac{41\pi}{4}$$

9. The curve goes through the pole when $\theta = \pi/4$, so we'll find the area for
$0 \le \theta \le \pi/4$ and multiply it by 4.

$$A = 4 \int_0^{\pi/4} \frac{1}{2} r^2 \, d\theta = 2 \int_0^{\pi/4} (4\cos 2\theta) \, d\theta$$

$$= 8 \int_0^{\pi/4} \cos 2\theta \, d\theta = 4\left[\sin 2\theta\right]_0^{\pi/4} = 4$$

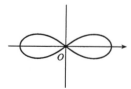

11. The curve is symmetric about the vertical line $\theta = \pi/2$, so we'll find the area of the right side and double it.

$$A = 2 \int_{-\pi/2}^{\pi/2} \frac{1}{2}(4 - \sin \theta)^2 \, d\theta = \int_{-\pi/2}^{\pi/2} (16 - 8\sin \theta + \sin^2 \theta) \, d\theta$$

$$= \int_{-\pi/2}^{\pi/2} (16 + \sin^2 \theta) \, d\theta \quad \text{[by Theorem 5.5.6(b)]}$$

$$= 2 \int_0^{\pi/2} (16 + \sin^2 \theta) \, d\theta \quad \text{[by Theorem 5.5.6(a)]}$$

$$= 2 \int_0^{\pi/2} \left[16 + \frac{1}{2}(1 - \cos 2\theta)\right] d\theta = 2\left[\frac{33}{2}\theta - \frac{1}{4}\sin 2\theta\right]_0^{\pi/2}$$

$$= \frac{33\pi}{2}$$

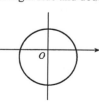

13. By symmetry, the total area is twice the area enclosed above the polar axis, so

$$A = 2 \int_0^\pi \tfrac{1}{2} r^2 \, d\theta = \int_0^\pi (2 + \cos 6\theta)^2 \, d\theta = \int_0^\pi \left(4 + 4 \cos 6\theta + \cos^2 6\theta \right) d\theta$$

$$= \int_0^\pi \left[4 + 4 \cos 6\theta + \tfrac{1}{2}(1 + \cos 12\theta) \right] d\theta$$

$$= \left[4\theta + 4\left(\tfrac{1}{6} \sin 6\theta \right) + \left(\tfrac{1}{24} \sin 12\theta + \tfrac{1}{2}\theta \right) \right]_0^\pi = 4\pi + \tfrac{\pi}{2} = \tfrac{9\pi}{2}$$

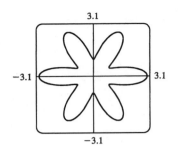

3.1

−3.1 ⎜ 3.1

−3.1

15. The shaded loop is traced out from $\theta = 0$ to $\theta = \pi/2$.

$$A = \int_0^{\pi/2} \tfrac{1}{2} r^2 \, d\theta = \tfrac{1}{2} \int_0^{\pi/2} \sin^2 2\theta \, d\theta$$

$$= \tfrac{1}{2} \int_0^{\pi/2} \tfrac{1}{2}(1 - \cos 4\theta) \, d\theta = \tfrac{1}{4} \left[\theta - \tfrac{1}{4} \sin 4\theta \right]_0^{\pi/2}$$

$$= \tfrac{1}{4} \left(\tfrac{\pi}{2} \right) = \tfrac{\pi}{8}$$

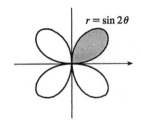

$r = \sin 2\theta$

17.

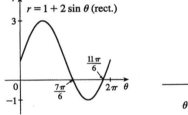

$r = 1 + 2 \sin \theta$ (rect.)

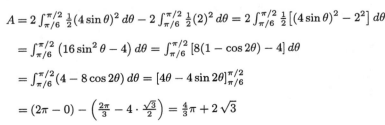

$\left(3, \tfrac{\pi}{2} \right)$ $r = 1 + 2 \sin \theta$

$\left(-1, \tfrac{3\pi}{2} \right)$

$\theta = \tfrac{7\pi}{6}$ $\theta = \tfrac{11\pi}{6}$

This is a limaçon, with inner loop traced out between $\theta = \tfrac{7\pi}{6}$ and $\tfrac{11\pi}{6}$ [found by solving $r = 0$].

$$A = 2 \int_{7\pi/6}^{3\pi/2} \tfrac{1}{2}(1 + 2 \sin \theta)^2 \, d\theta = \int_{7\pi/6}^{3\pi/2} \left(1 + 4 \sin \theta + 4 \sin^2 \theta \right) d\theta = \int_{7\pi/6}^{3\pi/2} \left[1 + 4 \sin \theta + 4 \cdot \tfrac{1}{2}(1 - \cos 2\theta) \right] d\theta$$

$$= \left[\theta - 4 \cos \theta + 2\theta - \sin 2\theta \right]_{7\pi/6}^{3\pi/2} = \left(\tfrac{9\pi}{2} \right) - \left(\tfrac{7\pi}{2} + 2\sqrt{3} - \tfrac{\sqrt{3}}{2} \right) = \pi - \tfrac{3\sqrt{3}}{2}$$

19. $4 \sin \theta = 2 \iff \sin \theta = \tfrac{1}{2} \iff \theta = \tfrac{\pi}{6}$ or $\tfrac{5\pi}{6}$ (for $0 \le \theta \le 2\pi$). We'll subtract the unshaded area from the shaded area for $\pi/6 \le \theta \le \pi/2$ and double that value.

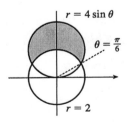

$r = 4 \sin \theta$

$\theta = \tfrac{\pi}{6}$

$r = 2$

$$A = 2 \int_{\pi/6}^{\pi/2} \tfrac{1}{2}(4 \sin \theta)^2 \, d\theta - 2 \int_{\pi/6}^{\pi/2} \tfrac{1}{2}(2)^2 \, d\theta = 2 \int_{\pi/6}^{\pi/2} \tfrac{1}{2} \left[(4 \sin \theta)^2 - 2^2 \right] d\theta$$

$$= \int_{\pi/6}^{\pi/2} \left(16 \sin^2 \theta - 4 \right) d\theta = \int_{\pi/6}^{\pi/2} \left[8(1 - \cos 2\theta) - 4 \right] d\theta$$

$$= \int_{\pi/6}^{\pi/2} (4 - 8 \cos 2\theta) \, d\theta = \left[4\theta - 4 \sin 2\theta \right]_{\pi/6}^{\pi/2}$$

$$= (2\pi - 0) - \left(\tfrac{2\pi}{3} - 4 \cdot \tfrac{\sqrt{3}}{2} \right) = \tfrac{4}{3}\pi + 2\sqrt{3}$$

21. $3\cos\theta = 1 + \cos\theta \iff \cos\theta = \frac{1}{2} \implies \theta = \frac{\pi}{3}$ or $-\frac{\pi}{3}$.

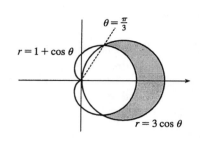

$$A = 2\int_0^{\pi/3} \frac{1}{2}\left[(3\cos\theta)^2 - (1+\cos\theta)^2\right]d\theta$$

$$= \int_0^{\pi/3}\left(8\cos^2\theta - 2\cos\theta - 1\right)d\theta$$

$$= \int_0^{\pi/3}\left[4(1+\cos 2\theta) - 2\cos\theta - 1\right]d\theta$$

$$= \int_0^{\pi/3}\left(3 + 4\cos 2\theta - 2\cos\theta\right)d\theta = \left[3\theta + 2\sin 2\theta - 2\sin\theta\right]_0^{\pi/3}$$

$$= \pi + \sqrt{3} - \sqrt{3} = \pi$$

23. $A = 2\int_0^{\pi/4}\frac{1}{2}\sin^2\theta\,d\theta = \int_0^{\pi/4}\frac{1}{2}(1 - \cos 2\theta)\,d\theta$

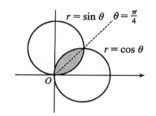

$$= \frac{1}{2}\left[\theta - \frac{1}{2}\sin 2\theta\right]_0^{\pi/4} = \frac{1}{2}\left[\left(\frac{\pi}{4} - \frac{1}{2}\cdot 1\right) - (0 - 0)\right]$$

$$= \frac{1}{8}\pi - \frac{1}{4}$$

25. $\sin 2\theta = \cos 2\theta \implies \dfrac{\sin 2\theta}{\cos 2\theta} = 1 \implies \tan 2\theta = 1 \implies 2\theta = \frac{\pi}{4} \implies$

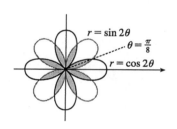

$\theta = \frac{\pi}{8} \implies$

$$A = 8\cdot 2\int_0^{\pi/8}\frac{1}{2}\sin^2 2\theta\,d\theta = 8\int_0^{\pi/8}\frac{1}{2}(1 - \cos 4\theta)\,d\theta$$

$$= 4\left[\theta - \frac{1}{4}\sin 4\theta\right]_0^{\pi/8} = 4\left(\frac{\pi}{8} - \frac{1}{4}\cdot 1\right) = \frac{1}{2}\pi - 1$$

27. The darker shaded region (from $\theta = 0$ to $\theta = 2\pi/3$) represents $\frac{1}{2}$ of the desired area plus $\frac{1}{2}$ of the area of the inner loop. From this area, we'll subtract $\frac{1}{2}$ of the area of the inner loop (the lighter shaded region from $\theta = 2\pi/3$ to $\theta = \pi$), and then double that difference to obtain the desired area.

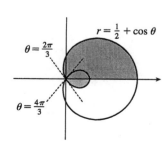

$$A = 2\left[\int_0^{2\pi/3}\frac{1}{2}\left(\frac{1}{2} + \cos\theta\right)^2 d\theta - \int_{2\pi/3}^{\pi}\frac{1}{2}\left(\frac{1}{2} + \cos\theta\right)^2 d\theta\right]$$

$$= \int_0^{2\pi/3}\left(\frac{1}{4} + \cos\theta + \cos^2\theta\right)d\theta - \int_{2\pi/3}^{\pi}\left(\frac{1}{4} + \cos\theta + \cos^2\theta\right)d\theta$$

$$= \int_0^{2\pi/3}\left[\frac{1}{4} + \cos\theta + \frac{1}{2}(1 + \cos 2\theta)\right]d\theta$$

$$\qquad - \int_{2\pi/3}^{\pi}\left[\frac{1}{4} + \cos\theta + \frac{1}{2}(1 + \cos 2\theta)\right]d\theta$$

$$= \left[\frac{\theta}{4} + \sin\theta + \frac{\theta}{2} + \frac{\sin 2\theta}{4}\right]_0^{2\pi/3} - \left[\frac{\theta}{4} + \sin\theta + \frac{\theta}{2} + \frac{\sin 2\theta}{4}\right]_{2\pi/3}^{\pi}$$

$$= \left(\frac{\pi}{6} + \frac{\sqrt{3}}{2} + \frac{\pi}{3} - \frac{\sqrt{3}}{8}\right) - \left(\frac{\pi}{4} + \frac{\pi}{2}\right) + \left(\frac{\pi}{6} + \frac{\sqrt{3}}{2} + \frac{\pi}{3} - \frac{\sqrt{3}}{8}\right)$$

$$= \frac{\pi}{4} + \frac{3}{4}\sqrt{3} = \frac{1}{4}\left(\pi + 3\sqrt{3}\right)$$

29. The curves intersect at the pole since $\left(0, \frac{\pi}{2}\right)$ satisfies $r = \cos\theta$ and

$(0,0)$ satisfies $r = 1 - \cos\theta$. Now $\cos\theta = 1 - \cos\theta \;\Rightarrow\;$

$2\cos\theta = 1 \;\Rightarrow\; \cos\theta = \frac{1}{2} \;\Rightarrow\; \theta = \frac{\pi}{3}$ or $\frac{5\pi}{3} \;\Rightarrow\;$

the other intersection points are $\left(\frac{1}{2}, \frac{\pi}{3}\right)$ and $\left(\frac{1}{2}, \frac{5\pi}{3}\right)$.

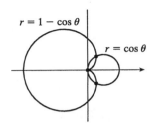

31. The pole is a point of intersection. $\sin\theta = \sin 2\theta = 2\sin\theta\,\cos\theta$

$\Leftrightarrow \sin\theta\,(1 - 2\cos\theta) = 0 \;\Leftrightarrow\; \sin\theta = 0$ or $\cos\theta = \frac{1}{2} \;\Rightarrow\;$

$\theta = 0, \pi, \frac{\pi}{3}, -\frac{\pi}{3} \;\Rightarrow\; \left(\frac{\sqrt{3}}{2}, \frac{\pi}{3}\right)$ and $\left(\frac{\sqrt{3}}{2}, \frac{2\pi}{3}\right)$ (by symmetry) are

the other intersection points.

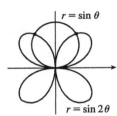

33.

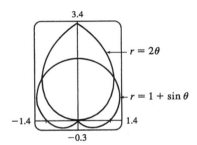

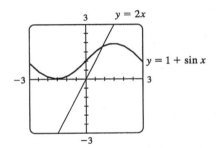

From the first graph, we see that the pole is one point of intersection. By zooming in or using the cursor, we find the θ-values of the intersection points to be $\alpha \approx 0.88786 \approx 0.89$ and $\pi - \alpha \approx 2.25$. (The first of these values may be more easily estimated by plotting $y = 1 + \sin x$ and $y = 2x$ in rectangular coordinates; see the second graph.) By symmetry, the total area contained is twice the area contained in the first quadrant, that is,

$$A = 2\int_0^\alpha \tfrac{1}{2}(2\theta)^2\,d\theta + 2\int_\alpha^{\pi/2} \tfrac{1}{2}(1 + \sin\theta)^2\,d\theta = \int_0^\alpha 4\theta^2\,d\theta + \int_\alpha^{\pi/2}\left[1 + 2\sin\theta + \tfrac{1}{2}(1 - \cos 2\theta)\right]d\theta$$

$$= \left[\tfrac{4}{3}\theta^3\right]_0^\alpha + \left[\theta - 2\cos\theta + \left(\tfrac{1}{2}\theta - \tfrac{1}{4}\sin 2\theta\right)\right]_\alpha^{\pi/2}$$

$$= \tfrac{4}{3}\alpha^3 + \left[\left(\tfrac{\pi}{2} + \tfrac{\pi}{4}\right) - \left(\alpha - 2\cos\alpha + \tfrac{1}{2}\alpha - \tfrac{1}{4}\sin 2\alpha\right)\right] \approx 3.4645$$

35. $L = \int_a^b \sqrt{r^2 + (dr/d\theta)^2}\,d\theta = \int_0^{3\pi/4}\sqrt{(5\cos\theta)^2 + (-5\sin\theta)^2}\,d\theta = \int_0^{3\pi/4}\sqrt{25\cos^2\theta + 25\sin^2\theta}\,d\theta$

$= 5\int_0^{3\pi/4}\sqrt{\cos^2\theta + \sin^2\theta}\,d\theta = 5\int_0^{3\pi/4}d\theta = 5[\theta]_0^{3\pi/4} = 5\left(\frac{3\pi}{4}\right) = \frac{15}{4}\pi$

37. $L = \int_a^b \sqrt{r^2 + (dr/d\theta)^2}\,d\theta = \int_0^{2\pi}\sqrt{(\theta^2)^2 + (2\theta)^2}\,d\theta = \int_0^{2\pi}\sqrt{\theta^4 + 4\theta^2}\,d\theta$

$= \int_0^{2\pi}\sqrt{\theta^2\,(\theta^2 + 4)}\,d\theta = \int_0^{2\pi}\theta\sqrt{\theta^2 + 4}\,d\theta$

Now let $u = \theta^2 + 4$, so that $du = 2\theta\,d\theta \;\left[\theta\,d\theta = \tfrac{1}{2}\,du\right]$ and

$$\int_0^{2\pi}\theta\sqrt{\theta^2 + 4}\,d\theta = \int_4^{4\pi^2+4}\tfrac{1}{2}\sqrt{u}\,du = \tfrac{1}{2}\cdot\tfrac{2}{3}\left[u^{3/2}\right]_4^{4(\pi^2+1)} = \tfrac{1}{3}\left[4^{3/2}(\pi^2 + 1)^{3/2} - 4^{3/2}\right]$$

$$= \tfrac{8}{3}\left[(\pi^2 + 1)^{3/2} - 1\right]$$

39. From Figure 4 in Example 1 with $r = \cos 2\theta$ and $r' = -2 \sin 2\theta$,

$$L = \int_{-\pi/4}^{\pi/4} \sqrt{r^2 + (r')^2} \, d\theta = 2 \int_0^{\pi/4} \sqrt{\cos^2 2\theta + 4 \sin^2 2\theta} \, d\theta \approx 2(1.211056) \approx 2.4221$$

 I **Complex Numbers** • • • • • • • • • • • • •

1. $(3 + 2i) + (7 - 3i) = (3 + 7) + (2 - 3)i = 10 - i$

3. $(3 - i)(4 + i) = 12 + 3i - 4i - (-1) \; [i^2 = -1] = 13 - i$

5. $\overline{12 + 7i} = 12 - 7i$

7. $\dfrac{2 + 3i}{1 - 5i} = \dfrac{2 + 3i}{1 - 5i} \cdot \dfrac{1 + 5i}{1 + 5i} = \dfrac{2 + 10i + 3i + 15(-1)}{1 - 25(-1)} = \dfrac{-13 + 13i}{26} = -\dfrac{1}{2} + \dfrac{1}{2}i$

9. $\dfrac{1}{1 + i} = \dfrac{1}{1 + i} \cdot \dfrac{1 - i}{1 - i} = \dfrac{1 - i}{1 - (-1)} = \dfrac{1 - i}{2} = \dfrac{1}{2} - \dfrac{1}{2}i$

11. $i^3 = i^2 \cdot i = (-1)i = -i$

13. $\sqrt{-25} = \sqrt{25}\,i = 5i$

15. $\overline{3 + 4i} = 3 - 4i$, $|3 + 4i| = \sqrt{3^2 + 4^2} = \sqrt{25} = 5$

17. $\overline{-4i} = \overline{0 - 4i} = 0 + 4i = 4i$, $|-4i| = \sqrt{0^2 + (-4)^2} = \sqrt{16} = 4$

19. $4x^2 + 9 = 0 \iff 4x^2 = -9 \iff x^2 = -\frac{9}{4} \iff x = \pm\sqrt{-\frac{9}{4}} = \pm\sqrt{\frac{9}{4}}\,i = \pm\frac{3}{2}i.$

21. By the quadratic formula, $x^2 - 8x + 17 = 0 \iff x = \dfrac{-(-8) \pm \sqrt{(-8)^2 - 4(1)(17)}}{2(1)} = \dfrac{8 \pm \sqrt{-4}}{2} = \dfrac{8 \pm 2i}{2} = 4 \pm i.$

23. By the quadratic formula, $z^2 + z + 2 = 0 \iff z = \dfrac{-1 \pm \sqrt{1^2 - 4(1)(2)}}{2(1)} = \dfrac{-1 \pm \sqrt{-7}}{2} = -\dfrac{1}{2} \pm \dfrac{\sqrt{7}}{2}i.$

25. For $z = -3 + 3i$, $r = \sqrt{(-3)^2 + 3^2} = 3\sqrt{2}$ and $\tan \theta = \frac{3}{-3} = -1 \Rightarrow \theta = \frac{3\pi}{4}$ (since z lies in the second quadrant). Therefore, $-3 + 3i = 3\sqrt{2}\left(\cos\frac{3\pi}{4} + i\sin\frac{3\pi}{4}\right).$

27. For $z = 3 + 4i$, $r = \sqrt{3^2 + 4^2} = 5$ and $\tan \theta = \frac{4}{3} \Rightarrow \theta = \tan^{-1}\left(\frac{4}{3}\right)$ (since z lies in the first quadrant). Therefore, $3 + 4i = 5\left[\cos\left(\tan^{-1}\frac{4}{3}\right) + i\sin\left(\tan^{-1}\frac{4}{3}\right)\right].$

29. For $z = \sqrt{3} + i$, $r = \sqrt{\left(\sqrt{3}\right)^2 + 1^2} = 2$ and $\tan \theta = \frac{1}{\sqrt{3}} \Rightarrow \theta = \frac{\pi}{6} \Rightarrow z = 2\left(\cos\frac{\pi}{6} + i\sin\frac{\pi}{6}\right).$
For $w = 1 + \sqrt{3}\,i$, $r = 2$ and $\tan \theta = \sqrt{3} \Rightarrow \theta = \frac{\pi}{3} \Rightarrow w = 2\left(\cos\frac{\pi}{3} + i\sin\frac{\pi}{3}\right).$
Therefore, $zw = 2 \cdot 2\left[\cos\left(\frac{\pi}{6} + \frac{\pi}{3}\right) + i\sin\left(\frac{\pi}{6} + \frac{\pi}{3}\right)\right] = 4\left(\cos\frac{\pi}{2} + i\sin\frac{\pi}{2}\right),$
$z/w = \frac{2}{2}\left[\cos\left(\frac{\pi}{6} - \frac{\pi}{3}\right) + i\sin\left(\frac{\pi}{6} - \frac{\pi}{3}\right)\right] = \cos\left(-\frac{\pi}{6}\right) + i\sin\left(-\frac{\pi}{6}\right),$ and $1 = 1 + 0i = 1(\cos 0 + i\sin 0) \Rightarrow$
$1/z = \frac{1}{2}\left[\cos\left(0 - \frac{\pi}{6}\right) + i\sin\left(0 - \frac{\pi}{6}\right)\right] = \frac{1}{2}\left[\cos\left(-\frac{\pi}{6}\right) + i\sin\left(-\frac{\pi}{6}\right)\right].$ For $1/z$, we could also use the formula that precedes Example 5 to obtain $1/z = \frac{1}{8}\left(\cos\frac{\pi}{6} - i\sin\frac{\pi}{6}\right).$

31. For $z = 2\sqrt{3} - 2i$, $r = \sqrt{\left(2\sqrt{3}\right)^2 + (-2)^2} = 4$ and $\tan\theta = \frac{-2}{2\sqrt{3}} = -\frac{1}{\sqrt{3}}$

$\Rightarrow \quad \theta = -\frac{\pi}{6} \quad \Rightarrow \quad z = 4\left[\cos\left(-\frac{\pi}{6}\right) + i\sin\left(-\frac{\pi}{6}\right)\right]$. For $w = -1 + i$, $r = \sqrt{2}$,

$\tan\theta = \frac{1}{-1} = -1 \quad \Rightarrow \quad \theta = \frac{3\pi}{4} \quad \Rightarrow \quad z = \sqrt{2}\left(\cos\frac{3\pi}{4} + i\sin\frac{3\pi}{4}\right)$. Therefore,

$zw = 4\sqrt{2}\left[\cos\left(-\frac{\pi}{6} + \frac{3\pi}{4}\right) + i\sin\left(-\frac{\pi}{6} + \frac{3\pi}{4}\right)\right] = 4\sqrt{2}\left(\cos\frac{7\pi}{12} + i\sin\frac{7\pi}{12}\right)$,

$z/w = \frac{4}{\sqrt{2}}\left[\cos\left(-\frac{\pi}{6} - \frac{3\pi}{4}\right) + i\sin\left(-\frac{\pi}{6} - \frac{3\pi}{4}\right)\right] = \frac{4}{\sqrt{2}}\left[\cos\left(-\frac{11\pi}{12}\right) + i\sin\left(-\frac{11\pi}{12}\right)\right]$

$\quad = 2\sqrt{2}\left(\cos\frac{13\pi}{12} + i\sin\frac{13\pi}{12}\right)$, and

$1/z = \frac{1}{4}\left[\cos\left(-\frac{\pi}{6}\right) - i\sin\left(-\frac{\pi}{6}\right)\right] = \frac{1}{4}\left(\cos\frac{\pi}{6} + i\sin\frac{\pi}{6}\right)$.

33. For $z = 1 + i$, $r = \sqrt{2}$ and $\tan\theta = \frac{1}{1} = 1 \quad \Rightarrow \quad \theta = \frac{\pi}{4} \quad \Rightarrow \quad z = \sqrt{2}\left(\cos\frac{\pi}{4} + i\sin\frac{\pi}{4}\right)$. So by

De Moivre's Theorem,

$$(1+i)^{20} = \left[\sqrt{2}\left(\cos\frac{\pi}{4} + i\sin\frac{\pi}{4}\right)\right]^{20} = \left(2^{1/2}\right)^{20}\left(\cos\frac{20\cdot\pi}{4} + i\sin\frac{20\cdot\pi}{4}\right)$$

$$= 2^{10}(\cos 5\pi + i\sin 5\pi) = 2^{10}[-1 + i(0)] = -2^{10} = -1024$$

35. For $z = 2\sqrt{3} + 2i$, $r = \sqrt{\left(2\sqrt{3}\right)^2 + 2^2} = \sqrt{16} = 4$ and $\tan\theta = \frac{2}{2\sqrt{3}} = \frac{1}{\sqrt{3}} \quad \Rightarrow \quad \theta = \frac{\pi}{6} \quad \Rightarrow$

$z = 4\left(\cos\frac{\pi}{6} + i\sin\frac{\pi}{6}\right)$. So by De Moivre's Theorem,

$$\left(2\sqrt{3} + 2i\right)^5 = \left[4\left(\cos\frac{\pi}{6} + i\sin\frac{\pi}{6}\right)\right]^5 = 4^5\left(\cos\frac{5\pi}{6} + i\sin\frac{5\pi}{6}\right) = 1024\left[-\frac{\sqrt{3}}{2} + \frac{1}{2}i\right] = -512\sqrt{3} + 512i$$

37. $1 = 1 + 0i = 1(\cos 0 + i\sin 0)$. Using Equation 3 with $r = 1$, $n = 8$, and $\theta = 0$, we have

$w_k = 1^{1/8}\left[\cos\left(\frac{0 + 2k\pi}{8}\right) + i\sin\left(\frac{0 + 2k\pi}{8}\right)\right] = \cos\frac{k\pi}{4} + i\sin\frac{k\pi}{4}$, where $k = 0, 1, 2, \ldots, 7$.

$w_0 = 1(\cos 0 + i\sin 0) = 1$, $w_1 = 1\left(\cos\frac{\pi}{4} + i\sin\frac{\pi}{4}\right) = \frac{1}{\sqrt{2}} + \frac{1}{\sqrt{2}}i$,

$w_2 = 1\left(\cos\frac{\pi}{2} + i\sin\frac{\pi}{2}\right) = i$, $w_3 = 1\left(\cos\frac{3\pi}{4} + i\sin\frac{3\pi}{4}\right) = -\frac{1}{\sqrt{2}} + \frac{1}{\sqrt{2}}i$,

$w_4 = 1(\cos\pi + i\sin\pi) = -1$, $w_5 = 1\left(\cos\frac{5\pi}{4} + i\sin\frac{5\pi}{4}\right) = -\frac{1}{\sqrt{2}} - \frac{1}{\sqrt{2}}i$,

$w_6 = 1\left(\cos\frac{3\pi}{2} + i\sin\frac{3\pi}{2}\right) = -i$, $w_7 = 1\left(\cos\frac{7\pi}{4} + i\sin\frac{7\pi}{4}\right) = \frac{1}{\sqrt{2}} - \frac{1}{\sqrt{2}}i$

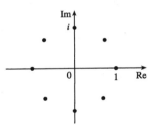

39. $i = 0 + i = 1\left(\cos\frac{\pi}{2} + i\sin\frac{\pi}{2}\right)$. Using Equation 3 with $r = 1$, $n = 3$, and $\theta = \frac{\pi}{2}$, we have

$w_k = 1^{1/3}\left[\cos\left(\frac{\frac{\pi}{2} + 2k\pi}{3}\right) + i\sin\left(\frac{\frac{\pi}{2} + 2k\pi}{3}\right)\right]$, where $k = 0, 1, 2$.

$w_0 = \left(\cos\frac{\pi}{6} + i\sin\frac{\pi}{6}\right) = \frac{\sqrt{3}}{2} + \frac{1}{2}i$

$w_1 = \left(\cos\frac{5\pi}{6} + i\sin\frac{5\pi}{6}\right) = -\frac{\sqrt{3}}{2} + \frac{1}{2}i$

$w_2 = \left(\cos\frac{9\pi}{6} + i\sin\frac{9\pi}{6}\right) = -i$

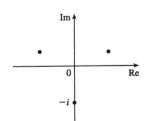

41. Using Euler's formula (6) with $y = \frac{\pi}{2}$, we have $e^{i\pi/2} = \cos\frac{\pi}{2} + i\sin\frac{\pi}{2} = 0 + 1i = i$.

43. Using Euler's formula (6) with $y = \frac{3\pi}{4}$, we have $e^{i3\pi/4} = \cos\frac{3\pi}{4} + i\sin\frac{3\pi}{4} = -\frac{1}{\sqrt{2}} + \frac{1}{\sqrt{2}}i$.

45. Using Equation 7 with $x = 2$ and $y = \pi$, we have $e^{2+i\pi} = e^2 e^{i\pi} = e^2(\cos\pi + i\sin\pi) = e^2(-1 + 0) = -e^2$.

47. Take $r = 1$ and $n = 3$ in De Moivre's Theorem to get

$$[1(\cos\theta + i\sin\theta)]^3 = 1^3(\cos 3\theta + i\sin 3\theta)$$

$$(\cos\theta + i\sin\theta)^3 = \cos 3\theta + i\sin 3\theta$$

$$\cos^3\theta + 3(\cos^2\theta)(i\sin\theta) + 3(\cos\theta)(i\sin\theta)^2 + (i\sin\theta)^3 = \cos 3\theta + i\sin 3\theta$$

$$\cos^3\theta + (3\cos^2\theta\,\sin\theta)i - 3\cos\theta\,\sin^2\theta - (\sin^3\theta)i = \cos 3\theta + i\sin 3\theta$$

$$(\cos^3\theta - 3\sin^2\theta\,\cos\theta) + (3\sin\theta\,\cos^2\theta - \sin^3\theta)i = \cos 3\theta + i\sin 3\theta$$

Equating real and imaginary parts gives

$$\cos 3\theta = \cos^3\theta - 3\sin^2\theta\,\cos\theta \quad \text{and} \quad \sin 3\theta = 3\sin\theta\,\cos^2\theta - \sin^3\theta$$

49. $F(x) = e^{rx} = e^{(a+bi)x} = e^{ax+bxi} = e^{ax}(\cos bx + i\sin bx) = e^{ax}\cos bx + i(e^{ax}\sin bx) \quad \Rightarrow$

$F'(x) = (e^{ax}\cos bx)' + i(e^{ax}\sin bx)' = (ae^{ax}\cos bx - be^{ax}\sin bx) + i(ae^{ax}\sin bx + be^{ax}\cos bx)$

$\qquad = a\left[e^{ax}(\cos bx + i\sin bx)\right] + b\left[e^{ax}(-\sin bx + i\cos bx)\right] = ae^{rx} + b\left[e^{ax}(i^2\sin bx + i\cos bx)\right]$

$\qquad = ae^{rx} + bi\left[e^{ax}(\cos bx + i\sin bx)\right] = ae^{rx} + bie^{rx} = (a+bi)e^{rx} = re^{rx}$